Wyoming Geological Association

33rd Annual Field Conference

GUIDEBOOK

1982

Geology of Yellowstone Park Area

—Editors—
STEVEN G. REID
DAVID J. FOOTE

Mammoth Hot Springs, Wyoming

September 15-18, 1982

C 720 MTN STATES LITHO—CASPER

Table of Contents

STRUCTURE

OIL & GAS

ECONOMIC MINERALS

ROAD LOGS

DEDICATION

Dr. Daniel N. Miller, Jr.

It is a unique privilege for the Wyoming Geological Association to dedicate this guidebook to our keynote speaker for the 1982 Field Conference, Dr. Daniel N. Miller, Jr. In his current position as Assistant Secretary for Energy and Minerals, U.S. Department of the Interior, Dr. Miller holds the highest appointed position in the United States government ever achieved by a W.G.A. member. He is directly responsible for the management of the U.S. Geological Survey, U.S. Bureau of Mines, Office of Surface Mining, and Office of Minerals Policy and Research Analysis; he has been instrumental in working toward the Interior Department's goals of the orderly development of the nation's mineral and energy resources thereby reducing dependence on foreign sources, and of the improvement of relationships between federal land-and-mineral management agencies and state and local governments.

Most W.G.A. members, however, know Dan Miller best for his service as Wyoming's State Geologist and Director of the Geological Survey of Wyoming, positions which he held from the time of reorganization of the State Survey in 1969 until his appointment to the Interior Department post by President Reagan in 1981. During his administration the Survey built and occupied its present office building in Laramie. With the phenomenal growth of the mineral industries in Wyoming, the Survey, under Dan's direction, grew steadily but modestly from three to fourteen full-time employees. Although still one of the smallest state geological staffs in the country, it became widely respected for being one of the most productive in terms of the quality work accomplished and number of reports published. In addition to his ex-officio position as a member of the Wyoming Oil and Gas Conservation Commission he was appointed by Governors Hathaway and Herschler to numerous federal and state bodies dealing with natural-resource issues.

Not so many of the younger W.G.A. members are familiar with Dans earlier careers. He was born, raised, and learned to fish in Missouri. World War II interrupted his formal education when he enlisted in the Army Air Force. His military experience was culminated by seven combat bombing missions over the Japanese mainland as navigator, co-pilot, and pilot of B-29's. He has the dubious distinction of being the co-pilot of the only B-29 ever to attempt a landing on an aircraft carrier! Following his discharge from military service he completed his education, earning B.S. and M.S. degrees in geology from Missouri School of Mines and a Ph.D. from the University of Texas where he specialized in sedimentary petrology, stratigraphy, and salt-water fishing.

Dan started his employment, like so many of us, with a major oil company, in his case, Stanolind (Pan American, AMOCO) for whom he worked in Texas, Billings, and Casper where he was introduced to rainbows, browns and mackinaws. After shorter stints with Lion Oil Division of Monsanto and Barlow & Haun, Inc. in Casper he switched careers to the academic world, and served for six years as professor and Chairman of the Geology Department at Southern Illinois University in Carbondale. While there he was elected President of the Illinois Geological Society, but more significantly, his participation in a multi-discipline research project at S.I.U.'s recreational lake led to his better understanding of the habits of catfish and bass. It was from Carbondale that he returned to Wyoming to embark upon his third career in government service as State Geologist.

Dan's professional contributions are as varied as his careers. In addition to his work for the W.G.A. which included editorship of the 1960 Guidebook on the Overthrust Belt, he has served on various committees of AAPG, AIPG and SEPM. He achieved national prominence through his presidency of the Association of American State Geologists in 1979-80 and his earlier chairmanship of that association's Federal Liason Committee. Not only did these positions lead to his familiarity with the operation of the U.S. Interior Department, but they also provided contacts which provided him with vital information on the best fishing spots in the other 49 states.

His wife, Esther, and children, Gwen and Jeff, have supported with more-than-reasonable patience and good humor the numerous moves, early breakfasts, late dinners, and missed plane connections that Dan's work has entailed. They have also tolerated countless pre-dawn departures and after-dark returns with muddy boots and smelly creel. We suspect that they will want to return to the Rocky Mountains when the big job in Washington is completed, and we hope that Dan will then have time to pursue that giant muskie that has eluded him for so many years.

Greetings

On behalf of the Wyoming Geological Association, we say "howdy" and extend a most enthusiastic welcome to each and every participant of the 1982 Field Conference. We are gathering within the boundaries of our first National Park, in the midst of spectacular surroundings and a diverse geologic setting which offers something for everyone.

In the thirty six years since the First Annual Field Conference in 1946, the W.G.A.'s field conferences and symposia have generally covered the length and breadth of the various basins, uplifts, rocks, and geology of the State of Wyoming with one notable exception, the Yellowstone Park area.

Yellowstone Park is a veritable outdoor laboratory with countless geologic features and phenomenon which we in Wyoming are fortunate to have at our disposal. Geologic studies within the boundaries of the Park will probably never lead to direct economic benefit there; however, geologic concepts and relationships seen within the Park can be applied to other areas, where natural resources can be developed.

We would like to thank everyone who has contributed their time and efforts toward this field conference. We hope this field conference will be of value as a geologic learning experience for everyone, and will also help to perpetuate close and friendly relations between all of us in the geologic profession.

Daniel D. Miller
Wyoming Geological Association
Co-Chairman

Jimmy E. Goolsby
Wyoming Geological Association
Co-Chairman

WGA Officers and Field Conference Committee

President	Ralph Specht	Maynard Oil Company
First Vice President	Jim Brown	Conoco Oil Company
Second Vice President	Roger Motten	Gulf Oil Company
Secretary	Mike Mendeck	Gulf Oil Company
Treasurer	Lyn George	Kirkwood Oil & Gas
Earth Science Bulletin Editor	Greg Parsons	Union Oil Company
Past President	Steve Champlin	Independent

FIELD CONFERENCE COMMITTEE

General Co-Chairmen	Dan Miller	Independent
	Jimmy Goolsby	Independent
Guidebook Co-Editors	Steve Reid	Independent
	David Foote	Independent
Road Log Editor	Nancy M. Doelger	Gulf Oil Company
Associate Editor	Jack Watson	Tooke Engineering
Program Chairman	Bill Purdon	Gulf Oil Company
Field Trip Chairmen	Jimmy Goolsby	Independent
	Dan Miller	Independent
Activities Chairman	Cliff Chew	Tooke Engineering
Advertising	Carl Babb	Union Oil Company
Registration & Housing	Herb Waterman	Tyrex Oil Company
	Ralph Pointer	Birdwell
Transportation	David Foote	Consultant

Previous Wyoming Geological Association Field Conferences

First Annual Field Conference, 1946	Central and Southwestern Wyoming
Second Annual Field Conference, 1947	Big Horn Basin, Wyoming
Third Annual Field Conference, 1948	Wind River Basin, Wyoming
Fourth Annual Field Conference, 1949	Powder River Basin, Wyoming
Fifth Annual Field Conference, 1950	Southwestern Wyoming
Sixth Annual Field Conference, 1951	Southcentral Wyoming
Seventh Annual Field Conference, 1952	Southern Big Horn Basin, Wyoming
Eighth Annual Field Conference, 1953	Laramie Basin and North Park, Wyoming and Colorado
Ninth Annual Field Conference, 1954	Casper Area, Wyoming
Tenth Annual Field Conference, 1955	Green River Basin, Wyoming
Eleventh Annual Field Conference, 1956	Jackson Hole Area, Wyoming
Twelfth Annual Field Conference, 1957	Southwest Wind River Basin, Wyoming
Thirteenth Annual Field Conference, 1958	Powder River Basin, Wyoming
Fourteenth Annual Field Conference, 1959	Big Horn Basin, Wyoming
Fifteenth Annual Field Conference, 1960	Overthrust Belt of Southwestern Wyoming and Adjacent Areas
Sixteenth Annual Field Conference, 1961	Symposium on Late Cretaceous Rock of Wyoming and Adjacent Areas
Seventeenth Annual Field Conference, 1962	Symposium on Early Cretaceous Rocks of Wyoming
Eighteenth Annual Field Conference, 1963	Joint Field Conference of the Billings Geological and Wyoming Geological Association, Northern Powder River Basin
Nineteenth Field Conference, 1965	Rock Springs Uplift, Wyoming
Twentieth Field Conference, 1968	Black Hills Area; Wyoming, South Dakota, and Montana
Twenty-first Field Conference, 1969	Symposium on Tertiary Rocks of Wyoming
Twenty-second Field Conference, 1970	Symposium on Wyoming Sandstones
Twenty-third Field Conference, 1971	Symposium on Wyoming Tectonics and their Economic Significance
Twenty-fourth Field Conference, 1972	Rawlins Area, Wyoming
Twenty-fifth Field Conference, 1973	Geology and Mineral Resources of the Greater Green River Basin, Wyoming
Twenty-sixth Field Conference, 1974	Muddy Sandstone, Wind River Basin, Wyoming
Twenty-seventh Field Conference, 1975	Bighorn Basin, Wyoming
Twenty-eighth Field Conference, 1976	Geology and Energy Resources of the Powder River Basin
Twenty-ninth Field Conference, 1977	Joint Wyoming - Montana - Utah Geological Associations Guidebook, Rocky Mountain Thrust Belt Geology and Resources
Thirtieth Field Conference, 1978	Resources of the Wind River Basin
Thirty-first Field Conference, 1980	Stratigraphy of Wyoming
Thirty-second Field Conference, 1981	Energy Resources of Wyoming

WYOMING GEOLOGICAL ASSOCIATION PUBLICATIONS

	WGA Members	Non Members
1981 GUIDEBOOK — ENERGY RESOURCES OF WYOMING	$45.00	$50.00
1981 OIL AND GAS FIELDS SYMPOSIUM - POWDER RIVER BASIN	$70.00	$75.00
1980 GUIDEBOOK - STRATIGRAPHY OF WYOMING	$32.00	$36.00
1979 OIL AND GAS FIELDS SYMPOSIUM - GREATER GREEN RIVER BASIN	$60.00	$65.00
1978 GUIDEBOOK - WIND RIVER BASIN	$28.00	$32.00
1977 GUIDEBOOK - ROCKY MOUNTAIN THRUSTBELT	$42.00	$47.00
1976 GUIDEBOOK - POWDER RIVER BASIN	$24.00	$27.00
1975 GUIDEBOOK - BIG HORN BASIN	$22.00	$25.00
1968 GUIDEBOOK - BLACK HILLS AREA	$14.00	$16.00
WELLS DRILLED FOR OIL AND GAS IN WYOMING TO JANUARY 1, 1950	$4.00	$5.00
BIBLIOGRAPHY OF WGA ARTICLES 1967 - 1977	2.00	$3.00
WYOMING STRATIGRAPHIC NOMENCLATURE CHART (1969)	$1.25	$2.00
CURRENT WGA DIRECTORY		$12.00
ALBANY CO. GEOLOGIC MAP	$1.00	$2.00
1974 APPLIED GEOLOGY AND ARCHAEOLOGY: HOLOCENE HISTORY OF WYOMING	$2.50	$3.50
1979 CORE BOOK AAPG-SEPM SEIMIC	$7.00	$10.00
JUNE 1979 ESB GUIDEBOOK FOR ROCKY MOUNTAIN AAPG-SEPM-EMD FIELD TRIPS	$5.50	$7.00
1946 - 1978 PUBLICATIONS ON MICROFICHE	$80.00	$90.00
WITH FILE BOX	$100.00	$110.00

****

ALL ORDERS MUST BE PRE-PAID

Wyoming Residents add 3% or 4% Sales Tax, depending on the tax rate for the County you reside in.

SEND ORDERS AND MAKE CHECKS PAYABLE TO:

WGA PUBLICATIONS
P.O. Box 545
Casper, Wyoming 82602

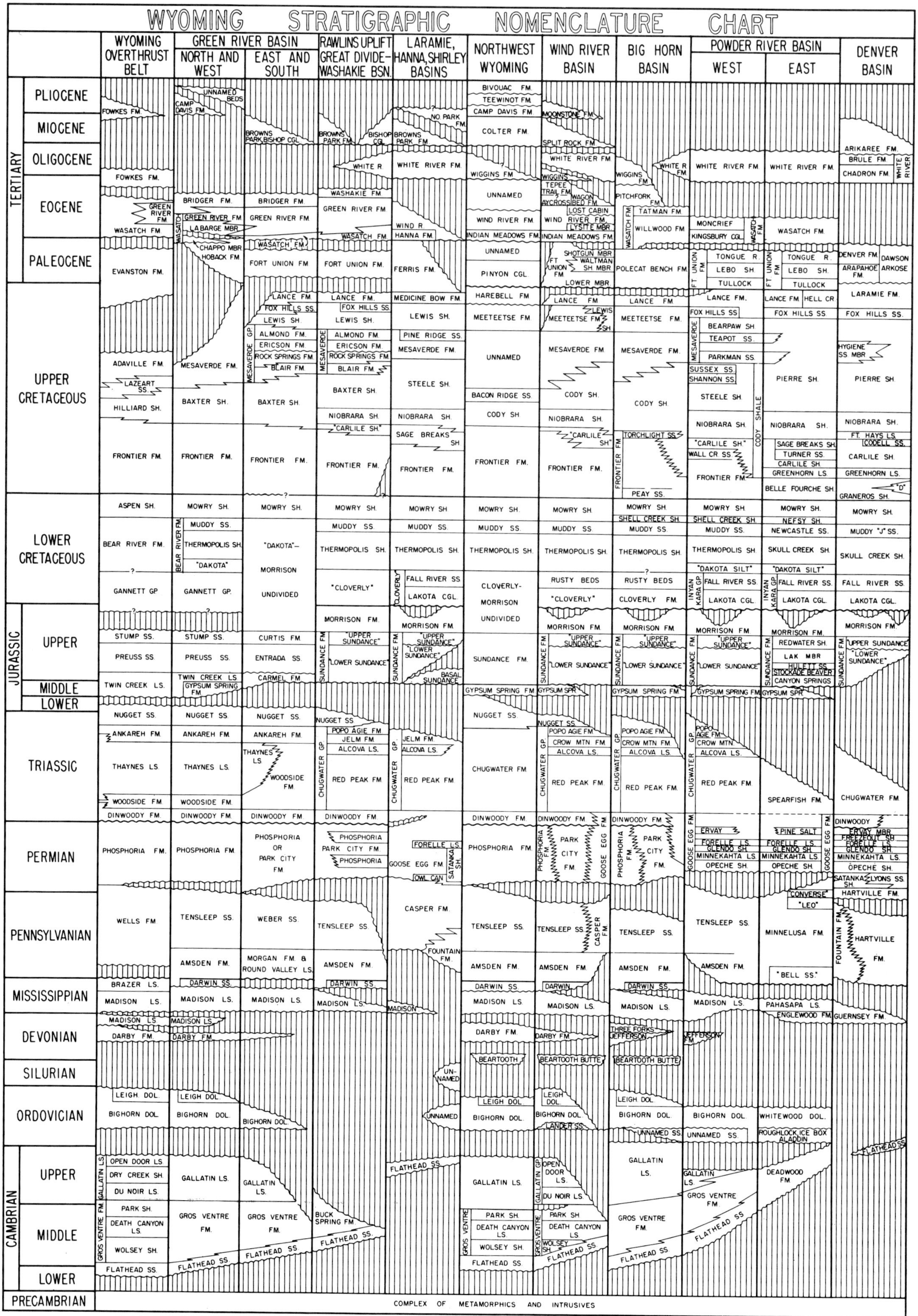

REVISED BY STRATIGRAPHIC NOMENCLATURE COMMITTEE, WYOMING GEOLOGICAL ASSOCIATION, 1969

Thirty-Third Annual Field Conference — 1982
Wyoming Geological Association Guidebook

ARCHEOLOGICAL RESEARCH IN YELLOWSTONE NATIONAL PARK

GARY A. WRIGHT

INTRODUCTION

A general consideration of recent archeological research in the Swan Lake Flat-Mammoth and Yellowstone Lake areas indicates that Yellowstone Park was clearly occupied by the Alttithermal. The hunting-gathering-fishing adaptation was maintained on a seasonal basis until ca. A.D. 1600 when the Park was apparently abandoned. It was reoccupied by the Shoshone shortly after A.D. 1800. Several new research problems are suggested.

Yellowstone National Park is one of the poorest known archeological areas of North America. Only 20% of the Park has been surveyed, and there is no adequate synthesis of the archeological data (cf. Wright et al. 1978). Over two decades ago a general surface survey was conducted by Montana State University, but the field crews were restricted to within about 50 to 100 m of the major highways (Taylor 1964). Since that time, most archeological work in the Park has been closely tied to mitigation projects (Wright et al. 1978). Our own recent research under the aegis of the Midwest Archeological Center has concentrated on two sections of the Park: Swan Lake Flat-Mammoth (Reeve et al. 1981; Smith 1979; Wright and Smith 1981) and Yellowstone Lake (Samuelson 1981; Wright et al. 1981). Some of the Park's archeological data has appeared elsewhere (e.g. Wright 1978, 1981, 1982; Wright and Reeve 1981), but no synthesis has been attempted. Here, I shall provide an outline of the results of our research.

SWAN LAKE FLAT-MAMMOTH

This subarea in extreme northern Yellowstone National Park is drained by the Gardner River and several other tributary streams of the Yellowstone River. Swan Lake Flat (ca. 2190 m) was apparently ice free by about 13,530 B.P. It contains one of the largest obsidian sources in the Central Rocky Mountains, Obsidian Cliff (48YE433), which is 12 km south of Swan Lake. Obsidian Cliff provided raw materials for tools to Native Americans not only locally but also over much of the Northwestern Plains (e.g. Frison et al. 1968) and even as far east as the Ohio Valley (Griffin et al. 1968). Obsidian is also distributed as glacial erratics about 2 km into Swan Lake Flat.

We still have only a poor understanding of post-glacial climatic changes in this area. There is but a single pollen core for this subregion (Gennett 1977). It comes from Blacktail Pond (2045 m) which lies east of Mammoth and about 12 km northeast of Swan Lake. However, it exhibits some major discrepancies when compared to the more extensive palynological data we have from Yellowstone Lake (2343 m). Most importantly for our purposes is the long period of maximum warmth which is bracketed by dates of 7300 to 1565 B.P. (calibrated, Zone B2) at Blacktail Pond. Zone B2 is dominated by *Pseudotsuga* pollen. Elsewhere on the Yellowstone Plateau this warmer period has ended by 4000 B.P. Following this there was a return to cooler and moister conditions. At Blacktail Pond this interval is marked by lodgepole pine forests.

We have intensively investigated 17 of the 53 currently known sites in this subarea. The summer of 1977 was spent excavating the Sheepeater Bridge site (48YE320) which is located just east of Mammoth on the Gardner River (Reeve et al. 1981). We have 26 obsidian hydration dates. They range in age from 7100 B.P. to 382 B.P., with 12 of them falling between 7100 and 5410 B.P. Thus, occupation at the site was clearly underway by the Altithermal. Our analysis of the artifacts indicates that it was a fishing station that was visited on a temporary (seasonal—spring) basis for over 6700 years.

In 1978 we concentrated our activities in Swan Lake Flat (Wright and Smith 1981). This work included a series of major excavations, most importantly at 48YE504. This site is located on a gravel knoll just north of both the Gardner River and Indian Creek Campground. It produced a wide variety of projectile point types. One of them, a lanceolate point, was dated by obsidian hydration analysis to 3335 B.P. This site also yielded a portion of a

Department of Anthropology
State University of New York
Albany, NY 1222

butchered mule deer (*Odocoileus heimonus*); an obsidian flake in associated with the faunal material was dated tp 2965 B.P. This site was a small hunting camp. The range of projectile point types suggests utiliazation of the site over several millennia, but this cannot be confirmed at present without more absolute dates.

There are several other sites of interest. One, now completely destroyed, was described a century ago by Park Superintendent P. W. Norris (1880). It was a game drive with log constructed wings that converged at Rustic Falls. One wing led toward Swan Lake and the other wound through the timber at the base of Bunsen Peak and continued on for nearly two miles. We do not know what species of animals were being driven into the trap and over Rustic Falls. Nearby to the southwest of Rustic Falls is site 48YE511. It is situated in a draw and is completely hidden from the trap. However, from the top of the draw both the trap area and northern Swan Lake Flat are clearly visible. The site is untested, but we have proposed that it functioned as a hunting-lookout station and for the manufacture of tools to be employed in hunting and butchering activities associated with the use of the game drive (Wright and Smith 1981 Neither the trap or 48YE511 are dated.

Site 24YE344 is located along the Gardner River in the canyon about 8 km downstream from Sheepeater Bridge (Lahren 1973). It has 3 stratigraphic levels. Based upon the typology of the points, the middle level should date in the range of ca. 1250 to 1450 B.P. This level produced remains of ungulates (still unidentified) and beaver (*Castor canadenis*). It undoubtedly functioned as a temporary camp for human groups on their way into and out of the Park area.

YELLOWSTONE LAKE

For our purposes the Yellowstone Lake region also includes the section of the Yellowstone River from its outlet to Chittenden Bridge. We have a much better data set for this subarea, including good stratified sites, 50 absolute dates, and a pollen sequence which shows that no significant climatic changes occurred over the past 4000 years (Beiswenger and Frison 1979). Here we have recorded 63 sites. They are heavily concentrated around West Thumb Bay, at the outlet of the Yellowstone River and on Fishing Bridge Peninsula, and along the river down to Chittenden Bridge (Taylor 1964; Samuelson 1981). Nine sites have been test excavated.

Although the recovered projectile point types suggest pre-5000 B.P. occupations, our earliest absolute dates are early post-Altithermal in age: a C-14 date of 4880+100 B.P. (TX-3764) from the Chittenden Bridge site and an obsidian hydration determination of 4675 B.P. from 48YE1 on Fishing Bridge Peninsula. Both dates relate to peoples using indented base points (McKean?) during what Frison (1978) terms the Middle Plains Archaic. Eight additional sites produced these points. Our most recent date on them is 2965 B.P. from 48YE650 on West Thumb Bay.

Corner notched points are extremely common (22 sites) and several types have been found. Two braod corner notched points from 48YE396 on West Thumb Bay have dates of 3890 and 3320 B.P., thus showing a temporal overlap with the indented base style. Later corner notched points which are Pelican Lake-like were discovered in good stratigraphic context at Chittenden Bridge with C-14 dates ranging from 1975+140 B.P. (TX-3726) to 1360+60 B.P. (TX-3727). Finally, another variety of corner notched points is a long, spikey form with serrated edges. A single example from 48YE1 was dated at 680 B.P. This type is common from the Big Horn Basin into the Central Rockies (including Jackson Hole) down to Utah where it ranges in age from 1900 to 675 B.P.

The latest point types is side-notched. Two varieties can be related to the Plains and Prairie types. At Chittenden Bridge C-14 dates on these types range from 1205+50 B.P. (TX-3763) to 615+50 B.P. (TX-3765). Late side-notched points are comparatively rare and have been found on only 8 sites.

One site needs to be noted in more detail. 48YE1 has produced the only known Native American burial from the Park. Originally missexed as a male (Condon 1948), it is in fact a female. She was about 40-45 years of age and about 154 cm tall. She exhibits several pathologies on the cranium, teeth, long bones, and the right iliac blade. The latter pathology may have derived from a wound which produced acute infection. No ethnic designation of the woman is possible.

The burial also apparently included two domesticated dogs (Condon 1948), but only one was present in the material that I studied. It was incomplete. According to its dentition, it was 6 to 7 months old (the permanent canines are visible anteriomedial to the corresponding deciduous canines in the maxilla). The dental measurements show that the specimen fits at the smaller end of the 8 size classes of prehistoric domesticated dogs defined by Haag (1948). Some artifacts are reported but were unavailable to me.

The burial, itself, was in an ovate pit, 100 cm in depth that began 13 to 21 cm below the present ground surface. It lay below an undisturbed charcoal layer, and the top of the pit was covered by 37 flat rocks (Condon 1948). A stratigraphically controlled series of obsidian hydration dates from 48YE1 indicates an age of ca. 3200 to 3100 B.P. for the construction of the burial pit.

GENERAL CONSIDERATIONS

We are only beginning to understand the prehistoric use of the area's resources. We have much better information for Jackson Hole just to the south of Yellowstone Park. For that region we have proposed a general model termed a High Country Adaptation (Wright etal. 1980). Briefly, it sees prehistoric hunting and gathering populations successively transversing a series of progressively higher (or lower) resource zones over the course of the yearly migration. In Jackson Hole proper, these resources included a variety of root crops including blue camas (*Camasia quamash*) (Wright 1981: Table 1). Double and triple cropping of the same species at different elevations was an adaptive alternative. Basically, from late spring to early fall, these groups concentrated on fishing and plant gathering in Jackson Hole and its surrounding mountains. Winter was spent hunting at lower elevations outside of Jackson Hole. Our current tests of the model suggest that it is valid and that this adaptation was a successful one for several millennia (Wright and Reeve 1981).

The Yellowstone Lake area may well have been drawn into this larger Jackson Hole sphere. For example, blue camas meadows are present on the west side of the Yellowstone Lake from Bridge Bay to Fishing Bridge. This root crop also grows extensively in the Thorofare which is still unsurveyed. At this elevation, harvesting would have taken place about 1 month later than on the valley floor of Jackson Hole. Yellowstone Lake, of course, contained abundant fish. Sites near West Thumb have heavy concentrations of projectile points suggesting some emphasis on hunting at these locations (Samuelson 1981). Following the early summer population congregations around Jackson Hole (e.g. the Lawrence site, 48TE509; Reeve et al. 1979), some families may have dispersed toward Yellowstone Lake. The several large sites near the outlet of the lake are ideally situated for easy access to both fish and plant (e.g. blue camas) resources.

For the Swan Lake Flat-Mammoth area, we hypothesize that the groups who exploited that region wintered in the sheltered canyons of the Yellowstone River north of the Park (Wright and Smith 1981). Several bison jumps are known between Gardner and Livingston. Movement into the Park was then timed to coincide with early access to fishing resources, particularly in the vicinity of 48YE320. The route was along the Gardiner River (e.g. site 24YE344). The blue camas equivalent in northern Yellowstone Park is bitterroot (*Lewisia redivia*) which is found east of Mammoth.

Swan Lake is summer range for elk (*Cervus canadensis*) and mule deer and is also near the Terrace Mountain migration route used by mule deer and bighorn sheep (*Ovis canadensis*). Hunting was undoubtedly an important activity too. Swan Lake Flat is adjacent to Obsidian Cliff, and new supplies of obsidian for tools could be obtained. From here, human populations could have continued into even higher elevations eastward along the Bannock Trail or toward Yellowstone Lake and the Absaroka Range for the summer. Exit back to the winter hunting grounds would have been along the Gardner River. We can detect no cultural change here that might be correlated with the environmental change detected in the Blacktail Pond pollen core (Wright 1981; Wright and Reeve 1981).

Several other problems may be noted briefly. First, we do not know the ethnic identity of the Park's prehistoric inhabitants. It seems clear that the Shoshone (or Sheepeaters) were not residents prior to ca. A.D. 1800 (Wright 1978). We have no evidence of the Shoshone in the Park, or Jackson Hole for that matter, before the 1830s. A pot of Shoshonean manufacture (=Intermountain Ware) was found at 48YE449 on West Thumb Bay, but the site is undated. The single Shoshone site in Jackson Hole (48YE576) has a date of about A.D. 1844. One artifact of presumed Shoshonean culture is the steatite bowl of which more than two dozen are known from northwestern Wyoming, but none are from a datable context. However, a recent study by Marceau (1981) concludes that there is no solid evidence to attribute steatite bowls to the Shoshone. The prehistoric inhabitants left the Park around A.D. 1600, and we are unable to trace them after this date.

A second problem is the date of the earliest human penetration of the Park. As noted, our earliest absolute date is 7100 B.P. Yet, deglaciation of the major areas of the Park was well underway by ca. 13,000 B.P. By 10,000 B.P., many of the root crop species whcih were to be key items in prehistoric diet had already invaded the Park (Wright and Reeve 1981). In regions adjacent to the Park, we have excellent data on human occupations by +11,000 B.P. (Frison 1978). But precisely when these human groups began to take advantage of the high country resources is still unknown.

Next, what role did mountainous northwestern Wyoming play during the Altithermal? Many archeologists have argued for a large scale abandonment of northwestern Plains during parts of the Altithermal, with a corresponding migration of those peoples into the high country. This proposition has never been adequately tested. We can, however, comment here that the latest pre-Altithermal groups on the adjacent Plains are identified by the use of lanceolate projectile points with parallel oblique flaking (Frison 1978). These points seem to disappear on the Plains by ca. 8000 B.P. In the high country this is the earliest point type that occurs in any frequency, but we do not have it in stratigraphic context prior to 7100 B.P. On the Plains, this type is replaced during the Altithermal by early side-notched forms (Frison 1978). This type is virtually absent in Yellowstone Park, but some have been recovered in Jackson Hole.

The hypothesis, thus, may be stated more explicitly. The earliest occupation of Yellowstone Park began during the Altithermal. The Park served, initially, as a refugium for prehistoric peoples who were abandoning the Plains. They developed a new land use pattern — the high country adaptation — which they maintained until around A.D. 1600 (Wright 1978). It is this hypothesis that should be tested.

Finally, considering Jackson Hole and the Yellowstone Lake subarea as a unit, we have 67 absolute dates falling between 3500 and 350 B.P. However, for Yellowstone Lake, 20 are in the 3500 to 1300 B.P. interval, and only 6 are at less than 1300 B.P. In contrast, 14 of the absolute dates from Jackson Hole fall between 3500 and 1300 B.P., while 27 are at 1300 B.P. or less. Unless we have a major sampling error, it suggests a decline in land use intensity around Yellowstone Lake by 1300 B.P. with an attendant shift toward greater exploitation of Jackson Lake.

CONCLUSION

It is obvious that Yellowstone Park is still poorly known archeologically, particularly in comparison with Jackson Hole just to the south. Considerably more surface survey and additional extensive excavations are required before we can offer any truly confident synthesis of the Park's prehistory. The Park's potential is, however, enormous for archeological research.

REFERENCES CITED

Beiswenger, J. M. and G. C. Frison, 1979. Vegetative History of Yellowstone Park as Determined by Pollen Analysis. Department of Anthropology, University of Wyoming, Project Number 2/78-79.

Condon, D. 1948. American Indian Burial Giving Evidence of Antiquity Discovered in Yellowstone National Park. Yellowstone Nature Notes, V. 22, No. 4, pp. 37-43.

Frison, G. C. 1978. Prehistoric Hunters of the High Plains. New York: Academic Press.

Frison, G. C., G. A. Wright, J. B. Griffin, and A. A. Gordus, 1968. Neutron Activation Analysis of Obsidian: An Example of its Relevance to Northwestern Plain Archeology. Plains Anthropologist, V. 13, No. 41, pp. 209-217.

Gennett, J. A. 1977. Palynology and Paleoecology of Sediments from Blacktail Pond, Northern Yellowstone National Park, Wyoming. Masters Thesis, Department of Geology, University of Iowa.

Griffin, J. B., A. A. Gordus, and G. A. Wright. 1968. Identification of the Sources of Hopewellian Obsidian in the Middle West. American Antiquity, V. 34, pp. 1-14.

Haag, W. G. 1948. An Osteometric Analysis of Some Aboriginal Dogs. University of Kentucky, Reports in Anthropology, V. 7, No. 3, pp. 107-264.

Lahren, L. L. 1973. Archeological Salvage Excavations at a Prehistoric Indian Campsite (24YE344) in Yellowstone National Park. Manuscript on file, Midwest Archeological Center, Lincoln.

Marceau, T. E. 1981. Steatite, Intermountain Pottery, and the Shoshone: Problems in Association. Paper presented at the 39th Annual Plains Conference, Bismarck.

Norris, ZP. W. 1880. Annual Report of the Superintendent of the Yellowstone National Park to the Secretary of the Interior for the Year 1880. Washington: United States Government Printing Ofiice.

Reeve, S. A., T. E. Marceau, and G. A. Wright. 1981. Mitigation of the Sheepeater Bridge Site, 48YE320, Yellowstone National Park. Midwest Archeological Center, Occasional Publications. in press.

Reeve, S. . G. A. Wright, T. E. Marceau, and P. Mecham. 1979. Archeological Investigations of the Lawrence Site (48TE509), Grand Teton National Park, Wyoming. Report on file, Midwest Archeological Center, Lincoln.

Samuelson, A. E. 1981. Archeological Investigations in the Grant Village-West Thumb Area of Yellowstone National Park, Wyoming. Masters Thesis, Department of Anthrology, State University of New York, Albany.

Smith, R. C. 1979. Archeological Investigations in Swan Lake Flat: Lithic Technology and Seasonal Patterns in Yellowstone National Park, Wyoming. Masters Thesis, Department of Anthropology, University of Wisconsin.

Taylor, D. C. 1964. Preliminary Archeological Investigations in Yellowstone National Park, Wyoming. Manuscript on file, Midwest Archeological Center, Lincoln.

Wright, G. A. 1982. Notes on Chronological Problems on the Northwestern Plains and Adjacent High Country. Plains Anthropologist, in press.

______. 1981. Environmental Change and Cultural Variation: The View from the High Country of Northwestern Wyoming. In: Proceedings of the Second Conference on Scientific Research in the National Parks, V. 1, pp.324-33.

______. 1978. The Shoshonean Migration Problem. Plains Anthropologist, V. 23, pp. 113-137.

Wright, G. A., T. E. Marceau, F. Calabrese, and M. Tune. 1978. Archeological Data for the Greater Yellowstone Cooperative Regional Transportation Study. Lincoln: Midwest Archeological Center. 91 p.

Wright, G. A., T. E. Marceau, T. Koenig, and A. E. Samuelson. 1981. Chronology of Occupations at Yellowstone Lake, Yellowstone National Park. Paper presented at the 39th Annual Plains Conference, Bismarck.

Wright, G. A. and S. A. Reeve. 1981. Prehistoric Resource Procurement and Climatic Change in Northwestern Wyoming. In Quaternary Paleoclimate, ed. W. Mahaney, pp. 423-448. Geoabstracts.

Wright, G. A. and R. C. Smith. 1981. Archeological Investigations in Swan Lake Flat: The Prehistory of Northern Yellowstone National Park. Manuscript on file, Midwest Archeological Center, Lincoln.

Thirty-Third Annual Field Conference — 1982
Wyoming Geological Association Guidebook

DISCOVERING YELLOWSTONE

BY WILLIAM L. LANG[1]

It was September 1867 on the Yellowstone River near Otter Creek when a party of itinerant Montana prospectors spurred their horses into the swift current, crossing from the east to the west bank. Bart Henderson, one of the group helped set up camp and then ambled down the river that late summer day to explore more of this new region. It was his first visit to Yellowstone. He and his friends had marvelled at Yellowstone Lake along their route north from the Snake River country, and they knew something of the hot springs and geysers to the west. He watched the smooth-flowing Yellowstone rush past verdant banks and rugged boulders on its northward course.

Tramping the bank, Henderson unknowingly approached the brink of the Upper Falls of the Yellowstone. There he watched "the water disappear from my sight. I walked out on a rock," he wrote in his journal, "and made two steps at the same time, one forward, the other backward, for I had unawares as it were looked down into the depth or bowels of the earth, into which the Yellow plunged as if to cool the infernal region that lay under all this wonderful country of lava and boiling springs."

The Yellowstone, as it slides and riffles toward the ledge-jutted drop at the Upper Falls, is silent in its duty and barely warns of the chasm hole ahead. It surprised Henderson in 1867 and it surprises us today as we watch the river's deception unfold. But what is most surprising about Henderson's discovery is the date — 1867. The guns of the Civil War had been silent for over two years, Montana had been a territory for three years, the first transcontinental railraod was only two years off, and six decades had elapsed since the first white recorded a visit to Yellowstone. It was late in the western exploration game.

High on a plateau at the top of western America, Yellowstone resisted discovery. Jim Bridger, the famous fur trapper, scout and teller of tall tales, knew much about Yellowstone and its wonders decades before Henderson's party stumbled on the Upper Falls, but most scoffed at Bridger's descriptions of geysers, boiling mud pots and a huge lake at mountain-top elevation. This disbelief only encouraged "old Gabe," as Bridger was known, to stretch his exaggerations, perhaps in hopes of convincing someone of the truths about Yellowstone wonderland. Bridger's stories of petrified animals in petrified forests and other travesties on the truth did not engender belief.

Mountain men in search of good beaver streams first penetrated the Yellowstone area in the late 1820s. Many followed the old Bannock Trail that crossed the northern portion of today's Park, fording the Yellowstone at the only safe spot north of the Grand Canyon at Tower Creek and proceeding eastward on one of the many branches of the trail to the Soda Butte Creek, Slough Creek or Lamar River drainages. In the 1830s, the fur men wandered back and forth across Yellowstone, including forays into the geyser basins, trampings up in the Cooke City area, steep reconnaissance in the mountains on the east boundary of the Park, and a near circumnavigation of the Yellowstone Lake. Some of their descriptions of Yellowstone, such as Osborne Russell's poetic portrait of the Lamar River valley near Soda Butte Creek, impress us today with the timelessness of the region. "There is something in the wild romantic scenery of this valley," Russell wrote, "the impressions made upon my mind while gazing from a high eminence on the surrounding landscape one evening as the sun was gently gliding behind the western mountain and casting its gigantic shadows across the vale were such as time can never efface from my memory."

It was Joe Meek, a nineteen-year-old trapper, who drifted into the Norris Geyser Basin in 1829 in flight from a skirmish with Indians some distance to the north. There he saw a sight that he "thought himself reminded of the city of Pittsburgh." Shocked at what belched from the ground, Meek described it as an unreal countryside: "The whole country beyond was smoking with the vapor from boiling springs, and burning with gases, issuing from small craters, each which was emitting a sharp, whistling sound." Meek, like other mountain men, prided themselves on accurate reporting of mountain ranges, drainages and landmarks; by necessity in an unmapped region, they observed with a keen eye for detail. Yet, Joe Meek characterized the Norris Geyser Basin as an area containing smoking craters four miles wide that spewed "molten brimstone." He was not the only visitor to Yellowstone who lost perspective when geysers blew and fumaroles hissed.

By the early 1860s, overland trail travellers, gold prospectors, and would be explorers had heard about

[1]William L. Lang is editor of *Montana the Magazine of Western History*, published by the Montana Historical Society. He is author of *Montana: Our Land and People* (1979) and numerous articles on the history of the West. Dr. Lang also teaches courses in the history of Yellowstone National Park for the Yellowstone Institute, the educational program of the Yellowstone Library and Museum Association.

Yellowstone, but no one had mapped it; no one had broken through Yellowstone's resistence to thorough exploration. Captain Willian F. Reynolds of the U.S. Topographical Engineers led an expedition in 1860, with Jim Bridger as guide, that was to proceed from the Wind River drainage to the source of the Yellowstone and down that river to the plains — a route that would have taken the party high on Two Ocean Plateau and through the heart of the Yellowstone caldera, the falls and the Grand Canyon. But it was May; too early to penetrate the Yellowstone's high-pass snow barrier. "I told you," Old Gabe jabbed at Reynolds, "you could not go through. A bird can't fly over that without taking a supply of grub along."

Three years later Walter Washington de Lacy, a surveyor and sometime prospector who had roamed all over the West and experienced more than his allottment of hazards and close scrapes, led an adventurous pack of gold hunters into the upper Snake River country. Not finding much color up the Snake, de Lacy and just over two dozen others headed north out of the Jackson Hole area, trying to make it back to Virginia City, Montana in a direct manner. Those familiar with the terrain soon recognized no landmarks and cautioned that they were in unknown territory. De Lacy's entorage ascended the Snake to the hot springs near the forks and then climbed up on to the Pitchstone Plateau, heading northwest. Rough going prompted a division of their group, one following de Lacy's lead and the other tagging along behind Charles Ream.

De Lacy discovered Shoshone Lake (at first called DeLacy's Lake) and skirted its east shore on their way over the divide and into the geyser basins. Ream's outfit struggled up the Lewis River drainage to Lewis Lake and pressed on to Shoshone Lake, passing on the west shore and through the Shoshone Geyser Basin. The two parties regrouped at the headwaters of the Madison River. As one of their number stated years later, we "never dreamed that we were on the threshold of a great discovery," but, of course, they were. De Lacy's map, published in 1865, provided the first observed cartographic description of Yellowstone Lake and the geyser basins. Seemingly, de Lacy was relatively unaware of the importance of his journey and discoveries. Only when the later Hayden Survey called some of his notations into question did de Lacy assert that he was the first to record Yellowstone's wonders.

Yellowstone continued to yield its secrets almost grudgingly. It had tortured many a trapper, frustrated explorers and gave up its information to only the hardy and determined. Small groups of prospectors combed the region for promising outcroppings or glittering sands on alluvial slopes, each one adding bits and pieces of knowledge about the shape of Yellowstone. Yet, a circuit of the high plateau remained unachieved until the last year of the decade when a most unassuming and unexpected expedition raced against the weather in an inspired dash through the heart of today's Park.

Planned expeditions into Yellowstone in 1867 and 1868 had failed through unexpected obstacles, poor planning or lack of determination. Knowledge of the area was current enough to spark adventurous interest. Too many prospectors and wanderers who traversed the region brought back snapshot spoken images of a land that rumbled and boiled, spewed forth and smoked. Two Quakers and a Dane, as historian Richard Bartlett characterized them, undertook the first genuine expedition in 1869. Charles W. Cook, David Folsom and William Peterson were the first to chart Yellowstone's wonderland.

The two Quakers, Charles Cook and David Folsom, lived remarkable parallel lives. Of the same age and general New England Quaker background, they attended school together in Rhode Island and in later life would be sheep ranchers and neighbors in Montana's Smith River Valley. In 1869 Cook and Folsom were in the business of supplying water to miners at Confederate Gulch, one of the great Montana gold strikes of the mid-1860s. Cook had made his way to the mining camp, which lay some thirty miles east of Helena near the Missouri River, by way of St. Joseph, Missouri, Denver, Virginia City and Helena. Folsom had attached himself to James L. Fisk's overland expedition from Minnesota in 1862 and came to Bannack first before moving on to Alder Gulch and finally Confederate Gulch in 1868.

Also at work at the ditch company was William Peterson, a Danish native who at thirty-five was five years older than Cook and Folsom. A seaman, Peterson had shipped to California in 1860, followed the mining excitement to Idaho in 1861 and operated a pack string for a season before stopping briefly in Montana in 1864 on his way back to British Columbia. By the summer of 1865, Peterson was at Diamond City in Confederate Gulch working for Charles Cook at moving water for the miners. Peterson, Folsom and Cook worked well together and the ditch company prospered, but there was a wanderlust and the experiences of adventuresome travel that equally linked these three bachelors and set them up for their moment in Yellowstone history.

It was an exasperated Charles Cook who said within earshot of Peterson: "If I could get one man to go with me, I'd go anyway." He had lost his chance to explore Yellowstone once again. This time he had signed up for what he believed was a well planned and well financed expedition. But key individuals pulled out by mid-summer 1869 and the whole organization collapsed. Bill Peterson left no doubt where he stood: "I can go as far as you can," he told his friend. Peterson had also signed up for the journey, as had David Folsom. Folsom quickly took up the challenge and threw in with his comrades. The three of them would explore the mysterious Yellowstone alone.

Friends cautioned them against trying such a journey so late in the season — it was the first week of September. Snow and low temperatures always came early in the Upper Yellowstone Valley. Some told them the trip was "next to suicide," and one hailed as they left: "Goodbye, boys, look out for your hair." But Cook, Folsom and Peterson had plenty of grit. They took three riding horses and two pack horses, enough stuff for six weeks in the woods, de Lacy's map of the region, a compass, and firearms. They had gathered their outfit within a day of their pledge to go it alone, and with little more thought they rode out of Diamond City, Montana Territory, on September 6, 1869.

They provisioned at the small town of Bozeman on their way to the Upper Yellowstone, made their way over the Gallatin divide and trekked up river, to the present-day Gardiner area by September 14. They continued up river, avoiding the deep canyons and finally struck the Bannock Trail, which took them to the crossing near Tower Creek. From there the explorers followed the trail up Specimen Ridge, up the Lamar River to Calfee Creek. From their vantage point, some distance from the Grand Canyon, they plotted the direction of the falls, or where they supposed the falls to be. Using de Lacy's map and sighting off distant mountains they continued overland, leaving the Bannock Trail, which continued east over the mountains.

"Turning in the opposite direction," Folsom wrote, "as far as we could see from the northeast around to the southeast mountain and valley were covered with timber. In contrast with the barren, snowcapped peaks behind us (Absaroka Mts.) the dark green foliage deepened in hue as it receded, till it terminated at the horizon in boundless black forest. In several places a few miles to the southward we could see dense clouds of steam rising above the tree tops, indicating the location of tremendous hot springs." Those springs were the Basin Group near Shallow Creek, which they described as looking "like pots of boiling yellow paint and one of the largest was black as ink." They called these springs "The Chemical Works," where "the ground echoed beneath our tread with a hollow sound and we could hear the waters surging below, sending up a dull, resonant roar like the break of the ocean surf into a cave." The next day, which was a continuation of difficult going through rough terrain and fallen timber, Cook was leading the group when his horse halted abruptly, startling the rider. "I turned and looked," Cook remembered years later, "forward from the brink of the great canyon, at a point just across the canyon from what is now called Inspiration Point. I sat there in amazement, while my companions came up, and after that, it seemed to me that it was five minutes before anyone spoke." Cook and his fellow explorers had not been prepared for the splendor of the Grand Canyon of the Yellowstone — its immensity, its color, its reflection of the sky's light and cloud patterned walls. None of the trappers' tales or the prospectors descriptions had told them of the canyon's awesomeness and power.

From their vantage point they could not see or hear the falls. Following the canyon rim up river, they camped along the lower falls of the Yellowstone on the east bank of the river just a short walk above today's Chittenden Memorial Bridge. The falls, which Cook described as a "masterpiece of nature's handiwork," captivated the explorers. "We spent a day at the falls," Folsom recorded, "a day that has been a succession of surprises, and we return to camp realizing as we never have before how utterly insignificant are man's mightiest efforts when compared with the fulfillment of the Omnipotent Will."

They stared and marveled at the power of the Yellowstone. With forked pole and weighted twine, Folsom and his pals plumbed the depths of the Upper Falls' drop, gauging the precipice to be 115 feet; with makeshift methods the explorers overestimated the actual height by only six feet. The Lower Falls seemed even more magnificent. "The ragged edge of the precipice tears the water into a thousand streams," Cook wrote, "changing it to the appearance of molten silver, all united together and yet separate." Air turbulence at pool level under the powerful Lower Falls made measuring difficult; this time they erred on the high side by over fifty feet.

The trio continued their journey up the Yellowstone, remarking at the strange sounds and sights at Mud Volcano and Dragon's Mouth Spring. Striking Yellowstone Lake on September 24, eighteen days after they had left Diamond City, they dallied for three days pursuing the shoreline before heading over the divide and into the geyser basins. The geyser displays overwhelmed them, particularly the Great Fountain Geyser. But it was in the Midway Geyser Basin that they stood in awe at the Excelsior Geyser eruption: "They (hot springs) were situated along the river bank, and discharged so much hot water that the river was blood warm a quarter of a mile below. One of the springs (Excelsior) was two hundred and fifty feet in diameter . . ."

Their route out of the Yellowstone area was out the Madison drainage, on to Virginia City and then by stage back to Diamond City. Cook, Folsom and Peterson returned to the disbelief of some friends just thirty-six days after they rode out on what many believed a foolhardy mission. They had done what none others had — they made a circuit of the Yellowstone and set into motion a chain of events that ended with the declaration of the area as the nation's and the world's first national park in 1872. The Yellowstone had held its secrets for decades, and like a tantalizing viand that lives up to its aroma the discovery of "wonderland" immediately suggested preservation, not development. Over fifty years later, when Charles Cook returned to Yellowstone as an honored guest at the semicentennial of the Park in 1926, he remembered a discussion at the group's last camp on the Firehole River:

> "In camp that night we were talking over the great array of natural marvels we had seen and the scenic beauty of the area we had traversed. David Folsom, William Peterson and myself made up the party. Peterson remarked that probably it would not be long before settlers and prospectors began coming into the district and taking up the land around the canyons and the geysers, and that it would soon be all in private hands. I said that I thought the place was too big to be all taken up, but that, anyway, something ought to be done to keep the settlers out, so that everyone who wanted to, in future years, could travel through as freely and enjoy the region as we had. Then Folsom said: 'The Government ought not to allow anyone to locate here at all.' 'That's right,' I said, 'It ought to be kept for the public some way.' "

Thirty-Third Annual Field Conference — 1982
Wyoming Geological Association Guidebook

GEOLOGICAL INVESTIGATIONS IN YELLOWSTONE NATIONAL PARK, 1976-1981

WAYNE L. HAMILTON[1]

ABSTRACT

The 1959 Hebgen Lake earthquake triggered increased interest in the geology of Yellowstone and the surrounding area, especially the relationship between seismic and hydrothermal activity. By the mid-1960's this work had moved beyond its original scope, as it became known that the area was blanketed by ash-flow tuffs, with the recognition of major volcanic calderas. In the mid-to-late 1960's, the U.S. Geological Survey became heavily involved, with NASA support, in a variety of investigations, including studies of Eocene and Quaternary volcanics, glacial and surficial geology, and hydrothermal geology and geochemistry. The major products of this period of field work included fullpark bedrock and surficial geological maps, revised mapping on most 15 minute quadrangles of the park, and many technical papers. Some are still forthcoming.

The accomplishments of the U.S. Geological Survey set the stage for and attracted the interest of other earth scientists who are now focusing upon some of the larger questions of the volcanic history of Yellowstone, its setting in the tectonic framework of the region, evolution of magmas, development of calderas, modeling of hydrothermal systems, and many other topics.

The findings of the past five years are briefly summarized here, together with a listing of studies in broad categories, identification of areas of special interest to National Park Service scientists and managers, identification of some trends, and a listing of publications for that period. The number of geological investigations increased steadily between 1975 and 1980, but decreased by 40 per cent last year [1981].

INTRODUCTION

Yellowstone National Park is a unique geological area that attracts millions of visitors annually to view the natural wonders represented here. It is the responsibility of the National Park Service to interpret and explain those wonders to inquisitive visitors so that they gain a better understanding not only of Yellowstone, but of geological processes and concepts that apply in other parts of the world as well.

The uniqueness of Yellowstone also attracts earth scientists, and the National Park Service understands very well that to accomplish our interpretive mission we must facilitate geological investigations that further our understanding of the park. Most proposed studies are authorized by our permitting process provided that they are non-manipulative, cause no permanent environmental impact, and seek to evaluate a problem represented in Yellowstone that could not be similarly addressed elsewhere.

In addition to assuring resource protection, the permitting process must be egalitarian so that the National Park Service provides the opportunity for the pursuit of understanding that includes but also transcends "schools of thought" and "currently accepted theory". Ideally, many researchers in Yellowstone are those on the frontiers of their fields.

Moreover, Yellowstone National Park is a unique geological laboratory in which earth scientists can develop models that can be applied elsewhere to such fields as hydrothermal mineralization and alteration and to

[1]Physical Science Coordinator, National Park Service, Yellowstone National Park.

geothermal exploration.

Yellowstone encourages research by the serious and the responsible earth scientist with the understanding that all of us share in the awesome responsibility of protecting Yellowstone for future generations. Yellowstone has been designated a Biosphere Reserve and a World Heritage Site by the United Nations, so that it is not only a "crown jewel" of America but a precious natural heirloom for all of earth's inhabitants.

INVESTIGATIONS, 1976 - 1981

Hydrothermal Investigations

Continuing long-term chemical and isotopic studies of hydrothermal systems in Yellowstone have been conducted by Robert O. Fournier, Alfred H. Truesdell, and J. M. Thompson of the U.S. Geological Survey Branch of Experimental Geochemistry and Mineralogy, Menlo Park, California. Their work involves analyses of thermal waters, gases, and hot spring deposits, and a continuing assessment of waters in U.S.G.S. experimental drill holes. The overall goal is development of models of hydrothermal systems, development of geothermometers, and modeling the mixing of waters and steam loss affecting water chemistry. They have been joined by Harmon Craig and co-workers in determination of isotopic ratios of C, H, N, O, and He in studies to distinguish between shallow and deep sources.

Donald E. White, of the U.S. Geological Survey Branch of Experimental Geochemistry and Mineralogy, Menlo Park, California, continues his long-term investigations of geology and thermal activity with studies of the Norris Geyser Basin and Recent hydrothermal explosion craters. M. H. Beeson and L. J. P. Muffler, also of U.S.G.S. in Menlo Park, California, are completing large-scale maps of the Upper and Lower geyser basins. They are also finalizing petrologic, mineralogic, and chemical investigations of drill hole cuttings from U.S.G.S. drill holes.

Earthquake prediction studies, through monitoring of helium and chloride in thermal discharges, are being carried out by Irving Friedman and Dan Norton, of the U.S.G.S. Isotope Branch in Denver. They have installed equipment for automated monitoring of discharge, conductivity, and temperature at the Upper Boundary Creek Thermal Area.

R. A. Hutchinson continues to carry out a large share of the National Park Service effort on benchmark measurements and monitoring of thermal features in many parts of the park, with important contributions from permanent and seasonal NPS personnel in the Division of Interpretation and this writer. Hutchinson is developing a computer data file of geyser activity, and the writer is similarly filing all available data for thermal discharges parkwide.

Seismic studies of geyser activity are being done by Susan W. Keiffer, U.S.G.S., Flagstaff, who is modeling eruption mechanisms by evaluating the noises produced during eruptions, and David H. Oppenheimer and H. M. Iyer, U.S.G.S., Menlo Park, who are evaluating the nature of wave emissions from geyser systems. R. A. Levinson and R. W. Marrs, of the University of Wyoming, completed an evaluation of thermally active areas in the southwestern part of Yellowstone and identified, from remote sensing imagery, linear features that suggest possible hydrologic connections with the Island Park Geothermal Area.

David S. Chapman, William Parry, and John Bowman at the University of Utah, are in the process of completing their analysis of data collected during a 1981 assessment of thermal features in the southwest corner of the park. It is hoped that they may be able to model hydrologic inputs into those thermal areas.

Other significant geochemical and hydrologic investigations have been carried out by graduate students from Arizona State University (boron and heavy metals), Montana State University (hydrologic controls), and other institutions.

Miscellaneous Geochemistry

A variety of other geochemical studies have been conducted in the past five years. These include a geochemical reconnaissance of the Ashton Quadrangle, showing concentrations of uranium and other trace metals in water and sediments, by Spencer Shannon, Jr., and co-workers at Los Alamos Scientific Laboratory; a study of microstructure and chemistry of rock "varnish" at Dragon's Mounth Spring, by Ronald I. Dorn of the University of California, Berkeley; a study of clays representing ore-forming environments, by Mike Palin, New Mexico Technical University; and hot spring chemical analysis by Duncan Foley, University of Utah Research Institute.

Igneous Geology Investigations

Robert L. Christiansen of the U.S. Geological Survey in Menlo Park, with E. W. Hildreth, is continuing his studies of Pleistocene volcanic history in the Yellowstone area with particular attention to identification and mapping of volcanic rocks, origin and evolution of magmas, their relation to regional evolution of the crust and upper mantle, and possible future eruptions.

The U.S.G.S. work has included thermoluminescence dating of silicic volcanic rocks, by Rodd May, and a paleomagnetic study of lava flows and sediments by Richard L. Reynolds. Important work on Precambrian intrusives continues, with a description and mapping of the southwestern Beartooth Mountains, by Clarence J. Casella and Carla W. Montgomery of Northern Illinois University.

Other university-supported work includes a number of thesis and dissertation studies of such topics as petrology of Sylvan Pass intrusives and volcanics (University of Montana), mixed rhyolite-basalt lavas (University of Montana), mineralization of the Jardine-Crevice gold district (Johns Hopkins University), detailed mapping, chemistry, and petrography of Electric Peak and Sepulcher Mountain (University of New Mexico), Rb-Sr dating of Beartooth Precambrian rocks (Northern Illinois University), and Eocene ash-flow tuff studies (Wesleyan University and University of Chicago).

Geophysics

Some of the most interesting results in recent years have come from geophysical studies of the Yellowstone Caldera (or calderas) and the eastern Snake River plain. H. M. Iyer and co-workers at the U.S.G.S in Menlo Park have used teleseisms to model a deep low-velocity body underlying the Yellowstone Caldera. Their results argue against a deep mantle plume but show that the low velocity body may extend to as much as 300 km depth.

Much of the latest work has been done by Robert B. Smith and co-workers at the University of Utah and L. W. Braile and co-workers at Purdue University, with the former group concentrating on Yellowstone and the latter, the Snake River plain.

The Utah group has reported on rapid uplift of the Yellowstone Caldera as shown by leveling measurements (up to 15 mm per year near LeHardy Rapids). Uplift occurs in an elongate area corresponding closely to the configuration of the Yellowstone Caldera. They have also reported on seismic wave delays, especially in an area beneath the extreme northeastern end of the caldera. They interpret this attenuation, together with a gravity low in that area, as consistent with a partial melt body extending to approximately 3 km depth beneath the Hot Springs Basin Group — the largest area of steam-dominated hydrothermal activity in the park. Smith and co-workers have also installed seismographs in the southern part of the park, significantly extending the coverage of active seismic zones.

Most earthquake monitoring in Yellowstone has been done by A. M. Pitt of the U.S.G.S. Office of Earthquake Studies in Menlo Park. The U.S.G.S. Yellowstone Earthquake Network grew to a total of 26 seismographs by 1976, but due to budget cuts that number was reduced to 2 in late 1981. Pitt has reduced much of the data to produce epicenter maps with tabulated magnitudes. Hypocenters become appreciably shallower within the Yellowstone Caldera, supporing the suggestion of hot plastic rock at relatively shallow depth. Pitt has been joined by R. A. Hutchinson in evaluating the effects of a seismic disturbance in the Mud Volcanoes area and by George Zandt in developing crustal models using refraction data. These data have, of course, provided the base for the investigations confirming the existence of a low velocity body.

An interesting method of dating fault scarps has been employed by David B. Nash, University of Cincinnati. Nash has investigated scarp morphology and calculated paleoseismic chronology in the western part of the park.

H. M. Iyer and David H. Oppenheimer have also investigated microseisms at Norris Geyser Basin, finding that body wave microseisms do not originate in a deep hydrothermal system.

Paleoecology, Paleobotany, Paleoenvironment

Building upon definitive earlier work by Ehrling Dorf at Princeton University there has been a major effort to resolve questions involving the Eocene fossil forests in Yellowstone. Most of this work has been carried out by Lanny H. Fisk and William J. Fritz at Walla Walla College and Arthur Chadwick and his students at Loma Linda University. Fisk's group has concentrated upon paleoecological investigations, leading to a revised stratigraphic evaluation of the fossil forest. Chadwick's people have also developed paleoecological reconstructions, with emphasis upon dendrochronological cross-matching. Robert Ammons and co-workers at the University of Montana have also done cross-identification of ring structures in assessing Eocene climatic cycles.

Most recently, Richard Yuretich, University of Massachusetts, Amherst, has been reevaluating the contrasting hypotheses: *in situ* burial of trees by ash-flow (or mud-flow), and mudflow transport of trees to the depositional site. Investigations of recent paleoenvironment include a vegetation history of Yellowstone (from 3,700 years B.P.) based on pollen studies, by George C. Frison and Jane M. Beiswenger of the University of Wyoming, and study of limnological changes indicated by diatoms in Yellowstone Lake (from 3,900 years B.P.), by Michael Parker and Brian Shero, also of the University of Wyoming. The diatom study showed decreasing productivity over the interval.

Glacial Geology

The earlier work during this period was done by Gerald Richmond and Kenneth L. Pierce, U.S.G.S.

Denver. Their studies of glacial geology, palynology, paleoclimate, and paleomagnetics have resulted in a much improved understanding of the history of glaciation in Yellowstone. Pierce's studies show that large parts of Yellowstone were covered with ice up to 3,000 feet thick. John Good, with the National Park Service, is now involved in further fieldwork to refine earlier estimates of ice cover.

Other Geological Studies

Thomas Hauge, University of Southern California, has reported recently on results of a reevaluation of the Heart Mountain detachment fault. He proposes an even greater amount of movement than previously recognized.

TRENDS AND AREAS OF SPECIAL INTEREST

As can be seen from the above summary, Yellowstone attracts a large number of geological investigators and a great variety of interests. The one common factor linking these studies is that they relate to problems that are relatively unique to Yellowstone. Geologists are looking at problems here that in most cases are best represented in Yellowstone, and that is as it should be.

We expect to see gradual development of unifying concepts that explain the presence of the Yellowstone hot spot at the leading edge of the Snake River Plain that now bisects a "kink" in the trend of regional earthquakes. Although U.S. Geological Survey budgets for earthquake studies and geothermal research have been greatly reduced, we hope to see resumption of earthquake monitoring and hydrothermal systems studies. Undoubtedly other sources of funding can be found, but amounts are likely to be small. Because of reduced budgets, the National Park Service should not be viewed as a funding source. However, a small amount of research funds is channeled through the University of Wyoming — National Park Service Research Center at the University of Wyoming.

We would particularly like to see proposals for funded studies that would monitor microearthquakes, uplift, and hydrothermal chemistry changes in the northeast part of the Yellowstone Caldera as a means of evaluating possible magma movements there. We would also welcome research that would model sub-surface hydrology of hydrothermal systems in park boundary areas, especially those adjacent to the Island Park Geothermal area, the Yellowstone Known Geothermal Resource area (KGRA), and the Corwin Springs KGRA.

Yellowstone is in the process of developing interpretive displays that more clearly explain Quaternary volcanic activity and other fascinating events such as glaciation, readjustment of drainage to lava-damming, and effects of uplift, and clearly much can be done to improve our effectiveness in telling the Yellowstone story. We ask for the continued help and cooperation of all Yellowstone researchers in this effort and gratefully acknowledge the foresight, perseverence, and assistance of those whose names may be found in the bibliography that follows.

The number of geology-related research projects in Yellowstone increased in almost a linear fashion from 21 in 1976 to 33 in 1980 with a precipitous drop to 20 in 1981. We think this can be explained by reduced research budgets, particularly within the U.S.G.S. I am sure we will all be challenged over the next five years to find new ways to do more with less, and I look forward to working with many of you to attempt to carry on in the excellent established tradition.

BIBLIOGRAPHY

HYDROTHERMAL

Barger, K. E. and M. H. Beeson, 1981, Hydrothermal alteration in in research drill hole Y-2, Lower Geyser Basin, Yellowstone National Park, Wyoming, Amer. Mineralogist *66*, 473-490.

Barger, K. E., M. H. Beeson, and T. E. C. Keith, 1981, Zeolites in Yellowstone National Park, The Mineralogical Record, Jan.-Feb., p. 29-38.

Barger, K. E. and M. H. Beeson, Hydrothermal alteration in research drill hole Y-6, Upper Firehole River, Yellowstone National Park, Wyoming, in press.

Barger, K. E. and L. J. P. Muffler, Hydrothermal alteration in research drill hole Y-11, from a vapor-dominated geothermal system, at Mud Volcano, Yellowstone National Park, Wyoming, in press.

Beeson, M. H., T. E. C. Keith, and D. E. White, 1978, Hydrothermal alteration and self-sealing in Y-7 and Y-8 drill holes in the northern part of Upper Geyser Basin, Yellowstone National Park, Wyoming, U.S. Geol. Survey Prof. Paper 1054-A.

Beeson, M. H., T. E. C. Keith, and D. E. White, in preparation, Hydrothermal minerals in U.S. Geological Survey research drill hole Y-13, Yellowstone National Park, Wyoming.

Craig, H., J. E. Lupton, J. A. Welham, and R. Poreda, 1978, Helium isotope ratios in Yellowstone and Lassen park volcanic gases, Geophys. Res. Letters *5*, 897-900.

Craig, H., J. A. Whelham, R. Poreda, and J. E. Lupton, Helium isotope variations in the Yellowstone Snake River plain region and the western U.S. (abstract), EOS Trans. AGU, *60*, p. 945.

Des Maris, D. J., J. H. Donchin, N. L. Nehring, and A. H. Truesdell, 1981, Molecular carbon isotopic evidence for the origin of geothermal hydrocarbons, Nature *22*, 826-828.

Doering, W. P. and I. Friedman, 1980, Survey of helium in natural water wells and springs in southwest Montana and vicinity, U.S. Geol. Surv. Open-File Report 80-181.

Duffield, W. A., 1981, the Geothermal Research Program of the Geol. Surv., U.S. Geol. Survey Open-File Report 81-564.

Evans, K. L., 1981, Ion Chromatographic analysis of geological and environmental samples, Ph.D. Dissertations, Arizona State University.

Fournier, R. O., D. E. White, and A. H. Truesdell, 1976, Convective heat flow in Yellowstone National Park, *in* Proceedings of the Second U.N. Symposium on Development and Use of Geothermal Resources, 1, 731-739.

Fournier, R. O., 1979, A revised equation for the Na/K geothermometer: Geothermal Resources Council, *Transactions*, vol. 3, Sept. 1979, p. 221-224.

Friedman, I. and D. R. Norton, 1981, Ground temperature measurements, part III, Ground temperature in and near Yellowstone National Park, U.S. Geol. Surv. Prof. Paper 1203.

Ghiorso, M. S., 1980, Studies in natural solid/liquid equilibrium, Ph.D. dissertation, University of California, Berkeley.

Hirschmann, F., 1982, *Yellowstone*, Graphic Arts Center Publishing Co., Portland, Oregon, 88 p.

Hutchinson, R. A., 1979, Geologic setting of Sylvan Springs geothermal area, Yellowstone National Park, Masters Thesis, Iowa State Univ., 43 p.

Hutchinson, R. A., 1980, Boundary Creek thermal areas of Yellowstone National Park, I: Thermal activity and geologic setting, Geothermal Resources Council Transactions, *4*, 129-132.

Keith, T. E. C., J. M. Thompson, and R. E. Mays, Selective Concentration of Cesium during hydrothermal alteration, Yellowstone National Park, Wyoming, in press.

Kieffer, S. W. and S. Ihnen, 1977, seismic and aseismic geyser eruptions, EOS Trans., A.G.U., *58*: 12, p. 1187.

Kieffer, S. W., 1977, Fluid dynamics during eruption of water — steam and magma-gas systems: geysers, maars, and diatremes, second International Kimberlite Conf., Abstr., September 1977, Santa Fe, N.M.

Levinson, R. A. and R. W. Marrs, 1979, Evaluation of thermally active areas in southwestern Yellowstone National Park, Univ. of Wyoming Remote Sensing Laboratory Report for contract CX-1200-7-Bo33.

Martinez, S., 1978, Report on geyser and hot spring activity during the field trip of 8/20/78 - 9/3/78, National Park Service report, 55 p.

McKenzie, W. F. and A. H. Truesdell, 1977, Geothermal reservoir temperatures estimated from the oxygen isotope compositions of dissolved sulfate and water from hot springs and shallow drillholes, Geothermics, *5*, 51-61.

Muffler, L. J. P., D. E. White, M. H. Beeson, and A. H. Truesdell, 1982, Geologic Map of Upper Geyser Basin, Yellowstone National Park, Wyoming, U.S. Geol. Surv. Misc. Invest. Ser. Map I-1371.

Nehring, N. L. and A. H. Truesdell, 1978, Hydrocarbon gases in some volcanic and geothermal systems, Geothermal Resources Council Transactions, *2*, 483-486.

Parry, W. T., D. S. Chapman, J. R. Bowman, and R. G. Allis, 1982, Quantitative assessment of thermal features of the southwest portion of Yellowstone National Park, 8 p.

Pearson, F. J., Jr., and A. H. Truesdell, 1978, Tritium in the waters of Yellowstone National Park, U.S. Geol. Survey Open-File Report 78-701.

Pitt, A. M. and R. A. Hutchinson, 1982, Hydrothermal changes related to earthquake activity at Mud Volcano, Yellowstone National Park, Wyoming, Journ. Geophys. Res., *87*, 2762-2766.

Rinehart, J. S., 1980, *Geysers and geothermal energy*, Springer-Verlag Inc., New York.

Robertson, E. C., R. O. Fournier, and C. P. Strong, 1976, Hydrothermal activity in Southwestern Montana, Proc. Second United Nations Symposium on Development and Use of Geothermal Resources, *1*, 553-561.

Schoen, R. and R. O. Rye, 1976, Sulfur isotope distributions in solfataras, Yellowstone National Park, Science *170*, 1082-1084.

Stauffer, R. E. and J. M. Thompson, 1976, Arsenic and Antimony in geothermal waters of Yellowstone National Park, Abstr. with Program, First Conference on Scientific Research in the National Parks, NPS-AIBS.

Stauffer, R. E. and J. M. Thompson, 1978, Phosphorus in hydrothermal waters of Yellowstone National Park, Wyoming, Jour. Research U.S. Geol. Survey, *6*:6, 755-763.

Stauffer, R. E., E. A. Jenne, and J. W. Ball, 1980, Chemical Studies of selected trace elements in hot spring drainages of Yellowstone National Park, U.S. Geol. Survey Prof. Paper 1044-F.

Strusacker, E. M., 1976, geothermal systems of the Corwin Springs-Gardiner areas, Montana: Possible structural and lithologic controls, Masters Thesis, Montana State University, 93 p.

Thompson, J. M., 1979, Arsenic and fluoride in the upper Madison River system: Firehole and Gibbon Rivers and their tributaries, Yellowstone National Park, Wyoming and southwest Montana, Environmental Geology, *3*, 13-21.

Thompson, J. M. and S. Yadav, 1979, Chemical analyses of waters from geysers, hot springs, and pools in Yellowstone National Park, Wyoming from 1974 to 1978, U.S. Geol. Survey Open-File Report 79-704.

Thompson, J. M. and R. A. Hutchinson, 1980, Boundary Creek thermal areas of Yellowstone National Park II: Thermal water analyses, Geothermal Resources Council Trans. *4*, 189-192.

Thompson, J. M. and R. A. Hutchinson, 1981, Chemical analysis of waters from the Boundary Creek Thermal Area, Yellowstone National Park, Wyoming, U.S. Geol. Survey Open-File Report 81-1310.

Truesdell, A. H. and R. O. Fournier, 1976, Conditions in the deeper parts of the hot spring system of Yellowstone National Park, Wyoming, U. S. Geol. Survey Open-File Report 76-428, 30 p.

Truesdell, A. H., 1976, Summary of Section III, geochemical techniques in explorations, Proceedings 2nd U.N. Symposium on the Development of Geothermal Resources, San Francisco, 1975, *1*, U.S. Govt. Printing Office.

Truesdell, A. H. and R. O. Fournier, 1976, Calculation of deep temperatures in geothermal systems from the chemistry of boiling spring waters of mixed origin, Proc. 2nd U.N. Symposium on the Development and Use of Geothermal Resources, San Francisco, v. 1.

Truesdell, A. H., 1976, Chemical evidence of subsurface structure and fluid flow in a geothermal system, Proc. for the International Symposium on Water-Rock Interactions, Czechoslavakia, 250-257.

Truesdell, A. H., M. Nathenson, and R. O. Rye, 1977, The effects of subsurface boiling and dilution on the isotopic compositions of Yellowstone thermal waters, Jour. Geophysical Res., *82*, 3694-3704.

Truesdell, A. H., R. O. Rye, F. F. Whelan, and J. M. Thompson, 1978, Sulfate chemical and isotopic patterns in thermal waters of Yellowstone National Park, Wyoming, U.S. Geol. Survey Open-File Report 78-701.

Truesdell, A. H., R. O. Rye, J. F. Whelan, and J. M. Thompson, 1978, Sulfate chemical and isotopic patterns in thermal waters of Yellowstone Park, Wyoming: Short papers of the Fourth International Conference, Geochronology, Cosmochronology, Isotope Geology, 1978, U.S. Geol. Survey Open-File Report 78-701, p. 435-436.

Truesdell, A. H. and J. R. Hulston, 1980, Isotopic evidence on environments of geothermal systems, *in Handbook of Environmental Isotope Geochemistry, Vol. 1, The Terrestrial Environment, A*, edited by P. Fritz & J. C. Fontes, (chapter 5), p. 179-226.

Wehlan, J. A., 1981, Carbon and hydrogen gases in hydrothermal systems: the search for a mantle source, Ph.D. Dissertation, University of California, San Diego, 194 pages.

White, D. E., 1978, Conductive heat flows in research drill holes in thermal areas of Yellowstone National Park, Wyoming, J. Res. U.S. Geological Survey, *6*:6, p. 765-774.

Wrenn, M., 1981, Tritium in hot spring water at Porcelain Basin. Unpublished report University of Utah.

OTHER GEOTHERMAL STUDIES

Dorn, R. I., 1981, Analysis of Dragon's Mouth hydrothermal deposits, memorandum in Yellowstone National Park Library.

Foley, D., D. L. Nielson, and C. R. Nichols, 1980, Geothermal systems of the Yellowstone Caldera, Field Trip Guide, Geothermal Resources Council, Sept. 6, 7, 8, 1980, 69 p.

Keith, T. E. C., and L. J. P. Muffler, 1978, Minerals produced during cooling and hydrothermal alteration of ash flow tuff from Yellowstone drill hole Y-5, Jour. Volcanol. Geothermal Res., *3*, p. 373-402.

Keith, T. E. C., D. E. White, and M. H. Beeson, 1978, Hydrothermal alteration and self-sealing in Y-7 and Y-8 drill holes in northern part of Upper Geyser Basin, Yellowstone National Park, Wyoming, U.S. Geol. Survey Prof. Paper 1054-A, 26 p.

Phelps, D., and P. R. Buseck, 1980, Distribution of soil Mercury and the development of soil Mercury anomalies in the Yellowstone geothermal area, Wyoming, Economic Geology *75*, p. 730-741.

Shannon, S. S., Jr., 1980, Uranium hydrogeochemical and stream sediment reconnaissance data release for the Ashton NTMS Quadrangle, Idaho/Montana/Wyoming, including concentrations of forty-two additional elements, Los Alamos Scientific Laboratory Informal Report LA-8009-MS, 168 p.

IGNEOUS GEOLOGY

Casella, C. J., 1979, The geology of the southwestern Beartooth Mountains, Yellowstone National Park, Montana and Wyoming , *in* Guide to the Precambrian rocks of the Beartooth Mountains, Univ. of Florida Press, p. 5-9.

______, 1979, Definition and extent of the Precambrian Beartooth Batholith, Montana and Wyoming, Geol. Soc. Amer. Abstr. with Programs *11*:6, p. 268.

Christiansen, R. L., and H. R. Blank, in press, Quaternary and Pliocene volcanism of the Yellowstone rhyolite plateau region of Wyoming, Idaho, and Montana, U.S. Geological Survey Prof. Paper 729-.

Christiansen, R. L., in press, Late Cenozoic volcanism of the Island Park area, eastern Idaho, for Cenozoic Geology of Idaho, Idaho Bur. of Mines and Geology.

Christiansen, R. L., in press, Yellowstone Magmatic evolution: its bearing on understanding large-volume explosive volcanism, for Symposium on Explosive Volcanism, National Academy of Sciences.

French, A. N., and R. Van Der Voo, 1978, Absaroka Mountains, Yellowstone National Park, a summary report of a paleomagnetic study of lower Eocene volcanics, in Yellowstone National Park Library, 10 p.

Hedenquist, J. W., 1979, Geologic history including mineralization of the Archean rocks of the Jardine-Crevice Au, As, W district, Park County, Montana, unpublished report, Yellowstone research file, 1 p.

Hildreth, W., 1981, Gradients in silicic magma chambers: Implications for lithospheric magmatism, J. Geophys. Res. *86:* B11, p. 10153-10192.

Izette, G. A., 1981, Volcanic ash beds: recorders of Upper Cenozoic silicic pyroclastic volcanism in the western United States, J. Geophys. Res. *86*: B11, p. 10200-10222.

Izette, G. A., and R. E. Wilcox, 1981, Map showing localities and inferred distributions of the Huckelberry Ridge, Mesa Falls, and Lava Creek ash beds (Perlette family ash beds) of Pliocene and Pleistocene age in the western United States and Canada, U.S. Geol. Surv. Misc. Invest. Map I-1325.

LaPointe, D. D., 1977, Geology of a portion of the Eocene Sylvan Pass volcanic center, Absaroka Range, Wyoming, Masters Thesis, University of Montana, 60 p.

Leeman, W. P., and D. W. Phelps, 1981, Partitioning of rare earths and other trace elements between sanidine and coexisting volcanic glass, J. Geophys. Res. *86*:, B11, p. 10193-10199.

Leeman, W. P., in press, Development of the Snake River Plain and Yellowstone Plateau Province: an overview petrologic model, Bull. Idaho Bur. Mines and Geology.

Love, L. L., Am. M. Judo, and D. W. Love, 1976, Dacites of Bunsen Peak, the Birch Hills, and the Washakie Needles, Northwestern Wyoming, and their relationship to the Absaroka volcanic field, Wyoming and Montana, G.S.A. Bull. *87*, p. 1455-1462.

Lytwyn, J. N., 1982, Geochemistry and ages of Precambrian plutonic rocks from the southwest Beartooth Mountains, Yellowstone Park, Wyoming and Montana, M.S. Thesis, Northern Illinois University.

Montgomery, C. W., 1981, Metamorphic history of Archean schists from the southwest Beartooth Mountains, Montana and Wyoming, Geol. Soc. America Abstr. with Programs *13*:4, p. 220.

Montgomery, C. W., and J. N. Lytwyn, 1981, Ages of Precambrian plutonic rocks and migmatites in the southeast Beartooth Mountains, Geol. Soc. America Abstr. with Programs *13*:7, p. 513.

Montgomery, C. W., 1982, Preliminary zircon U-Pb dating of biotite grandodiorite from the South Snow Block, Beartooth Mountains, in press Montana Bur. Mines Special Publication.

Reynolds, R. L., 1979, Comparison of the TRM of the Yellowstone Group and the DRM of some Perlette ash beds, J. of Geophys. Res., *84*: B9.

Struhsacker, D. W., 1978, Mixed basalt-rhyolite assemblages in Yellowstone National Park, the petrogenic significance of magma mixing, Masters Thesis, University of Montana.

Wooden, J. L., C. W. Montgomery, and C. J. Casella, 1979, Rb-Sr and zircon U-Pb ages of the granitic rocks of the southwest Beartooth Mountains within Yellowstone Park, Montana and Wyoming, Geol. Soc. America Abstr. with Programs *11*:6, p. 306.

GEOPHYSICS

Benz, H. M., and R. B. Smith, 1981, Simultaneous inversion for lateral velocity variations and hypocenters in the Yellowstone region using earthquake and controlled-source data, EOS, Transactions, *62*, 470.

Braile, L. W., R. B. Smith, *et al.*, 1982, The Yellowstone — Snake River Plain seismic profiling experiment: Crustal structure of the eastern Snake River Plain, J. Geophys. Res. *87*, 2597-2609.

Clawson, S. R., and R. B. Smith, 1981, The lateral Q-structure of the upper crust in Yellowstone National Park from seismic refraction data, EOS, Transactions, *62*, 970.

Daniel, R. G., and D. M. Boore, 1982, Anomalous shear wave delays and surface wave velocities at Yellowstone Caldera, Wyoming, J. Geophys. Res. *87*, 2731-2744.

Friedline, R. A., R. B. Smith, and D. D. Blackwell, 1976, Seismicity and contemporary tectonics of the Helena, Montana area, 1976, Bull. of the Seismol. Soc. of America, *66*:1, 81-95.

Iyer, H. M., J. R. Evans, G. Zandt, R. M. Stewart, J. M. Coakley, and J. N. Roloff, 1981, A deep low-velocity body under the Yellowstone Caldera, Wyoming: Delineation using teleseismic P-wave residuals and tectonic interpretation: Summary, Geol. Soc. of America Bull., Part 1, *92*, p. 792-798.

Iyer, H. M., J. R. Evans, G. Zandt, R. M. Stewart, J. M. Coakley, and J. N. Roloff, 1981, A deep low-velocity body under the Yellowstone Caldera, Wyoming: Delineation using teleseismic P-wave residuals and tectonic interpretation, Geol. Soc. of America Bull., Part II, *92*:11, p. 1471-1646.

Lehman, J. A., R. B. Smith, M. M. Schilly, and L. W. Braile, 1982, Upper crustal structure of the Yellowstone Caldera from delay time analyses and gravity correlations, J. Geophys. Res. *87*, 2713-2730.

Morgan, P., D. D. Blackwell, R. E. Spafford, and R. B. Smith, 1977, Heat flow measurements in Yellowstone Lake and the thermal structure of the Yellowstone Caldera, J. Geophys. Res. *82*, 3710-3732.

Oppenheimer, D. H., and H. M. Iyer, 1979, Microseisms analysis at Norris Geyser Basin, Yellowstone National Park, Wyoming, Geothermal Resources Council Trans. *3*, p. 523-525.

Oppenheimer, D. H., and H. M. Iyer, 1980, Frequency-wavenumber analysis of geothermal microseisms at Norris Geyser Basin, Yellowstone National Park, Wyoming, Geophysics *45*:5, p. 952-963.

Otis, R. M., and R. B. Smith, 1977, Homomorphic deconvolution by log spectral averaging, Geophys. *42*:6, p. 1146-1157.

Otis, R. M., R. B. Smith, and R. J. Wold, 1977, Geophysical surveys of Yellowstone Lake, Wyoming, J. Geophys. Res. *82:26*, 3705-3717.

Pelton, J. R., and R. B. Smith, 1979, Recent crustal uplift in Yellowstone National Park, Science *206*, 1179-1182.

Pelton, J. R., and R. B. Smith, 1982, Contemporary vertical surface displacements in Yellowstone National Park, J. Geophys. Res. *87:*B4.

Pitt, A. M., 1979, Earthquake epicenters in Yellowstone Park and vicinity, 1973-1978, U.S. Geol. Survey Open-File Report 79-717.

Pitt, A. M., 1979, Preliminary map of earthquake epicenters in Yellowstone Park and vicinity, 1973-1978, U.S. Geol. Survey Open-File Report 79-717.

Pitt, A. M., C. S. Weaver, and W. Spence, 1979, The Yellowstone Park earthquake of June 30, 1975, Bull. of Seismol. Soc. of America, *69:*1, 187-205.

Pitt, A. M., 1980, Catalogue of earthquakes in the Yellowstone Park — Hebgen Lake region, from Nov. 1972 to Dec. 1975, U.S. Geol. Survey, Open-File Report 80-2006.

Pitt, A. M., 1981, Preliminary catalogue of earthquakes in the Yellowstone Park — Hebgen Lake region, from January to March, 1981, unpubl. rept.

Pitt, A. M., 1981, Preliminary catalogue of earthquakes in the Yellowstone Park — Hebgen Lake region, from April to June, 1981, unpubl. rept.

Pitt, A. M., 1981, Preliminary catalogue of earthquakes in the Yellowstone Park — Hebgen Lake region, from October to December, 1980, unpubl. rept.

Schilly, M. M., R. B. Smith, J. A. Lehman, L. W. Braile, and J. Ansorge, 1982, The Yellowstone-eastern Snake River Plain seismic profiling experiment: Data and detailed crustal structure of the Yellowstone region, J. Geophys. Res. *87*, 2692-2704.

Shuey, R. T., R. O. Ugland, and C. R. Schmit, 1977, Geophysical study of a hydrothermal variation in cores from Yellowstone and Jackson lakes, Wyoming, J. Geophys. Res. *82:26*, 3739-3746.

Smith, R. B., in press, Crustal structure and evolution of an explosive silicic volcanic system at Yellowstone, for National Academy of Sciences Symposium on Explosive Volcanism.

Smith, R. B., M. M. Schilly, L. W. Braile, J. A. Lehman, J. Ansorge, M. R. Baker, C. Prodehl, J. H. Healy, S. Muller, and R. W. Greenfelder, 1982, The 1978 Yellowstone-eastern Snake River Plain seismic profiling experiment: Crustal structure of the Yellowstone region and experiment design, J. Geophys. Res. *87*.

Smith, R. B., 1977, Intraplate tectonics of the western North American plate, Tectonophysics *37*, p. 323-336.

Smith, R. B., and R. T. Shuey, 1977, Earthquake and magnetic evidence for the Yellowstone hot spot, J. Geophys. Res. *82:26*.

Smith, R. B., R. T. Shuey, J. R. Pelton, and J. P. Bailey, 1977, Yellowstone hot spot: Contemporary tectonics and crustal properties from earthquakes and aeromagnetic data, J. Geophys. Res. *82*, 3665-3676.

Smith, R. B., and R. L. Christiansen, 1980, Yellowstone Park as a window on the earth's interior, Scientific American *242*, 104-117.

Stanley, W. D., J. E. Boehl, F. X. Bostick, and H. W. Smith, 1977, Geothermal significance of magnetotelluric sounding in the eastern Snake River Plain — Yellowstone region, J. Geophys. Res. *82*, 2501-2514.

Ugland, R. O., 1976, An investigation of paleomagnetic secular variation in the sediments of Yellowstone Lake and Jackson Lake Wyoming, Masters Thesis, University of Utah.

Wold, R. J., M. A. Mayhew, and R. B. Smith, 1977, Geophysical study of a hydrothermal explosion crater in Mary Bay, Yellowstone Lake, Wyoming, J. Geophys. Res. *82:26*.

Zandt, G., 1978, Study of three-dimensional heterogeneity beneath seismic arrays in central California and Yellowstone, Wyoming, Ph.D. Dissertation, Massachusetts Institute of Technology, 489 p.

PALEOBOTANY, PALEOECOLOGY, PALEOCLIMATE

Aguirre, M. R., 1980, An Eocene leaf florule from the Amethyst Mountain "fossil forest", Yellowstone National Park, Wyoming, Masters Thesis, Walla Walla College, 82 p.

Ammons, R., A. Ammons, and R. B. Ammons, 1980, Cross-identification of ring structures in apparently contemporary Eocene trees of Yellowstone: preliminary report of methodology and evidence, Montana Academy of Science, April 1980 meeting.

Arct, M. J., 1979, Dendrochronology in the Yellowstone fossil forests, Masters Thesis, Biology, Loma Linda University, 65 p.

Baker, R. G., 1976, Late Quaternary vegetation history of the Yellowstone Lake Basin, Wyoming, U.S. Geol. Survey Prof. Paper 729-E.

Beiswenger, J. M. and G. C. Frison, 1979, Vegetational history of Yellowstone Park as determined by pollen analysis, Dept. of Anthropology, Univ. of Wyoming Proj. 2/78-79, 30 p.

Chadwick, A. and T. Yamamoto, submitted for publication, A paleoecological analysis of the petrified trees in the Specimen Creek area of the Gallatin petrified forest, Yellowstone National Park, Montana, U.S.A.

Coffin, H. G., 1976, Orientation of trees in the Yellowstone Petrified Forest, Jour. of Paleontology, *50:*3, 539-543.

DeBord, P. C., 1976, Palynology of the Gallatin Mountain "fossil forests" of Yellowstone National Park, Montana: preliminary report, Abstr., First Conference on Scientific Research in the National Parks, New Orleans, NPS-AIBS.

DeBord, P. C., 1977, Gallatin Mountain "petrified forests": a palynological investigation of the *in situ* model, Ph.D. Dissertation, Loma Linda University, 86 p.

Fisk, L. H., 1976, Palynology of the Amethyst Mountain "fossil forest", Yellowstone National Park, Wyoming, Ph.D. Dissertation, Loma Linda University, 357 p.

Fritz, W. J., 1977, Paleoecology of petrified woods from the Amethyst Mountain "fossil forests", Yellowstone National Park, Wyoming, Masters Thesis, Walla Walla College, 57 p.

Fritz, W. J. and L. H. Fisk, 1978, Eocene petrified woods from one unit of the Amethyst Mountain "fossil forest", Northwest Geology, 7, 10-19.

Fritz, W. J. and L. H. Fisk, 1979, Paleoecology of the petrified woods from the Amethyst Mountain "fossil forests", Yellowstone National Park, Wyoming, *in* National Park Service Proceedings, Ser. 5, *2*, 743-749.

Fritz, W. J., 1980, Depositional environments of the Eocene Lamar River Formation in Yellowstone National Park, Ph.D. Dissertation, Univ. of Montana, 113 p.

Fritz, W. J., 1980, Stratigraphic framework of the Lamar River Formation in Yellowstone National Park, Northwest Geology, *9*, 1-18.

Fritz, W. J., 1980, Reinterpretation of the depositional environment of the Yellowstone "fossil forests", Geology, *8*, 309-313.

Fritz, W. J., 1981, Reinterpretation of the depositional environment of the Yellowstone "fossil forests"; Reply, Geology, *9*, 53-54.

Gennett, J. A., 1977, Palynology and paleoecology of sediments from Blacktail Pond, northern Yellowstone Park, Wyoming, Masters Thesis, University of Iowa, 74 p.

Retallack, G., 1981, Reinterpretation of the depositional environment of the Yellowstone "fossil forests"; Comment, Geology *9*, 52-53.

Richmond, G. M., W. M. Mullenders, and M. Coremans, 1978, Climatic implications of two pollen analyses from newly recognized rocks of latest Pliocene age in the Washburn Range, Yellowstone National Park, Wyoming, U. S. Geological Survey Bull. 1455, 13 p.

Schmit, C. R., 1977, Sedimentology and geochemistry of piston cores from Yellowstone and Jackson Lakes, Wyoming, Ph.D. Dissertation, University of Utah.

Shero, B. and M. Parker, 1976, Limnological changes in Yellowstone Lake as indicated by sediment diatoms, University of Wyoming, Dept. of Zoology and Physiology Contract 43-8329 for the National Park Service, 114 p.

Yamamoto, T. and A. Chadwick, 1980, Identification of fossil wood from the Specimen Creek area of the Gallatin petrified forest, Yellowstone National Park, Montana, in press.

Yuretich, R. F., 1981, Comment and reply on 'Reinterpretation of the depositional environment of the Yellowstone "fossil forests" ' and 'Stumps transported and deposited upright by Mount St. Helens mud flows'. *Geology* G.S.A., *9*, p. 146-147. Reply by Wm. J. Fritz.

GLACIAL GEOLOGY

Pierce, K. L., J. D. Obradovich, and I. Friedman, 1976, Obsidian hydration dating and correlation of Bull Lake and Pinedale glaciations near West Yellowstone, Montana, Geol. Soc. of America Bull. *87*, 703-710.

Pierce, K. L., 1979, History and dynamics of glaciation in the northern Yellowstone National Park area, U.S. Geological Survey Professional Paper 729-F, 90 p.

Porter, S. C., K. L. Pierce, and T. Hamilton, in press, Glaciation of the Western United States, *in Late Wisconsin Glaciation of the United States*, S. C. Porter, ed.

Richmond, G. M., 1976, Surficial geologic history of the Canyon Village Quadrangle, Yellowstone National Park, Wyoming, for use with Map I-652, U.S. Geol. Surv. Bull. 1427, 35 p.

STRUCTURAL AND GENERAL GEOLOGY

Hauge, T. A., 1982, The Heart Mountain detachment fault: Role of Absaroka volcanic rock, Abstract with Program, Rocky Mountain Section, Geol. Soc. of America meeting.

Montagne, J., and R. A. Chadwick, 1982, Cenozoic history of the Yellowstone valley south of Livingston, Montana, Fieldtrip Guidebook in conjunction with the 35th Ann. Meeting of the Rocky Mountain Section, Geol. Soc. of America, 67 p.

Nash, D. B., 1981, Fault scarp morphology: Indicator of paleoseismic chronology, U.S. Geol. Surv. Final Tech. Report, Contract No. 14-08-0001-19109.

Thirty-Third Annual Field Conference — 1982
Wyoming Geological Association Guidebook

HISTORY AND DYNAMICS OF GLACIATION IN THE NORTHERN YELLOWSTONE NATIONAL PARK AREA*

BY KENNETH L. PIERCE[1]

Large masses of ice formed in northern Yellowstone National Park during the Bull Lake and Pinedale Glaciations. Bull Lake terminal moraines of the northern Yellowstone icemasses occur only on the park's western margin, in the West Yellowstone area. The Bull Lake Glaciation here is about 150,000 years old as dated by combined obsidian hydration and K-Ar methods. Bull Lake ice advanced into the West Yellowstone basin largely through lowlands later filled by the West Yellowstone rhyolite flow. The moraines deposited by this advance are bulky, loess mantled, and have the muted but clearly glacial constructional morphology typical of the Bull Lake elsewhere in the Rocky Mountains. Bull Lake glaciers extended 20 kilometers beyond Pinedale glaciers west of the park; but north of the park, Pinedale glaciers overrode and obliterated Bull Lake terminal moraines. This difference is attributed to emplacement of large rhyolite flows in the western part of the park in the interval between the Bull Lake and the Pinedale maxima. These rhyolite flows have K-Ar ages between 70,000 and 120,000 years old. They did not affect the flow of Bull Lake ice but diverted Pinedale flow by blocking westward flow and shunting it to the north.

Just downvalley from the Pinedale terminus of the outlet glacier in the Yellowstone valley are moraines of both Bull Lake and Pinedale age deposited by *local* glaciers. The location and degree of preservation of these moraines demonstrate that if any outlet glacier existed larger than the Pinedale one, it was of pre-Bull Lake age.

At some time between the Pinedale and Bull Lake glacial maxima, ice from the Beartooth uplift advanced southward beyond the Washburn Range and dammed the Yellowstone River, leaving erratices of Precambrian crystalline rocks.

During Pinedale time, major ice streams from four sources converged near Gardiner, Montana, to form the northern Yellowstone outlet glacier. This outlet glacier then flowed 60 km down the Yellowstone valley to the Eightmile terminal moraines, and although it was joined by several tributary glaciers along the way, the main outlet glacier dominated ice flow.

The northern Yellowstone outlet glacier and its accumulation area together constituted the northern Yellowstone glacier. This glacier covered 3,400 km^2, averaged more than 700 m in thickness, and included a flow line 146 km long. Five interconnected icecap sources contributed to the northern Yellowstone glacier: an icecap on the Gallatin Range, an icecap on the plateau between the Gallatin and Washburn Ranges, an icecap on the plateau south of the Washburn Range, an icecap on the Beartooth uplift, and an icecap in the upper Lamar drainage basin.

The ice divides along the crest of these icecaps defined the outer boundary of the northern Yellowstone glacier. Across these divides, from the northern Yellowstone glacier, glacier flow was toward the other glacial termini, generally tens of kilometers away. The ice divides were generally offset toward the northern Yellowstone glacier from the present drainage divides. Thus, some ice flowed outward from the Yellowstone drainage into adjacent basins where ice levels were lower.

Obsidian hydration methods date Pinedale terminal moraines near West Yellowstone as about 30,000 years old. Pinedale glaciers probably remained near full-glacial size until 15,000 to 20,000 years ago, and deglaciation of the Pinedale icecap on the Yellowstone plateau was largely completed by 13,000 to 14,000 radiocarbon years ago.

During Pinedale recession, ice from different sources did not diminish at the same rate. The northern Yellowstone outlet glacier was still near its maximum volume when tributary mountain-valley glaciers had receded and were contributing little or no ice to it. Lakes formed where the outlet glacier blocked these tributary valleys. In Tom Miner Basin the outlet glacier advanced upvalley and deposited moraines and erratices within a few kilometers of the local cirques. Elsewhere it transported erratics at high levels across the mouths of valleys, such as Mol Heron and Bear Creeks, defining boundaries across which local ice did not advance.

At a later phase in the Pinedale recession, here termed the Deckard Flats readjustment, glaciers from the Gallatin Range and Beartooth uplift advanced into terrain previously occupied by the icecap on the Yellowstone

*Abstract taken from U.S. Geological Survey Professional Paper 729-F (90 pages). Reprinted with permission of the author and the director of the U.S.G.S.

[1]U.S. Geological Survey, Denver, Colorado 80255.

plateau. Crossed striations and distribution of erratics indicate that their direction of flow was at angles of 60°-130° to the flow under full-glacial conditions.

A late Pinedale readvance of mountain-valley glaciers in the Gallatin Range left as many as three moraines 0.5-8 km from the cirques. At the same time, the icecap on the Beartooth uplift paused in its recession and left extensive moraines near Junction Butte.

Along the Yellowstone River in the Gardiner area, longitudinal and midchannel bars and flood-ripped fronts of alluvial fans indicate the occurrence of at least two separate floods, with waters 60 m and 45 m deep. Two additional, younger floods down the valley of the Yellowstone River resulted from the breach of a landslide dam in Yankee Jim Canyon and from release of a lake in the Grand Canyon of the Yellowstone.

Calculation of glaciological parameters, especially basal shear stress and mass balance, provides independent means for evaluating the glacial reconstruction based on field studies. Basal shear stress was computed using the formula $T^b = Foghsin\alpha$. Because this formula ignores the effect of longitudinal stress gradients, its calculated value is called primary basal shear stress. Calculations were made for 50 contiguous reaches of lengths of 5-15 km along the main flow paths of the northern Yellowstone glacier and of several other glaciers heading on the same ice divide. The calculated values are all between 0.6 and 1.5 bars and are consistent with the range of 0.5-1.5 bars observed for modern glaciers. Much of the variation in primary basal shear stress is related to differing flow patterns. Primary basal stress averaged 1.21 $\pm$ 0.12 bars for reaches with strongly extending flow lines, 1.04 $\pm$ 0.16 bars for reaches with nearly uniform flow, and 0.84 $\pm$ 0.14 bars for reaches with strongly compressing flow. These calculated values probably do not indicate actual differences in shear stress at the base of the glacier, but do suggest that the variation in primary basal shear stress is strongly related to flow pattern, especially convergence in plan view, and to ablation.

Basal shear stress calculations can also be used as an independent parameter in evaluation of alternate glacial reconstructions. In particular, an alternate reconstruction published for Yellowstone shows Pinedale ice levels 430 to 730m lower on the average than previously mapped* in the Washburn Range, but at the same altitudes towards the ice margin. This reconstruction results in basal shear stress values between 0.05 and 0.15 bars, as much as an order of magnitude lower than that observed for modern glaciers. Thus, in addition to being incompatible with the field observations reported in this paper, this alternate reconstruction appears glaciologically unreasonable.

The mass balance of the reconstructed northern

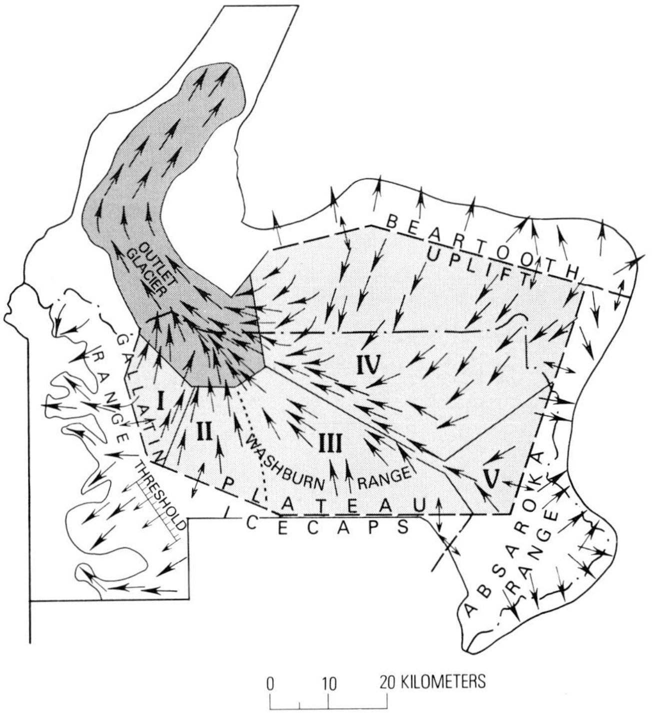

Figure 1: The northern Yellowstone glacier, showing its major source areas and its outlet galcier. Heavy dashed line with double-headed arrows indicates ice divides. Smaller arrows, direction of ice-flow. Boundary between areas II and III is dotted to indicate these two areas are both part of the plateau icecap.

Yellowstone glacier is calculated using (1) the mapped* area-altitude distribution, (2) an equilibrium-line altitude of 2,835 m as approximated by the glaciation limit, and (3) curves of specific net balance from modern glaciers. These values result in an annual net balance in which both net accumulation and ablation are equal to about 3 km^3. Because the calculated values of accumulation and ablation are approximately equal, this glacial reconstruction can meet mass-balance requirements.

The accumulation-area ratio (AAR) of the reconstructed northern Yellowstone glacier is 0.75. Although this value is high, the mass-balance calculations show that because of its area-altitude distribution, a high AAR is required for the northern Yellowstone glacier.

The discharge of the outlet glacier was approximated by three flow models. The discharge required by mass-balance data suggests that for the outlet glacier, flow by bed-slip was much greater than flowage within the glacier.

The mass-balance data when combined with very rapid rates of recession, or meltback, indicate that the increase in discharge of the Yellowstone River owing to meltback would be less than 9 percent of the discharge during the Pinedale glacial maximum. The paucity of both outwash and recessional moraines upvalley from the

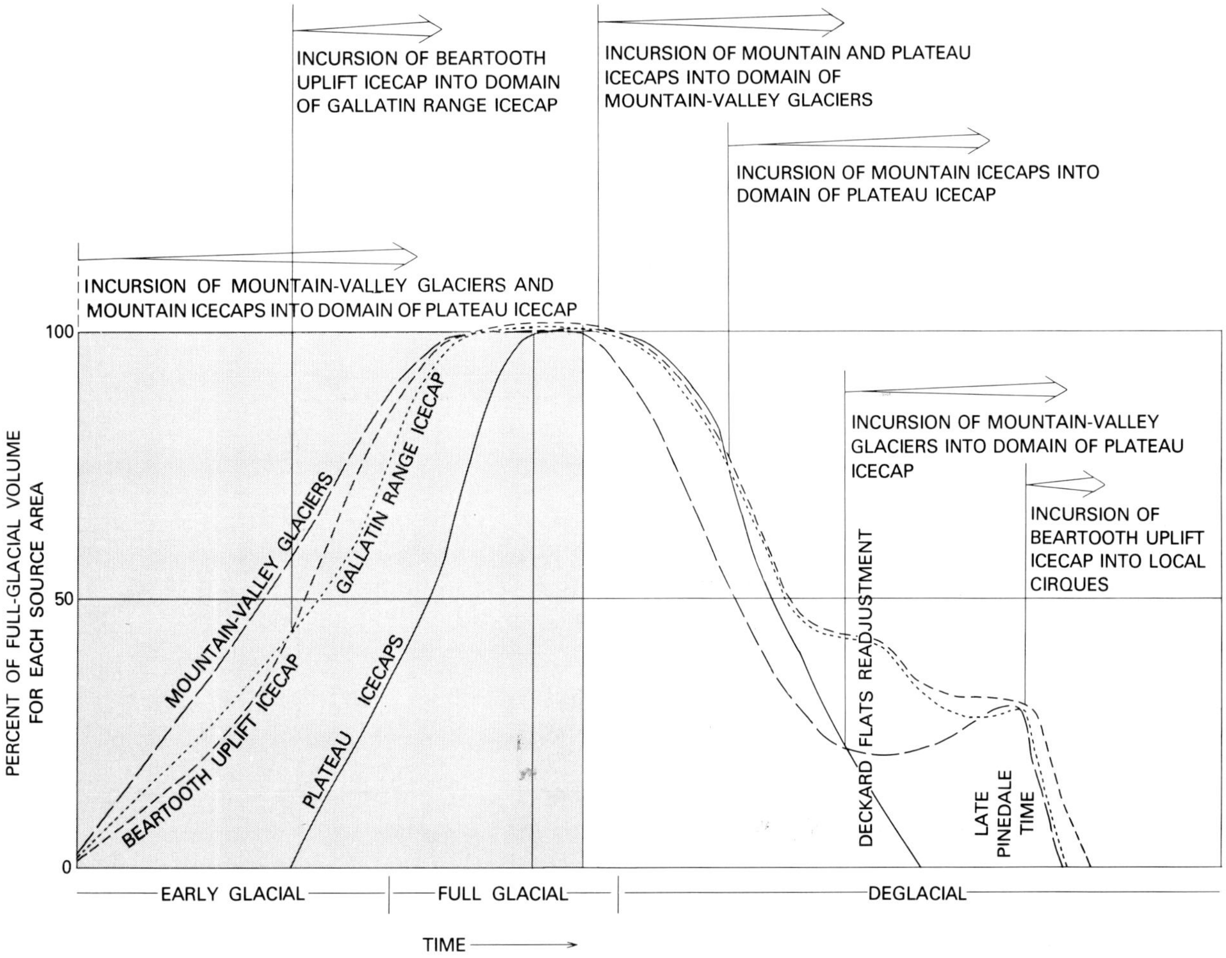

Figure 2: Inferred phase relations between the several ice masses in the northern Yellowstone area. The vertical axis is for each ice mass relative to its full-glacial volume. The point where one curve crosses another is determined from field observations of incursions of ice from one source area into the full-glacial domain of another source area, as noted by the labeled arrows above the curves. The plateau icecap had a stronger "peaking" response than the other ice sources, as indicated by the narrower base of this curve. Left half of the diagram is shaded to emphasize paucity of field relations to control interrelations of curves there.

northern Yellowstone terminal moraines also indicates that extensive reasearch did not result in outwash deposition due to greatly increased discharge and sediment load.

General models, devised to show the relations in space and time among the various source areas of the northern Yellowstone outlet glacier, demonstrate the differing size and efficiency of these source areas for generating ice under full- and waning-glacial conditions (figs. 1 & 2). The waxing and waning of each source area did not occur in unison, and field relations documenting incursions of ice from some sources into areas previously occupied by ice from other sources demonstrate this lack of unison. This lack of simultaneous phase relations for the growth and recession of northern Yellowstone ice masses implies that end moraines deposited elsewhere — for example, by large icecaps and small mountain-valley glaciers — are not of the same age.

*The map referred to here is plate 1A in the Professional Paper from which this abstract was taken.

Thirty-Third Annual Field Conference — 1982
Wyoming Geological Association Guidebook

CENOZOIC HISTORY OF THE YELLOWSTONE VALLEY BETWEEN YELLOWSTONE PARK AND LIVINGSTON, MONTANA

BY JOHN MONTAGNE[1]
ROBERT A. CHADWICK[1]

PREFACE

This paper is designed to provide a glimpse of the major Cenozoic geologic events that built, shaped, and modified the spectacular upper Yellowstone Valley during Cenozoic time. For clarity and chronologic presentation, the authors have divided the text into three essays which outline the Cenozoic structural, volcanic, and geomorphic-glacial history. R. A. Chadwick was responsible for the volcanic discussion while J. Montagne took the lead in the reamining two subjects.

The essay was originally prepared to accompany Field Trip 1, for the Rocky Mountain Section of the Geological Society of America in May, 1982, and is presented here through the courtesy of the Montana Bureau of Mines and Geology whose pending Publication 86 will formally include the manuscript.

Studies of this area over the past twenty years have been aided by the support of two grants from the National Science Foundation, G-24326, and GP-2912, for which the authors are grateful. In addition, the Montana State University Endowment and Research Foundation contributed financial aid and support. We are also indebted to many graduate students for their aid in the field, among them W. A. Van Voast, J. A. Bush, E. M. Struhsacker, S. E. Personius, and Albert Basler whose theses contributed importantly to this treatise. Visits in the field with John Good, Clifford Montague, K. L. Pierce, J. D. Love, and our late colleague, W. J. McMannis have stimulated progress in a variety of geologic ways. The Bruce Weeter, James and William Warfield, and Nelson families have been particularly helpful in giving access to critical areas. The cooperation of the MSU Earth Sciences Department and its former Head, M. J. Edie are gratefully acknowledged. We are indebted to D. L. Smith for critical reading and editing of the manuscript.

SETTING

Since the early explorations of Dr. F. W. Hayden and his associates (1872-73). the upper Yellowstone Valley has attracted scientist, tourist, and rancher alike by its beauty, dramatic landforms, structural contrasts and rich range lands. Called "Paradise Valley" by some, the important waterway stretches some fifty miles north of Yellowstone Park toward Livingston, Montana. The Yellowstone River and its tributaries are directed and redirected by various structural, volcanic, and geomorphic influences, forming a deep narrow canyon through Precambrian rocks ten miles north of Gardiner, and opening once again to a wide valley from where the river passes northward beyond the portal of Yankee Jim Canyon (Figs. 1, 2, 4). North of the portal the valley is basically a half graben bounded on the east by the upfaulted Beartooth Mountains, and on the west by the eastward-tilted volcanic-capped Gallatin Range. To the north, the valley narrows once again as the river passes through folds and thrust faults formed along the westward continuation of the Nye-Bowler lineament (Wilson, 1936).

The southern extremity of the Yellowstone Valley is

[1]Montana State University

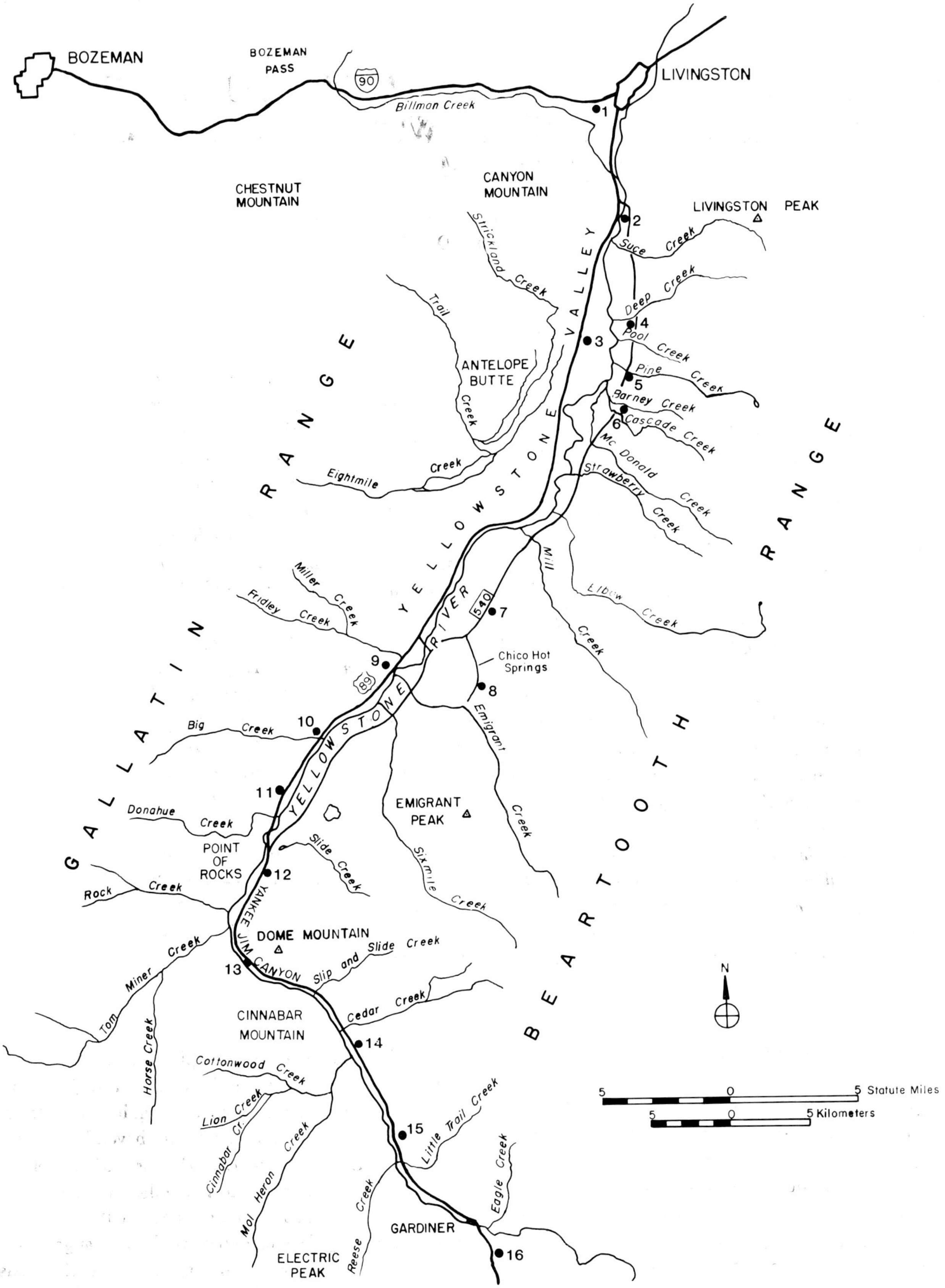

Figure 1: Map 1 showing field trip road log stops and highways and drainage of the Upper Yellowstone Valley and adjacent areas, Gallatin and Park Counties, Montana.

roughly coincident with the southeast-trending Gardiner fault and associated highlands which merge southward with volcanic realms of the Yellowstone Plateau. The entire area lies within the northwest boundary of the Middle Rocky Mountains Physiographic Province (Fenneman, 1931).

STRUCTURAL HISTORY

The modern Yellowstone Valley separates the Beartooth and Gallatin ranges both structurally and topographically. In Paleocene time, however, the two ranges were topographically unified along common northwest-striking compressional structures. Several of these folds are *en echelon* (Fig. 2) and continue in stepped fashion westward from the north end of Yellowstone Valley, veering from west to northwest along strike, and finally plunging northward along the continuation of the Nye-Bowler lineament (Wilson, 1936; Roberts, 1972).

The Suce Creek fault is one of these structures and occupies a position on the south flank of the Canyon Mountain anticline, five miles south of Livingston. Its hanging wall forms the prominent northern limit of the upper Yellowstone Valley where the fault transects the Yellowstone River and ramps up the northern extremities of the Beartooth and Gallatin ranges. A few miles farther south, a related disturbed belt carries the south-directed compression into a prominent fold called "The Hogback", which is slightly thrust on its south flank. Mild folds south of The Hogback also bring Paleozoic rocks to the surface near Antelope Butte (Figs. 1, 2).

Although the Beartooth and Gallatin Ranges were originally linked by Laramide compressional complexes, several authors cite evidence for at least some early Cenozoic tensional separation of these two foreland blocks. For instance, the Deep Creek-Luccock Park fault zone (Lammers, 1937, p. 278; Reid and others, 1975, p. 120), bounds the Beartooth Range in a series of normal faults that also mark the east margin of the Yellowstone Valley (Figs. 2, 6, 7). Richards (1957, p. 430) claimed that tension along this fault system preceded the final compressional phase of Laramide folding, since the Suce Creek Thrust truncates the Deep Creek fault in the valley of Suce Creek. This argument is refuted by Foose and others (1961, p. 1162) who noted that local alluvium obscures the relationship. Other hints of early tensional activity along the Deep Creek-Luccock Park fault zone appear in studies by Chadwick (1969, p. 152) who observed that the earliest Eocene volcanic rocks of the northern part of the Gallatin Range dip eastward toward the Yellowstone Valley at 10 degrees whereas the upper units of the same sequence are nearly horizontal. McMannis and Chadwick (1964, p. 21-22) also demonstrated that pre-volcanic drainage of the northern Gallatin Range was to the east, presumably in response to eastward tilting of the Yellowstone Valley-Gallatin Range block.

Important movement along the Deep Creek-Luccock Park fault zone has taken place since mid-Cenozoic time in the segment that extends 30 miles southward from Pool Creek to the mouth of Yankee Jim Canyon (Figs. 7, 8). Along this front, the Deep Creek fault has displaced Pleistocene to Recent morraines, terraces, and loess along its trend. The fault bifurcates, offsets *en echelon*, dies out, and develops anew along its strike. Associated with these movements, late Miocene beds at White Cliffs (Fig. 9) are tilted eastward 10 degrees (Horberg, 1940, p. 285; Van Voast, 1964, p. 25) but overlying Pliocene basalts on Hepburn's Mesa are apparently not affected. The Miocene beds are relatively fine grained and tuffaceous, and probably were deposited in response to recurrent movement along the Deep Creek fault (Fig. 8).

Using gravimetric data, Bonini and others (1972, p. 125-5) produced an interpretive cross section transecting the Yellowstone Valley and Beartooth front near Chico which shows, conservatively, 18,000 ft of displacement along the Deep Creek fault. In addition, the gravity models bring out the probability that several thousand feet of Tertiary sediments are trapped on the downdropped wedge. The assembled data suggest that the Yellowstone Valley north of Yankee Jim Canyon is a major fault trough and that uplift along the west-facing Beartooth front is also of considerable magnitude.

The Mill Creek fault system, which includes the Elbow Creek fault (Wilson, 1936), enters the Yellowstone Valley at Mill Creek and Elbow Creek. The south side of both of these faults is downthrown. According to Wilson (1936), the eastern extension of the system curves northward and projects toward the Nye-Bowler lineament near the mouth of the Stillwater River. This system divides the Beartooth uplift into two units which Wilson named the North and South Snowy blocks, respectively.

The North Snowy block is dominated by Precambrain rocks, whereas the South Snowy block has a greater variety of intrusive and extrusive rocks within and on the basement complex, tying in more directly to the Absaroka volcanic field in the Cooke City area, northeast of Yellowstone Park.

Ten miles south of the Mill Creek fault zone, the Deep Creek fault apparently dies out at the mouth of Tom Miner Basin. The Precambrian crystalline rocks continue across the Beartooth Mountains to the Gallatin Range, forming the uplift through which Yankee Jim Canyon is cut.

South of Yankee Jim Canyon, the Gardiner fault, a high-angle reverse fault, projects southeastward from the north side of Cinnabar Mountain and extends diagonally across the Yellowstone Valley. This fault, first described by

Fig. 2 MAP II - MAJOR IGNEOUS, STRATIGRAPHIC AND STRUCTURAL FEATURES OF THE UPPER YELLOWSTONE VALLEY AND ADJACENT AREAS

GALLATIN AND PARK COUNTIES, MONTANA

Mapped or Compiled by R. A. Chadwick and J. Montagne

1982

-Note: Only major drainage shown: for detailed drainage, highway and road log stops, see Map I (Fig. 1).

LEGEND

AGE	SYMBOL	FORMATION OR UNIT
Cenozoic		
Quaternary	Qb	Basalt flows
Tertiary	Tb	Basalt flows
Tertiary	Td	Dacites Td_3 youngest Td_2 Td_1 oldest
Tertiary	Tvh	Hyalite Peak volcanics
	Tvhv	Hyalite Peak volcanic vent
	Tvg	Golmeyer Creek volcanics
Cenozoic undifferentiated	C	Sedimentary rock units including alluvium
Mesozoic	M	Mesozoic rock units
Paleozoic	Pal	Paleozoic rock units
Precambrian	P€	Precambrian rock units

STANDARD SYMBOLS

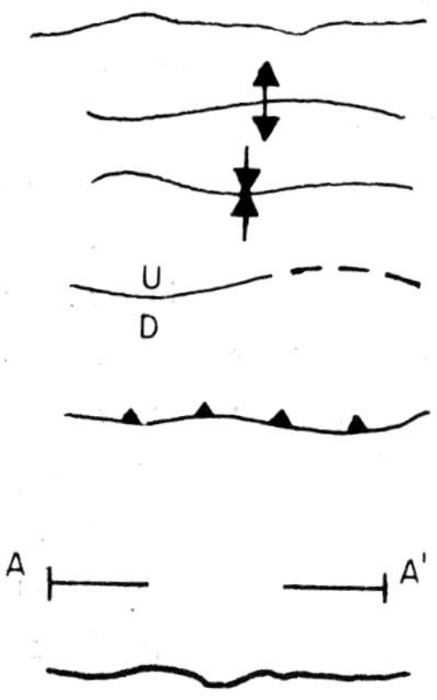

Formation boundary

Axial surface, anticline

Axial surface, syncline

Normal fault, dashed were indefinite, dotted where covered, U (up) D (down)

Reverse fault, dashed where indefinite, dotted where covered, sawteeth on upper plate

Igneous dike, andesite or dacite

Line of cross section (shown in Fig. 3)

Major stream

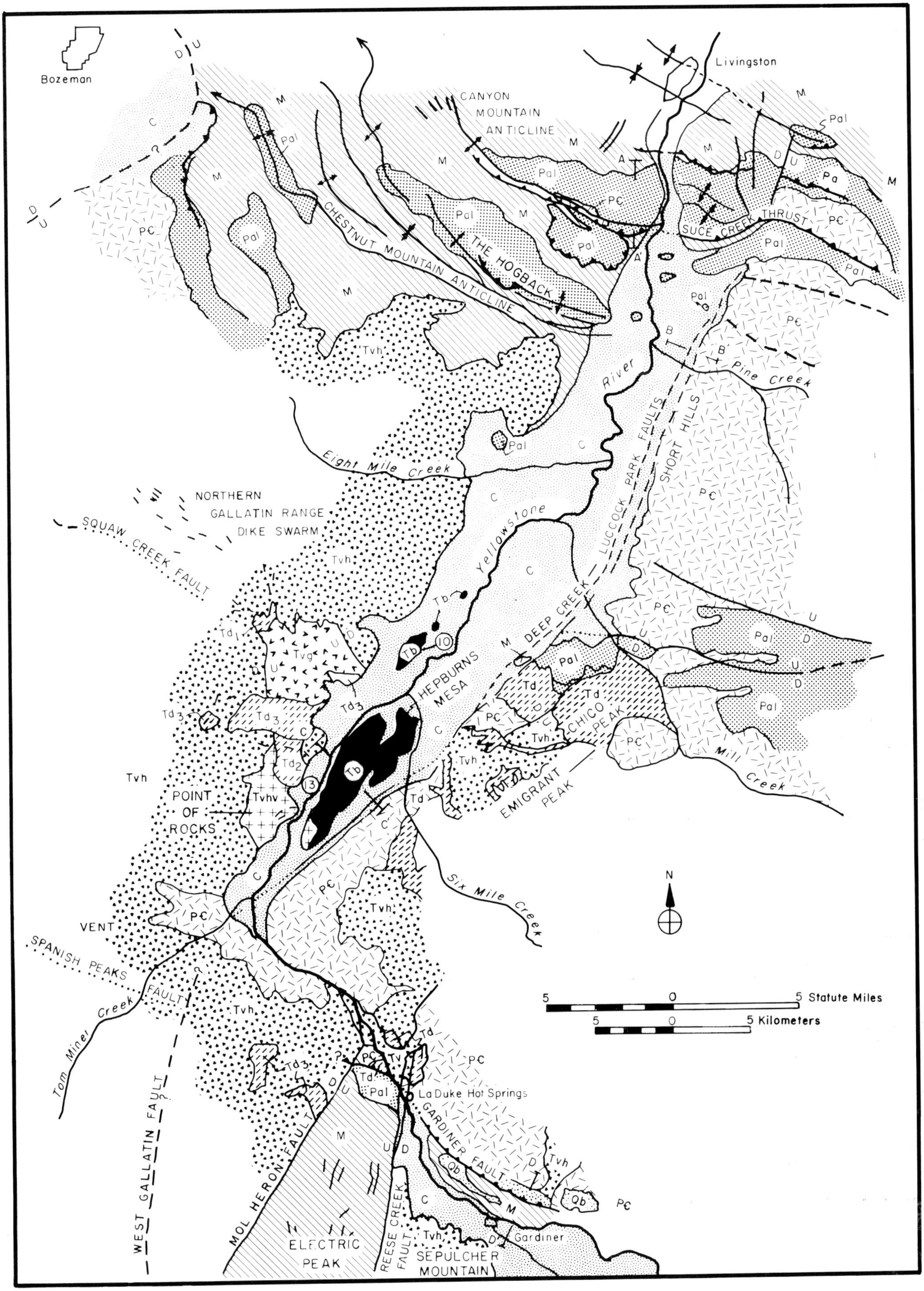
Bozeman
Livingston
CANYON MOUNTAIN ANTICLINE
CHESTNUT MOUNTAIN ANTICLINE
THE HOGBACK
SUCE CREEK THRUST
Pine Creek
River
Eight Mile Creek
NORTHERN GALLATIN RANGE DIKE SWARM
SQUAW CREEK FAULT
Yellowstone
LUCCOCK PARK FAULTS
SHORT HILLS
DEEP CREEK
HEPBURNS MESA
CHICO PEAK
EMIGRANT PEAK
Mill Creek
POINT OF ROCKS
Six Mile Creek
VENT
SPANISH PEAKS FAULT
Tom Miner Creek
WEST GALLATIN FAULT
MOL HERON FAULT
REESE CREEK FAULT
GARDINER FAULT
LaDuke Hot Springs
Gardiner
ELECTRIC PEAK
SEPULCHER MOUNTAIN
5 0 5 Statute Miles
5 0 5 Kilometers

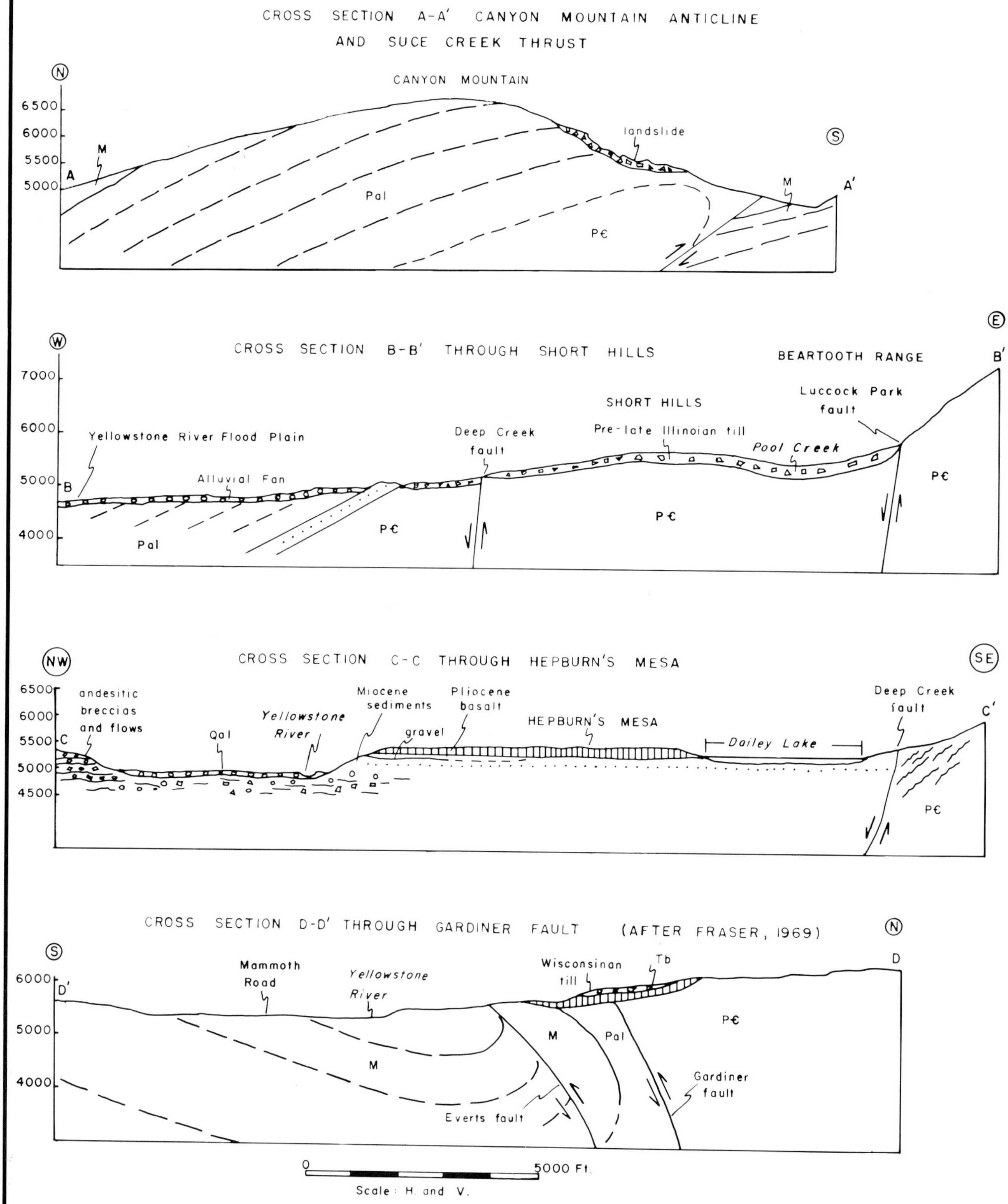

FIG.3 - CROSS SECTIONS ILLUSTRATING STRUCTURE SHOWN ON FIGURE 2, MAP 2.

Holmes (1883, p. 56), forms the southwest edge of the uplifted Beartooth Mountains and strikes perpendicular to the Deep Creek fault. The Gardiner fault dips steeply to the northeast (Wilson, 1934, p. 649) and has a throw of 15,000 ft. The Yellowstone River follows this structural trend northward from Gardiner to Yankee Jim Canyon. Paleozoic and Mesozoic rocks are preserved in synclinal relationship along the footwall of the fault and are displayed in peculiar "reverse drag" relationship at the "Devil's Slide", a part of Cinnaber Mountain.

Where the Gardiner fault crosses the Yellowstone Valley just east of the town of Gardiner, it is offset by incremental uplifts along a number of north-trending, high-angle faults that extend through and beyond the northern Yellowstone Park area. They are part of a system bounded by the Mammoth fault on the east (Brown, 1961, p. 1190) and the east Gallatin fault on the west (Holmes, 1883, p. 27). The latter forms the imposing front of the southern Gallatin Range at Electric Peak, west of Gardiner.

All of the north-trending, high-angle faults tend to die out in the Precambrian complex of the Beartooth Mountains as they cross and offset the Gardiner fault (Struhsacker, 1976, p. 21). La Duke Hot Spring is located at the intersection of the Gardiner and Reese Creek faults.

As discussed in this guidebook by R. A. Chadwick, the Gardiner fault may be part of the northwest-trending system of faults and igneous centers that includes the Spanish Peaks fault which emerges from volcanic cover in the western Gallatin Range (Fig. 2). This fault extends across the Gallatin River at Big Sky, crosses the Madison Range, and continues as the North Meadow Creek fault along the northern margin of the Madison Valley at Ennis Lake. The fault system appears to continue still farther west, crossing the Tobacco Root Mountains (Reid, 1957; Chadwick, 1970) where it is represented by the Bismark fault. In the other direction, the Gardiner fault may extend southeastward as far as the flank of the Big Horn Basin, south of Cody, Wyoming (Wilson, 1934, p. 653), although evidence in support of this seems vague.

Ruppel (1972, p. A-49-A53) described repeated Cenozoic movements on the Gardiner fault, the connection between the east and west Gallatin faults and the Teton fault system south of the Park, and various interrelations of the Snake River volcanic system which are beyond the scope of this treatise. However, Fraser, and others (1969, p. 49-53) clarified some locally important earlier misconceptions about the relationship of the Gardiner fault to the offset of basalt flows in the Gardiner area.

Several basalt flows mantle the trace of the Gardiner fault on a high terrace east of Gardiner. Chadwick (1978, p. 26) interpreted the age of flows, from radiometric and paleomagnetic evidence, to be about 0.6 million years. Fraser (1969, p. 50) believed that the flows originated upstream from this locality. One of the flows may be offset 400 ft along the Gardiner fault, although it is uncertain whether the offset is due to movement on the fault or to recent slumping. If due to faulting, Holocene movement has taken place on the Gardiner fault. Since late Tertiary compressional movement is not known in this part of the Rocky Mountains, an alternative possibility may be proposed: normal fault movements associated with splays of the Mammoth fault exploited earlier weakness along the Gardiner trend and locally uplifted the hanging wall of the Gardiner fault in the process.

Summary of Structural Relationships

The upper Yellowstone Valley incorporates a number of divergent structural trends. Its north-northeast topographic grain north of Yankee Jim Canyon approximates the strike of Precambrian foliation in many places. The valley is controlled more directly by movement on the Deep Creek-Luccock Park fault system that exploited this structural weakness locally. Important Quaternary tensional movement on this fault (Fig. 8) suggests an affinity with Basin and Range extensional tectonics.

The structural and topographic grain of the valley is also influenced by northwest- to west-trending Laramide compressional structures which are widely displayed along the Gardiner, Mill Creek, and Suce Creek fault systems.

VOLCANIC HISTORY

During the Eocene Epoch, widespread and intensive volcanism erupted across southwestern Montana and northwestern Wyoming. Two sub-parallel, northwest-trending chains of volcanic centers and related intrusives developed along pre-existing zones of weakness. These belts of igneous centers, 15 to 35 miles apart, extend from the southern Absaroka Mountains of Wyoming northwestward across the Yellowstone National Park-upper Yellowstone Valley area through the Gallatin Range (Chadwick, 1970). The northeastern belt, which has somewhat more potash-rich igneous rocks, includes in the Montana portion, the Cooke City, Independence, Emigrant Peak-Chico Peak, and Point of Rocks volcanic-intrusive centers and a dike swarm in the northern Gallatin Range. The Montana portion of the southwestern belt includes the Electric Peak-Sepulcher Mountain, Sheep Mountain, and Porcupine Creek igneous bodies. Figure 2 illustrates the location of the Emigrant Peak-Chico Peak, Point of Rocks,

Fig. 4 - MAP III - GENERALIZED GEOMORPHIC MAP OF THE UPPER YELLOWSTONE VALLEY AND ADJACENT AREAS, GALLATIN AND PARK COUNTIES, MONTANA

Mapped or Compiled by John Montagne, 1982

Principal References: Horberg, 1940; Van Voast, 1964; and Pierce, 1979

Note: Only major drainage is shown: for detailed drainage, highway, and road log stops, see Map I (Fig. 1).

LEGEND

Late Quaternary	Qal	Stream alluvium, mostly post-glacial; includes terrace gravel
Late Quaternary	Ls	Major landslides
Pinedale (Wisconsinan)	Qow	Glacial outwash
Pinedale II (Late Wisc)	PII	Piedmont glacial moraine complex (locally the Chico advance)
Pinedale I (Early Wisc)	PI	Piedmont glacial moraine complex (locally the Eightmile advance)
Pinedale (Wisconsinan)	P	Alpine glacial moraine complex at canyon mouths
Pinedale (Wisconsinan)	L	Proglacial lake or stream deposits
Bull Lake (Early Wisconsinan-Late Illinoian)	BL	Alpine glacial moraine complex at canyon mouths
Pre-Wisconsinan	Ped	Pediments graded to phantom levels at least 50 feet higher than present drainage
Pre-Late Illinoian	⊗	Sheet till or isolated erratics beyond Pinedale limit

STANDARD SYMBOLS

	Rock or unit boundary
U / D	Mid-Tertiary to late Holocene fault; dashed where indefinite, dotted where covered
A — A'	Line of cross section (shown on Fig. 5)

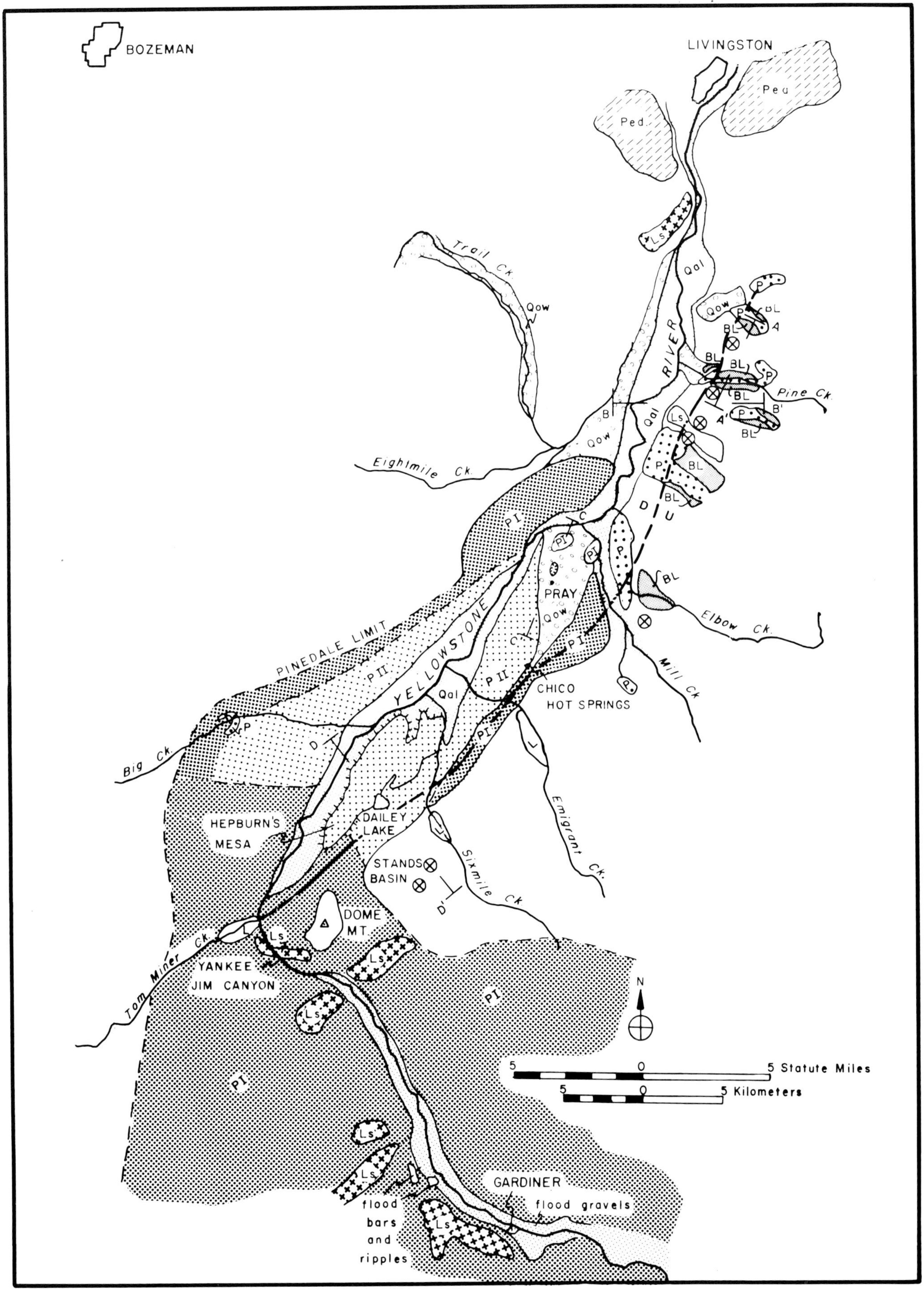
BOZEMAN
LIVINGSTON
Ped.
Peu
Ls
Trail Ck
Qow
Qal
RIVER
Pine Ck.
Eightmile Ck.
YELLOWSTONE
PINEDALE LIMIT
PRAY
Elbow Ck.
Mill Ck
CHICO
HOT SPRINGS
Big Ck.
Emigrant Ck.
HEPBURN'S
MESA
DAILEY
LAKE
STANDS
BASIN
Sixmile Ck
DOME
MT.
Tom Miner Ck.
YANKEE
JIM CANYON
N
5 0 5 Statute Miles
5 0 5 Kilometers
GARDINER
flood gravels
flood
bars
and
ripples

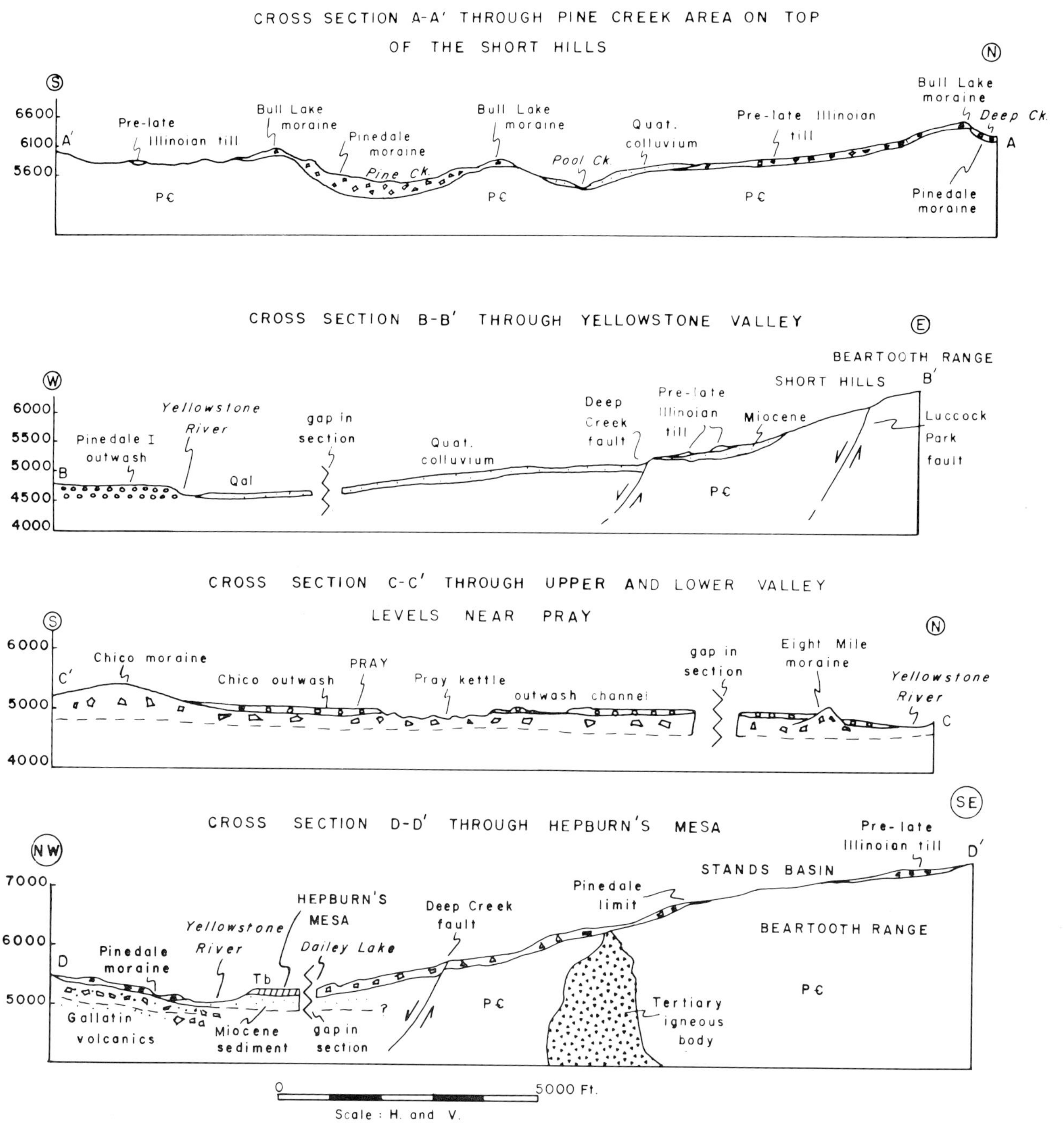

FIG.5 CROSS SECTIONS ILLUSTRATING QUATERNARY RELATIONS SHOWN ON FIG.4, MAP 3.

Figure 6: Panoramic view southeast along west margin of North Snowy Range block, Beartooth Range. Deep Creek is in foreground. Pine Creek Canyon shows as two storied glacial trough discharging through the Short Hills. The latter extend in high grassy meadows in distance and are capped by scattered boulders of the pre-Illinoian piedmont ice sheet.

Figure 7: Deep Creek fault scarp along base of Emigrant Peak, east of the Yellowstone Valley. Above the battered scarp are ice marginal troughs and lateral moraines of several piedmont glacial advances, the latest and most prominent being the Pinedale advance. Low terrace is part of the dissected fan of Emigrant Creek.

and Electric Peak igneous centers with respect to the Yellowstone Valley.

Compositionally, the eruptive centers consist of andesitic flows, flow breccias, vent breccias, mudflow and fluvial breccias, dikes, and agglomerates which have been intruded by dacitic to rhyodacitic plugs, stocks, and dikes. Across most of the range, the andesites are medium gray to black and contain crystals of sodic labradorite to andesine, augite, hypersthene, and locally, hornblende. The dacites are light gray to pink, aphanitic to glassy, locally flow banded and/or autobrecciated, and characterized by biotite and andesine phenocrysts. Some bodies contain megascopic quartz crystals.

Traced to the northwest, the northeastern belt of igneous centers is aligned with and appears to continue as the Squaw Creek fault in the western Gallatin Range (McMannis and Chadwick, 1964), which becomes the Cherry Creek fault in the Madison Range and the London

Figure 8: Mouth of Pool Creek (left) at Beartooth front. Deep Creek fault shows as dark escarpment cutting up through Precambrian Short Hill block. Top of Short Hills is capped by lateral moraine V of Bull Lake age (north lateral of Pine Creek glacier).

Figure 9: Yellowstone River gorge through west edge of Hepburn's Mesa at White Cliffs. Whitish deposit is Late Miocene tuffaceous sequence overlain by Pliocene(?) Yellowstone gravel. Sequence is capped by 5.5 m.y. Pliocene basalt.

Hills fault in the northern Tobacco Root Mountains (Schmidt and Hendrix, 1981).

The southwestern belt lines up northwestward with the Spanish Peaks-North Meadow Creek-Bismark fault system. This system is traceable across the Gallatin River, Madison Range, and Tobacco Root Mountains (Reid, 1957; Chadwick, 1970; Schmidt and Hendrix, 1981).

These major fault zones appear to represent major northwest-trending zones of weakness in the crystalline crust (Chadwick, 1981; Garihan and others, 1982) which perhaps originated in Precambrian time (Reid, 1957) and were reactivated at later times to provide conduits for magmas of the Tobacco Root batholith and for Eocene volcanic-intrusive centers. Kavanagh (1965) described Eocene rhyolitic to rhyodacitic bodies along the zones near Norris.

The Gallatin Range (Fig. 10) is capped by gently eastward-dipping andesitic flows and breccias which lie unconformably on a complex of Precambrian crystalline rocks and infolded Paleozoic-Mesozoic sedimentary strata. The volcanic sequence attains an estimated thickness of 6,000 ft including two principal sequences. The volcanic deposits comprise the high peaks and rolling foothills of the Gallatin Range as seen from the Yellowstone Valley.

The older of the two principal volcanic sequences is the Golmeyer Creek Volcanics (Chadwick, 1969). This sequence includes pyroxene-hornblende andesitic flows and breccias, and a sill. The rocks are commonly propylitized and locally pyritized. The younger of the two sequences overlies the Golmeyer Creek Volcanics unconformably, is more widespread, and consists mainly of augite-hypersthene andesitic flows and breccias. This sequence is the Hyalite Peak Volcanics.

Figure 10: View west from Yellowstone River to crest of Gallatin Range near Point of Rocks. Prominent dike in middle ground intrudes nearly horizontal, stratified andesitic volcaniclastics.

Figure 11: View west toward dissected stratiform volcanic body at Point of Rocks. Yellowstone River is in trees in middleground. Volcanic rocks here are andesitic and dacitic.

Figure 12: Burlington Northern quarry showing 5.5 m.y. basalt flow resting on Pliocene(?) river gravel.

Dacites intruded the volcanic pile at three major times: (1) post-Golmeyer Creek, pre-Hyalite Peak; (2) Hyalite Peak; and (3) post-Hyalite Peak. The third group of dacites intruded during a magnetically reversed epoch; the earlier dacites and both andesite sequences are paleomagnetically normal (Shaver, 1978).

Potassium-argon dates bracket the andesitic and dacitic activity in the 53 to 48 m.y. range. These rocks are believed to correlate in large part with the Washburn sequence of Yellowstone National Park (Smedes and Prostka, 1972).

Andesitic units of the Hyalite Peak Volcanics are

exposed along most of the foothills west of the Yellowstone Valley from Antelope Butte southward to Tom Miner basin. Individual lava flows range from a few feet to about 200 ft thick and may exhibit blocky or columnar jointing.

Flow breccias are generally near-monolithologic, containing principally pyroxene andesite fragments which, however, may be fresh to highly oxidized. Commonly graduations from massive andesite to flow-brecciated material may be observed. Gases have altered the fragmental matrix of the breccia and some of the fragments, giving the rock a characteristic red color.

Mudflow and fluvial breccias are heterolithologic: fragments are dominantly pyroxene andesite, but some are hornblende-bearing andesite or dacite. The greater variety suggests transport from lithologically more varied source areas. Boulders may exceed 6 ft in diameter. Sorting is better than in flow breccias, and stratification is typically developed on a crude scale. Lenses of fluvial origin within mudflow sequences commonly have better sorting and finer stratification than the enclosing mudflow deposits: some zones could be termed volcanic sandstone and siltstone.

Mudflow sequences in the northern Gallatin Range attain hundreds of meters in thickness and extend laterally for over 9 miles. These mudflows evidently derived much of their material from autobrecciated lava flows rather than vent breccia complexes. Flows in the northern Gallatin Range were extruded from parallel fissures rather than central vents.

The Point of Rocks vent complex (Fig. 11) is particularly interesting. At this locality, a sequence of andesitic breccia units was laid down, probably on the slopes of a cone. The sequence consists of alternating red, monolithologic units and light gray, heterolithographic beds. Evidence for vent origin includes (1) quaquaversal dips of the breccia units, (2) increase in stratification of the heterolothographic units outward from the center, (3) numerous crosscutting andesite and dacite dikes, and (4) intrusive breccia units, including some highly vesicular vent-type breccias.

Dacitic plugs and dikes intrude through the andesitic units at Point of Rocks. At the mouth of Donahue Creek, an autobrecciated zone within a dacite plug grades laterally into epiclastic breccia which becomes better stratified and acquires a greater variety of dacite and andesite clasts away from this plug. The change takes place over about 2,000 ft. The relationships imply that the autobrecciating plug neared or reached the surface and crumbled and eroded to contribute material to nearby epiclastic breccia deposits. The relationships suggest that the Eocene land surface was at approximately the same position as the present land surface at Point of Rocks during dacite plug extrusion. However, 3 miles to the west, the previous andesitic

Figure 13: Closeup of contact zone of 5.5 m.y. basalt and Pliocene(?) gravel at Burlington Northern quarry (see Fig. 12 above).

volcanism had built the land surface to a much higher level, since Eocene andesitic flows are stacked to a height of 2,900 ft above the level of the plugs. Therefore, during Eocene Hyalite Peak volcanism, pre-dacite erosion greatly dissected the volcanic pile and evidently established major drainage to the east, perhaps into an ancient Yellowstone Valley in the vicinity of the present valley. The dacite plugs then erupted into this valley.

In Pliocene time, basalts from an unknown source, possibly the Deep Creek fault, flowed into the Upper Yellowstone Valley. Two sequences of olivine basalt are exposed in erosional remnants along the valley floor and on Hepburn's Mesa (Figs. 9, 12, 13) in the Emigrant area (Bush, 1967).

The older sequence, dated at 8.4 m.y., lies on Eocene andesites south of Hepburn's Mesa. It consists of two flows, has black, aphanitic to glassy texture, and has considerable magnetite content in the matrix. The younger sequence, also consisting of two flows, is medium gray, contains conspicuous olivine and plagioclase (about An_{64}) phenocrysts, and has ophitic texture of augite and plagioclase. This sequence is dated at 5.4 m.y. and overlies gravel and white siltstones of 17.1 m.y. age on Hepburn's Mesa. It is well exposed in a railroad ballast quarry south of Emigrant. The older sequence is paleomagnetically normal and the younger is reversed (Chadwick, 1972).

These basalts are much older than the basalt flows at Gardiner for which an age of 0.6 m.y. seems most reasonable. The Emigrant area basalts probably did not flow down the Yellowstone Valley from the Gardiner area as evidenced by the difference in age and lack of basalt remnants upstream. Local origin for the Emigrant area basalts along the Deep Creek fault seems likely because the fault may have extended deeply enough to serve as a conduit. However, no feeder dikes have been located in the area. Since the basalts are so much older than volcanic rocks of the Yellowstone Park-Island Park caldera complex, they do not fit into the pattern of northeastward-migrating volcanism postulated for the Snake River-Yellowstone rift system (Christiansen and Blank, 1969).

GEOMORPHIC AND GLACIAL HISTORY

Introduction

Erosional and depositional processes have etched the previously described bedrock and structural framework and contributed greatly to the complex of geomorphic features in the Yellowstone Valley. Few valleys in the Rocky Mountains can match the Yellowstone for vividness and drama of these forms, particularly when viewed in the early morning or late afternoon light. This treatise can add only minor embellishment to the elegant descriptions already published by Holmes (1883), Weed (1893), and Horberg (1940).

Paleogeomorphology

Pleistocene events have obviously been most influential on current topographic expression. Nevertheless, pre-Pleistocene topography should not be overlooked even if interpreted from indirect evidence. As indicated in the previous structural description, the Yellowstone Valley became somewhat trough-like early in Tertiary time in response to movement of the Deep Creek-Luccock Park fault system along the western flank of the Beartooth Range. Thick, gently east-dipping, volcanic and volcaniclastic rocks of mid- to late-Eocene age capping the Gallatin Range extend into the Yellowstone Valley and do not occur in the Beartooth Range north of Mill Creek. This simple evidence suggests that the North Snowy block of the Beartooth Range remained topographically higher than the adjacent Gallatin Range and Yellowstone Valley during early Cenozoic time.

Topography upon which Eocene volcanic rocks rest in the Yellowstone Valley was created by erosion of bedrock affected by Laramide structural deformation. In a few places, such as in the vicinity of Antelope Butte on the north and Cinnabar Mountain on the south, exhumation of such pre-existing topography indicates that post-Laramide, pre-Eocene relief was at least as great as that of today and that resistant rocks stood high in contrast to non-resistant rocks which tended to erode to valleys or draws. Modern processes have served to enhance the etching of these exhumed topographic forms.

A hint of Eocene topography in the valley is given by the occurrence of external volcanic forms, including dacitic domes and stratified cone flank deposits near Point of Rocks as described previously. Occurrence of a thick sequence of andesite flows farther west in the Gallatin Range implies that the topographic surface of the Gallatin Range during Eocene time was relatively higher in elevation at the time of Hyalite Peak igneous activity than was the Yellowstone Valley. Chadwick (1969) also noted that northeastward drainage into the Yellowstone River Valley near Antelope Butte is implied by buried valleys containing thick volcaniclastic deposits.

The rolling upland flats of the "Short Hills" (Figs. 2, 6) just east of the Deep Creek-Luccock fault are roughly 1,000 ft above the main valley floor and contain "fossil" topography developed during mid-Tertiary time. (The name, Short Hills, has been used colloquially by residents in recognition of the fact that the Shorthill family homesteaded here.) The Short Hills are bounded on the east by the Luccock Park fault and on the west by the Deep Creek fault. The block is either a foundered remnant of the Beartooth Range proper, just to the east, or a partly uplifted remnant of the Yellowstone Valley to the west. Upon its gently rolling surface, patches of tuffaceous sandstone and claystone are preserved in north-trending, partly exhumed topographic depressions. Since the light-colored Tertiary sediments compare favorably with known late Miocene deposits at nearby White Cliffs (Fig. 9), they are also probably of Miocene age. Thus, the benchland on top of the Short Hills is partly an exhumed pre-late Miocene topographic form.

Because the coarser fractions of the above sediments are locally derived angular Precambrian rocks and not rounded stream gravels, it is impractical to predict whether the Short Hills surface was once accordant with the Yellowstone River floor. However, a conglomerate consisting of angular Precambrian boulders and smaller particles contained in a white calcareous matrix is downfaulted along an extension of the Deep Creek fault north of the falls of Cascade Creek (Fig. 1) and suggests the Miocene sediments were once more widespread at this location.

Some sense of absolute movement of the Short Hills, and the timing of the major increment of such movement may be approached through geomorphic reasoning. Post-late Miocene movement on the block along the Deep Creek fault is proved by scarplets at the base of the Short Hills (Fig. 8). If the Short Hills achieved their major elevation above the main Yellowstone Valley floor prior to late Miocene time, a considerably greater thickness of Miocene sediments than is presently known would be necessary to reach high enough to cover the Short Hills.

Since such excessive thicknesses are not known except possibly beneath the Yellowstone Valley floor, it is suggested that major downfaulting of the valley block or uplift of the Short Hill block, or both, continued through post-Miocene time. This concept is enhanced by the fact that early to mid-Pleistocene piedmont glacial debris on the Short Hills appears to be lacking on the west side of the Yellowstone Valley at similar elevations.

At the White Cliffs (Fig. 9), Pliocene (5.4 m.y. old) basalt described in the preceding article of the guidebook, overlies a gravel containing boulders with Yellowstone River affinities. This gravel in turn overlies Miocene (17.1 m.y.) (Hughes, 1980) tuffaceous sediments. At a nearby rock quarry, the gravel contains cobbles of 8.4 m.y. old basalt. The gravel is thus evidently of Pliocene age and it is certain that the Yellowstone River existed at this time and that basalt flowed along the ancient valley floor which today is less than a hundred feet higher than the present valley floor.

Glaciation

The major characteristics of glaciation in the Yellowstone Valley have been known for nearly a hundred years, as attested by the excellent observations of Holmes, (1883), and later work of Horberg (1940) and Alden (1932, 1953). Pierce (1979) recently added important insights to some of the hitherto anomalous relations of the Wisconsinan and Illinoian glacial sequence. The original observations described in the present paper are mostly details augmenting these earlier studies. During these studies, I have had the benefit of frequent conferences with Dr. K. L. Pierce in developing many of the new concepts of piedmont glaciation in the Yellowstone Valley. Without his collaboration, the new ideas and synthesis of this interesting area would hardly have been possible, and it is satisfying to fit the glacial geology of the Yellowstone Valley into the larger context of the Yellowstone Icesheet.

Basically, the Yellowstone Valley was the gathering point for two types of glaciers. One type, a piedmont outlet lobe of the Yellowstone Ice sheet, accumulated mostly within the Yellowstone Plateau and flowed northward down the Yellowstone at least twice during the Pleistocene Epoch (Fig. 4). The mountains that surround the Yellowstone Valley were also gathering grounds for valley glaciers (Fig.s 4, 6) which discharged terminal moraines at canyon mouths on the main Yellowstone Valley floor. Where these latter glaciers projected into the piedmont ice, the morainal debris was swept down-glacier, leaving no trace of concentrated deposition on the valley floor. Various ages of glaciation are represented, as described below.

Pre-Illionoian Stade(s?) of the Yellowstone Glacier

One or more times a pre-Illinoian piedmont ice sheet descended the axis of Yellowstone Valley from the Yellowstone Plateau and surrounding mountains. Horberg (1940) recognized that scattered boulders on the crest and west flank of the Short Hills were of possible glacial origin, although these are far beyond the outer limit of the Pinedale age piedmont ice sheet that occupied the valley (Fig. 14). Among the lithologic types represented are Paleozoic dolomite boulders up to five feet in diameter, Flathead Quartzite, Precambrian gneiss, and Eocene andesite porphyry.

The author has worked extensively with these sheet-like deposits that display no semblance of morainic form. They are present on most interstream divides between Stands Basin on the south and Pool Creek on the north. The boulders are not associated with a recognizable and distinctive deep soil.

The most convincing evidence for glacial origin of these boulder deposits is the inclusion of angular andesite prophyry rock types within the deposits. The nearest source for these is the volcanic area south of Millcreek. The erratics of Flathead quartzite found among these deposits are quite abundant and were likewise transported from areas south of Mill Creek.

On the rim of Stands Basin, several hundred feet above the Pinedale piedmont limit, angular, corner-rounded Precambrian erratics occur in a sheet-like deposit which quite convincingly qualifies as an up-glacier correlative of the previosuly described boulder deposits.

Suspect erratics with rounded shapes have been found close to the main valley floor, resting on Paleozoic outliers beyond the Pinedale piedmont limit. The latter abraded bedrock forms are located west of the mouth of Pine Creek and east of the mouth of Strickland Creek. A few roundstones, in this case andesitic in composition, also have been found on the Cambrian limestone cliffs west of the Yellowstone River near Livingston. If ice were as thick as the Short Hills erratic train suggests, the ice lobe should have reached the outskirts of Livingston. Evidence for the latter supposition is not available today, however.

The strange fact is that no unequivocal glacial erratics have yet been located on the hillsides west of the Yellowstone River beyond the Pinedale limit in the northern part of the main valley. Quadrant Qyartzite and Madison Group boulders are present on the crest of Antelope Butte at elevations that correspond to those of the crest of the Short Hills, but nearby andesitic breccias of the Gallatin volcanic field also contain such boulders and raise the question of whether Eocene lahars were responsible for their emplacement. Such an ice sheet, as implied by the eastern erratic train, would presumably leave some recognizable trace on the western valley margin, but an alternative hypothesis, such as a thin ice sheet at this location would also explain the enigma. In the latter case, boulder deposits could have been spread at low levels and subsequently differentially raised on the Short Hills by activity of the deep Creek-Luccock Park fault. Erratics on the short Hills today rests on Miocene and Precambrian rocks 1,500 ft above the main lower valley floor in the vicinity of Pine Creek.

Pre-Illinoian ice sheets of greater magnitude than late Illinoian or Wisconsinan glaciers are clearly evident in nearby areas. For example, the Buffalo Glaciation of Jackson Hole (Blackwelder, 1915; Fryxell, 1930; Love, 1968) or the several larger glaciers of the Taylor Fork of the West Gallatin River along the crest of the Madison Range (Hall, 1960, p. 197; and Walsh, 1975, p. 649) certainly qualify. At least one of the latter tills is known to occur beneath the 1.9 m.y. old Huckleberry Ridge Tuff at Wapiti Creek in the Madison Range (Walsh, 1975, pers. comm.). Broad valleys which now contain post-glacially notched streams and Pinedale age moraines within new canyons suggest previous glacial activity of much greater magnitude.

The Late Illinoian Piedmont Glacial Enigma In Yellowstone Valley

The age of fresh young end moraine discharged by a piedmont ice sheet in Yellowstone Valley near the mouth of Eightmile Creek has long been argued. There is no question as to the origin of the material, for it is distributed abundantly in various morainal forms back to the source area of the Yellowstone Ice sheet. The best reasoning, which will be described below, relates these features to a Pinedale advance. Granting this, the question arises why the older Bull Lake piedmont glaciation, which outflanks the Pinedale in many Rocky Mountain localities, is not evident here. The Bull Lake sequence is present in the West Yellowstone area, on the west side of Yellowstone Plateau, however (Richmond, 1964, p. 225). The Pinedale moraines in this area are restricted to relatively inconspicuous forms several miles upstream from the Bull Lake moraines. Further, the Bull Lake moraines at West Yellowstone, although large and bulky, are very subdued in topographic expression.

Pierce (1979, p. F70) proposed that the Bull Lake advance through West Yellowstone was the principal discharge glacier for the northern part of the Yellowstone Ice sheet because it had been an easy exit route from the Plateau at that time. The 117,000 year old West Yellowstone rhyolite flow, which covered the Bull Lake moraines south of West Yellowstone, blocked the advance of later Pinedale piedmont ice, shunting it northward along the east side of the Gallatin Range and finally down the Yellowstone to its present, relatively distal position. If Bull Lake debris lay in its path, the Pinedale ice overwhelmed it. The Pinedale advance in the Yellowstone Valley was thus of greater magnitude than that of the Bull Lake, whereas in the West Yellowstone Basin the Bull Lake was of greater magnitude than the Pinedale.

Additional arguments favoring a Pinedale age for Yellowstone Valley piedmont moraines are outlined as follows. Although early workers such as Weed (1893), Horberg (1940), and Alden (1932) proposed that the outer piedmont moraine near Eightmile Creek in Yellowstone Valley correlates with the Bull Lake glaciation of early Wisconsinan or late Illinoian age, several lines of evidence point more logically to a Pinedale age for this feature:

1. Soils developed on the Eightmile moraine northwest of the Mill Creek bridge and on top of the Yellowstone River bluffs compare quite favorably with those developed on the Chico moraine west of the junction of Highway 540 and the Chico Hot Springs access road. Both soils are thin and poorly developed with depths to calcium carbonate reaction varying from five to 12 in (C. Montagne, 1970, pers. comm.).
2. The outer moraine originally classified as Bull Lake by Horberg (1940) is temporally related to the inner piedmont moraine (the Chico moraine of Horberg). The critical evidence is a large elongate kettle lying between the two moraines at Pray (Figs. 4, 5, 15). The bottom of this depression is nearly the same

Figure 14: View northward from top of Short Hills showing erratic boulder remnant of Bighorn Dolomite deposited by piedmont glacier in pre-Illinoian time. South lateral moraine of Pine Creek valley glacier is at farthest row of trees.

Figure 15: Fresh morainal surface forms within the "Pray Kettle". Base of kettle is approximately the same level as that of the Yellowstone River in middle distance.

elevation as the inner gorge of the Yellowstone River a mile to the west and its surface is veneered with fresh stagnation moraine (Fig. 5 C-C'). Outwash channels from the Chico moraine extend to the brink of the depression but not across it. These channels continue northward beyond the kettle and finally plunge off the higher outwash bench to the inner gorge of the Yellowstone. The best interpretation is that the kettle was occupied by stagnant recessional ice from the outer advance while the Chico moraine was actively being built; thus, the two moraines represent substades of the same glaciation (ie: Pinedale I and Pinedale II).

3. Pierce (1979, p. F33) correctly observed that terrain beyond the proposed limit of Pinedale I piedmont moraine on the Yellowstone Valley floor does not show any affinity to a Bull Lake piedmont ice sheet.

Dual Topographic Levels on Yellowstone Valley

Two topographic levels were formed in the Yellowstone Valley during Pinedale glaciation. Alden (1932, p. 23) and Horberg (1940, p. 297) observed that the Chico moraine descends northward from the higher level to the inner gorge level of the present Yellowstone River at the mouth of Mill Creek. Late Pinedale floods appear to have flushed the terminal lobe away at this location.

Other evidence in support of two levels of activity is the fact that the high outwash terrace that extends northward from the Pinedale I piedmont terminus gradually descends several miles to the north practically to the level of the present flood plain. This outwash terrace and related moraine has also been deeply trenched by outwash streams discharging laterally near Millcreek from the receding Pinedale I lobe, cutting once again to the lower level which is basically the present flood plain of the river. As the ice retreated, these trenches were progressively abandoned. The diverted channels of Emigrant Creek also show progressive migration southward to the present channel as the blocking piedmont ice retreated.

Finally, behind the Chico moraine, a void space extending to present river level indicates the rapid retreat of ice from the Chico position, and further establishes the two-level idea.

Based on observations in Yellowstone Valley and Jackson Hole, the most logical model for the establishment of two levels of geomorphic activity in a glacial valley can be developed from a situation in which outwash is deposited on the downstream side of an arcuate piedmont terminal moraine. Hydraulic requirements demand the maintenance of grade, hence a sloping plain is created which gradually descends to the downstream pre-glacial river level. As the ice lobe recedes from its terminus, outlet streams pond in the void behind the inner morainal wall where overflow quickly causes dissection and downcutting through the moraine and related outwash plain. The recession of the ice to a new stable position upstream, with possible partial or full readvance, leaves a glacial floor on which the master outwash stream continues to flow, building deposits at this lower level as the situation demands. The new position of the piedmont ice then drapes and drains over its own deposits and those of the former stade while both higher and lower levels of drainage and deposition continue to be simultaneously active.

The concept of two simultaneously active levels in a major valley during a single glacial episode is well illustrated in Jackson Hole, Wyoming, where the Burned Ridge moraine (Pinedale I) is separated from the Jackson Lake moraine (Pinedale II) by fresh pitted outwash. The latter relationship is developed in an area which also served as a southern distributary of the same ice sheet that nourished the Yellowstone Valley piedmont sequence.

Late Illinoian-Wisconsinan Glaciation of Canyons Tributary to Yellowstone Valley

The canyon mouths of the Beartooth Range from Elbow Creek northward to Deep Creek contain a "normal" sequence of late-Illinoian through late-Wisconsinan morainal deposits. That is, a relatively subdued but still recognizable Bull Lake moraine lies several hundred yards beyond the fresher Pinedale terminus and lateral remnants wing out above the Pinedale laterals in the frontal zone, just outside the confining bed-rock canyon walls (Fig. 4). This relationship is particularly well developed at Pine Creek,

Strawberry Creek, and McDonald Creek. Bull Lake lateral moraines at Pine Creek and Deep Creek occupy the crest of a flat bench on the Precambrian bedrock of the Short Hills (Figs. 6, 8). The westward projection of this bench and its morainal and Tertiary cover is truncated by the Deep Creek fault. Consequently, Bull Lake moraines and some Pinedale moraines have been uplifted as much as a hundred feet on this bolck and are not accordant with equivalent deposits on the down-thrown side of the fault.

At many canyon mouths, Pinedale moraines drape around the front of dissected Bull Lake laterals. Horberg (1940, p. 298) noted that the Pinedale glacier at Pine Creek extended northward through a gap between the Beartooth Front and the westward-projecting Bull Lake laterals, hinting that some erosion or faulting along the Luccock Park fault had taken place between Bull Lake and Pinedale advances. In spite of an intensive search, no interstratified soils have been located to prove that a period of erosion did, in fact, occur between the two advances.

Canyon-mouth moraines do not occur south of Elbow Creek because the piedmont glacier swept them downglacier. Perhaps this fact also accounts for the relatively mild expression of glacial canyon abrasion in the valleys of Emigrant and Sixmile Creeks. It also appears as though the Piedmont glacier lingered longer in the main Yellowstone Valley than did the canyon glaciers descending west from the Beartooth Range. This is indicated by lacustrine and fluvial plug deposits, particularly near the mouth of Sixmile Creek where such deposits have reached a thickness of over 200 ft. Because there is no morainal block on their down-stream side, piedmont ice must have served as the blocking mechanism. Emigrant Canyon, likewise, contains such deposits, though they are not as well developed as those of Sixmile Canyon.

On the west side of the Yellowstone Valley, canyon mouth fluvial and lacustrine plugs also occur, indicating that here, as on the east side, piedmont ice was longer lasting than valley ice. However, the canyon ice at Big Creek deposited arcuate moraines of pinedale age on the canyon floor far below the level of maximum piedmont glaciation.

As indicated by Weed (1893, p. 19) the Pinedale piedmont ice occupied the existing floor of Yankee Jim Canyon and must have been as thick as the highest canyon shoulders in order to have extended northward as far as Eightmile Creek. The ice did not extend completely over Dome Mountain at this location, contrary to the indications of Weed's map, for no definitive evidence was found by the writer to corroborate this.

The level and gradient of the piedmont ice during the Pinedale stage is indicated by the ice marginal canyon on the west side of the valley which serves as the practical limit of erratics there (Fig. 16). The canyon shoulders and erratic train drop at a rate of 154 ft per mile toward the north, discharging on the high terrace level near the mouth of Eightmile Creek. A very impressive display of other ice marginal features occurs all along the west side area from the mouth of Yankee Jim Canyon to Eightmile Creek, and were described aptly by Weed (1893).

Post-Glacial Mass Movements of the Yellowstone Valley Area

Several extensive landslides occurred along Yellowstone Valley following the Pinedale glaciation. One notable landslide or mudflow body is located between McDonald and Cascade Creeks on the east side of the main valley along the flank of the Short Hills (Figs. 1, 4). This body has a smooth surface and covers a pre-existing normal fault scarp which cuts the north side of the McDonald Creek alluvial fan. Material in the flow consists of boundary Precambrain rocks in a clayey matrix. The flow probably originated from ground saturation during the Pinedale glaciation or just following that event.

An impressive slide discovered by Good (1965) occurs just south of the mouth of Yankee Jim Canyon (Figs. 4, 17). Large cuts of U.S. Highway 89 extend through this slide. A large mass of Precambrian rock slid intact from the east side of the Canyon wall and crossed the Yellowstone River, damming it to form a lake. Lacustrine deposits upstream from the slide dam are visible along Highway 89 and can be traced as far south as Corwin Springs where a hearth site developed in the deltaic deposits at the head of

Figure 16: View north through ice-marginal canyon of the Pinedale I piedmont glacier along the west margin of Yellowstone Valley. Marginal stream discharged on valley floor here at the forward edge of the ice advance. Bedrock in foreground is andesitic breccia.

Figure 17: Highway 89 along the Yellowstone River at Yankee Jim Canyon cuts through post-Pinedale landslide involving Precambrian rock.

the lake has been dated as 4,900 years b.p. Good observed that "giant ripple marks" extending north of the slide on the west side of the river attest to the bursting of the slide dam with consequent flooding of the Yellowstone Valley for some miles northward. Flood gravels occur near Carbella bridge on the first terrace above river level and downstream to the vicinity of Point of Rocks.

Two earthflows consisting largely of volcanic rock debris descend to river level, one on each side of the river, near the mouth of Slip and Slide Creek (Fig. 18). They are several miles in length and show evidence of recent movement, trapping water within them to form small ponds. These slides apparently did not ever move rapidly enough to effectively block the main Yellowstone River.

Figure 18: Slump-earthflow on floor of Yellowstone Valley at Slip and Slide Creek. Material consists of volcanic debris from high flanking hills of the south Snowy block of the Beartooth Range on right skyline.

Small rockslides and slumps occur upstream from the Yankee Jim landslide as fas as Cinnabar Mountain and involve local deposits of both Precambrian and Paleozoic rocks. At the base of Cinnabar Mountain, the largest of these forms a mound that appears to have diverted the Yellowstone River. Waldrop (1963, p. E11) also described blockglides, slumps, and earthflows along the north side of Sepulcher Mountain, making a complex geomorphic picture there.

Pierce (1979, p. 65) concluded that flood bars and "giant ripple marks" within the northern boundary of Yellowstone Park near the mouth of Reese Creek on the west side of Yellowstone River were the result of catastrophic flooding (Figs. 4, 19). Similar features in the valley floor near Gardiner are part of the same complex. In the latter case, giant boulders formed tongue-like bars that terminate in lobate deposits that arc through the Gardiner High School area. Bar-like deposits descend northwestward in terrac-like form on the north side of the Yellowstone River at this locality.

Figure 19: View southeast on top of flood bar near mouth of Reese Creek, Yellowstone Valley. Flood surge was from right to left, leaving a slightly imbricated slurry that formed these "ripple marks". Dark talus-producing rocks in the background are Pleistocene basalt flows erupted near the trace of the Gardiner Fault.

Pierce proposed that the escape of proglacial lakes, such as Hayden Lake in the Yellowstone Plateau area, could account for such floods. In addition, the effects of geothermal melting and saturation of ice could contribute to the sudden water fluxes. The coursing of such water scoured the Yellowstone Valley, eradicating the Pinedale II (Chico) terminal moraine where it crosses the inner valley floor near Emigrant. These floods were similar to the "Spokane Floods" but of smaller magnitude. The imbricated, unsorted deposits of these floods superficially resemble glacial moraine but close inspection usually reveals such features as imbrication in the boulder deposits, regular crest-to-crest wave forms, and other features peculiar to flood deposits.

During the latter part of Pleistocene time and extending into the Recent, floods and mudflows modified nearly all inner canyon floors of streams descending to the Yellowstone River from surrounding highlands. These features are particularly well displayed where no late-stage valley glaciers occupied the canyons. The lower extremities of both Cascade and McDonald Creeks are good examples of these features. Whether these are the result of cloudburst activity, the release of peri-glacial debris, or both is not clear, but the deposits are certainly an important contribution to the geomorphic characteristics of these stream bottoms.

GENERAL CONCLUSIONS

This paper attempts to integrate the Cenozoic structural, volcanic, and geomorphic history of an impressive Rocky Mountain valley. Insofar as the broad scene is concerned, structural frameworks and compositional differences of bedrock types, including volcanics, are about as influential in determining the shape of the landscape as the specific geomorphic processes. Although Quaternary events climaxed the shaping of the topography, the initial development of many of these features was related to Laramide and early post-Laramide events. The following are highlights of structural, volcanic, and geomorphic principals illustrated by the Yellowstone Valley.

1. The upper Yellowstone Valley appears to have undergone an early tensional history, either contemporaneous with or slightly later than compressional activity of the Laramide orogeny. If tensional activity commenced early in Cenozoic time, the climax of such activity was not reached until near the very end of the Cenozoic. Hence, structural influence such as that shown by episodic movement on the Deep Creek-Luccock Park fault zone is an important contributing geomorphic factor.
2. Volcanism and related intrusive activity served principally as a constructional influence during the early Cenozoic history of the valley and surrounding highlands. Some evidence suggests that Eocene volcanism helped distinguish the valley from the surrounding highlands, whereas the less prolific late Cenozoic volcanic deposits merely conformed to the already existent valley. Both early and late Cenozoic volcanism took place concurrently with faulting, uplift, and tilting of crustal blocks.
3. In most cases, Bull Lake glaciation in the canyons of the Beartooth Range was of greater magnitude than that of the Pinedale. However, the reverse is true of the glacial record of the Yellowstone outlet glacier which formed the late glacial piedmont lobe in the Yellowstone Valley.
4. The hydraulic systems of the Wisconsinan piedmont glacier provided at least two simultaneously active topographic levels in the Yellowstone Valley. The highest central valley level ranges from a few feet to over a hundred feet above the Wisconsinan and Holocene flood plain of the Yellowstone River.
5. Tensional faulting has offset glacial deposits along the Beartooth front and must be reckoned with in rationalizing the history and magnitude of glaciation at this location.
6. The stamp of glacial activity, its abrasional, fluvial, and depositional aspects, even to the extent of providing the impetus for giant landslides and catastrophic floods, remains the most dominant of all geomorphic influences on the Yellowstone Valley area.

Early workers did a remarkable job of interpreting the true meaning of the physical geologic history of the Yellowstone Valley. We stand on their shoulders to carry out studies of the finer details.

REFERENCES CITED

Alden, W. C., 1932, Physiography and glacial geology of eastern Montana and adjacent areas: United States Geological Survey Professional Paper 174, 133 p.

______, 1953, Physiography and glacial geology of western Montana and adjacent areas: United States Geological Survey Professional Paper 231, 200 p.

Blackwelder, E., 1915, Post-Cretaceous history of the mountains in central western Wyoming: Journal of Geology, v. 23, p. 97-117, 193-217, 307-340.

Bonini, W. E., Kelley, W. N., Jr., and Hughes, D. W., 1972, Gravity studies of the Crazy Mountains and the west flank of the Beartooth Mountains, Montana: Guidebook, Montana Geological Society, 21st Annual Geological Conference, Crazy Mountains Basin, p. 119-127.

Brown, C. W., 1961, Cenozoic stratigraphy and structural geology, northeast Yellowstone National Park, Wyoming and Montana: Geological Society America Bulletin, v. 72, p. 1173-1194.

Bush, J. H., 1967, The basalts of Yellowstone Valley, southwestern Montana: unpublished M.S. thesis, Montana state University, Bozeman, 66 p.

Chadwick, R. A., 1969, The Northern Gallatin Range, Montana; northwestern part of the Absaroka-Gallatin Volcanic Field: Contributions to Geology, University of Wyoming, Laramie, v. 8, pt. 2, p. 150-166.

______, 1970, Belts of eruptive centers in the Absaroka-Gallatin volcanic province, Wyoming-Montana: Geological Society America Bulletin, v. 81, p. 267-273.

______, 1972, Igneous activity south and west of the Crazy Mountains basin, Montana: Montana Geological Society 21st Annual Field Conference Guidebook, p. 141-148.

______, 1978, Geochronology of post-Eocene rhyolite and basaltic volcanism in southwestern Montana: Isochron/West, no. 22, p. 25-28.

______, 1981, Chronology and structural setting of volcanism in southwestern and central Montana: Montana Geological Society, 1981 Field Conference and Symposium Guidebook, p. 301-310.

Christiansen, R. L., and Blank, H. R., 1969, Volcanic evolution of the Yellowstone Rhyolite Plateau and eastern Snake River Plain, U.S.A. (abstract): International Association Volcanology Symposium on Volcanoes and their Roots, Oxford, England, p. 220-221.

Fenneman, N. M., 1931, Physiography of western United States: McGraw-Hill Book Company Incorporated, New York, 534 p.

Foose, R. M., Wise, D. U., and Garbarini, G. S., 1961, Structural geology of the Beartooth Mountains, Montana and Wyoming: Geological Society America Bulletin, v. 72, p. 1143-1172.

Fraser, G. D., Waldrop, H. A., and Hayden, H. J., 1969, Geology of the Gardiner area, Park County, Montana: United States Geological Survey Bulletin 1277, 118 p.

Fryxell, F. M., 1930, Glacial features of Jackson Hole, Wyoming: Augustana Library Publication 13, Rock Island, Illinois, 129 p.

Garihan, J. M., Schmidt, C. J., Williams, M. A., and Young, S. W., 1982, Tectonic history of the Bismark-Spanish Peaks-Gardiner fault system, southwest Montana (abstract): Geological Society America Abstracts with Programs, v. 14, no. 6, p. 311.

Good, J. M., 1965, Prehistoric landslide in Yankee Jim Canyon, Park County, Montana (abstract): Geological Society America Special Paper 82, p. 327-328.

Hall, W. B., 1960, Multiple glaciation in the Madison and Gallatin Ranges, southwestern Montana: *in* West Yellowstone-Eartquake area, Billings Geological Society Guidebook, 11th Annual Field Conference, p. 191-206.

Hayden, F. V., 1872, Preliminary report of the United States Geological Survey of Montana and portions of adjacent territories: Fifth Report of Progress, Chapters III and IV, p. 44-66.

______, 1872, Sixth Annual Report of the United States Geological Survey of the Territories: Chapter II, p. 26-44.

Horberg, Leland, 1940, Geomorphic problems and glacial geology of the Yellowstone Valley, Park County, Montana: Journal of Geology, v. 48, no. 3, p. 275-303.

Hughes, G. C., 1980, Cenozoic geology and geomorphology of the Dry Creek Valley, Gallatin County, Montana: Unpublished M.S. thesis, Montana State University, Bozeman, 145 p.

Iddings, J. P. and Weed, W. H., 1894, Livingston Folio, United States Geological Survey Folio 1, Geological Atlas of the United States, 5 p.

Kavanaugh, J., 1965, A study of selected igneous bodies of the Norris-Red Bluff area, Madison County, Montana: unpublished M.S. thesis, Montana State University, Bozeman, 60 p.

Lammers, E. C. H., 1937, The structural geology of the Livingston Peak area, Montana: Journal of Geology, v. 45, p. 268-296.

Love, J. D., 1968, Creation of the Teton landscape: Grand Teton Natural History Association, Moose, Wyoming, 120 p.

McMannis, W. J., and Chadwick, R. A., 1964, Geology of the Garnet Mountain Quadrangle, Gallatin County, Montana: Bulletin 43, Montana Bureau of Mines and Geology, 47 p.

Pierce, K. L., 1979, History and dynamics of glaciation in the northern Yellowstone National Park area: United States Geological Survey Professional Paper 729-F, 90 p.

Reid, R. R., McMannis, W. J., and Palmquist, J. C., 1975, Precambrian geology of the North Snowy block, Beartooth Mountains, Montana: Geological Society of America Special Paper 157, 135 p.

Reid, R. R., 1957, Bedrock geology of the north end of the Tobacco Root Mountains, Madison County, Montana: Montana Bureau of Mines and Geology Memoir 36, 24 p.

Richards, P. W., 1957, Geology of the area east and southeast of Livingston, Park County, Montana: United States Geological Survey Bulletin 1021-L, 438 p.

Richmond, G. M., 1964, Glacial geology of the West Yellowstone Basin and adjacent parts of Yellowstone National Park: United States Geological Survey Professional Paper 435-T, p. 223-236.

Roberts, A. E., 1972, Cretaceous and early Tertiary depositional and tectonic history of the Livingston area, southwestern Montana: United States Geological Survey Professional Paper 526-C, 120 p.

Ruppel, E. T., 1972, Geology of pre-Tertiary rocks in the northern part of Yellowstone National Park, Wyoming; Geology of Yellowstone Park: United States Geological Survey Professional Paper 729-A, 66 p.

Schmidt, C. J. and Hendrix, T. E., 1981, Tectonic controls for thrust belt and Rocky Mountains foreland structures in the northern Tobacco Root Mountains-Jefferson Canyon area, southwestern Montana: Montana Geological Society, 1981 Field Conference and Symposium, p. 167-180.

Shaver, K., 1978, Chronology of dacite intrusion in the northwest end of the Absaroka-Gallatin volcanic province, Montana: northwest Geology, v. 7, p. 21-26.

Skeels, D. C., 1939, Structural geology of the Trail Creek-Canyon Mountain area, Montana: Journal of Geology, v. 47, p. 816-840.

Smedes, H. W. and Prostka, H. J., 1972, Stratigraphic framework of the Absaroka Volcanic Supergroup in the Yellowstone National Park region: United states Geological Survey Professional Paper 729-C, 33 p.

Struhsacker, E. M., 1976, Geothermal systems of the Corwin Springs-Gardiner area, Montana: possible structural and lithologic controls: unpublished M.S. thesis, Montana State University, 93.

Van Voast, W. A., 1964, General geology and geomorphology of the Emigrant Gulch-Mill Creek area, Park County, Montana: unpublished M.S. thesis, Montana State University, Bozeman, 65 p.

Waldrop, H. A. and Hayden, H. J., 1963, Landslides near Gardiner, montana: United States Geological Survey Professional Paper 450-E, p. E11-E14.

Walsh, T. H., 1975, Evidence for two pre-Wisconsin glaciations in the Madison Range, Gallatin and Madison Counties, Montana (abstract) Geological Society of America, Abstracts with Programs, v. 7, no. 5, p. 649.

Weed, W. H., 1893, Glaciation of the Yellowstone Valley north of the Park: United States Geological Survey Bulletin 104, 41 p.

Wilson, C. W., Jr., 1934, Geology of the thrust fault near Gardiner, Montana: Journal of Geology, v. 42, p. 649-663.

______, 1936, Geology of the Nye-Bowler lineament, Stillwater and Carbon Counties, Montana: American Association of Petroleum Geologists Bulletin, v. 20, no. 9, p. 1161-1188.

Wilson, J. T., 1936, The geology of the Mill Creek-Stillwater area, Montana: unpublished Ph.D. thesis, Princeton University, Princeton, New Jersey.

Thirty-Third Annual Field Conference — 1982
Wyoming Geological Association Guidebook

THE YANKEE JIM CANYON LANDSLIDE — PARK COUNTY, MONTANA

JOHN M. GOOD[1]

ABSTRACT

More than 5,000 years ago a landslide in Yankee Jim Canyon formed a dam 200 feet high across the Yellowstone River forming a lake which extended upstream to Gardiner, Montana. Indians camped along the lake shore 5,000 years ago. Failure of the dam was sudden as evidenced by lag boulders on the slide and giant ripples immediately downstream produced by catastrophic flooding.

The Yankee Jim Canyon landslide was recognized during investigations of an archeological site approximately one mile northeast of Corwin Springs, Montana.

The site consisted of two campsites approximately 200 yards apart in the bluffs on the east side of the Yellowstone River. It is known as the Rigler Bluffs site, 24Pa68P. Gullies, 16 or more feet deep and extending eastward from the river through alluvial deposits, expose the campsites.

One of these was a hearth lined with fire-blackened granite cobblestones of glacial origin containing disturbed sand and fragments of charcoal. The hearth is overlain by eight feet of thin-bedded sand and silt. Beneath the hearth, extending to river level some 25-feet below, are inter-bedded silts and fine sands containing a few thin lenses of coarse sand and fine gravel. Thin streaks of carbon separated by sterile silt layers occur at the same elevation as the hearth and below it in the section. Plant material in growth position constitutes significant parts of these carbon streaks. Charcoal from the hearth was dated by C-14 methods at about 5,000 yrs.b.p.(1)(2). The hearth's elevation is 5,096 feet. The other site was a simple fire pit in the same silt and sand section at an elevation of 5,088 feet.

Because the thin-bedded sands and silts of Rigler Bluffs represented 16-feet of sediments radically different from modern deposits of the Yellowstone River, a search was initiated for something which may have dammed the stream. The answer was found in Yankee Jim Canyon, 14.5 miles northwest of Gardiner, Montana.

[1]Present address: P.O. Box 354, Grand Canyon, AZ 86023

Yankee Jim Canyon is a major constriction in the valley of the Yellowstone River. Here the stream has cut a narrow gorge through Pre-Cambrain gneiss which was subsequently scoured by glacial ice. Figure 1 is a view of a landslide which moved northwesterly from the south slope of Dome Montain more than 2,000 feet across the Yellowstone River and well beyond the photo point.

More than 99% of the slide material consists of Pre-Cambrian crystallines from sand-size particles to house-sized boulders with the finer fractions making up a very small percentage of the total. Distorted individual masses of fine sands and gravel, as seen in Figure 2, are

Figure 1: View of Yankee Jim Canyon landslide looking north. The slide extended beyond the photo point.

Figure 2: Pinedale kame or outwash gravel and sand masses exposed in landslide debris.

conspicuous although they form a small part of the slide debris. A few striated cobbles and pebbles were found on the crest of the slide.

It is difficult to establish the lowest elevation on the slide crest which controlled the maximum lake level. Glacially striated stones on the crest were originally considered to have been ice rafted into position, but subsequent work by Pierce (3) demonstrating the presence of 2,800 feet of Pinedale ice in Yankee Jim Canyon renders glacial deposition much more likely.

On the western flank of the slide (downstream) stream-deposited slide debris at an elevation of 5,100 feet indicates the dam crest was higher. The low point of the existing crest at 5,225 feet establishes maximum height of the dam. It is believed the latter figure is closer to the landslide dam's true crest, in which case the dam was approximately 200 feet high and would have formed a lake extending upstream to Gardiner.

The undulating surface of the landslide west of the crest consists of numerous angular cobbles and boulders projecting above finer material. South of the river, immediately west of the landslide debris, is a series of giant flood ripples composed of locally derived angular gneiss fragments and whose crests are dotted with boulders of the same material. These bars overlie glacial outwash.

Synthesis of existing data suggests that extensive Pinedale glaciation left the walls of Yankee Jim Canyon and the walls of the Yellowstone Valley upstream in an unstable, steep condition. In Yankee Jim, the fractured gneiss standing in vertical walls failed, perhaps in an earthquake, more than 5,000 years ago producing a landslide which formed a dam in the canyon two-hundred feet high. The slide incorporated isolated lenses of glacial gravel and sand of Pinedale age. The lack of sand and silt-sized particles in the slide's debris suggests the dam leaked. Repeated streaks of plant material in growth position separated by thin sterile silt layers also suggest cyclical changes in lake level which might have been caused by the lake filling in spring runoff followed by partial drainage in fall.

The dam did not exist very long. Nowhere in the valley above Yankee Jim Canyon is there evidence of landscape modification by the lake, and lake sediments are neither extensive nor thick. In fact, increasing amounts of coarser material in the sands and silts overlying the campsites suggest a shift away from lake deposition.

When the dam failed less than 5,000 years ago, it collapsed suddenly as is evidenced by flood bars of landslide debris immediately downstream. Also, the extraordinary number of boulders on the downstream flanks of the slide suggest lag deposits in the wake of catastrophic flooding. The cause of dam failure is unknown, however, Pierce (3) describes post-Pinedale floods which originated in Yellowstone National Park. Perhaps one of the younger of these provided the volume of water necessary for dam failure.

SELECTED REFERENCES

1. Rubin, M., 1962, Personal communication, U.S. Geol. Surv. Lab. No. W1135, 4,900-3 yrs. b.p.
2. Gray, D. Sheridan College. Sample 24Pa68P, 5040-150 yrs.b.p. (Personal Communication, 1962)
3. Pierce, K. L., 1979, History and dynamics of glaciation in the northern Yellowstone National Park area: U.S. Geol. Surv. Professional Paper 729-F, Plate 1, p. F67.

Thirty-Third Annual Field Conference — 1982
Wyoming Geological Association Guidebook

GEOLOGY OF THE SLOUGH CREEK TUFF, NORTHERN ABSAROKA VOLCANIC FIELD, PARK COUNTY, MONTANA

BY JOHN W. HICKENLOOPER, JR.[1], JAMES T. GUTMANN[2]

INTRODUCTION

The Slough Creek Tuff is a quartz latitic, welded ash-flow tuff of early middle Eocene age located in the northern part of the Absaroka volcanic field, north of Yellowstone National Park (Fig. 1). This unit, which is about 80 m thick in many places, is composed chiefly of two compound cooling units of densely welded tuff. Total original volume evidently was of the order of 20 km^3. These resistant rocks cap the ridges across more than 500 km^2 of the volcanic field and form a distinctive stratigraphic unit within the brown andesitic volcanic and volcaniclastic rocks of the Mount Wallace Formation, of which it is a member.

The Slough Creek Tuff was formally named by Smedes and Prostaka (1972) from exposure along Slough Creek near the northern boundary of the Park (Fig. 2). Rocks of this unit had been mapped as a flow of "trachytic rhyolite" on Folio No. 1 of the U.S. Geological Survey Geologic Atlas series ((Iddings and Weed, 1894). J.P. Iddings (in Hague and others, 1899) described the petrography of a trachytic rhyolite located two miles west of the mouth of Slough Creek and compared it to similar rocks to the west near Ash Mountain (Fig. 2), then known as Sunset Peak. Howard (1937) named the rocks two miles west of Slough Creek the Lost Creek Trachyte. Brown (1961) pointed out the tuffaceous nature of this unit and renamed it the Lost Creek Tuff. Fraser and others (1969) suggested correlation between the Lost Creek Tuff and rocks southwest of Ash Mountain, near Jardine.

Smedes and Prostka (1972) recognized that the Lost Creek Tuff was stratigraphically beneath the Slough Creek Tuff, which they established as the medial member of the Mount Wallace Formation, Sunlight Group, Absaroka Volcanic Supergroup. Wedow and others (1975) mapped the tuffaceous rocks near Ash Mountain, and many others to the east, as the Slough Creek Tuff. Stratigraphic relationships and K-Ar ages cited by Smedes and Prostka (1972) suggest an age bracket for the Slough Creek Tuff of 48.0 to 49.2 m.y.

[1]Buckhorn Petroleum Co., 1625 Broadway, Denver, Colorado 80217
[2]Dept. of Earth & Environmental Sciences, Wesleyan University, Middletown, Connecticut 06457

In the study area, the Mount Wallace Formation consists of a succession of andesitic, basaltic and dacitic flows and volcaniclastic rocks and is divided into lower and upper flow sequences by the Slough Creek Tuff. The Mount Wallace Formation is about 700 m thick near Mount Wallace and the Slough Creek Tuff lies near the middle of that section (Smedes and Prostka, 1972). South of Mount Wallace this formation thins, the lower flows pinch out, and the Slough Creek Tuff rests on Paleozoic limestone at Slough Creek.

Wedow and others (1975) provided a reconnaissance geologic map at 1:62,500 of the region directly north of the Park and including most known outcrops of the Slough Creek Tuff. This they mapped as a separate unit and the accuracy of their regional mapping greatly facilitated the present study.

LITHOLOGY AND STRATIGRAPHY

Christiansen and others (1968) noted repeated association of ash-flow tuffs with compositionally similar lava flows and proposed that such associated rock units be mapped together as co-genetic sequences. The Slough Creek Tuff is such a sequence and includes basal lava flows and two compound cooling units of ash-flow tuff. The basal flows have been informally designated the Kristan flows and the overlying welded tuffs the Hildan and Aslan tuffs.

The two ash-flow units are similar in overall appearance but petrographically distinct. Typically, the tuffs are densely welded (bulk density 2.3-2.4 gr/cc) and pinkish tan to reddish brown in color. Nonwelded to partially welded tuff occurs at the top and bottom of both ash-flow sheets; in most places these zones of friable rock are thin and poorly exposed. The densely welded tuffs generally are lithoidal as a result of devitrification and alteration although zones of black vitrophyre are common and, in the Hildan tuff, locally are very thick. All of these pyroclastic materials contain 15 to 30 percent of feldspar phenocrysts about 1 to 2 mm across. Feldspar phenocrysts in the Hildan tuff are solely plagioclase whereas, in the overlaying Aslan unit, sanidine generally is more abundant than plagioclase.

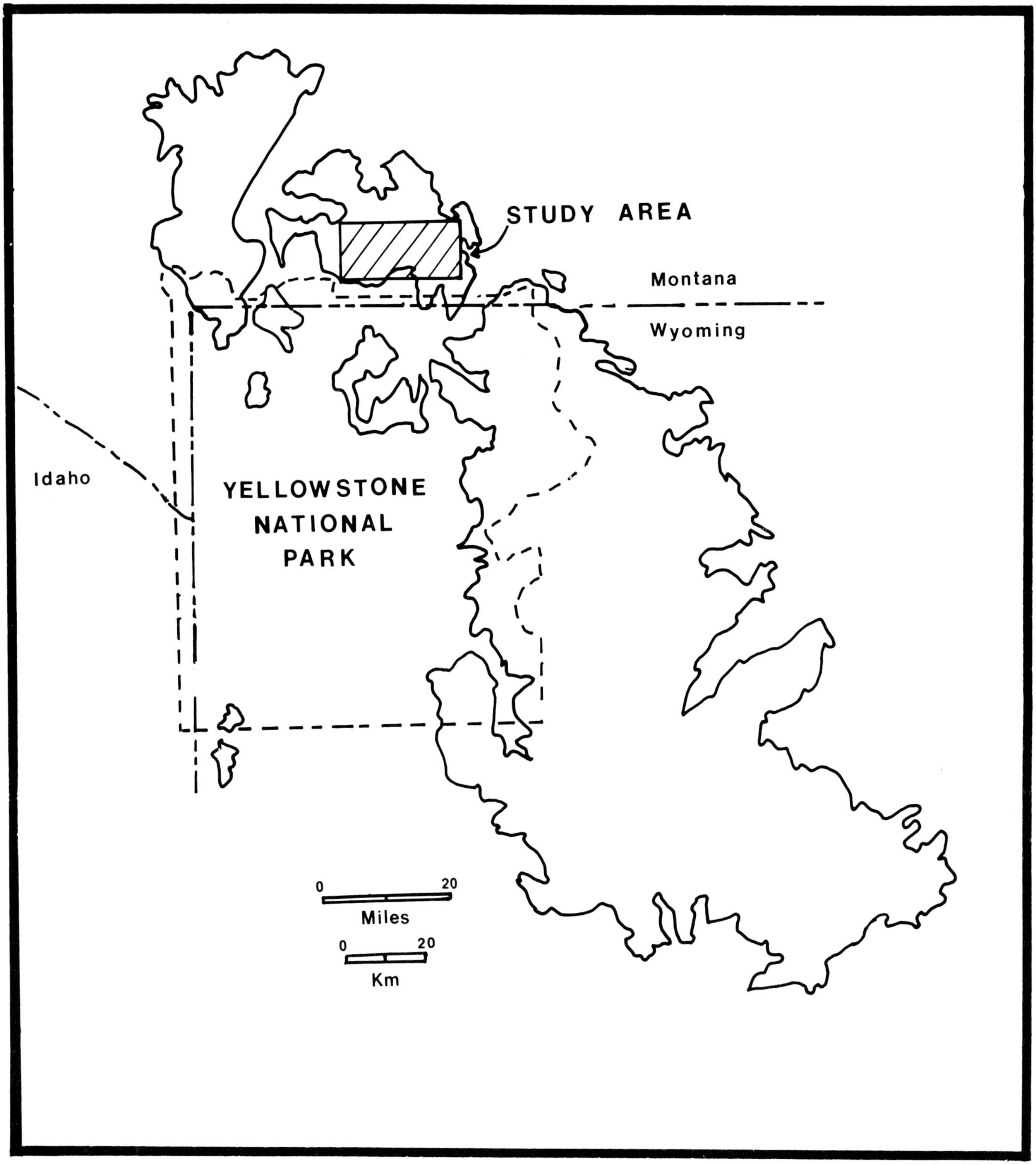

Figure 1: Generalized map of the Absaroka volcanic field.

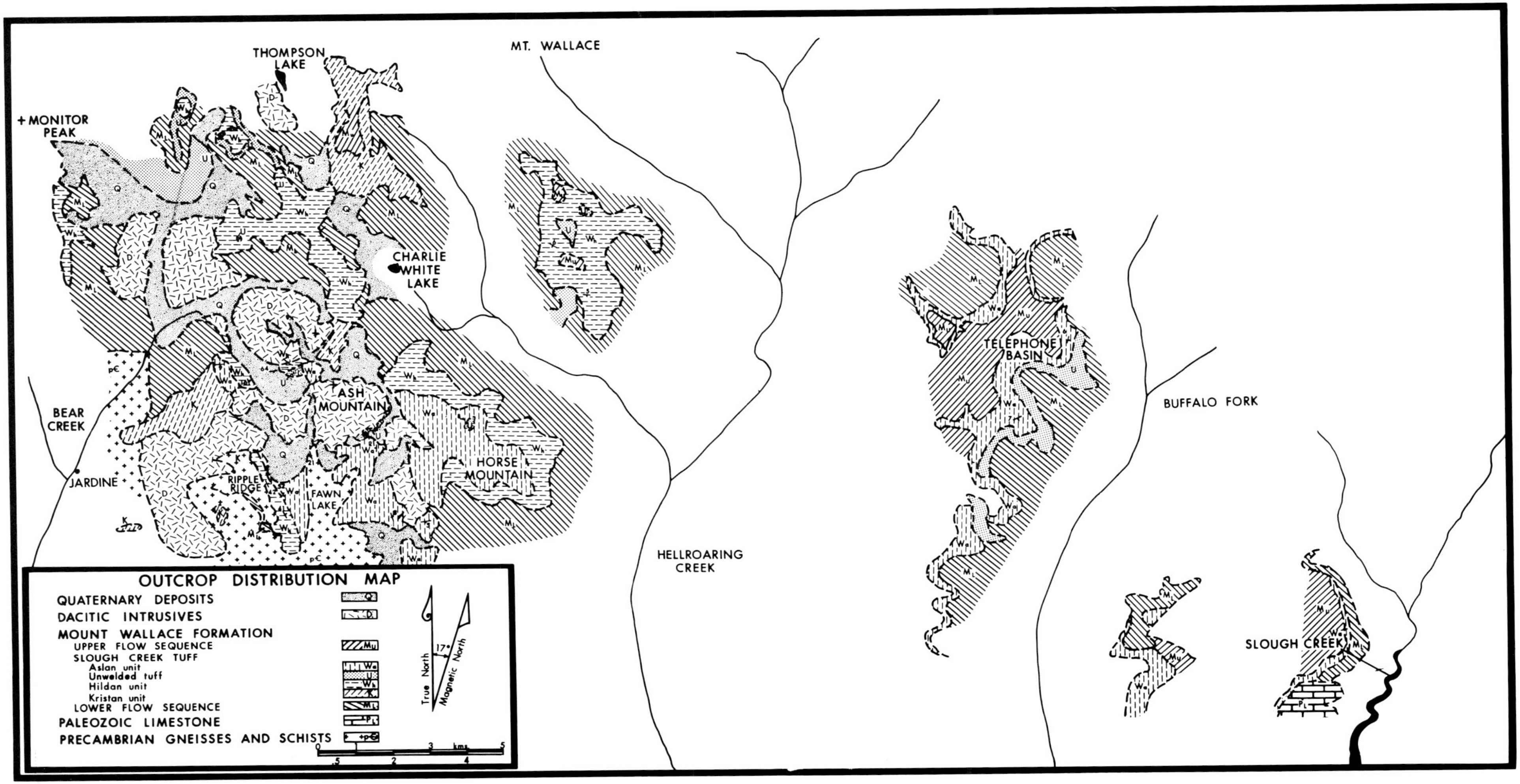

Figure 2: Outcrop distribution map for the Slough Creek Tuff.

Lithic fragments, almost exclusively of volcanic rocks, are present in variable amounts up to about six percent.

Distribution of the three units of the Slough Creek Tuff is shown in Fig. 2. The Kristan flows have so far been identified only in the western part of the area, around Ash Mountain and to the north near the head of Thompson Creek. The Hildan tuff also occurs only in the western part of the area, west of Hellroaring Creek. In contrast, the overlying Aslan tuff occurs in the western and eastern parts of the area, and is the only unit present in the exposures along Slough Creek from which the tuff was named.

Approximately 40 to 80 m of densely welded tuff are present at most localities. Original thickness plainly varied somewhat in response to topographic irregularities but evidently was rarely in excess of 85 m in the case of the Hildan tuff. A maximum thickness of the Aslan tuff in the study area occurs in the exposures at Slough Creek, where approximately 225 m of tuff, with several intercalated lava flows, accumulated in a paleovalley cut in Paleozoic rocks.

The Kristan unit is best exposed on a ridge southwest of Ash Mountain (Ripple Ridge, Fig. 2), where several tens of metres of flinty, reddish dark gray porphyritic lava flows and associated breccias crop out below the Hildan tuff. The contact between these units is not exposed in the study area.

The Hildan tuff is best exposed in the northwestern part of the area, southwest of Thompson Lake. Here the unit rests on laharic breccias although lithic inclusions in the tuff evidently are fragments of Kristan lavas. Approximately 8 m of nonwelded to partially welded tuff at the base grades within 2 m into dense vitrophyre. The lowermost 2 m contain well sorted airfall beds a few mm to a few cm thick; no cross bedding was observed. Bedding becomes less distinct upward and minute grains of black glass become larger and more abundant. In the second and third metres above the base, degree of welding increases rapidly and vitric lapilli increase in abundance. Across the upper four metres of the partially welded tuff, degree of flattening of lapilli increases as well as their abundance. Between 8 and 10 m above the base, the deposit grades into dense, black vitrophyre.

The vitrophyre at this locality is only a few metres thick and passes upward into lithoidal tuff in which the original glass is now turbid and feebly birefringent owing to exceedingly fine grained products of devitrification and alteration. About 2 km to the south, however, virtually the entire exposed section of Hildan tuff is a zone of black vitrophyre nearly 70 m thick; this grades upward into friable, apparently nonwelded tuff across only several cm. Nonwelded tuff at the top of the Hildan unit near Ash Mountain is anomalously rich in phenocrysts (about 40 percent after correction for porosity) and is well sorted ash nearly devoid of vitric lapilli and lithic fragments.

The section of Hildan tuff exposed east of Horse Creek can be seen from a distance to comprise at least six flow units separated by layers of light-colored rock. At the outcrop, only a few of these separations could be discerned; these consisted of about 1 m of more friable tuff within which pumice lumps evidently were as flattened as in the denser tuff above and below. Welding and compaction in the Hildan unit, wherein pumice lumps typically are flattened until thickness-to-length ratios are 1:10 to 1:15 (maximum), effectively obliterated discernible contacts between flow units in most places.

The basal Aslan tuff is best exposed on the western slope of Ash Mountain where it evidently rests in direct and conformable contact (never precisely exposed) on the top of the Hildan tuff. The lower most 30 to 35 m of the Aslan unit at this locality consists of massive, yellowish to faintly greenish buff or tan, nonwelded to weakly welded tuff. This tuff contains much pumice and abundant lithic clasts, up to 3 cm across, apparently derived from the Kristan flows. Here, as at other localities, vitrophyre is rare in the Aslan tuff, especially in contrast with the Hildan unit.

In the "type" locality of the Slough Creek Tuff, along Slough Creek, only the Aslan unit is present; its base is poorly exposed but nonwelded tuff there, if it occurs, must be very thin and succeeded rapidly upward by about 7 m of weakly to moderately welded, locally silicified tuff. This grades up into 100 m of densely welded material, including a layer of vitrophyre 1 m thick near the top and capped by a few m of virtually nonwelded tuff. Four mafic lava flows separate this nonwelded tuff from an overlying section of massive, pale tan, nonwelded to weakly welded tuff about 100 m thick. This unit is rich in pumice lumps as much as 20 cm long and is characterized by the presence of abundant phenocrysts of biotite up to 3 mm across. It is overlain disconformably by upper flows of the Mount Wallace Formation.

PETROGRAPHY

The two principal units of the Slough Creek Tuff are readily differentiated in thin section. The Hildan tuff generally carries 14 to 25 percent phenocrysts and crystal fragments of plagioclase, 1 to 8 percent of hypersthene crystals, and, in many samples, a few small phenocrysts of augite; sanidine is absent. In contrast, the Aslan tuff contains only 4 to 10 percent of plagioclase, less than one-half percent hypersthene if any, and, in nearly all samples, no augite; sanidine is invaribly present in amounts ranging from 6 to nearly 20 percent and is plainly more abundant than plagioclase in most samples.

Mineralogy of the Kirstan flows on Ripple Ridge is

similar to that of the overlying Hildan tuff in that they contain 10 to 15 percent of plagioclase phenocrysts as much as 5 m long and scattered phencrysts of hypersthene set in a fluidal, cryptocrystalline to glassy groundmass. Refractive indices of plagioclase clevage flakes indicate An_{45-49} for Kristan phenocrysts and An_{44-47} for Hildan phenocrysts.

The vitroclastic matrix of the Hildan tuff shows some evidence of devitrification in many samples, especially within pumice limps, but regular devitrification textures (e.g., axiolites) seem remarkably rare and the devitrification products which are present seem notably fine grained. Likewise, lithophysae and recognizable vapor-phase crystallization products are rare. The bulk of these rocks consists of cryptocrystalline to glassy(?), turbid material which may reflect not primary devitrification so much as the process of alteration and oxidation that must have accompanied hydration of the vitric matrices. Especially in the more densely welded rocks, plagioclase phenocrysts have undergone patchy alteration to fine aggregates of sericite, chlorite and calcite, and pyroxenes are entirely replaced by yellowish phyllosilicates. In addition to plagioclase and pyroxenes, phenocrysts in the Hildan unit include titanomagnetite, zircon, and rare apatite.

Sixteen samples were collected from a well exposed section of Hildan tuff 80 m thick east of Horse Creek. Bulk density and phenocryst modes were measured in all samples (data in Hickenlooper, 1980). Variations in bulk density reflect the compound nature of the cooling unit but show no systematic decrease upward through the sheet, which remains densely welded nearly to its top. Abundance of plagioclase phenocrysts tends to decrease slightly (perhaps five percent) up through the sheet, although this trend barely shows through the scatter arising from variations within Hildan flow units.

In the Aslan tuff, sanidine phenocrysts occur and are generally more abundant than those of plagioclase. The plagioclase phenocrysts here are notably more sodic (An_{23-27}) than in the Hildan unit (44-47). Also in contrast to the Hildan unit, plagioclase and especially sanidine phenocrysts in the Aslan tuff commonly are rounded and deeply embayed. Plagioclase cores in several instances were surrounded by sanidine rims. Pyroxene phenocrysts are relatively rare in the Aslan unit but hornblende and biotite occur in some samples and biotite is especially coarse and abundant in the thick, weakly to nonwelded tuff forming the upper half of the unit at Slough Creek.

Compositions of sanidine in the Aslan tuff were determined by X-ray diffraction using the "three peak" method of Wright (1968). The composition thus obtained for sanidine from each of several welded rocks was Or_{66-67}; $(\bar{2}01)$ peaks were sharp, indicating absence of cryptoperthite. Interpreted via the two-feldspar geothermometer of Stormer (1975), phenocryst compositions of these rocks indicate temperature of 690°C (assuming 1 kb pressure) to 800°C (at 10 kb). In the thick section of nonwelded tuff at Slough Creek, plagioclase composition (An_{25}) is similar to that in the weld rocks but the sanidine is more potassic (Or_{70}). The two-feldspar magmatic temperatures implied by these data are 650°C (at 1 kb) to 775°C (at 10 kb), or 25° to 40°C cooler than in the welded unit.

GEOCHEMISTRY

Ten whole-rock analyses representing all major units of the Slough Creek Tuff were performed by atomic absorption spectrophotometry and are presented in Table 1 (analytical procedures in Hickenlooper, 1980). The analyses are arranged in the table in order of decreasing age from left to right. Sample H-76 is from a Kristan flow on the west flank of Ash Mountain, H-138 is basal Hildan vitrophyre at Horse Mountain, and H-191 through 175 are from a prominent cliff of Hildan tuff northeast of Horse Mountain. H-65 represents densely welded Aslan tuff near its base at Ripple Ridge, H-86 is densely welded Aslan tuff from Slough Creek, and H-89 is from the thick, nonwelded Aslan tuff at Slough Creek.

Judging from the relative amounts of orthoclase, plagioclase and quartz in their norms, these rocks would be classified as rhyolites following Streckeisen (1979), or alternatively as rhyodacites in view of the excess of plagioclase over orthoclase in many samples. These rocks might well be termed rhyodacite but it should be recognized that they are very positive (average Na_20 /Na_20 = 1.F) and rich in alkalies relative to silica (average Na_20 + k_20 = 8.2 percent). Although somewhat less rich in soda, their anhydrous compositions are notably similar to those of nonhydrated quartz latite glasses from the Eureka Valley Tuff, California, the type locality of quartz latite (Noble and others, 1976). We believe that the compositions of a number of the Slough Creek rocks have been significantly altered by alkali migration and their analyses must be interpreted with caution. Nevertheless, it is clear that their potassic character is much stronger than that of most rhyodacites. Accordingly, the term quartz latite seems more appropriate for those rocks despite the fact that their ratio of normative quartz to quartz plus feldspar is slightly greater than 1 to 5.

The data indicate that most if not all samples have undergone considerable hydration since deposition. Furthermore, the behavior of sodium and potassium point to depletion and local concentration of these elements within the tuffs. Studies of post-depositional alkali migration in these rocks are in progress and will be reported elsewhere.

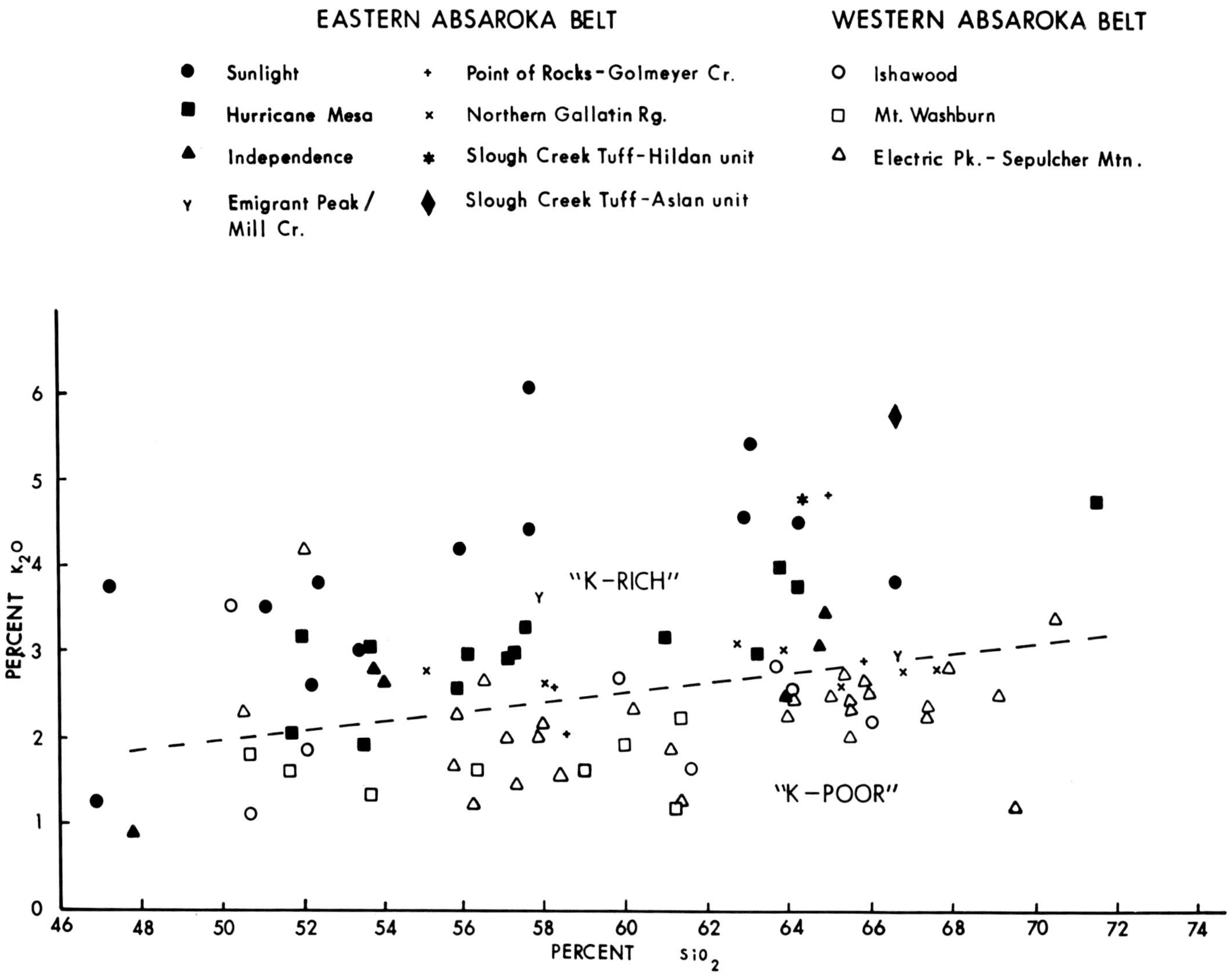

Figure 3: K_2O-SiO_2 diagram for rocks of the Absaroka-Gallatin province, from Chadwick, 1970, showing average anhydrous compositions of Hildan and Aslan units of the Slough Creek Tuff.

However, nothing in presently available chemical data suggests the occurrence of primary compositional magmatic zonation within either the Hildan unit or the Slough Creek Tuff as a whole.

DISCUSSION

Chadwick (1970) divided the Absaroka volcanic field into two northwest-trending belts of eruptive centers, of which the eastern contains more potassic rocks than does the western. Although its source is not precisely known, the highly potassic nature of the Slough Creek rocks indicates their chemical affinity with volcanics of the eastern belt. Average values for the Hildan Aslan welded tuffs, recalculated anhydrous, are shown in Fig. 3 on the K_2O-SiO_2 diagram compiled by Chadwick (1970); evidently these are among the most potassic of the "K-rich" Absaroka-Gallatin rocks.

The Hildan tuff plots on an Ab+An - Or - Q triangle between the position of the Aslan welded tuff and the plagioclase corner, inviting the suggestion that the Aslan was derived from Hildan magma by fractionation of the plagioclase so abundant in the Hildan unit. The connection between these units almost certainly is not that simple. The compositional range of plagioclase phenocrysts in each is greatly restricted and these ranges differ from one another by twenty percent anorthite with no trace of feldspar of intermediate composition. Biotite, hornblende and abundant sanidine occur in the Aslan unit but never in the Hildan. It seems improbable that these changes could have occurred so rapidly that there is no discernible stratigraphic record of the hiatus, especially between the uppermost Hildan and basal Aslan nonwelded tuffs on Ash Mountain. A more

Figure 4: Steeply dipping Slough Creek Tuff on the west flank of Ash Mountain, (symbols the same as in Figure 2).

plausible hypothesis would derive the two units from separate, albeit perhaps closely related, magma chambers (c.f., Noble and others, 1976).

Whatever its origin, dense welding within and evidently very close to the top of the Hildan unit suggests high emplacement temperature. According to the model of Sparks and Wilson (1976) for ignimbrite formation through gravitational column collapse, high temperature ash flows are more likely when the magma has low water content, thereby reducing column expansion and convective admixing of air. Preservation of sections of vitrophyre as much as 70 m thick in the Hildan unit may also indicate low water content inasmuch as water should promote devitrification (Ross and Smith, 1961). Absence of hydrous phenocrystic phases, fine grain size of devitrification products, and rarity of lithophysae and vaporphase crystallization likewise suggest eruption of relatively dry, quartz latitic magma to form the Hildan unit.

The source of the Slough Creek Tuff has yet to be identified. Ash Mountain is underlain by a large, hypabyssal intrusion of porphyritic dacite. Exposed on its western slope is a tabular mass of steeply dipping Slough Creek tuff about 300 m wide (Fig. 4). The outermost exposures of this mass consist of densely welded vitroclastic rock, whereas its core is composed of nonwelded tuffs. Elevation of the mass is about 200 m higher than that of flat-lying Slough Creek tuff surrounding Ash Mountain. Contact with the dacite intrusion is not exposed but must cut steeply inclined layering of the tuff. Wedow and others (1975) suggested that this mass is a feeder for the Slough Creek Tuff. We believe it represents a section of originally flat-lying Slough Creek strata which was rotated and carried up within the intrusion of dacite.

Rotation of the mass from a nearly flat-lying original position is confirmed by paleomagnetic studies. More than 40 oriented samples were collected from eight localities and their magnetic intensities and directions were measured using a four-probe spinner magnetometer. Figure 5 shows the resulting mean directions for samples from the mass of tuff on Ash Mountain (solid triangles) and samples from relatively flatlying Slough Creek Tuff (open circles) and younger and older flows (open triangles) elsewhere. Also

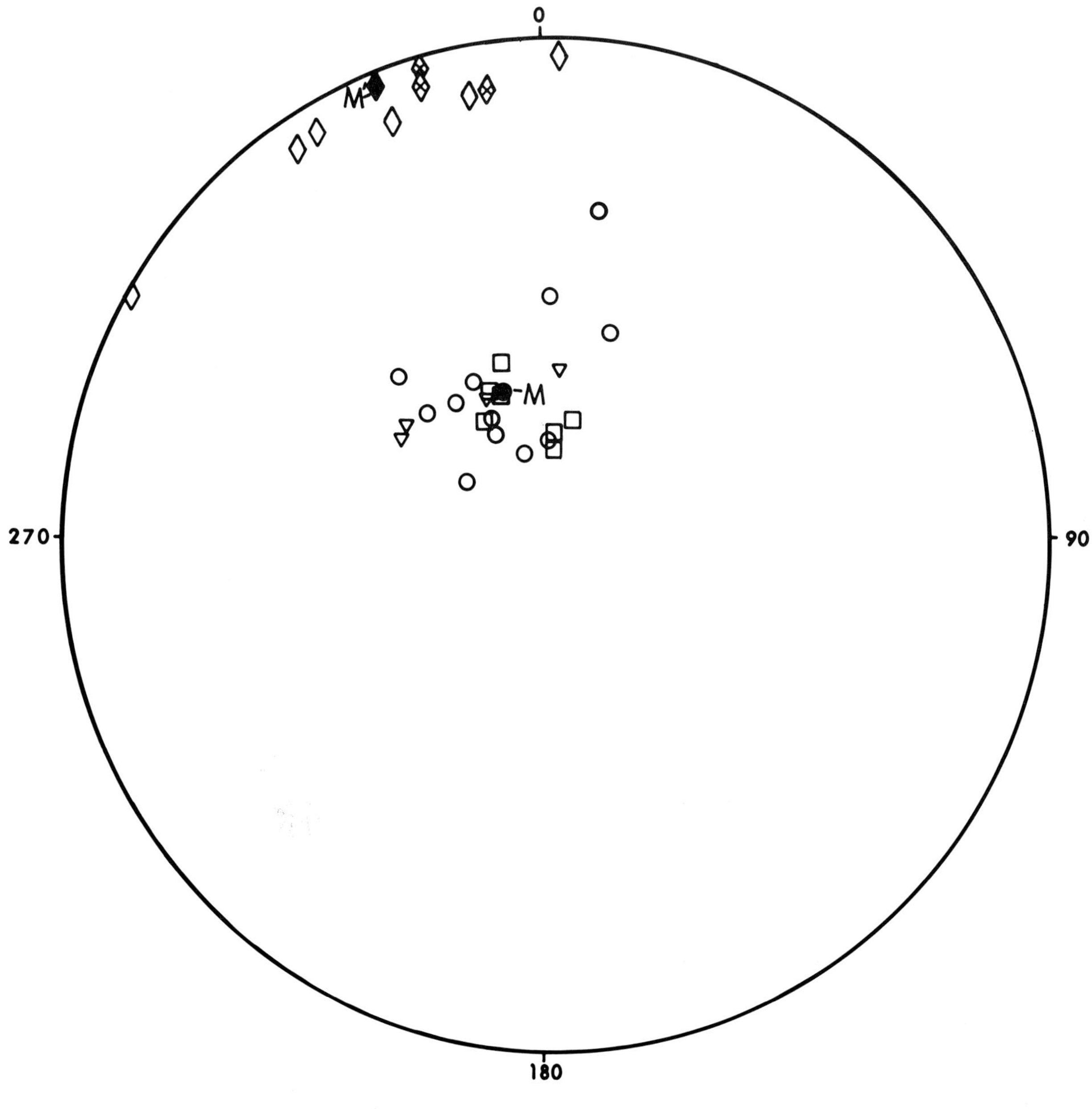

Figure 5: Magnetic directions for the Slough Creek Tuff and associated rocks. Open circles represent Slough Creek Tuff, open triangles represent Mount Wallace Fm. flows, open squares represent the dacite intrusion at Ash Mountain, and the solid circle represents the mean of these data (M). Open diamonds represent the steeply dipping Slough Creek Tuff at Ash Mountain, (cross inside the diamonds represent negative inclinations), and the solid diamond (M') represents the mean of these data.

shown in Fig. 5 is magnetic orientation of the dacite intrusion (open squares). The data cluster with a cone of confidence ($\alpha 95$) of 6°. Subhorizontal inclinations of the mass of tuff on Ash Mountain must be rotated 62° about a nearly horizontal axis trending N70E in order to bring their mean orientation (m') into coincidence with that of the other rocks (M). This trend and angular rotation is essentially identical to the strike and dip of foliation in the mass, which, therefore, must have been originally flay-lying. This mass provides the best single exposure of the combined Hildan and Aslan units of the Slough Creek Tuff, although the thick, nonwelded unit of the upper Aslan at Slough Creek is absent here.

The Slough Creek Tuff is less oxidized and more rich in vitrophyre north and northwest of Ash Mountain than it is there or elsewhere. Lenses of lithic fragments occur in the Hildan tuff near Thompson Lake, north of Ash Mountain, but have not been observed to the south or east. Elevations of the base of the tuff suggest a southeasterly regional paleoslope in the northwest part of the study area and a southerly slope elsewhere in the area. The southeasterly slope could represent the flank of a stratovolcano possibly centered at the intrusive complex of Emigrant Peak, 25 km north-northwest of Ash Mountain. Compositions of some rocks from that complex are rather similar to those of the Slough Creek Tuff, especially as regards high potassium content (R. Moore, personal communication, 1982). Accordingly, a vent located at or satellitic to this potassic center currently would appear to be the most likely source of the Slough Creek Tuff.

ACKNOWLEDGEMENTS

We thank Jelle de Boer for his assistance with the paleomagnetic data and Ken Vigil of the Drafting Department at Buckhorn Petroleum Co., for assistance in preparing the figures.

REFERENCES CITED

Brown, C.W., 1961, Cenozoic stratigraphy and structural geology, northeast Yellowstone National Park, Wyoming and Montana: Geol. Soc, America Bull., v. 72, p. 1173-1193.

Chadwick, R.A., 1970, Belts of eruptive centers in the Absaroka-Gallatin volcanic province, Wyoming-Montana: Geol. Soc. America Bull., v. 81, p. 267-274.

Christiansen, R.L., Noble, D.C., Orkild, P.O., and Sargent, K.A., 1968, Cogenetic sequences and source areas in silicic volcanic terranes: Geol. Soc. America, Abs, for 1966, Special Paper 101, p. 39.

Fraser, G.D., Waldrop, H.A., and Hayden, J.J., 1969, Geology of the Gardiner area, Park County, Montana: U.S. Geol. Survey Bull. 1277, 118p.

Hague, A., Iddings, J.P., Weed, W.J., Walcott, C.D., Girty, G.H., Stanton, T.W., and Knowlton, F.H., 1899, Descriptive geology, petrography, and paleontology, part II of Geology of the Yellowstone National Park: U.S. Geol. Survey Mon. 32, 893 p.

Hickenlooper, J.W., 1980, Geology of the Slough Creek Tuff, Northeastern Absaroka volcanic field, Southwestern Montana: unpubl. M.S. theiss, Wesleyan University, 89p.

Howard, A.D. 1937, History of the Grand Canyon of the Yellowstone: Geo. Soc. America Spec. Paper 6, 159 p.

Iddings, J.P., and Weed, W.H., 1894, Description of the Livingston quadrangle: U.S. Geol. Survey Geol. Atlas, Folio 1.

Noble, D.C., Korrings, M.K., Church, S.E., Bowman, H.R., Silberman, M.L., and Heropoulos, C.E., 1976, Elemental and isotopic geochemistry of nonhydrated quartz latite glasses from the Eureka Valley Tuff, east-central California: Geol. Soc. America Bull., v. 87, p. 754-762.

Ross, C.S., and Smith, R.L., 1961, Ash-flow Tuffs: Their origin, geologic relations, and identification: U.S. Geol. Survey Prof. Paper 366, 81 p.

Smedes, H.W., and Prostka, H.H., 1972 Stratigraphic framework of the Absaroka Volcanic Supergroup in the Yellowstone National Park regions: U.S. Geol. Survey Prof. Paper 729-C, 33 p.

Sparks, R.S.J., and Wilson, L., 1976, A model for the formation of ignimbrites by gravitational column collapse: Journal Geol. Soc. Lon., v. 132, p. 441-451.

Stormer, J.C., Jr., 1975, A practical two-feldspar geothermometer: Amer. Mineral., v. 60, p. 667-674.

Streckeisen, A., 1979, Classification and nomenclature of volcanic rocks. lamprophyres, carbonatites, and melitic rocks: Recommendations and suggestions of the I.U.G.S. subcommission of the systematics of igneous rocks: Geology, v. 7, p. 331-335.

Wedow, H., Jr., Gaskill, D.L., Banister, D.P., and Platte, E.C., 1975, Mineral resources of the Absaroka Primitive Area and vicinity, Park and Sweet Grass Counties, Montana: U.S. Geol. Survey Bull. 1391-B, 115 p.

Wright, T.L., 1968, X-ray and optical study of alkali fedspar; II. An X-ray method for determining the composition and structural state from measurement of 20 values for three reflections: Amer. Mineral., v. 53, p. 88-104.

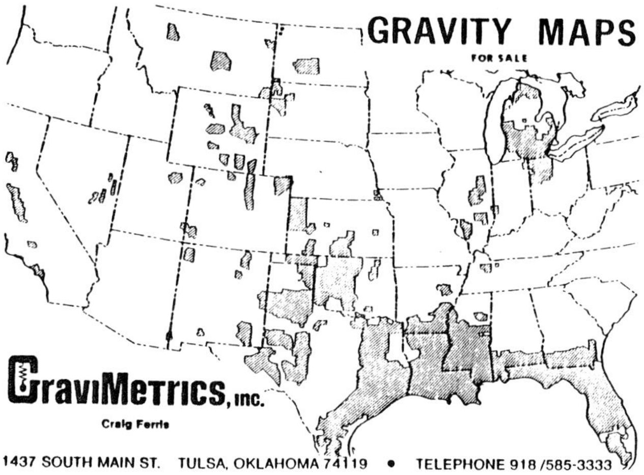
GRAVITY MAPS
FOR SALE
GraviMetrics, inc.
Craig Ferris

Thirty-Third Annual Field Conference — 1982
Wyoming Geological Association Guidebook

STRATIGRAPHIC RELATIONS WITHIN THE SOUTHEASTERN ABSAROKA VOLCANIC SEQUENCE, NORTHWESTERN WYOMING

KENT A. SUNDELL,[1] JEFFREY G. EATON[2]

INTRODUCTION

The southeastern Absaroka volcanic sequence is composed primarily of middle Eocene volcaniclastic rocks, including volcanic sandstones, siltstones, conglomerates, and breccias. Lava flows and pyroclastic deposits are commonly interbedded with epiclastic sediments near major vents in the upper Wood-Greybull River area and westward into the head of the South Fork of the Shoshone River (Fig. 1). The lack of extensive easily recognizable marker beds hinders correlation and contributes to the confusion in stratigraphic nomenclature.

The purpose of this paper is to clarify stratigraphic relationships and to point out an important rock stratigraphic marker bed. This is particularly important in light of recent exploration activities for oil and gas in Mesozoic and Paleozoic rocks underlying the Absaroka volcanic rocks in a westward extension of the Bighorn Basin.

STRATIGRAPHIC PROBLEMS INHERENT IN ABSAROKA VOLCANIC TERRAIN

The Absaroka volcanic sequence was formed by numerous volcanic eruptions and subsequent rapid weathering and transport of the ejected material. Most of these eruptions did not temporarily or spatially coincide. Because of the resultant complexity, many stratigraphic problems occur within the Absarokas. The following is a list of the major stratigraphic problems encountered.

1) Many of the major rock units are lithologically similar.
2) Laterally continuous marker beds are rare.
3) Correlation of the primary vent facies and the surrounding reworked alluvial facies is difficult.
4) The rocks contained within any group or formation are derived from many different vents or source areas.
   a. Complex lithologies are formed by the rework- and mixing of rock types.
   b. Compositions within a formation may vary greatly because of multiple source areas.
5) Rocks within any group or formation were deposited in different depositional environments.
6) Contemporaneous and post-depositional deformation, including widespread detachment faulting and gravity mass-movements, commonly obscures the stratigraphic record.

[1]University of California, Santa Barbara, California 93106
[2]Jeffrey G. Eaton, University of Wyoming, Laramie, Wyoming 82071

STRATIGRAPHY AND CORRELATION

Much confusion currently exists in the literature concerning nomenclature and stratigraphic correlation in the southeastern Absarokas. Recent mapping and lateral tracing of formations and marker beds by the authors have linked the previously isolated study areas of Love (1939), Wilson (1963, 1964, *in press*), Hay (1956), and Nelson and Pierce (1968). Figure 2 shows our interpretation of regional stratigraphic relations.

Previous problems in stratigraphic interpretation resulted from insufficient mapping between study areas forcing correlations to be based on similarities in gross lithologies and assumed ages. Detailed mapping and lateral tracing of key marker beds was accomplished as part of this study; radiometric dates, magnetostratigraphy, and biostratigraphy were used to test rock-stratigraphic correlations.

Description of Major Rock Units

The following brief descriptions are summarized from the original papers.

The Aycross Formation (Love, 1939) is comprised of middle Eocene rocks lying unconformably between the Wind River Formation (and older rocks) and the Tepee Trail Formation in the northwestern Wind River Basin. The rocks consist of nonresident variegated clays, shales, sandstones, minor andesitic conglomerates, and contain andesitic tuffaceous material. The formation's lithologic character changes rapidly over short distances (Smedes and Prostka, 1972; Sundell, 1982).

The Pitchfork Formation (Hay, 1956) consists of andesitic and dacitic volcanic sandstones, siltstones, and conglomerates overlying the early Eocene Willwood Formation and underlying the "early basalt sheets" (of Hague *et al.*, 1896, 1899) in the Greybull River area (Wilson, 1963, 1964).

The Wapiti Formation (Nelson and Pierce, 1968) replaced the informal term "early basic breccia" of Hague *et al.* (1896, 1899) and consists of volcanic breccias, sandstones, siltstones, conglomerates, and lava flows.

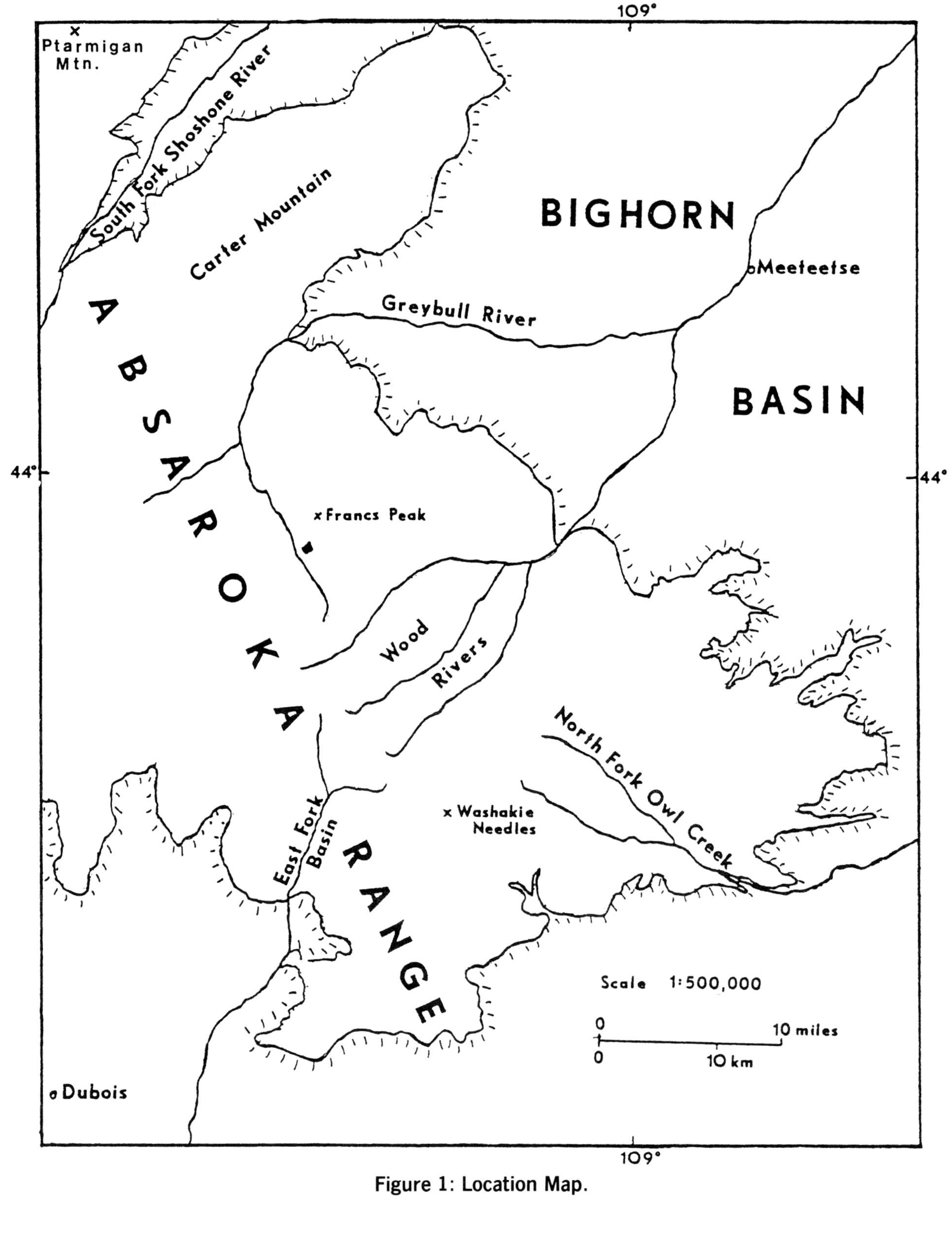

Figure 1: Location Map.

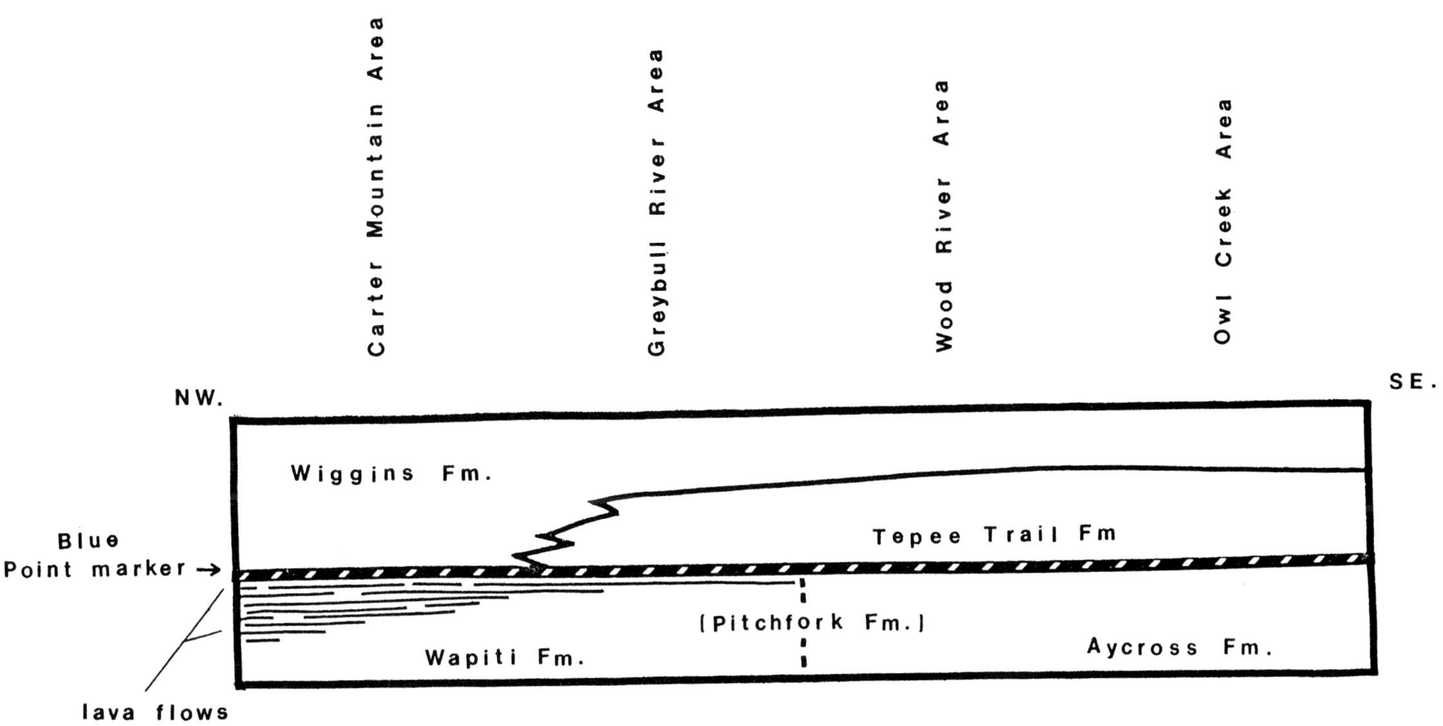

Figure 2: Stratigraphic relations within the southeastern Absaroka volcanic sequence.

Trachyandesite flows are common with lesser dacitic flows.

The Trout Peak Trachyandesite (Nelson and Pierce, 1968) overlies the Wapiti Formation and consists of lava flows interbedded with subordinate amounts of volcaniclastic rocks. The flows are dominantly trachyandesitic in composition. The Trout Peak Trachyandesite replaces the informal name "early basalt sheets" of Hague *et al.* (1896, 1899).

The Tepee Trail Formation (Love, 1939) consists of olive-drab, well-bedded tuffaceous claystones, siltstones, sandstones, conglomerates, breccias, and thin flows. It overlies the Aycross Formation and underlies the Wiggins Formation (in the area of the type section). Biotite-augite and hypersthene-augite andesite clasts are common (Smedes and Prostka, 1972; Sundell, 1982).

The Wiggins Formation (Love, 1939) is a tuffaceous light colored sequence unconformably overlying the Tepee Trail Formation in the East Fork Basin. No type section was designated because of great lateral variation in lithology, but a measured section provided by Love (1939) on Steamboat Rock includes volcanic conglomerates, tuffs, and thin flows. Biotite-hornblende andesites predominate in the East Fork Basin but locally other andesites, dacites, rhyolites, and basalts are important (Sundell, 1982; Wilson, 1963, 1964).

Description of the Blue Point Marker

The Blue Point marker includes part of the Blue Point Member of the Wiggins Formation as described by Wilson (1963). The Blue Point Member consists of two bluish-green andesitic cobble conglomerates separated by light greenish-gray clay beds and laterally equivalent white trachytic ash beds (Wilson, 1963, 1964; Dunrud, 1962; McGrew, 1965). The Blue Point marker, as used here, consists of a sequence (20-200 feet thick) of light colored volcanic claystones, siltstones, tuff, and bentonites originally included in the Blue Point Member (volcaniclastic terminology follows Fisher 1961). The Blue Point marker is commonly altered to bentonite and locally contains abundant secondary calcite. The abundance of bentonite in the Blue Point marker commonly creates a perched water table and unstable slopes, characterized by abundant springs and spectacular mass-movements including rockslides, earthflows, and slumps where it crops out throughout the southeastern Absarokas.

The Blue Point marker is interpreted as a nearly isochronous unit containing primary and reworked airfall ash which blanketed the southeastern Absarokas. The source and total extent of the marker are not presently known; however, it is traceable along the eastern margin of the Absarokas from Ptarmigan Mountain to the North Fork of Owl Creek.

DISCUSSION

Wilson (1963) first recognized the importance of the Blue Point Member as a marker bed. The present report clarifies and substantiates many of his correlations. However, the informal name "Blue Point marker" is preferable to the Blue Point Member of the Wiggins Formation for two reasons. First, the locally derived conglomerates within the Blue Point Member are not as laterally continuous as the tuffaceous beds. Secondly, recent mapping and tracing of the Tepee Trail Formation from the type section in the East Fork Basin to the Wood Rivers by Sundell (1982, and unpublished data) indicates that much of the lower Wiggins as used and mapped by Wilson (1963, 1964, 1970, *in press*) above the Blue Point Member is laterally equivalent and lithogically identical to the Tepee Trail Formation.

The Blue Point marker is used here to separate formations at a discrete boundary (see Fig. 2). This boundary is approximately equivalent to previously mapped formational contacts. Pierce (1978) included the Blue Point marker at the base of the Wiggins Formation in the Carter Mountain-Greybull River area. Sundell (1982) used the Blue Point marker as the top of the Aycross Formation in the Owl Creek area.

Wilson (1963, 1964, 1970, *in press*) mapped the Blue Point marker as the base of the Wiggins Formation in the Wood River area, but he did not recognize the overlying Tepee Trail Formation as a valid formation. Therefore, the type section of the Blue Point Member of the Wiggins Formation (Wilson, 1963), as interpreted here (also see Bown, 1982), lies below the Tepee Trail Formation and above the Aycross Formation. The writers believe assignment of the Blue Point marker bed to a formation is premature at this time because much still needs to be known about its total lateral extent, continuity, and source.

Bown (1982) did not believe that the Blue Point Member at its type section is the same unit Wilson (1963, 1964) and McGrew (1965) identified as the Blue Point Member to the north in the area of Carter Mountain. He based his conclusion on the occurrence of the Blue Point Member at a lower elevation (by 550 m) at the type section than to the north in the Greybull Valley area. Present elevations are useless for correlating rock-stratigraphic units in the Absarokas. Wilson (1964, 1975) correctly mapped the type section of the Blue Point Member on the downthrown (east) side of a fault. The unit is present at a higher elevation on the upthrown side of the fault (to the west) and is traceable to its occurrence in the Greybull River Valley.

Below the Blue Point marker bed the Aycross, Pitchfork, and Wapiti formations are considered to be lateral and approximately time equivalent rock units. A significant lithologic change occurs between the coarse grained, contorted volcanic breccias and flows of the Wapiti Formation and the generally fine-grained volcanic claystones, siltstones, sandstones, and minor conglomerates of the Aycross Formation. This transition is believed to represent a change between a breccia dominated, near vent facies (in the Wapiti Formation), and distal alluvial facies (in the Aycross Formation). The Pitchfork Formation, as previously mapped, is transitional between end member facies. The Pitchfork Formation could be retained to identify the transitional facies and requires two arbitrary boundaries between it and the Aycross and Wapiti formations. Alternatively, the name Pitchfork Formation could be abandoned and one arbitrary boundary established. The writers prefer the latter and suggest the boundary between the Aycross and Wapiti formations be defined as a vertical plane at the southeastern edge of the last continuous flows within the Wapiti Formation. Based on this definition, the boundary between the Aycross and Wapiti formations would be located in the Timber Creek area where the last flows of the Wapiti Formation pinch out to the southeast (Fig. 3).

Bown (1982) also advocated abandoning the name Pitchfork Formation. He suggested placement of the Aycross-Wapiti boundary at the northeast tip of Carter Mountain. However, the east end of Carter Mountain is dominated by breccias and flows of the Wapiti Formation (Eaton, 1982, *in press*) and we feel it is inappropriate to assign these beds to the Aycross Formation.

The lava flows on Carter Mountain have been previously mapped as the Trout Peak Trachyandesite by Pierce and Nelson (1969) and Pierce (1970, 1980). However, paleomagnetic data (Shive and Pruss, 1977) and recent work by Eaton (1982, *in press*) cast doubt upon this correlation. Because the Trout Peak Trachyandesite is not traceable to Ptarmigan and Carter mountains, and no criteria have been reported to differentiate between flows of the Trout Peak Trachyandesite and flows of the Wapiti Formation, all flows on Carter Mountain below the Blue Point marker are tentatively assigned to the Wapiti Formation and an unnamed sequence (Eaton, 1982, *in press*).

Above the Blue Point marker the Tepee Trail Formation is overlain by, and interfingers to the northwest with, the Wiggins Formation. This lateral transition in lithologies between the Tepee Trail and Wiggins Formations probably reflects differing source areas.

The Blue Point marker is invaluable to those interested in stratigraphy and structural interpretations within the southwestern Absaroka volcanic sequence. It should be easily recognized and mappable by geophysical

Figure 3: West side of Timber Creek, part of Sec. 21, 22, 27 and 28, T. 47 N., R. 103 W. (Sunshine Reservoir 15 Minute Quadrangle). Tt — Tepee Trail Formation; Bpm — Blue Point marker; Wf — last southern occurrence of lava flows within the Wapiti Formation.

and surface mapping techniques. There are gross lithologic changes above and below the marker bed and the marker itself possesses distinct lithologic characters.

AGE RELATIONS

Virtually all layered rocks in the southeastern Absaroka volcanic sequence are middle Eocene in age on the basis of vertebrate fossils and available radiometric (K/Ar) dates. The Caldwell Canyon Volcanics of Pliocene-Pleistocene(?) age described by Love (1939) and Wilson (1964) may be an exception; however, the exact age and stratigraphic relations of these rocks are not well known. The time scale of Berggren, et al. (1978), is used in this report.

The Wapiti and Aycross formations are thought to be equivalent in age. Bridgerian vertebrate fossils have been recovered from the Wapiti Formation in the Carter Mountain area (Eaton, 1982, *in press*), from the Aycross Formation in the Owl Creek area (Bown, 1982) and the East Fork Basin (Love, 1939), indicating an early middle Eocene age for both formations. One radiometric date of 49.2 m.y. from the Aycross Formation in the Wind River Basin supports this conclusion (Love, 1978).

Numerous Bridgerian taxa have been collected from the Blue Point marker bed on Carter Mountain (Eaton, 1980, 1982, *in press*) and appear to be transitional between faunas of the lower ("A-B") and upper ("C-D") Bridger Formation in the Bridger Basin.

A middle Eocene fauna collected from the lower Tepee Trail strata in the North Fork of Owl Creek is latest Bridgerian(?) to early Uintan (Eaton, 1980). An early Uintan fauna has been collected from the middle part of the Tepee Trail Formation at the type section along the East Fork of the Wind River (McKenna, 1980; McFadden, 1980). The middle and upper Tepee Trail Formation in the North Fork of Owl Creek has also yielded a Uintan fauna

(Eaton, 1980). Rocks assigned to the Tepee Trail Formation contain a Bridgerian flora in the Kisinger Lakes area (Rohrer, 1974) and a Bridgerian fauna in the Togwotee Pass area (McKenna, 1980). The Tepee Trail Formation is believed to be temporarily equivalent to the lower part of the Wiggins Formation on Carter Mountain (Eaton, 1982, *in press*).

Vertebrate fossils and a radiometric date (47.9+0.5 m.y.) indicate an early middle Eocene age (Bridgerian) for the Blue Point marker on Carter Mountain (Eaton, 1982, *in press*; Bown, 1982). The Wiggins Formation in the North Fork of Owl Creek and the East Fork Basin contains Uintan faunas and has yielded radiometric dates ranging from 43-47 million years (Bown, 1982; Eaton, 1980; Love, 1978; McKenna, 1980), suggesting a middle to late Eocene age. As Rohrer and Obradovich (1969) suggested the Tepee Trail/Wiggins contact is time transgressive, becoming younger to the southeast.

CONCLUSIONS

The Blue Point marker is a distinctive bed traceable throughout much of the southeastern Absarokas and can be used to clarify previous stratigraphic and nomenclature problems. Below the marker, the Aycross Formation is the lateral equivalent of the Wapiti Formation. The use of the transitional Pitchfork Formation is abandoned. The Tepee Trail Formation is present above the marker to the south, and the Wiggins Formation overlies the marker to the north. The Tepee Trail Formation is overlain by, and interfingers to the north, with the Wiggins Formation. The Aycross and Wapiti formations are early middle Eocene (Bridgerian) in age, the Tepee Formation is considered to be of middle Eocene age (Bridgerian?-early Uintan), and the Wiggins Formation ranges from early middle Eocene (Bridgerian-early Uintan) to possibly early late Eocene (late Uintan).

ACKNOWLEDGEMENTS

The writers would like to thank Dr. J. D. Love and Dr. W. H. Wilson for their interest and valuable discussions concerning the Absarokas. Dr. Brainerd Mears, Jr. and Harry G. Glicken kindly reviewed the manuscript. We acknowledge financial support from the National Science Foundation (Grants EAR 7709472 and EAR 7904369), the University of Wyoming, the University of California at Santa Barbara, and the Geological Society of America (Grants 2766-80 and 2926-81).

REFERENCES CITED

Berggren W. A., McKenna, M. C., Hardenbol, J., and Obradovich, J. D., 1978, Revised Paleogene polarity time scale: Journal of Geology, v. 86, p. 67-81.

Bown, T. M., 1982, Geology, paleontology, and correlation of Eocene volcaniclastic rocks, southeast Absaroka Range, Hot Springs County, Wyoming: U.S. Geological Survey, Professional Paper 1201-A, 75 p.

Dunrud, C. R., 1962, Volcanic rocks of the Jack Creek area, southeastern Absaroka Range, Park County, Wyoming (M.S. thesis): Laramie, Wyoming, The University of Wyoming, 92 p.

Eaton, J. G., 1980, Preliminary report on paleontological explorations of the southeastern Absaroka Range, Wyoming, *in* Gingerich, P. D., ed., Early Cenozoic paleontology and stratigraphy of the Bighorn Basin: University of Michigan, Papers on Paleontology, no. 24, p. 139-142.

______, 1982, Paleontology and correlation of Eocene volcanic rocks in the Carter Mountain area, Park County, southeastern Absaroka Range, Wyoming (M.S. thesis): Laramie, Wyoming, The University of Wyoming, 153 p.

______, *in press*, Paleontology and correlation of Eocene volcanic rocks in the Carter Mountain area, Park County, southeastern Absaroka Range, Wyoming: Contributions to Geology, The University of Wyoming.

Fisher, R. V., 1961, Proposed classification of volcaniclastic sediments and rocks: Geological Society of America Bulletin, v. 72, p. 1409-1414.

Hague, A., Iddings, J. P., Weed, W. H., Walcott, C. D., Girty, G. H. Stanton, T. W., and Knowlton, F. H., 1899, Descriptive geology, petrography, and paleontology, Part II *of* Geology of the Yellowstone National Park: U.S. Geological Survey Monograph 32, 893 p.

Hague, A., Weed, W. H., Iddings, J. P., 1896, Descriptions of the Yellowstone Park Quadrangle (Wyoming): U.S. Geological Survey Atlas Folio 30.

Hay, R. L., 1956, Pitchfork Formation detrital facies of Early Basic Breccia, Absaroka Range, Wyoming: American Association of Petroleum Geologists Bulletin, v. 40, p. 1863-1898.

Love, J. D., 1939, Geology along the southern margin of the Absaroka range, Wyoming: Geological Society of America Special Paper no. 20, 134 p.

______, 1978, Eocene rocks, fossils, and geologic history, Teton Range, northwestern Wyoming: U.S. Geological Survey Professional Paper 932-B, 40 p.

MacFadden, B. J., 1980, Eocene perissodactyls from the type section of the Tepee Trail Formation of northwestern Wyoming: University of Wyoming Contributions to Geology, v. 18, no. 2, p. 135-143.

McGrew, A. R., 1965, Stratigraphy and mineralogy of the Blue Point Member of the Wiggins Formation, southeast Absaroka Range, Park County, Wyoming (M.S. thesis): Laramie, Wyoming, The University of Wyoming, 74 p.

McKenna, M. C., 1980, Late Cretaceous and early Tertiary vertebrate paleontological reconnaissance, Togwotee Pass area, northwestern Wyoming *in* Jacobs, L. L., ed., Aspects of vertebrate history: Museum of Northern Arizona Press, Flagstaff, p. 321-343.

Nelson, W. H., and Pierce, W. G., 1968, Wapiti Formation and Trout Peak Trachyandesite, northwestern Wyoming: U.S. Geological Survey, Bulletin 1254-H, 11 p.

Pierce, W. G., 1970, Geologic map of the Devil's Tooth Quadrangle, Park County, Wyoming: U.S. Geological Survey, Map GQ-817, scale 1:62,500.

______, 1978, Geologic map of the Cody 1x2 degree Quadrangle, northwestern Wyoming: U.S. Geological Survey, Map MF-963, scale 1:250,000.

Pierce, W. G., and Nelson, W. H., 1969, Geologic map of the Wapiti Quadrangle, Park County, Wyoming: U.S. Geological Survey Map GQ-778, scale 1:24,000.

Rohrer, W. L., 1974, An early middle Eocene flora from the Yellowstone-Absaroka volcanic province, northwestern Wind River Basin, Wyoming *in* MacGinitie, H. D., Stratigraphy and stratgraphic relations of the fossil floras: California University Publications in the Geological Sciences, v. 108, p. 10-18.

Rohrer, W. L., and Obradovich, J. D., 1969, Age and stratigraphic relations of the Tepee Trail and Wiggins formations, northwestern Wyoming: U.S. Geological Survey Professional Paper 650-B, p. B57-B62.

Shive, P. N., and Pruss, E. F., A paleomagnetic study of basalt flows from the Absaroka Mountains, Wyoming: Journal of Geophysical Research, v. 82, p. 3039-3048.

Smedes, H. W., and Prostka, H. J., 1972, Stratigraphic framework of the Absaroka Volcanic Supergroup in the Yellowstone National Park region: U.S. Geological Survey Professional Paper 729-C, p. C1-C33.

Sundell, K. A., 1982, Geology of the headwaters of the North Fork of Owl Creek: Geological Survey of Wyoming, Report of Investigations, no. 15, 60 p.

Wilson, W. H., 1963, Correlation of volcanic rock units in the southern Absaroka Mountains, Wyoming: University of Wyoming Contributions to Geology, v. 2, p. 13-20.

______, 1964, Geologic reconnaissance of the Absaroka Mountains, northwest Wyoming: Part I — the Wood River-Greybull River area: University of Wyoming Contributions to Geology, v. 3, p. 60-77.

______, 1970, Geologic map of the Soapy Dale Peak Quadrangle, Hot Springs County, Wyoming: Wyoming Geological Survey map.

______, 1975, Detachment faulting in volcanic rocks, Wood River area, Park County, Wyoming: Wyoming Geological Association, 27th Annual Field Conference, Guidebook, p. 167-171.

______, *in press*, Geology of the Dick Creek Lakes, Dunrud Peak, Franc's Peak, Noon Point, and Twin Peaks Quadrangles Fremont, Hot Springs, and Park counties, Wyoming: Geological Survey of Wyoming map series no. 10.

Thirty-Third Annual Field Conference — 1982
Wyoming Geological Association Guidebook

GEOLOGY OF THE LAMAR RIVER FORMATION, NORTHEAST YELLOWSTONE NATIONAL PARK

WILLIAM J. FRITZ[1]

ABSTRACT

The Eocene Lamar River Formation exposed in the Lamar River Valley of Yellowstone National Park consists chiefly of volcaniclastic sediments deposited between two sub-parallel chains of andesitic stratovolcanoes. Recognizable facies include mudflows, braided streams, meandering streams, and lakes. Less than 10% of the formation consists of basaltic dikes, sills, and flows, trachyandesite flows, and ignimbrites. The depositional environment was, thus, very similar to that of the Mount St. Helens, Washington area.

A very diverse flora of temperate to tropical logs, stumps, leaves, needles, cones, and pollen is preserved throughout the formation. The mixture of habitat indicators was caused by transportation of plant parts, including some upright stumps, from various elevations within the depositional basin.

INTRODUCTION

Geological Setting:

Eocene volcaniclastic sediments of the Absaroka Volcanic Supergroup in Yellowstone National Park preserve some of the most unique petrified wood deposits found anywhere in the world (Fig. 1). Although several formations in the supergroup contain petrified trees, the best known and studied are in the Lamar River Formation in the northeastern part of the park (Fig. 2).

Rocks of the Absaroka volcanics in Yellowstone National Park were first described in detail by Hague and others (1899). Although they named no formations, they informally referred rocks preserving petrified wood in the Lamar River Valley to the "early basic breccia". Interfingering rock units containing petrified wood in the northwest portion of the park were called the "early acid breccia" (Fisk, 1976a; Fig. 3). Chadwick (1970) and Parsons (1974) proposed that rocks in the Absaroka Volcanic Supergroup were produced by two chains of stratovolcanoes with peaks eight to ten thousand feet above the valley floor. These vents existed as two northeast-southwest trending sub-parallel chains 25 to 60 Kilometers apart (Fig. 2). During Eocene time, this configuration would have produced a long narrow intermontane basin that received volcaniclastic sediment from surrounding volcanoes.

[1]Department of Geology, Georgia State University, Atlanta, GA 30303

Smedes and Prostka (1972) revised the stratigraphy of Hague and others (1899) by placing all Eocene volcanic and volcaniclastic rocks in the vicinity of Yellowstone into the Absaroka Volcanic Supergroup. The supergroup was then divided (in order of decreasing age) into the Washburn, Sunlight, and Thorofare Creek groups (Fig. 3). The basal Washburn Group contains the Cathedral Cliffs Formation in the northeast portion of the Park, the Lamar River Formation ("early basic breccia") in the Lamar River drainage, and the Sepulcher Formation ("early acid breccia") in northwestern Yellowstone Park. The petrified wood at Specimen Creek and Tom Miner Basin in northwestern Yellowstone Park and the Petrified Tree at old Yancey's Ranch west of Roosevelt Lodge belong to the Sepulcher Formation and are not covered in this report.

The Lamar River Formation unconformably overlies undifferentiated Paleozoic sedimentary rocks and Precambrian crystalline rocks (Smedes and Prostka, 1972) and underlies rocks of the Sunlight and Thorofare Creek groups. The Lamar River Formation is well-exposed along the Lamar River and its tributaries in the Tower Junction and Abiathar Peak quadrangles, Yellowstone National Park, Wyoming and Montana (Proska, Blank, and others, 1975; Prostka, Ruppel, and others, 1975) and has been assigned a Middle Eocene age based on sparse radiometric dating and by floral comparisons (Smedes and Prostka, 1972).

Figure 1: Thee classical petrified stumps with roots at the 75-100 m interval on Specimen Ridge. Unlike some upright stumps in the Lamar River Formation, these were probably preserved in place.

Paleobotany and Paleoecology:

Plant remains preserved in the Lamar River Formation have received considerable scientific attention starting with initial interpretations of Lesquereux (1872) and Holmes (1878, 1879). The area most extensively studied and described is the "Fossil Forest" locality on Amethyst Mountain (Fig. 2). The first leaf fossils collected from the area by the F. V. Hayden U.S. Geological and Geographical Survey of the Territories of 1871 were described by Lesquereux (1872). Holmes (1878), also a member of the 1871 Hayden expedition, gave a more detailed report of the Amethyst Mountain "Fossil Forest", and first interpreted petrified wood in the formation to represent successively buried forests.

A few years later Knowlton (1899) reported on petrified wood, leaves, and the geology of the "Fossil Forest" in his comprehensive monograph of the Yellowstone flora. More recently, Dorf (1960, 1964a, 1964b, 1974, 1980) suggested that the formation represents 27 or more successive living forests buried in place.

More recent studies on the leaves were conducted by Dorf (1960, 1964a) and Aguirre (1977). Petrified woods from the Lamar River Formation have been studied by Knowlton (1899), Conrad (1939), Read (1939), Beyer (1954), Wheeler, Scott, and Barghoorn (1977, 1978), Fritz (1977, 1980e), Fritz and Fisk (1978, 1979), Fisk and others (1977, 1978). Pollen and spores from the Amethyst Mountain locality were examined and described by Fisk and DeBord (1974a, 1974b) and Fisk (1976b, 1976c). These previous investigations of the numerous well-preserved leaves, needles, cones, pollen, spores, and wood from the Lamar River Formation describe speicmens of widely differing ecologic habitats (Fig. 4). The plant parts represented by nearly 200 taxa include cool temperate forms such as spruce (*Picea*) and fir (*Abies*) along with tropical to paratropical species like breadfruit (*Artocarpus*), avocado (*Persea*), mangroves (?*Terminalia* and ?*Rhizophora*), tropical magnolias, laurels, and aralias. Although it is not my intention to explore the

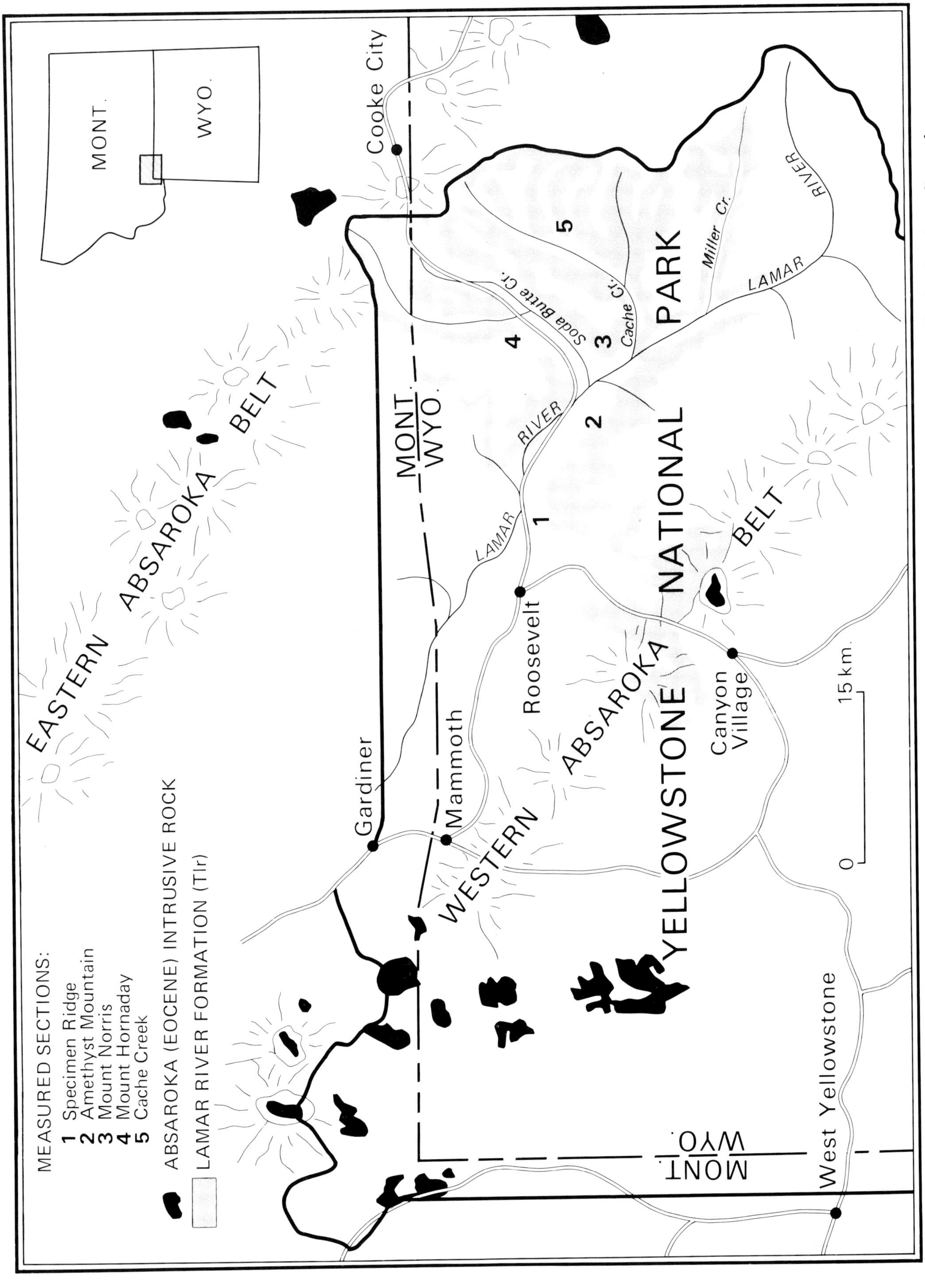

Figure 2: Geologic map of the Lamar River Formation in northern Yellowstone National Park showing sites of measured sections and their relation to Eocene eruptive centers and stylized Eocene topography. Modified from Fritz, 1980a; Tlr from USGS, 1972).

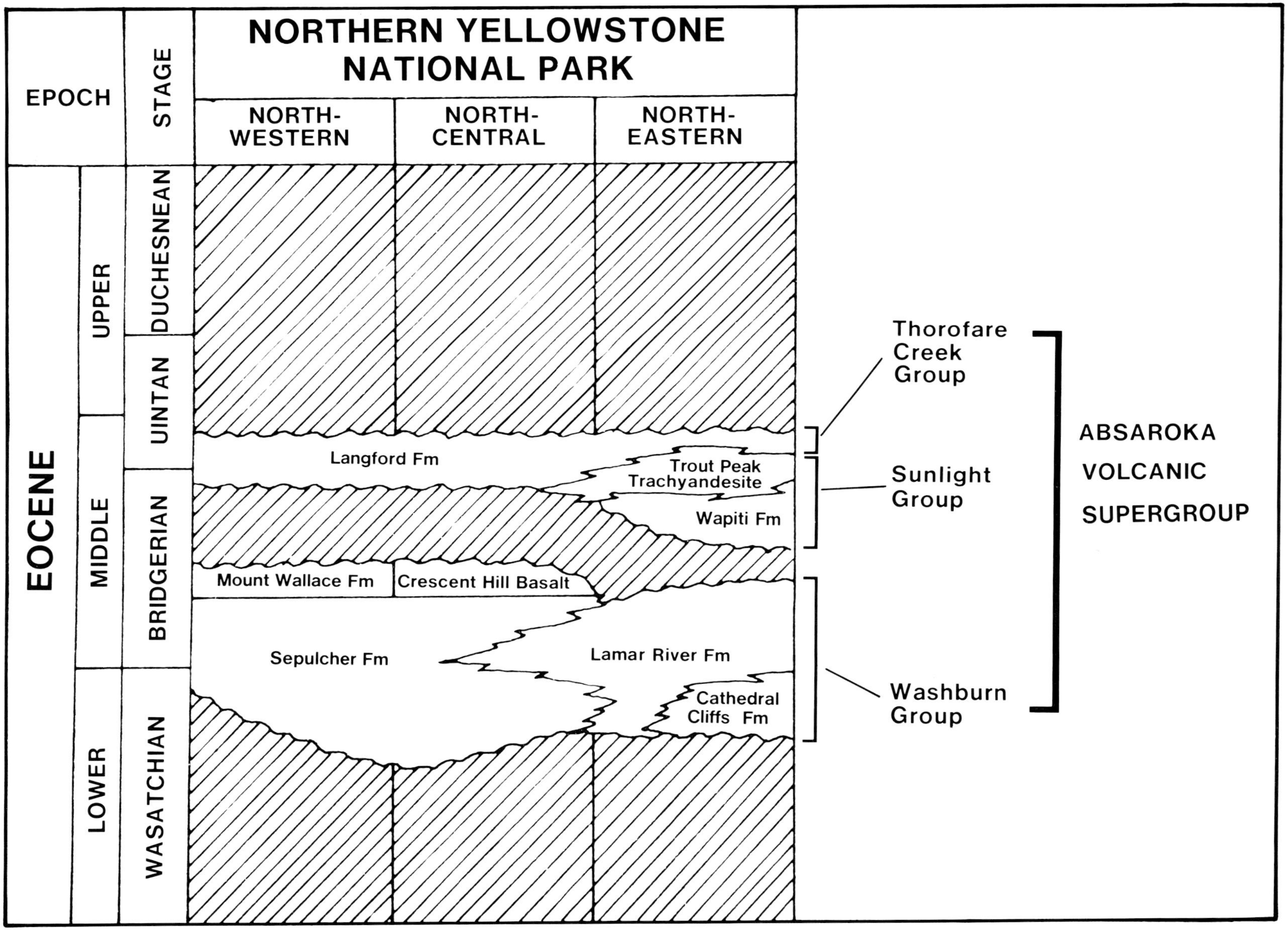

Figure 3: Generalized stratigraphic relations of the Absaroka Volcanic Supergroup in northern Yellowstone National Park. (Modified from Fisk, 1976b).

paleobotanical problems and errors of identification that may well exist in the Yellowstone flora, all previous paleobotanists working here have commented on the strange admixture (Fisk, 1976b). The previous workers reported that dominant species in the vegetation based on their relative abundance as fossils were sycamores (*Plananus* and *Platanophyllum*), magnolias (*Magnolia*), chestnuts (*Castanea*), oaks (*Quercus*), maples (*Acer*), dogwoods (*Cornus*), walnuts (*Juglans*), and redwoods (*Sequoia*). Also fairly common were figs (*Ficus*), laurels (*Laurus*, *Laurophyllum*, and *Ocotea*), and avocado (*Persea*). Other significant components of the vegetation were climbing ferns (*Lygodium*), pines (*Pinus*), soapberries (*Sapindus*), hickories (*Carya*), bay-berries (*Myrica*), elms (*Ulmus*), and willows (*Salix*). Exotic tropical forms include the katsura tree (*Cercidiphyllum*), breadfruit (*Artocarpus*), chinquapin (*Castanopsis*). There is apparently no single forest in the world where species comprising the floral assemblage are presently living together, however, most of the taxa can be found in small geographical areas that include several thousand meters of elevation change. Studies of the pollen (Fisk, 1976b) added 28 new families and 38 genera including *Ephedta*, *Podocarpus*, *Larix*, *Picea*, *Ilex*, *Alnus*, *Carpinus*, *?Terminalia*, *?Rhizophora*, *Moprus*, *Tilia*, *Taxus*, *Nyssa*, and *Fraxinus*.

This diverse assemblage of cool temperate and tropical taxa from one locality is not strange for a Paleogene flora. However, since the mixtures are generally attributed to mixing from several habitats, it is surprising "forests" interpreted as having been preserved in place with no transportation of plant remains. Since pollen and spores are easily transported it was first assumed that the mixture was present only in these plant parts and did not involve stumps or leaves of the forest. However, studies of petrified wood by previously cited authors indicate that mixtures of plants now growing in several habitats exist in the logs and stumps.

Evidence for a mixture of ecological types is present not only in taxonomic lists, but in the structure of the wood as well. Wheeler and others (1977, 1978) reported woods with both temperate and tropical affinities based on the presence of well-defined growth rings in some specimens and the lack of rings in others, and by affinities with extant taxa. They found inflated oil cells in the margins of the rays in the tropical members of the Magnolicaeae. Also reported were woods of *Alnus* and *Carpinus*. Since modern species of these two genera are temperate in distribution, their occurrence with tropical magnoloas seems to comfirm the mixture of temperate and tropical plants in the Yellowstone "fossil forests". In a separate study, I reported a mixture of ring and diffuse porous wood along with a mixture of temperate and tropical taxa from the same

Figure 4: Cross sections of wood from the 248.5-253 m interval on Amthyst Mountain illustrating the range of growth ring types and distinctiveness seen in the petrified stumps and logs, a,b,c, Sequoia magnifica **showing growth rings which range from a few microns to over 7 mm., d,** ?Podocarpus **showing less distinct rings over 2 cm wide, e,** Pinus. **f,** Picea. **g,h,i,** ?Aralia **showing the variety of ring distinctness seen in this one type. j,** ?Artocarpus **type showing an indistinct to non-existant ring. Bar is 2 mm long.**

stratigraphic unit on Amthyst Mountain (Fritz, 1977; Fritz and Fisk, 1978, 1979; Fig. 4). Thus, since the mixture was present at the same stratigraphic horizon, it could not be attributed to changing climatic conditions during disposition of the formation. Also stratigraphic analysis of the

Figure 5: Five stratigraphic sections measured across the depositional basin of the Lamar River Formation. Refer to Fig. 2 for exact locations.

pollen from Amethyst Mountain shows no significant vertical change and thus, does not support the idea that changing climates during time of deposition caused the mixture. Wheeler and others (1977, 1978) interpreted this mixture as having resulted from changing climatic tolerances for plants since Eocene. Another possibility is transportation of plant parts from a variety of habitats, as in other paleobotanical localities of a similar age. However, this idea was rejected by most paleobotanists since the classical interpretation of *in situ* forests as given by Holmes (1878) and Dorf (1960, 1964a) seemed to allow little or no transportation of any plant parts; especially wood. Typical of this thought is the following statement by Wheeler and others (1977; p. 298): "Long distance transport of the Yellowstone woods is unlikely; it is thus possible that these Eocene species of *Alnus* and *Carpinus* may have had different ecological requirements than are now characteristic of the genera."

In the following section of this paper, I describe the stratigraphic framework and sedimentology of volcaniclastic sedimentary rocks of the Lamar River Formation. I will argue that the Yellowstone area is not a critical test of the hypothesis that plants have changed their climatic tolerances since the Eocene because the depositional environment of the Lamar River Formation is one where plant parts, including some stumps and logs, could have easily been transported.

RESULTS

Description of Stratigraphic Sections:

To describe the stratigraphic framework, I measured five sections totaling 1699 m in the Lamar River drainage (Fig. 5) I chose the locations because they were located on the most accessible areas (other good exposures are not accessible due to treacherous cliffs) and they gave a cross section through the basin of deposition so that I could describe facies changes. Following is a general description of each section.

Specimen Ridge:

The 159 meter-thick Specimen Ridge Section contains the three classic vertical trees often seen by park visitors on Ranger-led hikes (Fig. 1) and is located 2.7 km SSW of the junction of Slough Creek and the Lamar River (figs. 2 and 6). Although this section contains no distinct marker beds, the well-rounded conglomerate of the 122-141 m interval correlates with a similar conglomerate exposed at the very top of the Amethyst Mountain and Mount Norris Sections. The top and bottom of the Lamar River Formation are not exposed at the Specimen Ridge locality. The Lamar River Formation interfingers with the Sepulcher Formation one kilometer west of the line of section. The Speicmen Ridge Section is farthest from known possible source areas (Fig. 2) and, along with Amethyst Mountain contains more fine grained tuffaceous sandstones than the other three sections. The section is 35% tuffaceous sandstone 59% conglomerate with 6% covered intervals (Table 1). More tall vertical trees with well-developed organic zones found near the roots occur here than on the other four sections. Retallack (1981) reported a moderately well-defined paleosol from the base of the three classic vertical trees at the 75-100 m interval (Fig. 1). The existence of paleosols is not surprising since this section is the farthest from a volcanic source area. Thus, it would represent deposition in the lowlands in low energy conditions — a situation more conducive to paleosol preservation than high energy conditions nearer to vent peaks. Unlike many short upright stumps with wide root systems, these tall vertical trees were probable preserved in place.

Amethyst Mountain Section:

The 430 m thick Amethyst Mountain Section located 3.5 km SW of the junctions of Soda Buttecreek and the Lamar River (Fig. 2) has been designated the type section for the Lamar River Formation (Fritz, 1980b) since most of the previous work both on the paleobotany and the geology was done here. The classic cliff exposures with excellent examples of cross cutting, channel cut and fill, and cross bedding 0.3 km southeast of the line of section was the site for the layer cake models depicted by Holmes (1878) and Dorf (1960, 1964a). Many of the plants described by Lesquereux (1872), Knowlton (1899), Conard (1930), Beyer (1954), Dorf (1960, 1964a), Fisk (1976b), Fritz '(1977, 1980f), and Wheeler and others (1977, 1978) were collected from Amethyst Mountain. Two previous sections were measured on Amrthyst Mountain (Brown, 1957; Fisk, 1976b). These sections were used to determine the number of petrified tree levels and to document a stratigraphic sequence of samples taken for pollen analysis. Thus, they did not contain rock descriptions of sufficient detail for interpreting the depositional environments.

The Amethyst Mountain Section probably represents almost the entire thickness of the Lamar River Formation. Overlying Trout Peak Trachyandesite flows are exposed at the top of the section and flows of the Elk Creek Basalt Member, that generally occur low in the section, were observed not far beneath the base of the section. The white ash of 138-151.5 m is the same unit as the white ash-ignimbrite of 240-242 m on Cache Creek. The white cross-bedded sandstone of 170-183.5 m on Mount Hornaday may represent stream reworking of this unit.

Figure 6: Exposures at the Specimen Ridge locality showing the line of section. Arrow points to the three classic vertical trees shown in Fig. 1.

The well-rounded conglomerate at the top of the section correlates with the top of the Mount Norris Section and with the 122-141 m interval on Specimen Ridge. The section is made of 36% tuffaceous sandstones and siltstones and 56% conglomerate (Table 1).

Mount Norris:

The Mount Norris Section is located on the southwest flank of Mount Norris 3.5 km ESE of the junction of Soda Butte Creek and the Lamar River (Figs. 2 and 8). Although the base of the Lamar River Formation is not exposed at this locality due to Quaternary landslides, the contact with the overlying Wapiti Formation was observed 36.5 m above the last outcrop (or at 216.5 m). The well-rounded conglomerate at the 159-180 m layer correlates with a similar conglomerate at the top of Amethyst Mountain Section and the unit exposed at the 121-141 m interval on Specimen Ridge. This section is coarser with a much smaller percentage of fine grained sandstone and siltstone than the Amethyst Mountain Section and Specimen Ridge Sections (Table 1). Those sandstones are bright blue-green. Based on field identification, all of the wood observed on this section is pine (Type II wood) with the exception of two horizontal angiosperm (Type III wood) logs.

Mount Hornaday:

The 547 m thick Mount Hornaday Section is located on the south flank of Mount Hornaday 3 km northeast of the junction of Soda Butte Creek and Pebble Creek (Fig. 2) and represents almost the entire thickness of the Lamar River Formation. The contact with the underlying Madison Limestone is approximately 30-35 m below the base of the section. The contact with the overlying Wapiti Formation at 547 m is 103 m (340 feet) higher than shown on the USGS map (Prostka, Ruppel, and others, 1975). I think this discrepancy results from the existence of up to 150 m of erosional relief on the Lamar River Formation below the Wapiti Formation. The erosional surface is well-

Table 1: Percentage and cumulative thickness of various rock types comprising the stratigraphic sections measured in the Lamar River Formation.

Rock Type	Specimen Ridge		Amethyst Mountain		Mount Norris		Mount Hornaday		Cache Creek	
	m	%	m	%	m	%	m	%	m	%
Tuffaceous sandstone and siltstone	56	35%	155	36%	23	13%	60	11%	35	9.5%
Conglomerate	94	59%	241	56%	157	87%	432	79%	335	90.0%
Volcanic Rock	0	0%	0	0%	0	0%	44	8%	2	0.5%
Covered Intervals	9	6%	34	8%	0	0%	11	2%	0	0%
Approximate Totals	159m	100%	430m	100%	180m	100%	547m	100%	372m	100%

FAR ------------Distance From Source--------------------NEAR

Figure 7: Exposures at the classic Amethyst Mountain "Fossil Forest" showing where the Amethyst Mountain Section was measured.

exposed in the nearly vertical cliffs southwest of the line of section. A large percentage of the section is well-bedded conglomerate; much of it is unaccessible due to the treacherous cliffs. One of the few fine-grained sandstone units is a white cross-bedded sandstone at 170-183.5 m that may represent a reworking of the white ashes of 240-242 m on Cache Creek and 138-151.5 m on the Amethyst Mountain Section.

Cache Creek:

The 372 m thick Cache Creek Section located 3.2 km northeast of the junction of South Cache Creek and Cache Creek (Fig. 2) contains better representations of all lithofacies than any of the other sections. Very fine grained silty shales below the base of the section correlate with lacustrine shales of the 15-17 m interval on Amethyst Mountain. The white ash-ignimbrite of 240-242 m correlates with the white ash of 138-151.5 m on Amethyst Mountain and was reworked into the white cross-bedded sandstone of 170-183.5 m on the Mount Hornaday Section. The Cache Creek Section contains less fine grained tuffaceous sandstones (9.5%) than any of the other sections (Table 1).

Description of the Lithofacies:

Since the initial interpretations (Holmes, 1879; Hague and others, 1899) rocks now placed in the Absaroka Volcanic Supergroup have typically been referred to as igneous breccias, vent breccias, air-fall ash, lava flows, and intrusions. Although volcaniclastic-sedimentary rocks are common in early Tertiary andesitic piles throughout the western United States (McKee, 1972; Taylor, 1977) this sedimentary aspect of the Absaroka Volcanics was long ignored. In fact, these rocks had been so widely called volcanic breccias that Jensen (1931) continued to use the term even though he reported a large percentage of rounded to sub-rounded clasts and personally documented that they should be more accurately called conglomerates. Because rocks of the Lamar River Formation were presumed to be volcanic, investigators interpreted the petrified wood deposits as having formed when "periodic breccia and ash eruptions buried successive forests during Eocene time" (Brown, 1961). Dorf (1964b) using this assumption estimated the number and frequency of local volcanic eruptions on the basis of successive "volcanic" layers containing petrified stumps and logs in the Lamar River

Figure 8: Exposures on the southwest flank of Mount Norris showing the line of section.

Formation.

A few investigators did, however, recognize the sedimentary nature of at least part of the Lamar River Formation. Holmes (1878) and Macomber (1956) described rounded, graded, and cross bedded conglomerate units, but interpreted them as being only a minor constituent of the Lamar River Formation. In describing the stratigraphic framework of Absaroka Volcanic Supergroup, Smedes and Proska (1972) along with Parsons (1974) referred to much of the rock as water deposited and called this material the alluvial facies, although they gave no detailed sedimentary analyses or rock descriptions. After describing the rocks in my measured sections, it is my opinion that most of the petrified wood occurs in the water laid alluvial facies, and, like the Eocene Clarno and Challis volcanics, most of the rocks in the Lamar River Formation were deposited by sedimentary processes, even though they originated from volcanic source areas. The alluvial facies are composed of sedimentary rock with 10% or less andesitic and basaltic volcanic flows. Following the suggestion of Parsons (1974), I refer to most of the rocks burying the wood as conglomerate and sandstone. I reserve the term breccia for a few minor, highly angular, volcanic units exposed on Mount Hornaday and Cache Creek. In this paper I use the term sandstone and siltstone in a textural sense that does not imply a quartz rich mineralogy. I believe that if the Lamar River Formation was deposited by airfall volcanic material, the entire section should look like the white ash-ignimbrite of 240-242 m on the Cache Creek Section. I will next describe the various lithofacies and propose depositional environments for each of the rock types present in the five stratigraphic sections that I measured for this study.

Mudflows:

Matrix-supported mudflow comglomerate units occur throughout the Lamar River Formation and comprise a large percentage of all of the sections. The mudflow conglomerate contains sub-angular to sub-round, poorly sorted clasts of homogeneous andesitic rock types. The clasts range in size from 1 cm to over 2 cm and are matrix supported by tan-brown silt and coarse sand. Occasionally units contain grain supported clasts, but most are matrix supported. Mudflow deposits are normally massive, structureless, and non-bedded, however, some are reversely graded. The top of the mudflow deposits are mostly moderately well-sorted, grain supported, pebble conglomerate with areas of imbricated clasts. This different lithology probably represents deposition in streams that formed by dewatering of the mudflows as they ceased to move. Only some clasts near tops of the units show imbrication since most are blocky in shape and would not become imbricated even if deposited in environments that normally imbricate clasts. Some mudflow units have erosional bases with relief of 1-20 m. Since mudflows do not usually produce scoured surfaces, the irregular basal contacts result from flows filling pre-existing stream-cut channels.

Jensen (1931) also interpreted some units with very large clasts to have formed as the result of mudflows since he felt that "It is preposterous to suppose that boulders six and eight feet in diameter could have been thrown from vents fifteen to twenty miles away. . ." I interprete these massive matrix supported conglomerates to have formed as a result of water saturated mudflows originating high on the flanks of surrounding strata volcanoes (Parsons, 1974) which can transport large clasts great distances.

Fluvial Deposits:

Much of the volcaniclastic material in the Lamar River Formation was deposited by streams. As previously mentioned, several workers recognized fluvial deposition in the Absaroka volcanics, but tended to down-play their importance (Holmes, 1898; Jensen, 1931; Macomber, 1956; Dorf, 1960, 1964b; Smedes and Prostka, 1972; Parsons, 1974). However, based on my stratigraphic frameworks, I believe streams reworking mudflow debris were the most important processes for depositing the Lamar River Formation. Well-rounded grain supported conglomerate of meandering streams and sub-rounded grain supported braided stream conglomerate along with interfingering tuffaceous (overbank) sandstone comprise the fluvial deposits.

Braided stream conglomerates are well-bedded and contain grain supported clasts with minor tan-brown silt and sand matrix. The clasts range in size from 1-30 cm (occasionally 1 m or larger) are sub-round to round, moderately well-sorted and contain a variety of clast rock types (Fig. 9). This heterogenous assemblage of rock types includes extrusive basalts and andesites, well-indurated tuffaceous sandstones, and, near the base of the formation, clasts of the underlying Precambrian granites and Paleozoic limestones (Macomber, 1956; Prostka, Ruppel, and others, 1975).

A variety of sedimentary structures occur within the braided stream conglomerate. Previously the only sedimentary structures reported from the Absaroka volcanics were local grading, cross bedding on Mount Hornaday (Macomber, 1956), and wide spread horizontal layering. Dorf (1964a) in his classic interpretation of the "fossil forests" depicts a layer cake stratigraphy with no

Figure 9: Braided stream conglomerate at 270 m on the Cache Creek Section showing the variety of rock types that make up clasts of this lithofacies. Camera lens cap in center of photo is 5.5 cm in diameter.

cross cutting or sedimentary structures. Fisk (1974) reported reversely graded beds along with normal and symmetrical grading in the Lamar River Formation. I found reverse graded beds throughout the braided stream conglomerate along with symmetrically and normally graded units. The reverse grading in fluvial deposits probably occurred due to increasing stream competency. These types of grading produce well-bedded units of conglomerate and was one criterion used to distinguish fluvial conglomerate from those produced by mudflows. However, these criteria may not always apply based on recent observations of well bedded Mount St. Helens mudflows (Harrison and Fritz, in preparation). Channel cut-and-fill that produces high and low tangential cross beds occur throughout all five sections (Figs. 10 and 11). Locally, clasts and pebbles of the stream conglomerate were imbricated — a condition that might have occurred more often if the majority of the clasts were not blocky in shape. The base of braided stream conglomerate units almost always occur on erosional surfaces with 1-20 m relief (Fig. 12). Previous streams cut channels of this depth that were subsequently filled with braided stream channel deposits and by mudflows.

Figure 10: High angle tangential cross beds in braided stream conglomerate at 150-155 m on the Cache Creek Section.

The widespread existence of braided streams producing deposits characteristic of arid lands at first seem surprising for humid conditions indicated by plants in the Lamar River Formation. However, Kuenzi, Horst, and McGehee (1979), describe similar braided streams in alluvial fan deposits in areas of high rainfall in Guatemala. There the slopes of nearby volcanoes are kept bare of vegetation by constant catastrophic mudflows that choke streams causing braiding. These streams were able to transport clasts form 30-120 cm in diameter (Kuenzi and others, 1979). Also, the mudflows of Mount St. Helens choked stream drainages causing widespread braiding (Fritz and Moore, 1981). Thus, the choking of streams by mudflows and sheet wash associated with dewatering of the flows caused numerous braided streams in a region of torrential rainfall.

Braided stream facies differ from other fluvial deposits in several ways. They are generally poorly sorted and contain a great predominance of angular to sub-angular clasts. A smaller percentage of fine grained sandstone and overbank mudstone overbank layers are associated with the braided streams when compared to meandering streams deposits. The braided streams units are thin and wide with low angle tangential cross bedded channel bar fill.

Figure 11: Low angle tangential cross bedding and channel fill of white pebbly sandstone in the braided stream conglomerate at 175 m on the Mount Hornaday Section.

Meandering streams, on the other hand, cut deeper channels, thus producing thick narrow channel fill deposits with high angle point bar cross beds. In contrast to the braided streams, meandering stream units in the formation are composed of well-rounded and sorted clasts with a large percentage of fine-grained sandstone and overbank mudstone. The meandering stream conglomerate is best exposed near the top of the Specimen Ridge, Amethyst Mountain, and the Mount Norris Sections. The clasts range in size from 1-50 cm and are composed of various rock types. The base of this lithofacies rests on an erosional contact of up to 50 m relief. High angle point bar cross beds in this unit grade into, or are overlain by low angle cross bedded tuffaceous sandstones (fig. 3). This change in cross bed angle and grain size might have occureed due to meander cut-offs and resulting deposition in oxbow lakes and from flood plain deposition. Smaller scale 'epsilon' high angle tangential cross beds are also common in the meandering stream lithofacies (Fig. 14). This stream conglomerate formed in parts of the streams that were not choked with mudflow debris, possibly due to steeper gradients in sections of the stream. The simultaneous existence of braided and meandering runs of streams depositing coarse conglomerate can be observed near Missoula, Montana, in the Cimarron River in Oklahoma (Shelton and Noble, 1974; Shelton and others, 1974), and in streams of the Toutle River drainage near Mount St. Helens. Meandering streams may also have formed near the end of periods of volcanic activity when input from mudflows ceased and streams were able to establish regular drainage networks. On Amethyst Mountain, Trout Peak Trachyandesite flowed down meandering stream channels before deposition of the overlying Langford Formation (Fig. 15).

As noted by Dorf (1964b) and Parsons (1974), conglomerates contain most of the petrified logs and stumps in the Lamar River Formation. Only on Amethyst Mountain do ash and sandstone units contain a significant portion of the petrified stumps and logs, but even there the conglomerate still contains over 50% of the fossil wood. Unlike braided stream and mudflow conglomerate, the meandering stream conglomerate contains few pieces of petrified wood. This is because meandering stream formed in the absence of mudflows that supplied logs and stumps to be buried. A similar situation occurs in the Mount St. Helens area. There little wood was carried by the meandering streams prior to May, 1980 mudflows. However, after mudflows choked streams with debris, including trees, many stumps and logs were buried in braided stream deposits (Fritz, 1980d; Fritz and Moore, 1981).

Sections far from possible highland sediment source areas (Specimen Ridge and Amethyst Mountain, see Fig. 2) have less conglomerate and more sandstone than sections near possible source areas (Mount Hornaday and Cache Creek) which have a high concentration of large boulders and few mudstone interbeds. This decrease in grain size results from deposition on a less steep gradient in the center of the basin of deposition.

Well-cemented blue-green to gray tuffaceous sandstone and siltstone was deposited by streams along with coarse conglomerate. Sandstone units are mostly well-sorted and highly silicified. Partially rounded volcanic ash-shard grains range in size from silt and clay to coarse sand and pebbles. As much as 10-20% of the grains are non-volcanic or of composition other than volcanic glass shards. The presence of 80-90% sub-angular volcanic sand grains is not surprising in a region with nearby volcanic peaks. An example is the recent work by Kuenzi and others (1979). The only prominant air-fall ash bed in any of the sections occurs at the 240 m interval on the Cache Creek section. Petrographic analyses thin sections show that many of the sedimentary ash-grains have altered to clay minerals after deposition.

In several cases, conglomerate grades vertically into tuffaceous sandstone units as blue-green ash replaces the brown sandstone matrix of the conglomerate. Eventually conglomerate clasts become smaller until only a pure tuffaceous sandstone remains. This change to deposition of finer grained material occurred during volcanic eruptions as air-fall ash choked streams providing only a fine grain source, or during periods of lower energy environments when streams were competent to move only fines. Often intraformational conglomerate occurs within tuffaceous

Figure 12: Bedded braided stream conglomerate showing an erosional base truncating fine grained blue-green tuffaceous sandstone units at 172-182 m on the Amethyst Mountain Section.

Figure 13: Cross bedding in well-rounded meandering stream conglomerate at the top of the Classic Cliff on Amethyst Mountain. Note change from high angle cross beds in coarse conglomerate to low angle cross beds in overlying tuffaceous sandstone lens, that represents overbank or lacustrine sedimentation.

sandstone layers. On the Amethyst Mountain Section and Specimen Ridge Sections, tuffaceous sandstone units are bright blue-green, while other sections are gray to tan. This color difference may reflect higher weathering rates producing tan and brown clays, hydrothermal alteration, or a different mineral composition possibly due to a different source vent for the volcaniclastic grains.

Tuffaceous sandstone units are well-laminated and display load casts, flame structures and ripple cross laminations. These sedimentary structures formed as sluggish streams reworked mudflow debris, air-fall ash, and detrital clasts. Many tuffaceous sandstone beds are discontinuous flat topped lenses within the mudflow and braided stream facies, thus, suggesting deposition as channel fill. Howell and Link (1979) describe channel fill lenses within Eocene conglomerate near San Diego, California that are identical in shape to those found in the Lamar River Formation. When tuffaceous sandstone units do form flat lying well-laminated beds, possibly from overbank flood plain deposition, they are cut into by overlying stream and mudflow conglomerate (Fig. 12).

Although tuffaceous sandstone units contain fewer pieces of petrified wood than conglomerate units, they preserve abundant pollen, leaves, needles, and some cones. These well-preserved organic remains occur in specific horizons sometimes containing roots of vertical stumps and have been called paleosols. I could find no A, B, or C horizon in organic zones that I observed. The organic zones that I saw were mostly very thin, well-laminated, have no decayed organic debris, and are often draped over large boulders (Fritz, 1981a). However, Retallack (1981) describes an apparent paleosol at the 90-94 m interval on Specimen Ridge. Although fossil soils could occur on sections far from source areas representing deposition in low energy conditions, caution should be used in making paleosol identifications (Fritz, 1981a). After examining the same organic zones reported on by Retallack, Jensen (1931, p. 46) states that "the soils in which the trees grew is never preserved." Jensen also found little or no organic material associated with the roots of trees and saw no evidence for

Figure 14: 'Epsilon' type tangential cross bedding in meandering stream lithofacies near the top of the Amethyst Mountain Section.

Figure 15: A flow of Trout Peak Trachyandesite (Tt) producing the final fill in a stream channel cut into the Lamar River Formation (Tlr).

lateritic soils that should have developed under the tropical conditions indicated by some plant remains. While I allow for fossil soils in the Lamar River Formation, I believe that many of the organic zones represent plant litter deposited by sheet wash, possibly during intense rainstorms associated with volcanic activity. This interpretation is strengthened by observations of fine organic debris associated with the roots of transported stumps around Mount St. Helens (Fritz, 1980d). Some of the organics also accumulated in slack water areas in the streams or in small mudflow dammed lakes. Vertical trees in the conglomerate facies normally have no organic zone or weathering profile associated with roots.

Lacustrine Deposits:

Much fine-grained siltstone and sandstone of the Lamar River Formation accumulated in small lakes. These lakes formed as mudflows choked, and temporarily dammed streams, allowing for slack water deposition of fine grained sediments. These lacustrine mudstones grade imperceptably into fluvial sandstone and mudstone. The lake deposits have more siltstone and claystone that the fluvial deposits and are horizontally laminated with no ripple cross laminations (Fig. 16). Many are indistinguishable from overbank and oxbow lake deposits except by stratigraphic succession. The shapes of the lakes that formed by braided streams were very irregular with ill-defined shorelines. The lakes probably existed for only short periods of time. Kuenzi and others (1979) describe similar dendritic lakes in mudflow dammed braided streams in an andesitic volcanic terrain in Guatemala. I have also observed many slack water ponds caused by flooding of the Toutle River in the Mount St. Helens area. Holmes (1879) and Dorf (1960) also report on the minor occurrence of lacustrine deposits.

Due to the lack of currents. leaves preserved in lacustrine sediments remain flat and show excellent preservation of organic tissue and cuticle. Leaves preserved in the stream-current deposited siltstone and mudstone, on the other hand, are generally rolled and folded. The lacustrine facies is best developed and exposed near the base of the Amethyst Mountain and Cache Creek Sections.

Figure 16: Well bedded lacustrine siltstone near the base (18-20 m interval) of the Amethyst Mountain Section.

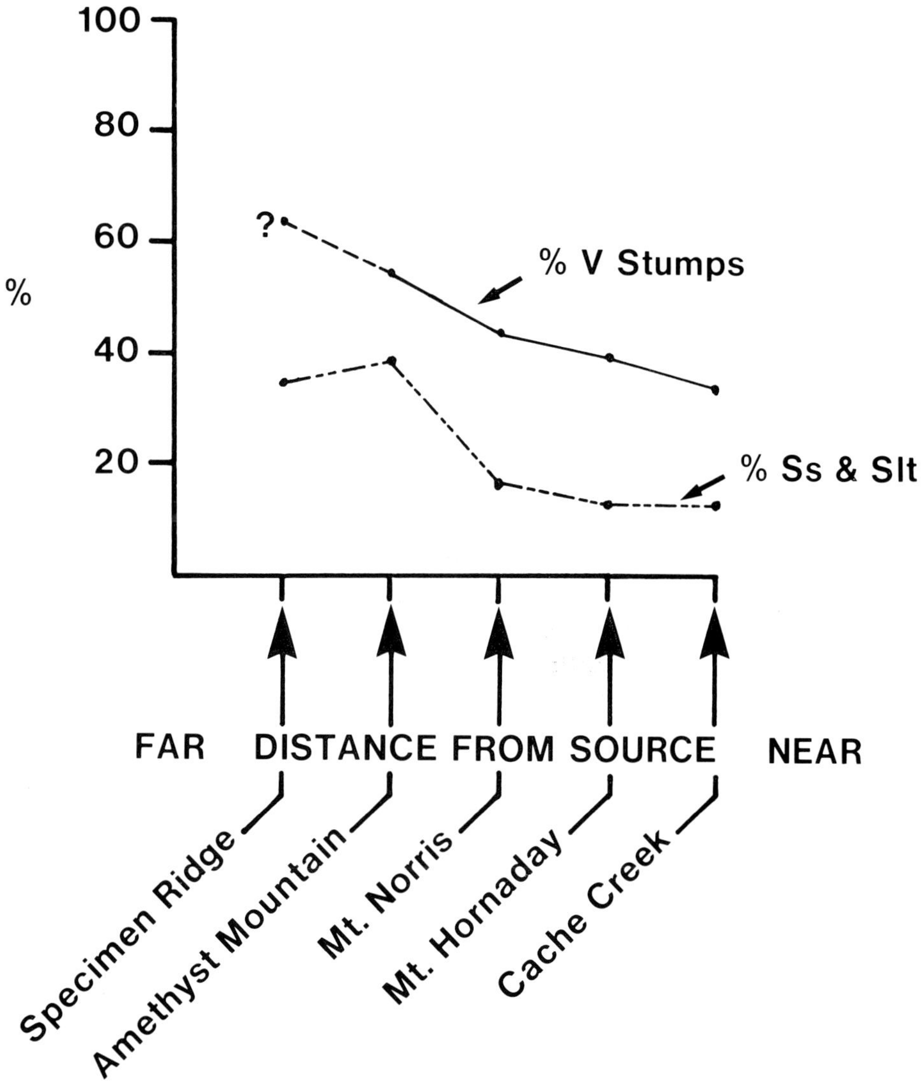

Figure 17: Graph showing fining of the Lamar River Formation away from sections deposited near a source. Also note the increase in horizontal logs near vent peaks.

Volcanic Rocks:

Volcanic rock forms less than 10% of the Lamar River Formation as shown in the stratigraphic sections (Fig. 5; Table 1). On Mount Hornaday there are three to four 8-25 m thick basalt units with an associated volcanic breccia. These flows become vesicular and scoraceous near the tops distinguishing them from dikes or sills. These basalt flows belong to the Elk Creek Basalt Member of the Lamar River Formation (Smedes and Prosatka, 1972) and occur throughout the lower part of the formation. A prominant two-meter thick white air-fall tuff is exposed on the Cache Creek Section. Portions of this white unit contain highly angular volcanic pebbles and hornblende crystals and may represent a volcanic breccia or ignimbrite. This white volcanic unit may correlate with a white, structureless, poorly indurated air-fall ash at the 138-151.5 m interval on Amethyst Mountain. As previously stated all rocks preserving the "fossil forests" should look much like this unit if they had been deposited according to many of the classic interpretations. Near the top of the Lamar River Formation flows of Trout Peak Trachyandesite produce the final fill in stream-cut channels (Fig. 15).

General Commenst on the Lithofacies:

Figure 5 shows a correlation of the five stratigraphic sections that I measured in the Lamar River Formation. I have found few key beds or layers in the formation that correlate between sections. As previously mentioned, there is a white volcanic ash unit that I can trace between the Amethyst Mountain, Cache Creek and possibly the Mount Hornaday sections. A well-rounded meandering stream conglomerate also exists near the top of the Specimen Ridge, Amethyst Mountain, and Mount Norris sections. Except for these, most beds lense or interfinger with other rock types and are not traceable for more than 100 meters.

In general, facies changes in the Lamar River Formation consist of a coarsening of the conglomerate and loss of sandstone towards the source areas (Figs. 7, 28). Conglomerates compose 55-60% of the Amethyst Mountain Section, the area farthest from a possible source, while the Mount Hornaday and Cache Creek section are 79-90% conglomerate, the nearest sections to a possible source (Table 1). Conglomerate units exposed on Mount Hornaday also contain a higher percentage of large boulder-size clasts than the other four sections. Northeast of Mount Hornaday the amount of volcanic flows increases towards what Smedes and Prostka (1972) termed the vent facies. Although I did not measure any sections in these vent areas, a large amount of sedimentary rock is inter-bedded with the volcanics on Mount Washburn and in exposures near Cooke City, Montana. Changes of composition of the floral elements can also be seen between sections. While taking counts of the horizontal, diagonal, and vertical trees, I also made field identifications of the petrified wood. Even though handlens identifications are at best crude, I could in most cases recognize three major wood types. The types I could best recognize were Type I, coniferous woods without resin ducts containing members of Taxodiaceae, Podocorpacae; Type II, coniferous wood with resin ducts, or the Pinaceae; and, finally Type III, wood with vessesl — the angiosperms. Note that most cool temperate species would fall in Types I and II with the exception of *Podocorpus* which would be lumped with Type I in the field. Most tropical trees and some of the temperate ones would be classified with Type III. Even though accurate identifications are still needed, trends can still be noted based on these field identifications. Sections near possible source areas have a larger percentage of coniferous wood with more logs in the horizontal position than sections far from source areas (Fig. 18). Sections farther from source areas have a mixture of temperate conifers along with many tropical trees, many of which are in the vertical stance. On Amethyst Mountain, a section far from possible source areas, the mixture was found even in the vertical stumps of one unit (Fritz, 1977; Fig. 4). These facies changes, along with changes in the forest composition support high energy, high elevation upland source areas with lower energy tropical sites of deposition.

In summary, as mentioned by Smedes and Prostka (1972), sections near source areas are coarser grained, have little ash, more volcanic rock, and contain predominately horizontally oriented temperate conifers. Sections farthest from known source areas are finer grained, have 30-40% tuffaceous sandstone, little volcanic rock, and a mixture of tropical and temperate trees with more in the vertical position than the other sections.

Sedimentology of the Petrified Stumps and Logs:

In addition to sedimentary structures found in the rocks, several important features of the petrified logs and stumps bear on interpretations of the depositional environment. Organic remains, such as leaves and logs, that might have been transported, should be considered and studied as any lithic clast in the conglomerate or sandstone.

Besides vertical stumps (Fig. 1), horizontal and diagnoal logs also occur in abundance (Figs. 19 and 20). Except for some intervals on Specimen Ridge, horizontal logs are more common than vertical stumps and comprise 60-100% of the total amount of petrified wood (Fig. 18). Typical is the 248-253 m layer on Amethyst Mountain (Unit 30 of Fisk, 1876b). This unit contains 39% vertical stumps, 53% horizontal logs, and 8% diagonal logs (Fritz and Fisk, 1978, 1979) (Fig. 18). In the field I recorded the orientation of the horizontal logs as shown in representative rose diagrams of Fig. 18. Since the roots are not exposed on all horizontal logs, a mirror image is given, indicating that true tree orientations might be in either of two 180° opposite directions. In some areas, such as the 272 m layer of the Cache Creek Section, all wood is horizontal and is preferentially oriented into log jams. The horizontal trees of Amethyst Mountain are also strongly oriented in a northwest - southeast orientation (Fritz and Fisk, 1978, 19798; Fig. 18). Jensen (1931) and Coffin (1976) reported oriented horizontal trees in the Lamar River Formation and suggested that they were oriented by entombing debris flows and might be useful in determing the flow direction and source area. I think the horizontal trees were oriented by mudflows and streams. However, since numerous source areas provided interfingering alluvial material, I could not determine the source area for a particular area or section.

Vertical stumps in the Lamar River Formation are mostly shorter than horizontal or diagonal logs and were broken above the root level by volcanic blasts or mudflows prior to burial and preservation (Fig. 21). Only on Specimen Ridge are significant numbers of tall vertical trees preserved. Many short vertical stumps have a very wide root system that forms a base wider than the broken stump is high (Fig. 21). These short stumps were probably broken by mudflows before they were buried and only a few

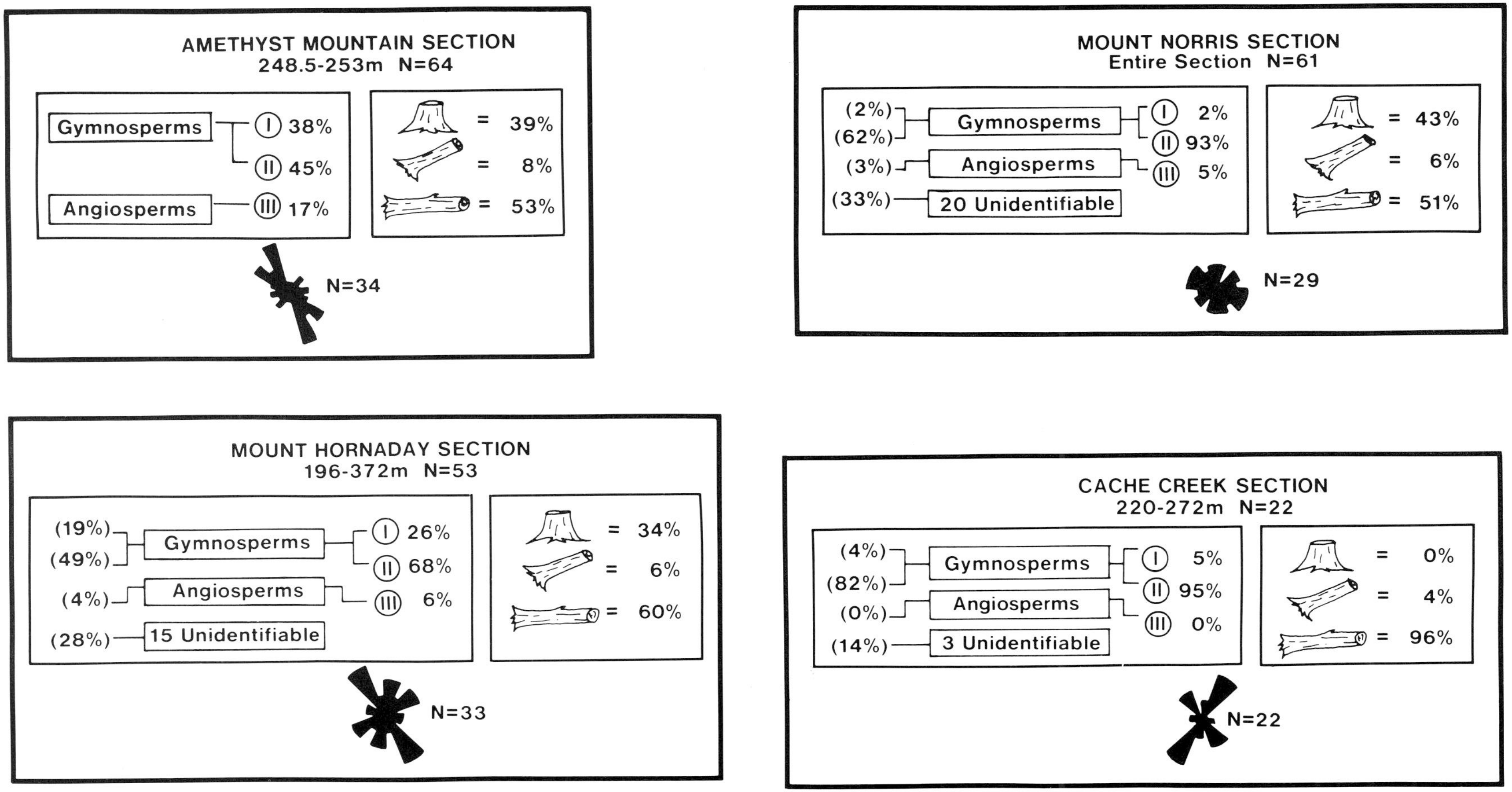

Figure 18: Percentage of horizontal, vertical, and diagonal trees, gymnosperms (Type I, wood without resin ducts; Type II, wood with resin ducts) and angiosperm (Type III wood with vessels) along with a rose diagram of the orientations of horizontal logs representative units in the Lamar River Formation.

locations are branches and rootlets present, both having been broken off before the trees were preserved. Some petrified trunks are coated with mud of a similar composition to the matrix of the mudflow conglomerate. This mud coating probably accumulated when mudflows engulfed and transported trees.

In only one or two known trees on Specimen Ridge is bark preserved. I think that the reports of bark on some of the vertical trunks by several early workers (Holmes, 1878; Weed, 1892) were caused by confusing the mud coating for bark. This mud coating often flakes off the trees after they have been exposed and could be easily mistaken for bark, but contains no cell structure of the woody tissue. None of the previous reports of bark were based on thin section analysis or a description of the woody tissue. All of the trees (with one or two questionable exceptions) appear to have had their bark removed prior to burial. This absence of bark is surprising especially for trees with thick resistant bark like *Sequoia*. Thus, the absence stands out in sharp contrast with the excellent preservation of other plant remains that show little or no decay. Since the other plant remains show no decay of even the fine structure, it is unlikely that the bark rotted away before preservation. Dorf (1964b) stated that much of the bark was removed from the trees by abrasion from air-fall ash, volcanic explosions, and other processes occurring during periods of volcanic eruptions. I think that much of the bark was removed during engulfment of the trees by the mudflows and from transport in the braided streams. It can be readily observed in most modern streams that the bark of stumps and logs is quickly abraided well before the wood starts to decay. These fluvial and mudflow processes appear to have broken off branches and fine rootlets in the horizontal logs as well as in many verical stumps. This accounts for their scarcity in the Lamar River fossil wood deposits.

DEPOSITIONAL SYSTEM OF THE YELLOWSTONE "FOSSIL FORESTS"

I believe that deposits and process operating around Mount St. Helens provide a ready modern analogue for the depositional environment of the Lamar River Formation (Fritz, 1980d). I here present my interpretation of the depositional environments for the Lamar River Formation (Fritz, 1980a, 1980b, 1980c).

As previously stated, Chadwick (1970) and Parsons (1974) proposed that rocks in the Absaroka Volcanic Supergroup were produced by stratovolcanoes with peaks eight to ten thousand feet above the valley floor. These vents existed as two northeast-southwest trending sub-parallel chains 25 to 60 kilometers apart (Fig. 2). During Eocene time, this configuration would have produced a long narrow intermontane basin that received sediment from surrounding volcanoes. All known fossil wood localities in the Lamar River Formation, and most of those in the supergroup, occur between these two chains of source areas (Fig. 2). Depositional environments would thus have been similar to those of many intermontane basins seen throughout the Tertiary in the western United States. Directly comparable modern environments exist in the andesitic volcanic terrain in Central America (Davies, and others, 1979; Kuenzi and others, 1979) and in the Cascade Mountain Range of western Oregon and Washington; in particular Mount St. Helens. Major depositional environments in both the modern and the Eocene include mudflows, braided and meandering stream deposits consisting of reworked mudflow debris, lacustrine, slack water deposits, and a small amount of volcanic rock.

During Eocene time, debris flows originated high on the flanks of volcanic vent areas. The flows were saturated with water, possibly from rain showers associated with volcanic eruptions or from snow cover. The flows uprooted trees growing in high elevations, cool temperate habitats near the rims of the volcanoes. During transport, roots and branches were broken off. The flows also debarked the trees, added a mud casing, and locally dammed streams. The resulting coalescing mudflow deposits were then reworked by braided streams. Braided streams further transported plant litter and wood, thus orienting the horizontal logs. Stream transportation also removed more bark, branches, and rootlets, and coated the trees with mud while producing flood plain deposits. Both mudflows and braided streams transported temperate plants from high elevation sites and mixed them with tropical plants growing in the lowlands. Even some vertical stumps, after being broken and uprooted by mudflows, were transported and deposited in the upright position by streams and subsequent mudflows. Trees with a wide root system and short broken trunks (Fig. 21) would behave as any such irregular clast with the vertical position being most stable. This interpretation explains the absence of soils for many of the vertical trees, since many of the stumps are not preserved where they grew and many soils were washed away by streams and mudflows. Many upright trees with long trunks and well-preserved root systems, such as those on Specimen Ridge (Fig. 1) were preserved in place as sediments buried them. Both transported vertical stumps and *in situ* vertical stumps co-exist in the Lamar River Formation, but distinguishing between them is often difficult. Very tall, vertical trees with relatively narrow root systems that contain unbroken roots and lie in a thick organic accumulation such as the paleosol described by

Figure 19: Horizontal log with roots on the Amethyst Mountain Section. Hammers are 33 cm long.

Retallack (1981) were probably preserved where they grew. Trees meeting these criteria are most numerous on Specimen Ridge (Fig. 2). Such trees are almost absent in the lower two-thirds of the Mount Hornaday Section. On the other hand, short vertical stumps broken prior to burial, wide root systems, no branches, rootlets, or bark, and the absence of organic zones indicate that the trees were transported by the depositional systems. However, many trees display only some of the criteria, making it difficult to interpret them. Vertical trees moved only a short distance might be impossible to distinguish from *in situ* ones. Modern vertical stumps transported by braided streams can have fine rootlets preserved along with an accumulation of organic debris at the base. At times large boulders entwined in the roots of stumps and logs are also transported. Additional detailed paleoecological studies conducted in a similar manner to that of Fritz (1977) are needed. These studies should concentrate on identifying and describing each vertical stump at individual stratigraphic horizons in an attempt to determine which have been moved and what percent of the total they comprise.

Simultaneous with the transportation of the wood, air-fall ash was washed into streams and lakes along with leaves and plant litter producing the laminated organic zones. as volcanic activity and resultant growth of vent peaks slowed near the end of deposition of the Lamar River Formation, mudflows gradually ceased. Braided streams established regular drainage networks and eventually meandering streams formed. These meandering streams produced erosional contacts and cross bedded conglomerates exposed near the top of Amethyst Mountain Section (Fig. 12).

Processes operating in the proposed depositional environment produced a complex series of sedimentary rocks (Fig. 22) where no "forest" or layer exists that can be traced for any distance. Thus, a layer cake model depicting continuous layers of even thickness alternating between air-fall tuff and breccia with 95% of the trees vertical (counts taken from the Figs. in Dorf 1960, 1964, 1964b, 1974, 1980) is not valid. Instead a complex typical alluvial deposit is present consisting of lenses, rapid facies changes and non-continuous layers. Dorf's model of layer-cake stratigraphy might be valid for a trailing continental

Figure 20: Eleven-meter long diagonal log on the Amethyst Mountain Section. Note horizontal logs protruding from cliff face of alluvial conglomerate.

margin with low relief such as the Gulf Coast which he uses as a modern analogue for Yellowstone. However, the Yellowstone area was not a trailing margin and contained considerable relief such as most andesitic mountain chains today. Thus, one should not expect to find layer cake stratigraphy. This model, presented in the generalized stratigraphic section of Figure 22 and the block diagram of Figure 23, predicts the observed loss of tropical indicators and increase in the number of horizontal logs and temperate indicators in sections near possible source areas.

In addition to Mount St. Helens, several modern environments can serve as analogues for my proposed depositional model for the Lamar River Formation. One of the closest is the andesitic volcanic region of Central and South America. There, in the tropics, with a five to ten thousand foot change in elevation, a full range of cool-temperate to tropical plants live in a fairly small geographic area. Numerous mudflows occur on these peaks (Kuenzi and others, 1979; Davies and Bonis, 1977; Davies and others, 1979; Vessel and others, 1977). These flows produce poorly sorted matrix supported conglomerates similar to those found in the mudflow facies of the Lamar River Formation. Recent mudflows in Guatemala choke streams with debris causing braiding and the formation of many small dendritic lakes (Kuenzi and others, 1979). These streams are able to transport clasts ranging in size from 30-170 cm — a size range comparable to clasts in the braided stream facies of the Lamar River Formation (Kuenzi and others, 1979). The sandstones deposited in this recent volcanic environment, like those preserved in the Lamar River Formation, are composed primarily of volcanic grains. These volcaniclastic sand and mud grains remain angular even though transported some distance by streams (Kuenzi and others, 1974). Kuenzi and others (1979, p. 836) describe the following facies changes that they observed in the recent volcanic terrain in Guatemala:

> "Individual fans would be composed of a complexly related assemblage of gravels, sands and organic-rich muds that record braided-stream deltaic, and lacustrine depositional environments,

respectively. . . The fluvial complex, in turn, would inter-tongue landward with lava flows, ash flows, and lahars."

I believe that nearly identical facies changes and sediment types occur in the Lamar River Formation.

Mudflows similar to those in the tropical region also occurred on Mount Rainer, Washington (Schmincke, 1967; Crandell, 1971). However, in a temperate region, the mixture of temperate to tropical trees does not occur or is of a much lesser variation than in the tropics.

More work is needed in these recent volcaniclastic sedimentary areas to determine if the mudflows do transport trees and plant debris. Kuenzi and others (1979) reported that the braided streams and mudflows buried trees. They did not, however, determine if any of the plant material thus preserved had been transported.

CONCLUSIONS

Due to the complex alluvial system responsible for deposition of the Lamar River Formation fossil wood deposits, it is impossible to determine the exact number of "layers" or "forests". Standing Eocene forests were not preserved, even though many upright trees were buried in place. It is also difficult to determine how many times surrounding volcanoes erupted.

Since mixing of plant remains occurred in this depositional system, the Lamar River Formation should not be used to argue that Eocene plants grew under climatic conditions unlike modern counterparts. Instead, the mixture is due to different habitats produced by abrupt elevation change in a fairly small geographic area and the transportation of plant parts from higher to lower elevations. Before paleoecological conclusions are drawn, the sedimentary geology of Tertiary paleobotanical localities should be studied in detail to reconstruct depositional environments and paleogeography. Even though I believe transportation of upright stumps to have occurred, I argue that many were preserved in place.

Mixtures of plants from a range of habitats seen in many Paleogene localities may be due to transportation from a variety of local habitats and not from changing

Figure 21: Short vertical stumps with a wide root system in a braided stream conglomerate on the Mount Horneday Section. No organic zone is associated with the roots of this stump. Hammer at base of stump is 33 cm long.

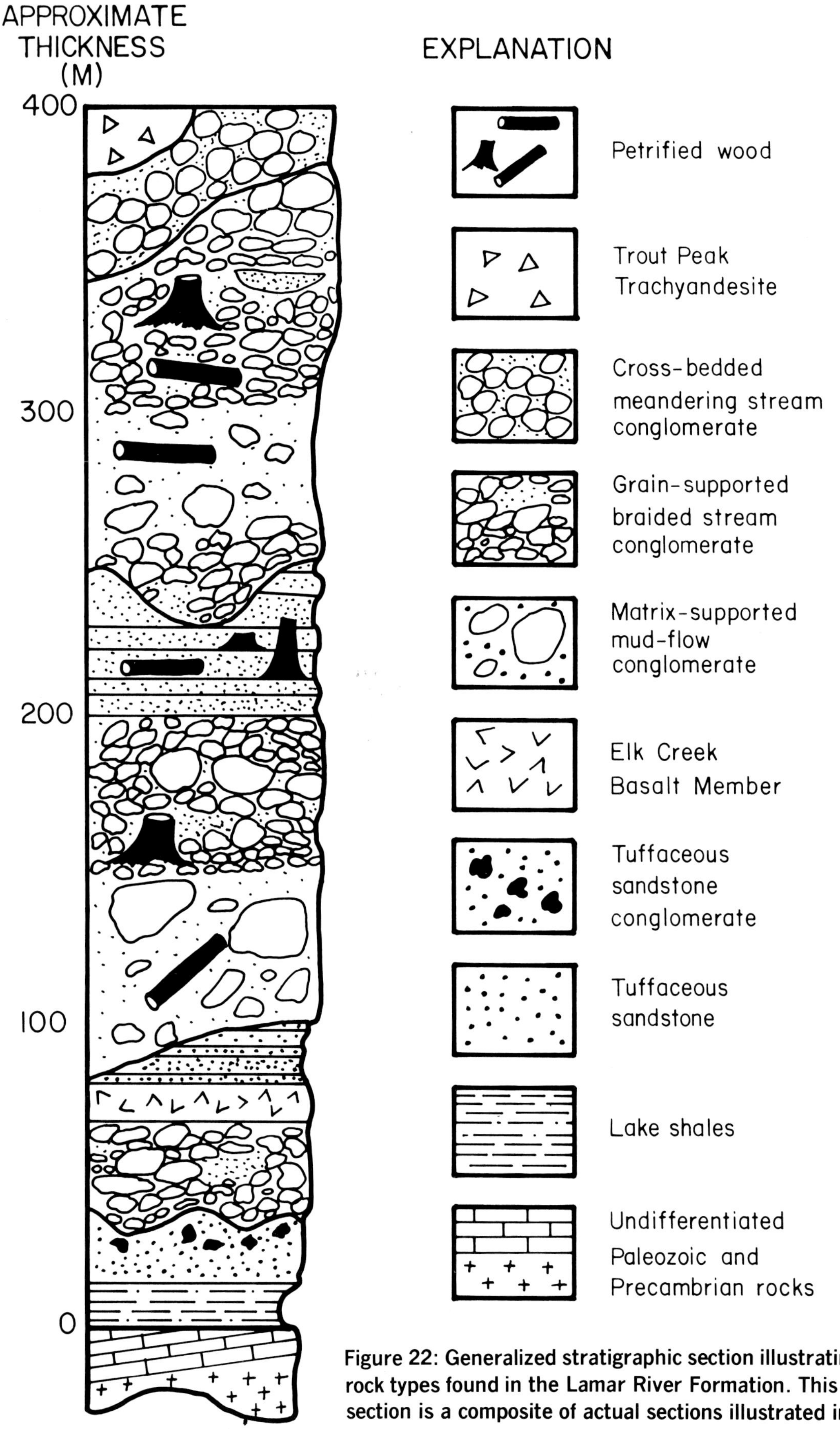

Figure 22: Generalized stratigraphic section illustrating various rock types found in the Lamar River Formation. This section is a composite of actual sections illustrated in Figure 5.

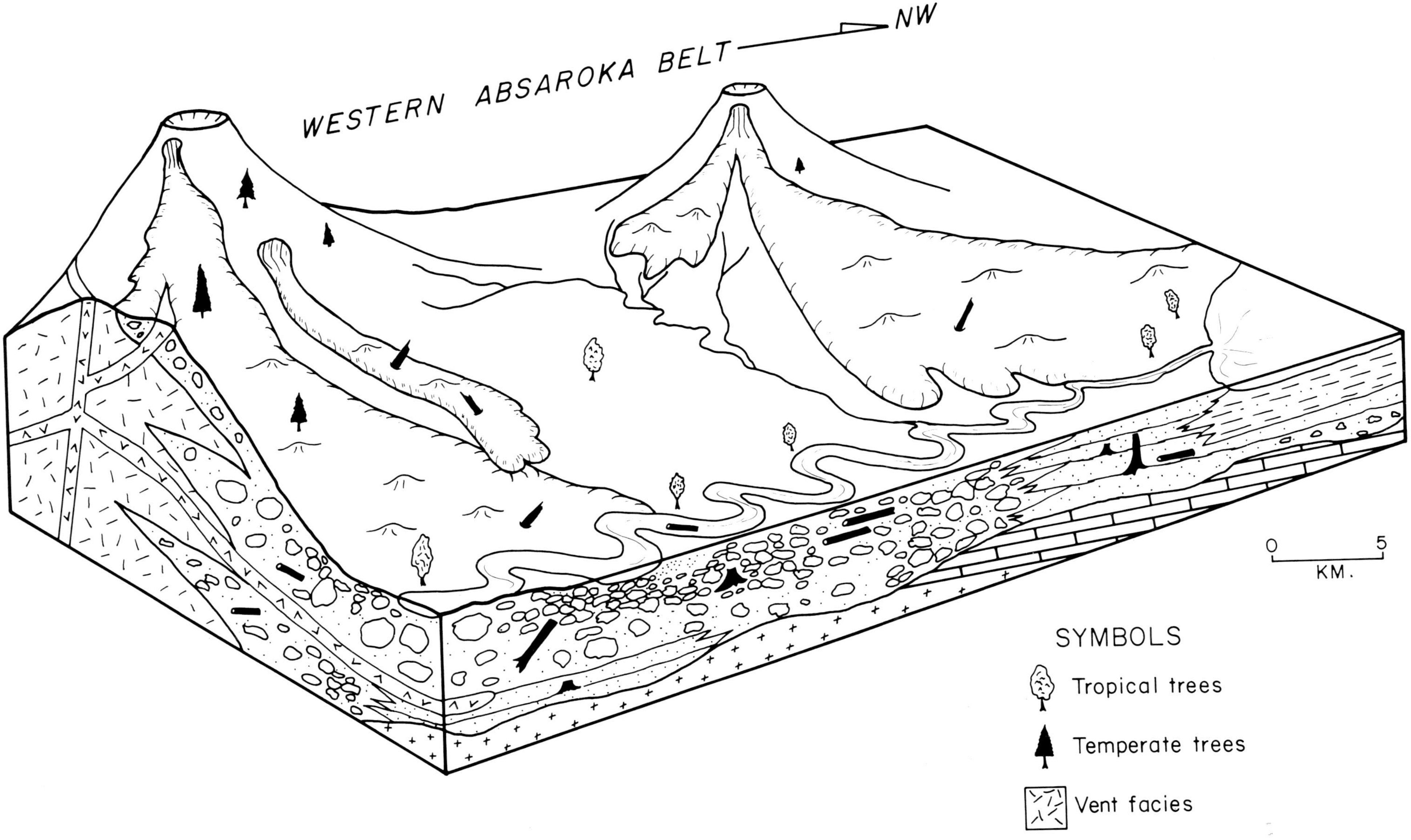

Figure 23: Block diagram showing depositional environments for "Fossil forests" in Lamar River Formation. From Fritz, 1980a.

climatic tolerances. In other localities, investigators should not use the presence of upright stumps to rule out transportation without first studying the stratigraphic framework.

ACKNOWLEDGEMENTS

Portions of this work formed the basis for my M.S. thesis at Walla Walla College and my Ph.D. dissertation at the University of Montana. I thank my advisors and committee members for their help. Special thanks go to Dr. R. W. Fields, Dr. L. H. Fisk, Dr. J. N. Moore, Dr. D. Winston, Dr. D. Alt and Dr. C. N. Miller. I thank the National Park Service for their cooperation and permission to collect samples and work in the Park. In addition to my advisors, C. A. LaPasha, B. Duncan, R. Nichols, M. Aquirre, R. Ammons, and the entire Ammons family provided welcome field assistance. M. Rucker kindly typed this version of the manuscript and F. Drago provided drafting assistance. Supported in part by GSA Research Grants Nos. 2131-76, 2448-79 and two Sigma-Xi Grants-In-Aid of Research.

REFERENCES CITED

Aquirre, M., 1977, Coniferous leaves from the Amethyst Mountain "Fossil Forest" Yellowstone National Park, Wyoming, (Senior Thesis): Walla Walla College, 16 p.

Beyer, A. F., 1954, Petrified wood from Yellowstone Park: American Midland Naturalist, v. 51, p. 553-567.

Brown, C. W., 1957, Stratigraphic and structural geology of north central-northeast Yellowstone National Park, Wyoming and Montana, (Ph.D. Dissertation): Princeton University, 178 p.

______, 1961, Cenozoic stratigraphy and structural geology, northeast Yellowstone National Park, Wyoming and Montana: Geological Society of America Bulletin, v. 72, p. 1173-1194.

Chadwick, R. A., 1970, Belts of eruptive centers in the Absaroka-Gallatin volcanic province, Wyoming-Montana: Geological Society of America Bulletin, v. 81, p. 267-274.

Coffin, H., 1976, Orientations of trees in the Yellowstone Petrified Forests: Journal of Paleontology, v. 50, p. 539-543.

Conrad, H. S., 1930, A Pityoxylon of Yellowstone National Park: American Journal of Botany, v. 17, p. 547-553.

Crandell, D. R., 1971, Postglacial lahars from Mount Rainer volcano, Washington: U.S. Geological Survey Professional Paper 677, 73 p.

Davies, D. K., and Bonis, S. B. 1977, Observations on glowing avalanches and lahars, at the Volcano Fuego, Guatemala: Geological Society of America Abstracts with Programs, v. 9, p. 943.

______, Almon, W. R., Bonis, S. B., and Hunter, B. E., 1979, Deposition and diagenesis of Tertiary-Holocene volcaniclastics, Guatemala: Society of Economic Mineralogists and Paleontologists Special Publication No. 26, p. 281-306.

Dorf, E. 1960, Tertiary fossil forests of Yellowstone National Park, Wyoming: Billings Geological Society, 11th Annual Field Conference, p. 253-260.

______, 1964a, The petrified forests of Yellowstone Park: Scientific American, v. 210, p. 106-114.

______, 1964b, The petrified forests of Yellowstone National Park: U.S. Government Printing Office, Washington, D.C., 12 p.

______, 1974, Early Tertiary fossil forests of Yellowstone Park: *in* Barry, V., ed., Rock Mechanics: The American Northwest, Third International Congress on Rock Mechanics, Expedition Guide, p. 108-110.

______, 1980, Petrified forests of Yellowstone: U.S. Government Printing Office, Washington, D.C., 31 p.

Fisk, L. H., 1974, Inverse grading as stratigraphic evidence of large floods: comment: Geology, v. 2, p. 613-615.

______, 1976a, The Gallatin "Petrified Forest": Montana Bureau of Mines and Geology Special Publication 73, p. 53-72.

______, 1976b, Palynology of the Amethyst Mountain "Fossil Forest", Yellowstone National Park, Wyoming, (Ph.D. Dissertation): Loma Linda University, 347 p. University Microfilms No. 77-1451.

______, 1976c, Paleoenvironmental interpretations of the Eocene "Fossil Forests" of Yellowstone National Park, Montana and Wyoming, U.S.A.: 25th International Geological Congress, Abstracts, V. 1, p. 303.

______, and De Bord, P., 1974a, Palynology of the "Fossil Forest" of Yellowstone National Park, Wyoming: American Journal of Botany (Supplement), v. 61, p. 15-16.

______, and De Bord, P., 1974b, Plant microfossils from the Yellowstone fossil forests: Preliminary report: Geological Society of America Abstracts with Programs, v. 6, p. 441-442.

______, Aguirre, M., and Fritz, W. J., 1978, Additional conifers from the Eocene Amethyst Mountain "Fossil Forest", Yellowstone National Park, Wyoming: Geological Society of America Abstracts with Programs, v. 10, p. 216.

______, and Fritz, W. J., 1977b, Pseudoborings in petrified woods from the Yellowstone "Fossil Forest". Botanical Society of America Abstracts, v. 154, p. 36.

Fritz, W. J., 1977, Paleoecology of petrified woods from the Amethyst Mountain "Fossil Forest", Yellowstone National park, Wyoming, (M.S. Thesis): Walla Walla College, 57 p.

______, 1980a, Reinterpretation of the depositional environments of the Yellowstone "Fossil Forests": Geology, V. 8, p. 309-313.

______, 1980b, Stratigraphic framework of the Lamar River Formation in Yellowstone National Park: Northwest Geology, v. 9, p. 1-18.

______, 1980c, Depositional Environment of the Eocene Lamar River Formation in Yellowstone National Park (Ph.D., Dissertation): Department of Geology, University of Montana, 113 pages.

______, 1980d, Stumps Transported and Deposited Upright by Mount St. Helens mud flows. Geology, v. 8, p. 586-588.

______, 1980e, A review of *Pinus* wood from the Eocene Lamar River Formation. Botanical Society of America Miscellaneous Series Publication, v. 158, p. 39.

______, 1981a, Reinterpretation of the depositional environment of the Yellowstone "Fossil Forests": Reply: Geology, v. 9, p. 53-54.

______, and Fisk, L. H., 1978, Eocene petrified woods from one unit of the Amethyst Mountain "Fossil Forest": Northwest Geology, v. 7, p. 10-19.

______, and Fisk, L. H., 1979, Paleoecology of petrified woods from the Amethyst Mountain "Fossil Forest", Yellowstone National Park, Wyoming: National Park Service Proceedings Series No. 5, v. 2, p. 743-749.

Fritz, W. J., and Moore, J. N., 1981, Mount St. Helens mud flows as related to the transport of trees. Geological Society of America Abstracts with Programs, v. 7, p. 455.

Hague, A., Iddings, J. P., Weed, W. H., Walcott, C. D., Gritty, G. H., Stanton, T. W., nad Knowlton, F. H., 1899, Descriptive geology, petrology, petrography, and paleontology, pt. II *of* Geology of the Yellowstone National Park: U.S. Geological Survey Monograph v. 32, 893 p. and atlas of 27 sheets folio.

Harrison, S., and Fritz, W. J., In Preparation, Depositional feature of March 1982 Mount St. Helens mudflows. Manuscript submitted to Nature.

Holmes, W. H., 1878, Report on the geology of the Yellowstone National Park: U.S. Geological and Geographic Survey, Territories of Wyoming and Idaho (1883 edition), Twelfth Annual Report, part 2, p. +—57.

______, 1879, Fossil forests of the volcanic Tertiary formations of Yellowstone National Park: U.S. Geological Survey of the Territories Bulletin v. 2, p. 125-132.

Howell, D. G., and Link, M. H., 1979, Eocene conglomerate sedimentology and basin analysis, San Diego and the southern California borderland: Journal of Sedimentary Petrology, v. 49, p. 0517-0540.

Jensen, C., 1931, Studies of the Early Basic Breccia of northeastern Yellowstone M.A. Thesis): University of California, 50p.

Kuenzi, W. D., Horst, O. H., and McGehee, R. V., 1979, Effect of volcanic activity on fluvial-deltaic sedimentation in a modern arc-trench gap, Southwestern Guatemala: Geological Society of America Bulletin, Part I, v. 90, p. 827-838.

Knowlton, F. H., 1899, Fossil flora of the Yellowstone National park: U.S. Geological Survey Monography, v. 32, p. 651-791.

Lesquerex, L., 1872, An enumeration with descriptions of some Tertiary fossil plants from specimens procurred in the explorations of Dr. F. V. Hayden in 1870: U.S. Geological and Geographical Survey of the Territories Annual Report, v. 5, p. 282-318.

Macomber, B. E., 1956, Geology of the Soda Butte are Yellowstone National Park, Wyoming, (M.S. Thesis): Northwestern University, 167 p.

McKee, B., 1972, Cascadia: McGraw-Hill Book Co., New York, 394 p.

Parsons, W. H., 1974, Volcanic rocks of the Absaroka-Yellowstone region: *In* Barry, V., ed., Rock Mechanics: The American northwest. Third International Congress on Rock Mechanics, Expedition Guide, p. 94-101.

Prostka, H. J., Blank, R. R. Christiansen, R. L., and Ruppel, E. T., 1975, Geologic map of the Tower Junction Quadrangle, Yellowstone National Park, Wyoming and Montana: U.S. Geological Survey, Map GQ-1247.

______, Ruppel, E. T., and Christiansen, R. L., 1975, Geologic map of the Abiathar Peak Quadrangle, Yellowstone National Park, Wyoming and Montana: U.S. Geological Survey, Map GQ-1277.

Read, C. B., 1930, Fossil floras of Yellowstone National Park, Part 1, Coniferous woods of Lamar River Flora: Carnegie Institue of Washington Publication, v. 461, p. 1-19.

Retallack, G., 1981, Reinterpretation of the depositional environment of the Yellowstone fossil forests: Comment: Geology, v. 9, p. 52-53.

Schmicke, H., 1967, Graded Lahars in the type sections of the Ellensburg Formation, south-central Washington: Journal of Sedimentary Petrology, v. 37, p. 438-448.

Shelton, J. W., Burman, H. R., Noble, R. L., 1974, Directional features in braided-meandering-stream deposits. Cimarron River, North-central Oklahoma: Journal of Sedimentary Petrology, v. 44, p. 114-117.

______, and Noble, R. L., 1974, Depositional features of braided-meandering stream deposits: The American Association of Petroleum Geologists Bulletin, v. 58, p. 741-752.

Smedes, H. W., and Prostka, H. J., 1972, Stratigraphic framework of the Absaroka Volcanic Supergroup in the Yellowstone National Park region: U.S. Geological Survey Professional Paper 729-C, 33 p.

Taylor, E. M., 1977, The Clarno Formation — A record of Early Tertiary volcanism in Central Oregon: Geological Society of America Abstracts with Programs, v. 9, p. 768.

USGS, 1972, Geologic map of Yellowstone National Park 1:125,000: Miscellaneous Geologic Investigations Map I-711.

Vessell, R. K., Davies, D. K., and Foley, M. G., 1977, Sedimentology and hydrology of flood flows on the active Volcano Fuego, Guatemala: Geological Society of America Abstracts with Programs, v. 9, p. 1210-1211.

Weed, W. H., 1892, The fossil forests of the Yellowstone: School of Mines Quarterly, v. 13, p. 230-236.

Wheeler, E., Scott, R. A., and Barghoorn, E. S., 1977, Fossil dicotyledonous woods from Yellowstone National Park: Journal of the Arnold Arboretum, v. 58, p. 280-302.

______, Scott, R. A., and Barghoorn, E. S., 1978, Fossil dicotyledonous woods from Yellowstone National Park, II: Journal of the Arnold Arboretum, v. 59, p. 1-26.

Thirty-Third Annual Field Conference — 1982
Wyoming Geological Association Guidebook

GENESIS OF EARLY TERTIARY EXOTIC METAQUARTZITE CONGLOMERATES IN THE ABSAROKA-BIGHORN REGION, NORTHWEST WYOMING

MARY J. KRAUS[1]

INTRODUCTION

Recent studies of early Tertiary sediments in the western Bighorn Basin of Wyoming have focused attention on thick and laterally extensive quartzite cobble conglomerates. Earlier workers interpreted these as alluvial fan deposits with an apex lying in the area presently covered by the Absaroka Range. However, a combination of field and laboratory work indicates that these units were deposited by gravelly braided streams that periodically introduced exotic metaquartzite clasts to the western Bighorn Basin from a distant western source. This report describes the occurrence, stratigraphy, and sedimentology of these conglomeratic bodies and is presented in advance of a more detailed and complete study of the depositional environments and paleohydraulics of these deposits and their significance to the Tertiary depositional and tectonic history of northwest Wyoming. Some of these conglomerates are gold bearing and, because a considerable volume probably lies buried beneath stratified volcanic rocks of the Absaroka Range, the sedimentologic origin of these conglomerates is of interest.

GENERAL GEOLOGY

Introduction

The Fort Union Formation is a Paleocene alluvial unit dominated by sandstones, drab mudstones, and thin coals. In the Grass Creek area of the western Bighorn Basin, the Fort Union Formation rests with apparent conformity on Upper Cretaceous Lance fluvial sediments. The two units are separated by angular unconformity in other areas of the western Bighorn Basin (*e.g.*, Hewett, 1914). The lower Eocene Willwood Formation is distinguished from the Fort Union by the absence of coals and the presence of brightly variegated, rather than drab, mudstones. Willwood sediments overlap all underlying rocks with an angular unconformity in the western part of the basin. The Willwood Formation intertongues with the overlying lacustrine Tatman Formation east and northeast of Meeteetse, and, along the western margin of the basin, Tatman or post-Tatman lacustrine sediments or the volcanic middle Eocene Aycross Formation rest unconformably on Willwood rocks (Bown, 1982).

Previous Investigations

Quartzite conglomerates were first described in the Grass Creek-Meeteetse region by Hewett (1926), who recognized three persistent beds of quartzite conglomerate in the Fort Union Formation as well as a zone of quartzite conglomerate near the local base of the Willwood Formation. Hewett ascribed deposition of the Willwood unit to alluvial fans spreading from the west. On the basis of Hewett's observations and measured sections, Rea and Barlow (1975) depicted a Paleocene alluvial fan in the Meeteetse area in their paleogeographic reconstructions for the lower part of the Fort Union Formation.

Pierce and Andrews (1941) and Neasham and Vondra (1972) recorded similar conglomerates in the Willwood Formation in other areas of the western Bighorn Basin. More detailed analysis of a single conglomerate body in the Willwood Formation was undertaken by Young (1972). He concurred with Hewett (1926) and Neasham and Vondra that the coarse detritus and crudely radial paleocurrent that he recognized for the Willwood conglomerates resulted from deposition as an alluvial fan that had an apex in the west under the area presently occupied by the eastern Absaroka Range.

Fort Union quartzite conglomerates also occur north of Cody and these were described by Lindsey (1972). His studies, however, suggested that these are unrelated to the other Fort Union quartzite conglomerates further south.

[1] Department of Geological Sciences, The University of Colorado, Boulder, Colorado 80309

Location and Distribution

The principal exposures of Fort Union quartzite conglomerates are restricted to the Grass Creek area of the western Bighorn Basin (Fig. 1), where the conglomeratic zone occurs in the basal part of the formation. A previously unreported outcrop of quartzite conglomerate was discovered during the course of this study farther to the east, in Secs. 22 and 23, T. 45 N., R. 95 W. This conglomerate, however, lies considerably higher in the Fort Union sequence and thus reflects a younger episode of

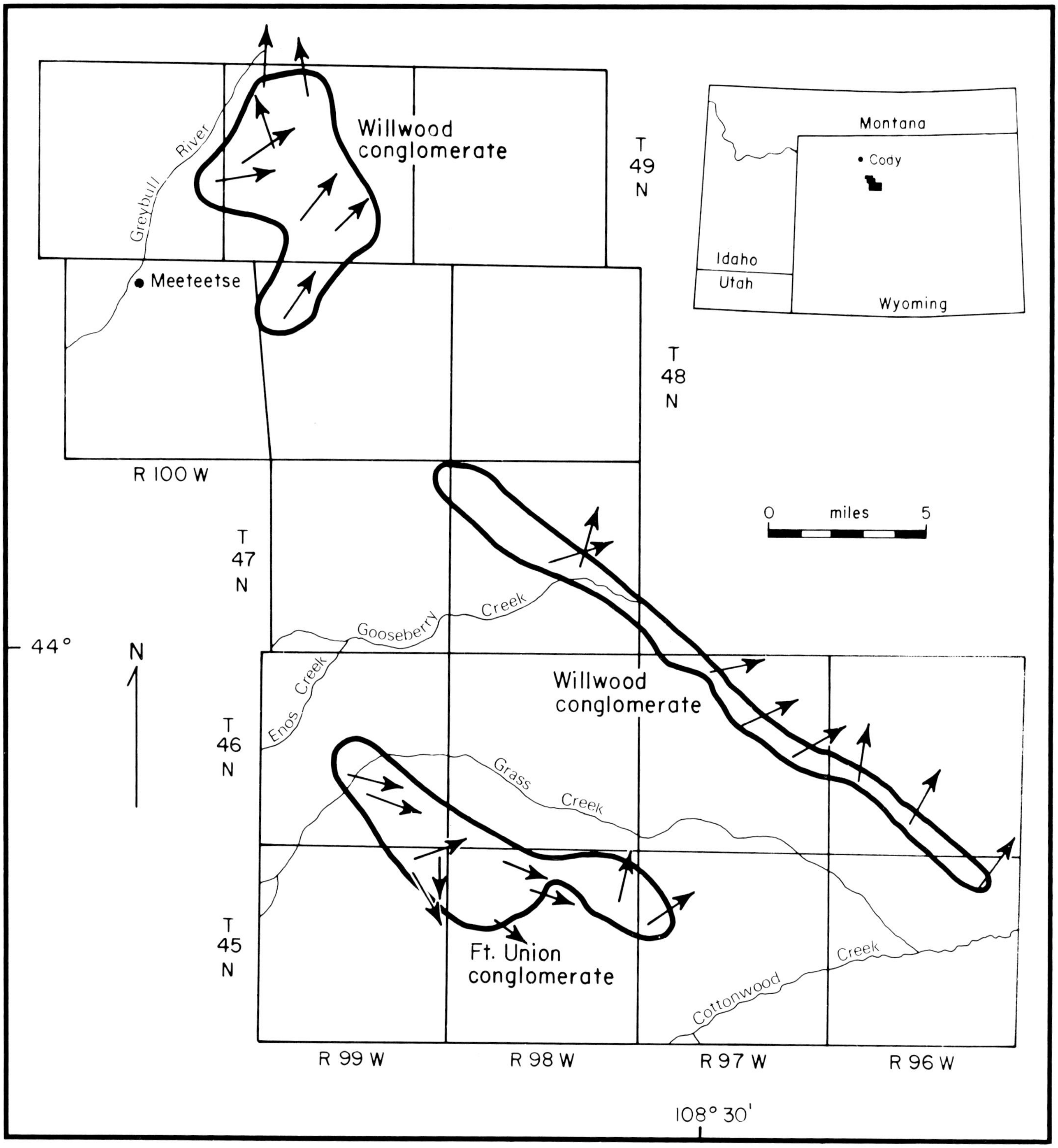

Figure 1: Map showing distributions of major Fort Union and Willwood quartzite conglomerate exposures, in western part of Bighorn Basin. Arrows indicate paleocurrent directions calculated from imbricated cobbles.

coarse quartzite deposition.

Similar conglomeratic zones are more widespread in the Willwood Formation. Extensive exposures of Willwood conglomerates occur throughout the Meeteetse area (Fig. 1) and in areas to the southeast along Blue Mesa. Areally smaller and less well exposed Willwood quartzitic conglomerate outcrops west of Meeteetse in T. 49 N., R. 102 W.

The conglomerates discussed in this report are confined to the major Fort Union and Willwood exposures outlined in Figure 1.

Stratigraphy

Hewett (1926) recorded three major zones of quartzite conglomerate in the Fort Union Formation. In some parts of the Grass Creek area, he recognized a conglomerate at the base of the formation, as well as a second horizon slightly higher in the section. Hewett also noted a quartzite conglomerate towards the top of the Fort Union Formation west of Meeteetse. My field investigations suggest that there is only one conglomerate in the Grass Creek area. This occurs at or near the base of the Fort Union, and in some localities it is not a single bed of conglomerate, but, rather, is a zone with several conglomeratic layers separated by intervening sandstones up to 6 m thick. Hewett's third conglomerate, in the Meeteetse area, lies not in the Fort Union but rather in the Willwood Formation.

In areas of good outcrop, two of four conglomerate sheets typically occur within a 53-76 m thick zone near the local base of the Willwood Formation. The conglomerate beds range from approximately 3 to 15 m in thickness, and are interbedded with mudstones and sandstones that are commonly 3-9 m thick. This conglomeratic zone can be traced from the Meeteetse area southeast as far as the southern tip of Blue Mesa in T. 45 N., R. 96 W. Hewett (1926) believed that the lowest occurring Willwood quartzite conglomerate marked the unconformable base of the formation, and his Fort Union-Willwood contact on geologic maps of the western Bighorn Basin reflects this concept. Field investigations during the course of this study, however, indicate that the lowest conglomerate occurs at least 15 m above the angular unconformity marking the Fort Union-Willwood contact. Diagnostic, brightly variegated Willwood mudstones occur beneatth this lowest conglomerate and an angular relationship with Fort Union beds is visible well below the conglomerate.

Mammal fossils are unknown from areas in which the Fort Union basal conglomerate is exposed. Fossil plants are known but have not been studied since Knowlton (*in* Hewett, 1926). The Lance Formation, which directly and apparently conformably underlies the quartzite conglomerate, is Late Cretaceous in age, and the Fort Union sequence in the Bighorn Basin is regarded to be wholly Paleocene. The conglomerate zone at the base of the Fort Union Formation is thus either latest Cretaceous or earliest Paleocene in age. Mammal fossils are well known from mudstones surrounding the Willwood conglomerates. On this basis, these conglomerates were deposited during the late Wasarchian (Lostcabinian) land mammal age and are late early Eocene (T. M. Bown and K. D. Rose, oral pers. comms., 1980).

Lithology

Fort Union and Willwood conglomerates possess a distinctive lithology. At least 75% of the cobbles and pebbles are quartzite and a sizeable proportion of these are metaquartzite. Young (1972) demonstrated that 90% of a random selection of quartzite clasts from Willwood conglomerate localities showed stretched quartz grains with sutured boundaries. There is a wide variety of quartzite colors including black, white, grey, green, pink, brown, and yellow.

Lesser amounts of chert, limestone, vein quartz, sandstone, and silicified wood clasts comprise the remainder of the cobble and pebble fraction. The conglomerates are also characterized by minute quantities of fine grained detrital gold that occurs as matrix particles (Antweiler and Love, 1967).

SEDIMENTOLOGY

Fort Union Formation Conglomerate

Geology: The basal Fort Union conglomerate in the western half of the Grass Creek area is a sheet-like body with a relatively uniform thickness ranging from 11 to 15 m. The conglomerate invariably overlies a sandstone with an irregular, erosional contact with up to 1 m of relief (Fig. 2). The quartzite conglomerate is itself overlain by sandstone with a sharp, planar contact (Fig. 3), and there is an abrupt textural transition from cobble conglomerate to medium sandstone.

The conglomerate zone in the eastern Grass Creek area (the eastern half of T.45N., R. 98 W) is thicker and more complex. One typical measured section consists of 18 to 21 m dominated by cobble conglomerate, followed by 6 m of sandstone and 10.5 m of interbedded conglomerate and sandstone.

Texture: Fort Union basal conglomerates are generally framework supported, indicating that the sandy matrix probably filtered into the interstices following deposition of the clasts. Boulders are rare, and mean clast size is in the cobble range (64-256 mm; Fig. 4). Most clasts are well

Figure 2: Basal erosional contact between Fort Union quartzite conglomerate and Lance sandstone; SE¼, Sec. 21, T. 46 N., R. 99 W. View to the northeast. Hammer gives scale.

rounded anf tabular in shape (Zingg, 1935 classification). Badly crushed clasts (Fig. 4), reflecting tectonic deformation (Clifton, 1965) are common; cobbles with pitted surfaces, resulting from pressure solution at points of grain to grain surfaces (Kuenen, 1942), are abundant. The matrix consists of moderately sorted, subangular to sub-rounded, medium to coarse sand.

Sedimentary Structures: Massive conglomerates or conglomerates exhibiting crude horizontal stratification are

Figure 3: Sharp, planar contact between Fort Union quartzite conglomerate and overlying Fort Union sandstone; N½, Sec. 9, T. 45 N., R. 98 W. Note cross-stratified set 5 m in thickness. View to the north.

common. Well developed clast imbrication is preserved in some cases (Fig. 5) and was utilized in determining paleocurrent directions. A second dominant sedimentary structure is large scale planar cross-stratification (Fig. 6). Individual sets are commonly 4-5 m thick but attain nearly 7 m, very unusual for cross-stratified conglomerate or gravel (Miall, 1977). Cross-stratified sets are either solitary or occur stacked in groups of two or at most three sets. Cross-strata are texturally heterogeneous (Fig. 6A), which enhances the cross-stratified appearance of the conglomerate. Strata dominated by cobbles (64-256 mm) alternate with pebbly (4-64 mm) strata. Cross grained cross-strata are typically 25 to 50 cm thick, whereas the pebbly cross-strata range from 15 to 40 cm in thickness. Both maintain a relatively uniform thickness down dip.

Figure 4: Cobbles in basal Fort Union conglomerate, NE¼, Sec. 5, T. 45 N., R. 98 W. Note several badly fractured clasts and pitted surfaces on clasts. 15 cm long pen shows scale.

Interbedded sandstones are neither common nor thick in the western and northwestern parts of the Grass Creek conglomerate. However, the eastern exposures exhibit zones of conglomerate interbedded with sandstones ranging from 0.3 to 2 m in thickness. Sandstone facies generally have a sharp, planar lower boundary and a sharp, erosional upper contact with overlying conglomerates. Some sandstones are horizontally stratified with well developed parting lineation, whereas other sandstone bodies exhibit large scale planar or trough cross-stratification. Sets are typically 0.5-1 m thick and may have some pebbly cross-strata.

Paleocurrents: Paleocurrent determinations were

Figure 5: Well preserved clast imbrication in Fort Union quartzite conglomerate; NW¼, Sec. 18, T. 45 N., R. 97 W. Flow was left to right. View to the north. Note tabular shape of many clasts.

made on imbricated clasts at a number of Fort Union conglomerate localities and are plotted on Figure 1. The mean azimuth for all localities is 103°, indicating a general paleoflow direction to the east-southeast.

Willwood Formation Conglomerates

Geometry: Individual Willwood conglomerates in the conglomeratic zone near the base of the formation are also sheet-like, with a much greater lateral extent than thickness. At their thickest, approximately 15 m, some of the Willwood conglomerate sheets are comparable to the basal Fort Union conglomerate at its maximum thickness. In places, however, Willwood quartzite conglomerate layers are no more than 3 m thick.

Willwood conglomerates have an irregular, erosional lower contact with underlying mudstones or sandstones. At their upper contact, Willwood conglomerates grade into finer grained facies.

Texture: Willwood conglomerates are texturally similar to Fort Union quartzite conglomerates. Boulders, however, are not uncommon at some Willwood localities, whereas they are rare throughout the Fort Union conglomerate. The coarsest Willwood conglomerates occur in Sec. 23, T. 47 N., R. 98 W., where clasts with long axes up to 40 cm were deposited. Quartzite conglomerates east and northeast of Meeteetse are generally less coarse than those to the southeast in the Gooseberry-Cottonwood Creek area.

Sedimentary Structures: Cobble conglomerates that are massive or that show crude horizontal bedding with preserved imbricated clasts are prevalent in the southeastern Willwood exposures between Gooseberry and Cottonwood Creeks. Thin, laterally discontinuous sandstone layers occur scattered through the conglomerate (Fig. 7). Some planar cross-stratified conglomerate is also present, but it is neither as common nor is it developed on as large a scale as in the Fort Union Formation. Sets are generally 1-2 m thick, markedly smaller than the 6 m Fort Union sets. With the exception of scale, Willwood large

Figure 6: Large scale cross-stratification in Fort Union conglomerate. 6A — Well developed alternating fine and coarse cross-strata in a relatively small cross-stratified set, SW¼, SW¼, Sec. 7, T. 45 N., R. 98 W. View to the southeast. 6B — Large scale cross-stratification at top of Fort Union conglomerate, NE¼, Sec. 2, T. 45 N., R. 99 W. This set is 3 m thick and cross-stratification is accentuated by sandstone layers. View to the northwest.

scale cross-stratification is similar to that in the Fort Union conglomerate.

Sandstone, pebbly sandstone, and pebble conglomerate facies are common at many Willwood localities, especially in the Meeteetse area where overall grain sizes are smaller than in Willwood conglomerate exposures to the southeast. Pebbly sandstone and pebble conglomerate facies typically exhibit large scale cross-stratification (Fig. 8). Sets range from approximately 0.5 to 1 m in average thickness and are either trough or planar in shape. Sandstones display horizontal stratification or large scale, usually planar, cross-stratification. Upper sandstone boundaries are scoured and followed in sequence by conglomerate or pebbly sandstone facies (Fig. 8).

Paleocurrents: Paleocurrent directions were determined from a number of Willwood localities on the basis of imbricated cobbles. The overall paleoflow trend was to the northeast. The mean azimuth for the conglomerates in the vicinity of Meeteetse is 29°, which corresponds closely with the mean azimuth of 45° calculated for Willwood conglomerates in the Gooseberry Creek-Cottonwood Creek region.

Figure 8: Conglomerate, large scale cross-stratified pebbly sandstone and horizontally stratified and cross-stratified sandstone facies in Willwood conglomerate, southeast of Meeteetse; N½, NW¼, Sec. 33, T. 49 N. , R. 99 W. Note upper erosional boundary of sandstone facies. Hammer shows scale. View to the northeast.

Figure 7: Thick, coarse Willwood conglomerate with thin, laterally discontinuous sandstone layers; SW¼, Sec. 15, T. 47 N., R. 98 W. View to the east.

DEPOSITIONAL ENVIRONMENT

Several lines of evidence demonstrate that the Willwood and Fort Union conglomerates were not deposited on alluvial fans. These include: the sheet geometry of Fort Union and Willwood conglomerates, the absence of disordered, matrix supported conglomerates, and the lack of significant grain size decreases in a downslope direction. Furthermore, there is no observable radial paleocurrent pattern to either conglomeratic deposit (Fig. 1). Alluvial fan deposits are usually thick wedges, reflecting deposition in tectonically active areas; they exhibit a rapid downfan decrease in grain size from gravel to sand; and they have debris flow deposits in proximal fan areas (Rust, 1979).

The very coarse sequences at many Fort Union and Willwood localities most closely resemble the braided stream succession of textures and sedimentary structures described from the Scott outwash gravels in Alaska (Boothroyd and Ashley, 1975; Maill, 1977). Massive or crude horizontally stratified gravel with preserved imbricate fabric is characteristic of proximal braided stream environments such as the Scott. Deposition occurs on low relief longitudinal bars and stratification parallels the upper bar surface. Massive-appearing deposits apparently result from deposition of texturally homogeneous gravels (Rust, 1975).

Gravel bars typically lack slip faces, and consequently

large scale cross-stratified gravels or conglomerates are not common. Church and Gilbert (1975) noted that foreset development at bar fronts is dependent on the relation between water depth and grain size. Where depth is not at least an order of magnitude larger than grain size, slip faces do not form.

The significance of water depth to foreset development in coarse material is also determined by other studies. Costello and Walker (1972), for example, reported large scale cross-stratified sandy gravel with foresets 2-10 feet in height and dipping 10-15°. Large scale cross-stratified gravels with sets up to 4 m thick were observed by Eynon and Walker (1974). In both cases, the foresets occurred downstream of thick sequences of massive or horizontally bedded gravel interpreted as longitundinal bars. The authors concluded that significant upward building of the bars resulted in rather deep basins on the lee sides of the bars and tall forests grew downstream from the bar fronts.

Unusually large scale, water deposited cross-strata were reported by Malde (1968) from the Snake River Plain in Idaho. There, large scale, dipping beds of sand and boulders, and sets of cross-stratified sand up to 30 m thick were deposited by the Late Pleistocene Lake Bonneville flood.

Cross-bedded sets in the Fort Union Formation basal conglomerate are significantly larger than those reported for any modern or ancient braided gravels known to me, but smaller than those recorded by Malde (1968) for the Lake Bonneville flood sediments. Minimum water depths necessary for deposition of the Fort Union foresets must have been comparable to set thickness, or at least 7 m for deposition of the thickest sets. Such deep flows in a braided stream environment could reflect either lateral confinement of flow in some areas, thus increasing the effective depth of the water, or exceptional, even catastrophic, flood events.

The abundant massive or horizontal bedding and the absence of gravel cross-bedding with foresets taller than 2 m in southeast Willwood exposures indicate that most deposition occurred on longitudinal bars and the braided streams were considerably more shallow than those depositing the Fort Union basal quartzite conglomerate.

The finer, more varied facies in the Willwood conglomerates near Meeteetse bear similarities to deposits from the Donjek River (Williams and Rust, 1969; Rust, 1972; Miall, 1977). Donjek braided stream deposits are developed more distally to source areas than are Scott type deposits, and they are typified by a variety of depositional facies, including small and large scale cross-stratified sand, gravels which are cross-stratified or horizontally stratified and which may have laterally discontinuous layers of sand or silt, and interbedded sands and silts. Gravels form a dominant part of the Willwood sequence in the Meeteetse area; however, lower energy facies, similar to those seen in the Donjek model, are also common. The grain sizes and sedimentary structures in these Willwood conglomerates near Meeteetse indicate shallow braided stream deposition with lower energy flows than occurred in the stream which deposited Willwood conglomerates in the Gooseberry Creek-Cottonwood Creek region.

PROVENANCE

Thick metaquartzite cobble and boulder conglomerates occur in other Upper Cretaceous through lower Tertiary sequences in northwest Wyoming, northeast Idaho, and south-central Montana. Examples include the Divide Quartzite Conglomerate (Upper Cretaceous-Paleocene) of southern Montana and northeast Idaho (Ryder, 1967), the Upper Cretaceous Harebell and Upper Cretaceous and Paleocene Pinyon Formations (Lindsey, 1972; Love, 1973). These conglomerates are characterized by abundant quartzite cobbles and boulders and, in the Harebell and Pinyon Formations, by small amounts of fine grained detrital gold (Antweiler and Love, 1967). similarities in trace element geochemistry of the gold and regional paleocurrent trends suggest a possible common stratigraphic source for all of these quartzite conglomerates. Exposed and known subsurface quartzite sources are too small to have supplied the enormous volume of quartzite conglomeratic debris, but the paleocurrent indicators, coupled with the absence of any obvious sources for the quartzite, led Love (1956), Eardley (1960), and Love and Reed (1968) to suggest that the quartzite clasts were derived from a now buried upland source. This was termed the Targhee uplift by Love and Reed. According to Love (1973), the Targhee uplift lies northwest of the Teton Range, and is now buried beneath upper Tertiary and Quaternary sediments in the Snake River plain in westernmost Wyoming and eastern Idaho.

Though Fort Union and Willwood quartzite conglomerates discussed in this paper lie 150 km east of the Jackson Hole area, they strongly resemble these other metaquartzite conglomerates. The presence of abundant metaquartzite cobbles and interstitial gold with similar trace element geochemistry (Antweiler and Love, 1967) suggests that the Bighorn Basin conglomerates were ultimately derived from the source shared by the Divide Quartzite, and Harebell and Pinyon Formations. It is also possible that some of them, especially the younger conglomerates, are second order deposits derived from reworking of the Harebell and Pinyon conglomerates. Either interpretation is supported by paleocurrent patterns of both the Fort Union and Willwood conglomerates which indicate a western source.

PALEOGEOGRAPHY

As outlined above, quartzite conglomerates in the Fort Union and Willwood Formations reflect episodic influxes of very coarse detritus into the Bighorn Basin from a distant western source. The nearest preserved exposures of the closely similar Harebell and Pinyon conglomerates lie approximately 150 km to the west. In the Absaroka-Bighorn area, deposition occurred in gravelly braided streams that were competent enough to transport cobbles and even boulders. The volumetrically large conglomeratic deposits of these west-to-east flowing streams in the western Bighorn Basin demonstrate that the intervening region (between the Teton-Bighorn areas), now covered by middle and later Eocene volcanic igneous and volcaniclastic rocks of the Absaroka Range, was deeply eroded and tectonically quiescent from at least earliest Paleocene through early Eocene time. This region was probably characterized by a broad, easterly dipping paleoslope with a gradient sufficent for the transport of very large volumes of coarse clastic material.

ACKNOWLEDGEMENTS

I am especially grateful to Thomas M. Bown (U.S.G.S., Denver) and J. D. Love (U.S.G.S., Laramie) for discussion and support during the course of this study. I also thank J. D. Love for introducing me to the Harebell and Pinyon Formations in the Jackson Hole area. Able field assistance was provided by Christine West and Mark Barton.

This research was supported in part by a Sigma Xi Grant-in-Aid of Research, Geological Society of America Research Grant, the Walker Van Riper Fund of the University of Colorado Museum, and a University of Colorado Doctoral Fellowship.

REFERENCES CITED

Antweiler, J. C., and Love, J. D., 1967, Gold-bearing sedimentary rocks in northwest Wyoming — A preliminary report: U.S. geol. Survey Circ. 541, 12 pp.

Boothroyd, J., and Ashley, G. M., 1975, Processes, bar morphology and sedimentary structures on braided outwash fans, northeastern Gulf of laska, *in* Jopling, A. V., and McDonald, B. C., eds., Glaciofluvial and glaciolacustrine sedimentation: Soc. Econ. Paleont. Mineral Spec. Pub., no. 23, p. 193-222.

Bown, T. M., 1982, Geology, paleontology, and correlation of Eocene volcaniclastic rocks, southeast Absaroka Range, Hot Springs County, Wyoming: U.S. Geol. Survey Prof. Paper 1201-A, 75 pp.

Church, M., and Gilbert, G., 1975, Proglacial fluvial and lacustrine environments, *in* Jopling, A. V., and McDonald, B. C., eds., Glaciofluvial and glaciolacustrine sedimentation: Soc. Econ. Paleont. Mineral. Spec. Pub., no. 23, p. 22-100.

Clifton, H. E., 1965, Tectonic polish of pebbles: Journ. Sed. Petrology, v. 35, p. 867-873.

Costello, W. R., and Walker, R. G., 1972, Pleistocene sedimentology, Credit River, southern Ontario: A new component of the braided river model: Jour. Sed. Petrology, v. 42, p. 389-400.

Eardley, A. J., 1960, Phases of orogeny in the deformed belt of southwestern Montana and adjacent areas of Idaho and Wyoming: Billings Geol. Soc. Guidebook, 11th Ann. Field Conf., p. 86-91.

Eynon, G. and Walker, R. G., 1974, Facies relationships in Pleistocene outwash gravels, southern Ontario: A model for bar growth in braided rivers: Sedimentology, v. 21, p. 43-70.

Hewett, D. F., 1914, The Shoshone River section: U.S. Geol. Survey Bull. 54, p. 89-113.

________, 1916, Geology and oil and coal resources of the Oregon Basin, Meeteetse, and Grass Creek Basin Quadrangles, Wyoming: U.S. Geol. Survey Prof. Paper 145, 111 pp.

Kuenen, P. H., 1942, Pitted pebbles: Leidse Geol. Meded., v. 13, p. 189-201.

Lindsey, D. A., 1972, Sedimentary petrology and paleocurrents of the Harebell Formation, Pinyon Conglomerate, and associated coarse clastic deposits, northwestern Wyoming: U.S. Geol. Survey Prof. Paper 734-B, 68 pp.

Love, J. D., 1956, New geologic formation names in Jackson Hole, Teton County, northwestern Wyoming: Am. Assoc. Petroleum Geol. Bull., v. 40, p. 1899-1914.

______, 1973, Harebell Formation (Upper Cretaceous) and Pinyon Conglomerate (Uppermost Cretaceous and Paleocene), northwestern Wyoming: U.S. Geol. Survey Prof. Paper 734-A, 54 pp.

______, and Reed, J. C., 1968, Creation of the Teton landscape — The geologic story of Grand Teton National Park: Moose, Wyo., Grand Teton Nat. Histroy Assoc., 120 pp.

Malde, H. E., 1968, The catastrophic Late Pleistocene Bonneville flood in the Snake River Plain, Idaho: U.S. Geol. Survey Prof. Paper 596, 52 pp.

Miall, A. D., 1977, A review of the braided-river depositional environment: Earth Sci. Rev., v. 13, p. 1-62.

Neasham, J. W., and Vondra, C. F., 1972, Stratigraphy and petrology of the lower Eocene Willwood Formation, Bighorn Basin, Wyoming: Geol. Soc. America Bull., v. 83, p. 2167-2180.

Pierce, W. G., and Andrews, D. A., 1941, Geology and oil and coal resources of the region south of Cody, Park County, Wyoming: U.S. Geol. Survey Bull. 921-B, p. 99-180.

Rea, B. D., and Barlow, J. A., 1975, Upper Cretaceous and Tertiary rocks, northern part of Bighorn Basin, Wyoming and Montana: Wyoming Geol. Assoc. Guidebook, 27th Ann. Field Conf., p. 63-71.

Rust, B. R., 1972, Structure and process in a braided river: Sedimentology, v. 18, p. 221-245.

______, 1975, Fabric and structure in glaciofluvial gravels, *in* Jopling, A. V., and McDonald, B. C., eds., Glaciofluvial and glaciolacustrine sedimentation: Soc. Econ. Paleont. Mineral. Spec. Pub., no. 23, p. 238-245.

______, 1979, Facies Models 2. Coarse alluvial deposits, *in* Walker, R. G., ed., Facies models: Geoscience Canada, Reprint Series 1, p. 9-21.

Ryder, R. T., 1967, Lithosomes in the Beaverhead Formation, Montana-Idaho: Montana Geol. Soc. Guidebook, 18th Ann. Field Conf., p. 63-70.

Williams, P. F., and Rust, B. R., 1969, The sedimentology of a braided river: Jour. Sed. Petrology, v. 39, p. 649-679.

Young, M., 1972, Willwood metaquartzite conglomerate in a southwestern portion of the Bighorn Basin, Wyoming: M. Sci. Thesis, Ames, Iowa, Iowa State University, 71 pp.

Zingg, T., 1935, Beitrage zur Schotteranalyse: Min. Petrog. Mitt. Schweiz., v. 15, p. 39-140.

Thirty-Third Annual Field Conference — 1982
Wyoming Geological Association Guidebook

GEOLOGY OF PRE-TERTIARY ROCKS IN THE NORTHERN PART OF YELLOWSTONE NATIONAL PARK, WYOMING*

BY EDWARD T. RUPPEL[1]

ABSTRACT

Metamorphic and sedimentary rocks of pre-Tertiary age are exposed over an area of about 200 square miles in the northern part of Yellowstone National Park, Wyo. and Mont., mostly in remote and seldom-visited parts of the park along its north boundary. The main outcrops are in the Gallatin Range in the northwest corner of the park, where the entire section of rocks from Precambrian to Late Cretaceous age is exposed; in the area around Mammoth Hot Springs and Mount Everts, where Jurassic and Cretaceous rocks are exposed; and in the north-central and northeastern parts of the park where rocks of Precambrian and Paleozoic age are exposed.

These outcrops consist of crystalline metamorphic rocks of Precambrian age and sedimentary rocks that represent every system except the Silurian. The Precambrian crystalline rocks are mainly granitic gneiss and quartz-biotite schist, but they also include amphibolite and pegmatite. These rocks are overlain by a mostly conformable sequence of sedimentary rocks about 8,000 feet thick, consisting of about 3,000 feet of marine rocks of Paleozoic age, a few hundred feet of Triassic and Jurassic nonmarine and marine rocks, and about 4,000 feet, or half the exposed sequence, of marine and nonmarine Cretaceous rocks. The Paleozoic rocks — limestone, dolomite, and, in subordinate amount, clastic rocks — range in age from Middle Cambrian to Permian.

The Mesozoic sedimentary rocks consist of marine deposits of Triassic and Jurassic age — mainly shale, sandstone, and, in subordinate amount, limestone — and overlying nonmarine mainly clastic rocks of later Jurassic and Early Cretaceous age and a thick sequence of marine and nonmarine clastic rocks of later Cretaceous age. Triassic and Jurassic rocks together are only a little more than 5000 feet thick. The 4,000 feet of Cretaceous rock is a monotonous sequence of shale, mudstone, and sandstone, broken by a few beds of conglomerate and limestone, and, in the upper part of the section, by a few beds of coal, carbonaceous sandstone, and shale. The youngest exposed pre-Tertiary sedimentary rocks are sandstone and conglomerate of the Landslide Creek Formation, which consists mainly of volcanic fragments.

The sedimentary rocks in the northwestern part of the park have been intruded by igneous rocks that form stocks, laccoliths, sills, and a few dikes. The oldest of these intrusives is the Mount Holmes stock of fine-grained biotite quartz monzonite, a steep-sided stock that domed its roof of Middle Cambrian sedimentary rocks and intensely hornfelsed both the roof rocks and its own chilled margin. It apparently was emplaced in early Tertiary time.

This paper is taken from U.S.G.S. Professional Paper 729-A. 66 pages reprinted with permission of the author and the director of the U.S.G.S.

[1]U.S. Geological Survey, Denver, Colorado 80225

The laccoliths, sills, and related dikes were emplaced after the Mount Holmes stock, but still in early Tertiary time, and these bodies of biotite rhyodacite porphyry and biotite quartz latite porphyry are the most widespread intrusive masses in the northern part of Yellowstone National Park. The Indian Creek laccolith, emplaced in Cambrian rocks near the south end of the Gallatin Range, is the largest of the group and is about 3 miles in diameter and more than 1,000 feet thick. The Snowshoe and Gray Peak laccoliths, emplaced in Jurassic and Cretaceous rocks in the Gallatin Range, are each about 1 mile in diameter and before erosion were probably each about 1,000 feet thick; they are surrounded by multiple sills that extend out for several miles from the laccolithic core to form an intrusive field 6-8 miles in diameter. The Snowshoe and Gray Peak laccoliths, thus, are the trunk, and the broad connected sills are the branches of a very large Christmastree laccolithic complex. The multiple thin sills and their interconnecting dikes on Electric Peak were emplaced in Cretaceous rocks. The laccoliths and sills were emplaced in shaly horizons. Their ultimate shape was possibly determined by depth — a comparatively simple laccolith if deep, multiple thin sills if shallow, and a combination of these, aChristmastree Laccolith, if intermediate in depth.
termediate in depth.

The Electric Peak stock, of Eocene age, is an irregular complex, about 1 mile across, on the east side of Electric Peak; it intrudes Upper Cretaceous sedimentary rocks and comprises several varieties of fine- to medium-grained rocks ranging in composition from calcic granodiorite to near granite, although most of the rock is granodiorite. The stock has long been considered the neck through which the volcanic rocks of Sepulcher Mountain were erupted.

The faults and folds in the pre-Tertiary rocks of northern Yellowstone National Park appear outwardly to have little relation to structural features in surrounding areas, probably because the park is in an unusual setting where several structural provinces come together. The Precambrian crystalline rocks reflect extreme deformation in Precambrian time. The sedimentary rocks are in parallel beds, which imply structural quiet throughout Paleozoic and most of Mesozoic time and to the latest Cretaceous, when they were folded at the beginning of a period of deformation that probably is still going on. The folding formed the broad, poorly defined northwest-trending Gallatin anticline, which was intruded in its axial part by both the Mount Holmes stock and the Indian Creek laccolith. Doming over the laccolith apparently caused the massive upper Paleozoic rocks to slide northward on the gently dipping Crowfoot fault zone.

In late Paleocene and Eocene time, the rocks in the northern part of Yellowstone Park were broken by steep northwest-trending reverse faults, which include (1) the faults of the north-dipping Gardiner fault zone and the associated reverse faults northeast of the Gallatin Range and (2) the south-dipping Grayling Greek fault southwest of the Gallatin Range. The Gallatin block was depressed between these bounding faults. Displacement on the Gardiner fault zone exceeds 10,000 feet, and that on the Grayling Creek fault is uncertain, but it probably exceeds 1,000 feet. The Gardiner fault zone is the southwest boundary of the uplifted Beartooth block, which is the dominant structural feature in the north-central and northeastern parts of the park.

In later Tertiary time, pre-Tertiary and younger rocks in the north-central and northeastern parts of the park were broken by northwest-trending normal faults, which now define the Lamar Valley and the south front of the Buffalo Plateau. Still later, perhaps in middle or late Pliocene time, the reverse faults and northwest-trending normal faults were broken by north-trending near-vertical normal faults, some of which have displacements of thousands of feet. These faults bound the

Gallatin Range and are represented in the north-central part of the park by the Buffalo Creek faults, and they probably control many of the thermal areas in the park. They may be reflected through younger volcanic rocks elsewhere in the park by a striking series of north-trending lineaments. The east and West Gallatin faults, major faults in the north-trending group, perhaps are continuous with somewhat similar faults south of Yellowstone National Park; north of the park they appear to merge into the northeast-trending Emigrant fault. The present Gallatin Range is an uplifted block much like the Teton Range and probably is genetically related to it. The structural features in the northern part of the park, however, do not appear to be related to most structural features in adjacent areas in Montana, and the structural pattern in and near the northern part of the park appears to reflect intersection and termination of the Snake River downwarp against the older, reverse-fault-bounded blocks of the Middle Rocky Mountains tectonic province.

INTRODUCTION

Although Yellowstone National Park is noted mainly for its widespread Cenozoic volcanic rocks, it also includes older (pre-Tertiary) sedimentary and metamorphic rocks that crop out in remote and seldom-visited parts of the park along its north and south borders. Such rocks in the northern part of the park (fig. 1) are the concern of this report.

The prevolcanic metamorphic and sedimentary rocks in the northern part of the park are exposed over an area of about 200 square miles (fig. 1). In the northwest corner of the park. Precambrian metamorphic rocks and Paleozoic and Mesozoic sedimentary rocks, intruded by sills, laccoliths, and stocks of younger igneous rocks, constitute the south end of the Gallatin Range and underlie all the high rugged peaks of this range. The Gallatin Range provides an almost complete section of sedimentary rocks ranging in age from Middle Cambrian to Late Cretaceous. Between the Gallatin Range and the west boundary of the park, Witkind (1964, 1969) mapped a few small outcrops of sedimentary rocks. East of the range, sedimentary rocks crop out near Mammoth Hot Springs and in several small areas from Bunsen Peak south to Whiterock Springs and Lemonade Creek. East of Mammoth Hot Springs, sedimentary and metamorphic rocks crop out almost continuously from Mount Everts across the Buffalo Plateau to the northeast corner of the park, an outcrop area almost as large as that of the southern part of the Gallatin Range.

These areas of metamorphic and sedimentary rocks are skirted by the major highways of the region (fig. 1): U.S. Highway 191, between Bozeman and West Yellowstone, Mont.; the West Entrance and Northeast Entrance Roads; and part of the Grand Loop Road. However, only a few outcrops can be reached easily from these highways, and most of the area is accessible only by foot or on horseback, using pack trails maintained by the National Park Service.

The geology was mapped on a topographic base (scale 1:62,500) in 1966 and 1967. About 5 months was spent in field mapping, mostly in the areas of sedimentary rocks; the metamorphic rocks were mapped only in reconnaissance.

PREVIOUS WORK

The principal early works on the metamorphic and sedimentary rocks in the northern part of the park are those by Iddings and Weed (1899) and by Weed (1896). Hayden (1873) had earlier noted many areas underlain by these rocks, and in 1878 Holmes (1883) had partly mapped most of the areas of pre-Tertiary rocks and had briefly described them. Holmes had also recognized and named the Indian Creek laccolith (Holmes, 1883, p. 24-26) and a major fault (Holmes, 1883, p. 5-6), now known as the Gardiner fault, along the valley of the Yellowstone River near its junction with the Gardner River. The studies in 1878 by Holmes (1883) and his colleagues, under the direction of Hayden, followed the earlier geological surveys made principally in thermal areas in the park in 1871 and 1872, also under Hayden's direction (Hayden, 1872, 1873). In turn, these studies were followed in the 1880's and 1890's by the complete geologic mapping of Yellowstone National Park by geologists of the U.S. Geological Survey, including Iddings and Weed (1899), under the direction of Arnold Hague.

The remoteness of most outcrops of pre-Tertiary rocks and the physical problems of working in such remote areas made the early explorations difficult (Hayden, 1872, p. 99) and also discouraged later mapping. To a large extent the work later than that by Iddings and Weed was limited to special stratigraphic studies on selected formations (Crickmay, 1936; Deiss, 1936; Cressman, 1955; Richards and Nieschmidt, 1961; McMannis, 1962; Grant, 1965; Benson, 1966). C.W. Brown (1961) published the only geologic map more recent than the maps by Iddings and Weed (1899) for the area between Mammoth Hot Springs and the Northeast Entrance; Brown's map shows the distribution of the pre-Tertiary rocks, but the emphasis of his study was on Cenozoic stratigraphy and structure, and he did not discuss the older rocks at any length. Love (1961) made a reconnaissance study of Quaternary movement on faults in Yellowstone National Park and examined many of the faults that cut the pre-Tertiary rocks in the northern part of the park.

Geologic reports on areas adjacent to the northern part of the park include maps and descriptions of pre-Tertiary

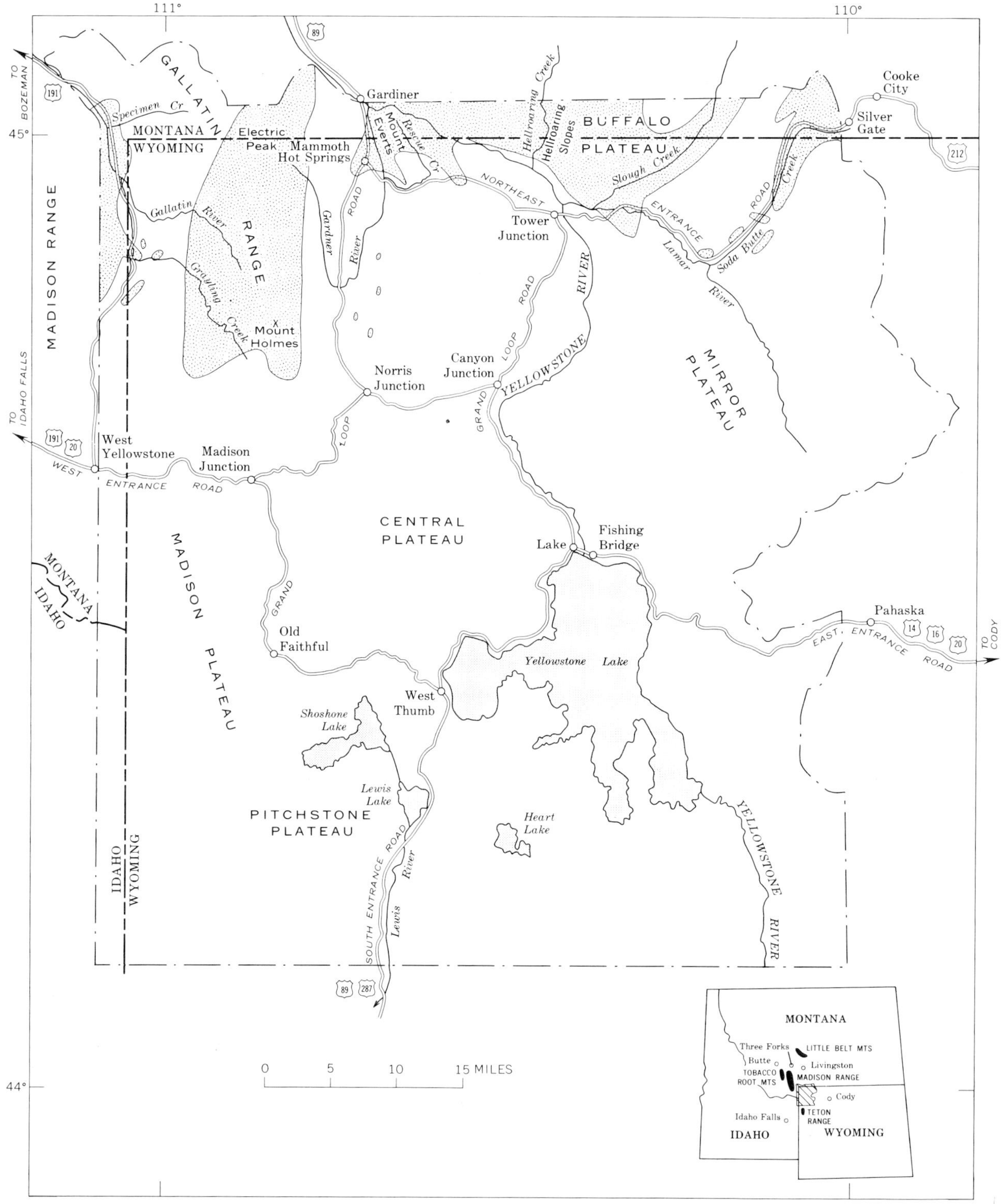

Figure 1: Areas of pre-Tertiary rocks (coarse stipple pattern), northern part of Yellowstone National Park, Wyo.

rocks that have been most helpful to me. Witkind (1964, 1969) mapped the Tepee Creek quadrangle, which includes a small part of Yellowstone National Park but is mostly in the Madison Range west of the park. Hall (1961) mapped the Gallatin Valley northwest of Yellowstone. Calvert (1912), Wilson (1934a, b), Seager (1944), L. E. Brown (1965), Brookins and Brown (1966), and Fraser, Waldrop, and Hyden (1969) have studied various aspects of the geology of the area from Cinnabar Mountain, just north of Electric Peak, to the Jardine-Crevasse Mountain mining district east of Gardiner, Mont. Lovering (1930) mapped the Cooke City, Mont., mining district near the Northern Entrance of the park, an area visited earlier by Hayden (1873, p. 47).

GEOLOGY

Pre-Tertiary rocks in the northern part of Yellowstone National Park range in age from Precambrian through Cretaceous and include rocks representative of every Paleozoic and Mesozoic system except the Silurian. Paleozoic sedimentary rocks unconformably overlie gneiss and schist of Precambrian age; the Precambrian Belt Supergroup that is so widespread in western Montana and Idaho is not present in or near the park. The sequence of sedimentary rocks is summarized in figure 2. The aggregate thickness of the sedimentary rocks is about 8,000 feet, of which about half are of Cretaceous age. All the Paleozoic rocks are marine, but the Mesozoic rocks consist of both marine and continental units.

The Precambrian rocks crop out in the southern part of the Gallatin Range and around the Buffalo Plateau; the Paleozoic rocks crop out in the Gallatin Range, and the Buffalo Plateau; the Paleozoic rocks crop out in the Gallatin Range, on the Buffalo Plateau, and in the northeast corner of the park; and the Mesozoic rocks crop out only in the Gallatin Range, near Mammoth Hot Springs, and on the just east of Mount Everts. The Precambrian rocks are strongly jointed and in most places foliated, and rocks of all ages are cut by faults of differing trends. In general, though, the sedimentary rocks are little deformed and dip gently toward the north boundary of the park.

In the Gallatin Range, the sedimentary rocks have been intruded by rhyodacitic and quartz latitic rocks that form laccoliths and sills, dioritic rocks that form dikes and a stock, and by a quartz monzonite stock at Mount Holmes. Despite the great volume of molten rock represented by these intrusive bodies, few contact metamorphic changes are visible except in the aureole surrounding the stock at Mount Holmes.

To simplify future work by others on the sedimentary

Age	Northern part of park (this report)	Terminology of Iddings and Weed (1899)	South-central part of park (J. D. Love and W. R. Keefer, written commun., 1971)
Quaternary or Tertiary			Heart Lake Conglomerate
Paleocene		Pinyon Conglomerate	Pinyon Conglomerate
Cretaceous	Landslide Creek Formation (part)	Laramie Formation	Harebell Formation
	Everts Formation		
	Eagle Sandstone	Montana Formation	Bacon Ridge Sandstone
	Telegraph Creek Formation		
	Cody Shale		Cody Shale
	Frontier Sandstone	Colorado Formation	Frontier Formation
	Mowry Shale		Mowry Shale
	Thermopolis Shale: Upper sandstone member		Muddy Sandstone Member
	Thermopolis Shale: Middle shale member		Thermopolis Shale
	Thermopolis Shale: Lower sandstone member	Dakota Formation	Rusty beds member
	Kootenai Formation		Cloverly and Morrison(?) Formations
Jurassic	Morrison Formation		
	Ellis Group: Swift Formation	Ellis Formation	Sundance Formation
	Ellis Group: Rierdon Formation		
	Ellis Group: Sawtooth Formation		Gypsum Spring Formation
Triassic	Thaynes(?) Formation	Teton Formation	Lower part Chugwater Formation
	Woodside Formation		
	Dinwoody Formation		Dinwoody Formation
Permian	Shedhorn Sandstone		Phosphoria Formation and related rocks
Pennsyl-vanian	Quadrant Sandstone	Quadrant Quartzite	Tensleep Sandstone
	Amsden Formation		Amsden Formation
Missis-sippian	Madison Group: Mission Canyon Limestone	Madison Limestone	Madison Limestone
	Madison Group: Lodgepole Limestone		
Devonian	Three Forks Formation	Three Forks Formation	Darby Formation
	Jefferson Formation	Jefferson Formation	
Ordovician	Bighorn Dolomite	Gallatin Formation	Not exposed but present in subsurface: Bighorn Dolomite
Cambrian	Snowy Range Formation: Grove Creek Ls. Mbr., Sage Ls. Mbr., Dry Creek Shale Mbr.		Gallatin Limestone
	Pilgrim Limestone		
	Park Shale	Flathead Formation	Gros Ventre Formation
	Meagher Limestone		
	Wolsey Shale		
	Flathead Sandstone		Flathead Sandstone

Figure 2: Nomenclature of sedimentary rock units in Yellowstone National Park.

rock units that are described in the text, the best exposed and least deformed sections noted in the course of my geologic mapping are identified in the text as suggested reference sections. Some of these sections have been measured and are described at the end of the report. Locations of other sections that could not be measured in the time available for fieldwork are given in the text.

The geology of pre-Tertiary sedimentary rocks in the south-central part of Yellowstone National Park has been studied by J. D. Love and W. R. Keefer. Their nomenclature differs somewhat from mine, and the stratigraphic units are correlated in figure 2.

PRECAMBRIAN METAMORPHIC ROCKS

Precambrian metamorphic rocks crop out discontinuously in the northwestern and north-central parts of the park, from The Crags on the west side of the Gallatin Range east to Slough Creek and Lamar Canyon. The Precambrian rocks in most of this area are mainly granitic biotite gneiss, but quartz-biotite schist predominates between Rescue Creek and Garnet Hill. The relations of the schist and gneiss in the Jardine — Crevasse Mountain area just north of the park were described by Seager (1944, p. 21-34, 40-42), L.E. Brown (1965), Brookins and Brown (1966, p. 613-615), and Brookins (1968, p. 5). They concluded that the schist was originally sedimentary rock, which was regionally metamorphosed and then intruded by granitic magma that formed a very large batholith. The granitic rocks were later metamorphosed to form the present granitic gneiss. Seager (1944, p. 28) pointed out that the granitic rocks crosscut the schist and occur as apophyses extending into the schist from the main granitic mass, evidence which supports his conclusion that the granitic gneiss was originally an intrusive igneous mass.

The uniformity of the Precambrian rocks exposed in the park contrasts with the variety of rocks of similar age reported in adjacent areas. In the Madison Range, west of the park, Witkind (1964, 1969) mapped six units of metamorphic rocks beneath the Flathead Sandstone: granite gneiss, dolomite, amphibolite, mica schist, tremolite marble, and quartzite. In the Jardine-Crevasse Mountain mining district, east of Gardiner and north of the area of schistose rocks in Yellowstone National Park, Seager (1944) and Brookins and Brown (1966) recognized biotite quartzite, quartz-biotite schist, quartz cummingtonite schist, quartz-horn-blende schist, gabbro diabase, biotite-muscovite granite, pegmatite, and aplite. This greater variety of rocks in areas adjacent to the park is partly a reflection of the greater detail of the studies made of the Precambrian crystalline rocks in these areas; for example, some of the varieties of schist reported in the Jardine-Crevasse Mountain district certainly extend into the park to the vicinity of Garnet Hill. But the quartzite and marble reported in these adjacent areas are not present in the park, and the dominant Precambrian crystalline rock is granitic gneiss.

PALEOZOIC SEDIMENTARY ROCKS

Paleozoic sedimentary rocks in the northern part of Yellowstone National Park have a total thickness of about 3,000 feet. More than two-thirds of the total thickness is limestone or dolomite — the rocks that form most of the conspicuous cliffs in the Gallatin Range, along Slough Creek, and on the lower slopes of Baronnette and Abiathar Peaks.

The oldest Paleozoic rocks, of Middle Cambrian age, are overlain by formations that represent every system except the Silurian. Although almost every system is represented, significant interruptions in deposition occurred, and no rock unit reflects deposition through more than part of a system.

In their discussion of the sedimentary rocks in the northern part of Yellowstone National Park, Iddings and Weed (1899, p. 6-8) described a measured section of Paleozoic rocks on Crowfoot Ridge and established this section as a reference sequence that has since been cited in many publications. Unfortunately, the Crowfoot Ridge section is cut by a zone of flat faults that destroy its usefulness as a reference sequence except for some of the Cambrian formations. To replace the Crowfoot Ridge section, other sections are suggested in the following pages.

CAMBRIAN

Cambrian rocks are among the most widespread sedimentary rocks in Yellowstone National Park; they crop out in the southern part of the Gallatin Range, on the Buffalo Plateau, and along Slough and Soda Butte Creeks. These rocks are separable into the Flathead Sandstone, Wolsey Shale, Meagher Limestone, Park Shale, Pilgrim Limestone, and the Dry Creek, Sage and Grove Creek Members of the Snowy Range Formation. Of these, the Flathead and Wolsey were mapped as a single unit and the Dry Creek, Sage, and Grove Creek were mapped together as the Snowy Range Formation. The aggregate thickness of the Cambrian rocks is about 1,100 feet.

The nomenclature of Cambrian rocks in Montana and Yellowstone National Park, long subject to controversy, has been reviewed by Deiss (1936), Dorf and Lochman (1940), Lochman-Balk (1950), Sloss and Mortiz (1951), Hanson (1952), and Robinson (1963). In the northern part of the park, the Cambrian rocks were originally divided into the Flathead Formation and the Gallatin Limestone by Weed (1896, p. 4), and the Crowfoot Ridge section was established as a reference sequence for these formations by Iddings and Weed (1899, p. 6-8). Deiss (1936) remeasured and described the Crowfoot Ridge section as part of a general study of Cambrian rocks in Yellowstone National Park and in Montana; he concluded (1936, p. 1325-1326) that the Flathead as originally used in the park had to be redefined and that the name Gallatin should not be used. To replace these names, Deiss proposed that the nomenclature used by Weed (1900, p. 285-286) for Cambrian rocks in the Little Belt Mountains, Mont., be adopted for Cambrian rocks in central and southern Montana and in Yellowstone National Park. Dorf and Lochman (1940) restudied the Cambrian rocks in southcentral Montana, and, though they agreed with Deiss that the name Gallatin Limestone should not be used in this area, they proposed three new formations — the Maurice, Snowy Range, and Grove Creek. They (Dorf and Lochman, 1940, p. 551) specifically stated that these new names were applied to the section redefined by Deiss (1936, p. 1318) on Crowfoot Ridge — the Maurice to replace the unit called Pilgrim, and the Snowy Range to include, in its lower part, rocks that Deiss called the Dry Creek Shale. Sloss and Moritz (1951, p. 2144-2145), Hanson (1952, p. 16), Lochman-Balk (1956, p. 603), and Robinson (1963, p. 15) pointed out that the Maurice is the lithologic equivalent of the Pilgrim and that, because the name Pilgrim was first applied to these rocks, the name Maurice should be dropped. Accordingly, the name Pilgrim is used in this paper, as are the names applied by Deiss (1936, p. 1318-1325) to the older Cambrian formations below the Pilgrim, which are, in ascending order, the Flathead Sandstone, Wolsey Shale, Meagher Limestone, and Park Shale.

The Cambrian rocks above the Pilgrim on Crowfoot Ridge were included in the Dry Creek Shale by Deiss (1936, p. 1318-1319), but the nomenclature of these uppermost Cambrian rocks has been extensively revised since then, principally by Dorf and Lochman (1940), by Lochman-Balk (1950, 1956), and most recently by Grant (1965). Dorf and Lochman (1940, p. 545-547) intended that the Snowy Range Formation and Grove Creek Formation include the Upper Cambrian rocks above the Maurice Formation (Pilgrim Limestone of this report), and they included the shale and limestone of Deiss' (1936) Dry Creek Shale on Crowfoot Ridge in the Snowy Range Formation. Lochman-Balk (1950, p. 2212), in a study of the Dry Creek Shale of central Montana, divided the Snowy Range Formation into a lower member (the Dry Creek Shale Member) and an upper member (the Sage Limestone Member). The name Dry Creek was reduced from formation to member rank. The Sage, a new name in her report, was derived from Lower Sage Creek and a stratigraphic section near Mill Creek, in the N½ SE¼ sec. 13, T. 6S., R. 9 E., in the Livingston quadrangle, Montana, was selected as the type locality. And in 1956 the Grove Creek Formation was reduced to a member of the Snowy Range Formation by Lochman-Balk (1956, p. 609-613). In this report the Cambrian rocks above the Pilgrim Limestone are included in the Snowy Range Formation and are divided in accordance with the terminology proposed by Dorf and Lochman (1940) and as revised by Lochman-Balk (1950, 1956) and Grant (1965). This revised terminology is herein adopted for Yellowstone National Park and the surrounding region.

The ages of the various Cambrian rocks have been well established in many of the reports referred to above.

The Cambrian rocks measured and described by Deiss (1936, p. 1318-1325) on Crowfoot Ridge crop out in a well-exposed, almost complete section, and Deiss' descriptions are so complete that the section was not remeasured. For Cambrian rocks between the base of the Flathead Sandstone and the top of the Pilgrim Limestone, the Crowfoot Ridge section is the most complete and best exposed reference sequence in the northern part of Yellowstone National Park. The Cambrian rocks above the Pilgrim — the Snowy Range Formation — are disrupted by faulting on Crowfoot Ridge, and the section is incomplete. However, both the Snowy Range Formation and the Pilgrim Limestone are well exposed on Three Rivers Peak, at the head of the Gallatin River, and this section (stratigraphic sections, see appendix) is here designated the reference section for the Pilgrim Limestone and Snowy Range Formation. Grant (1965, p. 67-69) measured sections of the Snowy Range Formation in the northeast corner of the park, and Lovering (1930, p. 18-24) described Cambrian rocks near Cooke City, Mont., a short distance from the Northeast Entrance to the park. Another excellent exposure of the Snowy Range Formation is half a mile southeast of the Buffalo Patrol Cabin on the west side of the Buffalo Plateau.

Flathead Sandstone

The Precambrian crystalline rocks are overlain by the Middle Cambrian Flathead Sandstone, which crops out in many places in the southern part of the Gallatin Range, on the Buffalo Plateau, and west of Bison Peak near Slough Creek, its easternmost exposure.

Most of the Flathead is quartzitic sandstone that is yellowish gray to very pale orange or grayish orange pink, fine to medium grained (0.1-0.3 mm), well sorted, and in beds 0.2-3 feet thick, although some beds on the Buffalo Plateau are as much as 5 feet thick. Locally, the Flathead weathers to pale red, grayish red, or pale reddish brown. Many beds are laminated or cross-laminated. The lower part of the formation is characterized by dark-reddish colors and coarse-grained rocks that are not common in the higher beds. On Crowfoot Ridge, conglomerate is especially plentiful in the lower 30 feet of the formation. Here, the basal 2-3 feet of the formation is grayish-red to light-gray very coarse (1-2 mm) poorly sorted sandstone interbedded with conglomerate that contains subangular to subrounded pebbles of glassy white quartz as much as 3 inches long. The overlying beds, 0.1-3 feet thick, are graded, with thin conglomeratic zones or very coarse sandstone zones at the base of each bed, grading upward into fine-grained (0.1 mm) clean, partly cross-laminated quartzitic sandstone. The upper part of the formation in the Gallatin Range contains thin interbeds of greenish-gray fissile shale similar to that in the overlying Wolsey Shale, and the contact is placed at the highest bed of quartzitic sandstone.

The Crowfoot Ridge section of the Flathead also includes a unique 1-foot-thick zone of sandy oolitic hematite beds that contain small brachiopods of Middle Cambrian age (Deiss, 1936, p. 1325), the only known fossils in the Flathead Sandstone.

On Crowfoot Ridge the formation is 160 feet thick (Deiss, 1936, p. 1315-1316), and on the Buffalo Plateau it is about 120 feet thick. At Cooke City it is about 100 feet thick (Lovering, 1930, p. 19), or a little less, depending on where the upper contact is placed. The formation is well exposed only on the crest of Crowfoot Ridge and on the Buffalo Plateau; elsewhere only the upper part of the formation crops out, in low rounded ledges.

Wolsey Shale

Although the Wolsey Shale of Middle Cambrian age is as widely distributed as the underlying Flathead Sandstone, it is exposed in only a few places in the Gallatin Range. On the Buffalo Plateau and near Slough Creek, its presence is suggested here and there by shale fragments in the soil, but no outcrops are known.

Most of the Wolsey is greenish-gray slightly micaceous sandy shale, but interbeds of soft rusty-weathering argillaceous calcareous thin-bedded sandstone and siltstone and of glauconitic sandy thin bedded limestone are common throughout the formation. The upper third of the formation contains the more abundant limestone interbeds, and the upper contact with the Meagher Limestone is gradational through a thickness of about 20 feet; the contact is placed at the first bed of yellow and gray mottled limestone, typical of the Meagher. Thus, both the upper and the lower contacts of the Wolsey are gradational in the Gallatin Range. The formation is characterized by its abundant grayish-green fissile shale, by rusty-weathering dirty calcareous sandstone, and by glauconitic sandy limestone interbeds. Many beds throughout the formation are marked by worm trails.

On Crowfoot Ridge, as elsewhere in the Gallatin Range, the Wolsey is 150 feet thick (Deiss, 1936, p. 1321-1322). It probably is only about 100 feet thick on the Buffalo Plateau, and farther east, near Cooke City, the lower shale member of the Gros Ventre Formation — the approximate equivalent of the Wolsey — is 90 feet thick (Lovering, 1930, p. 20).

Meagher Limestone

The Meagher Limestone, of Middle Cambrian age, is the lowest thick unit of carbonate rock in the northern part of Yellowstone National Park, and it forms conspicuous light-colored massive cliffs and ledges. The formation is widely exposed in the southern part of the Gallatin Range, on the Buffalo Plateau, and west of Slough Creek.

The Meagher is a distinctive formation that everywhere includes the mottled limestone beds so characteristic of the formation in southwestern Montana. The limestone that makes up most of the formation is medium gray to medium dark gray, is very fine grained, occurs in beds 0.1-3 feet thick, and is mottled with light-gray, yellowish-gray, or yellowish-orange slightly coarser grained limestone or dolomitic limestone. The upper part of the formation is not so conspicuously mottled as the middle and lower parts. The rocks weather medium light gray or light gray mottled with yellowish orange, pinkish gray, or lighter shades of gray. The origin of the mottling in the Meagher Limestone, and in similar rocks of the Upper Cambrian Pilgrim Limestone, was discussed by Hanson (1952, p. 14) and, most recently, by Robinson (1963, p. 19-20). The limestone commonly contains small twiggy calcite crystals and, in many places, is cut by thin (1-2 mm) veinlets of crystalline calcite that both parallel and cut across the bedding. Most beds in the lower part of the formation are 0.1-0.2 foot thick and are not so well exposed as the thicker, overlying beds. Shaly partings between limestone beds are common throughout the formation, and thin interbeds of grayish-green shale are widespread in the lower part of the formation in the Gallatin Range and throughout the entire formation in the north-central and northeastern part of the park. The top of the formation in the Gallatin Range is a unit, about 10 feet thick, of limonitic sandy calcareous shale containing thin beds and abundant concretions 3-5 feet in diameter of dark-gray to brownish-gray very finely crystalline ferruginous limestone (Deiss, 1936, p. 1320, 1323-1325). As Deiss pointed out, the unit is not found in the Meagher except in the Gallatin Range in Yellowstone National Park.

On Crowfoot Ridge, the Meagher is 388 feet thick in the section described by Deiss (1936, p. 1320-1321), and is about the same at Dome Mountain and south of Mount Holmes. On the Buffalo Plateau and near Buffalo and Slough Creeks, it appears to be 175-200 feet thick. In these areas, 25-50 feet of thinbedded limestone in the lowermost part of the formation is generally concealed, and the thicker, upper limestone beds form a cliff or series of ledges about 150 feet high. At Cooke City, rocks about equivalent to the Meagher were included in the Gros Ventre Formation by Lovering (1930, p. 19-21) as a middle limestone member, 200 feet thick. However, Lovering included shaly and sandy beds in the base of the middle limestone member of the Gros Ventre that I would include with the Wolsey Shale, so the part of the middle limestone member equivalent to the Meagher is somewhat less than 200 feet thick, and perhaps is a little thinner than the Meagher near Slough Creek. West of the park, in the Madison Range, the formation is about 450 feet thick (Witkind, 1964, 1969). The Meagher Limestone is only half as thick in the northeastern part of the park as it is in the northwestern part, and it clearly thickens westward.

Park Shale

The Park Shale conformably overlies the Meagher Limestone and is the uppermost unit of Middle Cambrian age. The formation is widespread but poorly exposed in the Gallatin Range. It is concealed by glacial deposits almost everywhere on the Buffalo Plateau, and only the uppermost few feet of the formation are exposed at its easternmost outcrops near Slough Creek.

The formation is made up mainly of grayish-green shale that, in the lower half, is papery and, in the upper half, becomes a thinly platy mudstone. Some medium-dark-gray to grayish-red beds occur throughout the formation, but the dominant color is grayish green. The lower half of the formation contains a few lenses, less than 0.5 foot thick, of limestone coquina and interbeds about 0.5 foot thick of greenish-gray rusty-weathering glauconitic finely crystalline limestone. Locally, a few thin beds of limestone pebble conglomerate are present in the basal part of the formation. In the Gallatin Range the upper 20-30 feet of the formation includes thin interbeds of yellowish gray argillaceous limestone and shale, but near Slough Creek the uppermost beds are grayish-yellow-green pisolitic glauconitic coarse-grained limestone, 0.1-0.3 foot thick, interbedded with greenish-gray shale and mudstone.

The only well-exposed sections in the Gallatin Range are on the crest of Crowfoot Ridge, where the formation is 120 feet thick (Deiss, 1936, p. 1320), and south of Mount Holmes, where it is about 100 feet thick. It appears to be about 100 feet thick on the Buffalo Plateau and near Slough Creek. Equivalent rocks in the Gros Ventre Formation near Cooke City, Mont., are about 90 feet thick (Lovering, 1930, p. 21).

Pilgrim Limestone

The Pilgrim Limestone, of Late Cambrian age, conformably overlies the Park Shale. The formation is well exposed and forms a conspicuous light-gray cliff at many places in the Gallatin Range. It forms cliffs on the upper part of the Buffalo Plateau, on the east canyon wall of Buffalo Creek near Slough Creek, and in the upper reaches of Soda Butte Creek near the Northeast Entrance to the park.

The Pilgrim includes three main lithologic units: (1) a lower unit of interbedded ribboned limestone, oolitic limestone, and less common limestone pebble conglomerate and shale; (2) a middle unit of limestone pebble conglomerate; and (3) an upper unit of mottled oolitic limestone. The three units are persistent throughout the northern part of the park.

The lower unit, which forms almost half the formation, is mostly medium gray very fine grained thin-bedded (0.2-0.5 ft) limestone that is ribboned with yellow-brown to grayish-orange silty limestone and irregular layers as much as 0.1 foot thick of silt and very fine sand. Interbeds as much as 3 feet thick of medium-gray glauconitic oolitic limestone rich in trilobite fragments are common in the ribboned limestone. The unit also contains a few interbeds, less than 1 foot thick, of medium-gray medium- to coarse-grained fossiliferous glauconitic limestone pebble conglomerate and, in the north-central and northeastern part of the park, a few thin interbeds of grayish-green shale.

The middle unit, about 60 feet thick, is medium-gray medium- to coarse-grained glauconitic limestone pebble conglomerate that contains abundant fossil fragments. The unit includes a few yellowish-gray sand pisolitic glauconitic limestone beds where it is exposed in Slough Creek and Soda Butte Creek.

The upper unit is medium-gray largely oolitic fine-to medium-grained limestone that is mottled yellowish gray to pale yellowish brown. Although most of the unit is massive or thick bedded, it also includes zones of thin-bedded to platy dolomite, particularly in the northeastern part of the park. It is 40-60 feet thick in the Gallatin Range and 100-150 feet thick on the Buffalo Plateau and near Soda Butte Creek. This is the "mottled limestone" of Iddings and Weed (1899, p. 8; Weed, 1896, p. 4), and the most conspicious part of the formation, for it forms prominent cliffs.

On Crowfoot Ridge, the Pilgrim Limestone is 172 feet thick (Deiss, 1936, p. 1320). On Three Rivers Peak it is

160-175 feet thick, but at that locality the upper part of the Park Shale and the lower part of the Pilgrim are intruded and metamorphosed by the Indian Creek laccolith, and placement of the contact is uncertain over a thickness of about 40 feet. On the Buffalo Plateau and near Buffalo Creek, Slough Creek, and Soda Butte Creek, the formation is about 250 feet thick, as are equivalent rocks near Cooke City, Mont. (Lovering, 1930, p. 21-24), where they were included by Lovering in the upper part of the Gros Ventre Formation and the lower part of the Gallatin Formation.

Snowy Range Formation

The Snowy Range Formation, of Late Cambrian age, conformably overlies the Pilgrim Limestone, and is the uppermost Cambrian unit in Yellowstone National Park. The formation typically consists of three members, the Dry Creek Shale (base), the Sage Limestone, and the Grove Creek Limestone (top), but only the Dry Creek and Sage Members are present in most of the northern part of the park. The Grove Creek Member crops out on Three Rivers Peak and also in the upper part of the canyon of Soda Butte Creek north of the Northeast Entrance.

The thickness of the Snowy Range Formation differs considerably from place to place. On the west side of Crowfoot Ridge, the Dry Creek Shale Member is about 40 feet thick, and the Sage Limestone Member is perhaps 75 feet thick, although its thickness can only be estimated because of crumpling associated with faulting. On the crest of Crowfoot Ridge, Grant (1965, p. 64) measured about 15 feet of Dry Creek and about 65 feet of Sage, but there the Dry Creek is cut by a fault. On Three Rivers Peak (stratigraphic sections, see appendix) the Dry creek is about 40 feet thick, the Sage is about 45 feet thick, and the Grove Creek is about 40 feet thick. Near the Buffalo Patrol Cabin on the Buffalo Plateau, the Dry Creek is about 50 feet thick, and the Sage is about 60 feet thick. North of the Northeast Entrance, Grant (1965, p. 67-69) measured about 50 feet of Dry Creek, 220 feet of Sage, and 25 feet of Grove Creek. Nearby on the southwest slope of Meridian Peak, the Dry Creek is about 50 feet thick, the Sage is about 150 feet thick, and the Grove Creek is absent. Grant (1965, p. 13-14, 17) suggested that the differences in thickness of the Sage and Grove Creek Members are due to erosion, which in most places in Yellowstone National Park had removed the Grove Creek Member and much of the Sage Limestone Member before deposition of the Bighorn Dolomite of Late Ordovician age.

Dry Creek Shale Member

The Dry Creek Shale Member of the Snowy Range Formation is mainly greenish-gray to light-olive-gray paper-thin shale, with thin (less than 1 ft thick) interbeds of light-olive-gray calcareous very fine grained sandstone and siltstone. The upper 10 feet or so of the member contains interbeds of medium-gray to medium-light-gray very fine grained to oolitic glauconitic limestone and limestone pebble conglomerate. The base of the member in the Gallatin Range is marked by a 1-foot-thick bed of yellowish-orange to yellowish to yellowish-brown platy calcareous very fine grained sandstone or siltstone, which overlies the mottled limestone at the top of the Pilgrim Limestone.

Sage Limestone Member

The thin and crinkly limestone beds of the Sage Limestone Member of the Snowy Range Formation are distinctive rocks that typify the member wherever it is exposed. These rocks, in beds about 1 foot thick, are pale yellowish brown to medium gray, partly glauconitic, very fine grained, and laminated and are irregularly, or "crinkly," ribboned with grayish-orange to dark-yellowish orange very fine grained sandy limestone and calcareous sandstone. Typically, they contain interbeds, as much as 3 feet thick, of similarly colored mottled limestone. From the Buffalo Plateau eastward to the Northeast Entrance, the base of the member is marked by a bed, 10-15 feet thick, of limestone that is made up of algal columns as much as 1 foot in diameter — the *Collenia magna* beds described by Grant (1965, p. 12-14) — but similar algal beds are not present in the Gallatin Range. In its eastern outcrops, the member also includes many thin interbeds of grayish-green shale and of yellowish-gray limestone pebble conglomerate that are absent at Three Rivers Peak (stratigraphic section, see appendix), and only a few occur in the Gallatin Range. The member includes many fossiliferous beds, some of them coquinas, and the upper part is generally more glauconitic than the lower.

Grove Creek Limestone Member

The Grove Creek Limestone Member of the Snowy Range Formation has been recognized in Yellowstone National Park only on Three Rivers Peak. It does crop out, however, south of Silver Gate, Mont., only a short distance from the Northeast Entrance and may well be present in some places locally along the east canyon wall of Soda Butte Creek, where the Snowy Range Formation is largely concealed by glacial and landslide deposits. At the Silver Gate outcrops, the Grove Creek Member (Grant, 1965, p. 67-69) is mostly yellowish-orange fine-to medium grained algal dolomitic and pyritic limestone interbedded with dusky-blue hard blocky calcareous or dolomitic shale. Typically, the member includes much limestone pebble conglomerate, but only a few thin beds are preserved in the thin remnants near Silver Gate.

On Three Rivers Peak (stratigraphic sections, see appendix) the member consists of light-brownish-gray very fine grained, partly algal, thin and irregularly bedded dolomite, dolomitic limestone, and limestone that form the base of the cliff above the softer rocks of the Dry Creek and Sage Members.

ORDOVICIAN

Although Ordovician rocks forming light-gray cliffs and ledges are widespread in the northern part of Yellowstone National Park, they were not recognized in the early geological studies by Iddings and Weed (1899) and apparently remained unknown until Lovering (1930, p. 24-26) identified the Upper Ordovician Bighorn Dolomite in the Cooke City mining district. Wilson (1934b, p. 374) later recognized the Bighorn north of the park at Cinnabar Mountain, and Richards and Nieschmidt (1961) measured a section of Bighorn Dolomite on Antler Peak in the Gallatin Range. C. W. Brown (1961, pl. 1, p. 1175) mapped the Bighorn on the Buffalo Plateau and farther east in the park. The principal stumbling block in recognition of the Ordovician rocks in the Gallatin Range has been the widespread acceptance of the Crowfoot Ridge reference sequence (Iddings and Weed, 1899, p. 6-8; Deiss, 1936, p. 1318-1325; McMannis, 1962, p. 8, pls. 4,5; Grant, 1965, p. 63-64). The Bighorn Dolomite, in fact, is not present on the crest of Crowfoot Ridge, as has been recognized by most who have measured this section, but its absence is due to faulting, rather than to erosion or nondeposition.

The Bighorn Dolomite is completely exposed on Three Rivers Peak, at the head of the Gallatin River, and is included in the Three Rivers Peak measured section (stratigraphic section,see appendix), which is designated as a new reference section for the Bighorn Dolomite in the northwestern part of Yellowstone National Park. The most complete exposures of the formation in the northeastern part of the park are on the southwest slopes of Meridian Peak, west of the Northeast Entrance.

Bighorn Dolomite

The rocks included in the Bighorn Dolomite unconformably overlie the Snowy Range Formation and are much alike wherever exposed. Most of the formation is light-brownish-gray dolomite that weathers pale yellowish brown to light gray and has a distinctive irregular or pitted surface. In the Gallatin Range, much of the formation is thin bedded (0.1-0.3 ft); these thin-bedded rocks (stratigraphic section, see appendix), are provisionally included in the Bighorn, but they are thinner bedded than is typical of the formation. On the Buffalo Plateau and near the Northeast Entrance, the formation is more massive, and many cliff outcrops show virtually no bedding throughout thicknesses of as much as 125 feet; the upper half of the formation at these localities includes more thin beds than the lower half, but thick-bedded or massive rocks are dominant even in the upper half. The dolomite is very fine grained to fine grained, partly laminated, and partly cherty, containing lenses that are as much as 0.1 foot thick and 5 feet long of light-gray to yellowish-orange chert. Lovering (1930, p. 25-26) described fetid rocks from the base of the Bighorn near Cooke City, Mont., but such rocks were not found at this stratigraphic position in Yellowstone National Park.

At Three Rivers Peak, the Bighorn is about 100 feet thick. On Antler Peak, only the lower part of the formation — some 55 feet — is exposed (Richards and Nieschmidt, 1961). East of Buffalo Creek and near Slough Creek and Soda Butte Creek, the formation is about 200 feet thick, and near Cooke City, it is about 175 feet thick (Lovering, 1930, p. 25).

DEVONIAN

Rocks of Late Devonian age, 300-350 feet thick, are separated into the Jefferson Formation and Three Forks Formation, a division that follows that of Iddings and Weed (1899, p. 7-8; Weed, 1896, p. 4). The regional stratigraphy of these rocks has been the subject of many reports (Sloss and Laird, 1947; McMannis, 1962; Robinson, 1963, p. 24-38; Sandberg, 1965; Benson, 1966). Devonian rocks in areas immediately adjacent to the northern part of Yellowstone National Park have been described by several geologists: in the Madison Range to the west by Witkind (1964, 1969), on Cinnabar Mountain to the north by Wilson (1934b, p. 373-374), and near Cooke City, Mont., to the northeast by Lovering (1930, p. 37-38).

The Crowfoot Ridge section is widely cited as a reference sequence for Devonian rocks in northwestern Yellowstone National Park (Iddings and Weed, 1899, p. 6-8; McMannis, 1962, pl. 4; Benson, 1966, p. 2578). However, the flat faults that cut across Crowfoot Ridge have removed most of the Jefferson and part of the Three Forks from the ridge crest. On the west side of the ridge, part of the Jefferson is apparently repeated by faulting, but much of it and the Three Forks are gone. The formations are well exposed on Three Rivers Peak (stratigraphic sections, see appendix), where the section is designated a reference section for the Jefferson and Three Forks Formations in northwestern Yellowstone National Park.

Jefferson Formation

The rocks of the Jefferson Formation in the northwestern part of the park differ considerably from those in the northeastern part. In the Gallatin Range the formation is mainly brownish-gray to pale brown, or various shades of gray, very fine grained to fine-grained sugary dolomite, in beds from 0.1 to 3 feet thick, with the thinnest beds commonly thinly laminated. A few beds are dark gray, and most of these darker rocks are fetid. Here and there, a few thin beds of dark-gray very fine grained limestone occur near the base. Thin and platy beds of yellowish-brown dolomitic siltstone and silty dolomite are intercalated throughout the formation, and a single 3-foot-thick bed of greenish-gray shale is near the top. The shale is overlain by a 30-foot-thick massive unit of yellowish-brown to brownish-gray very fine grained dolomite that marks the top of the formation and is correlated with the Birdbear Member of the Jefferson Dolomite (Sandberg, 1965, p. 7).

In the northeast corner of the park, the Jefferson is pale-brown to pale-yellowish-brown fine-grained sugary dolomite and dolomitic limestone in beds 0.2-4 feet thick. The thicker beds are in the lower third of the formation, and the uppermost beds are thinner and finer grained than those in the rest of the formation. In this area, there are few, if any, dark fetid rocks, and the section differs not only from that in the Gallatin Range, but also from that of the Devonian rocks measured by Lovering (1930, p. 27) near Cooke City, Mont., which was dominantly dark-gray fetid limestone. Similar light-colored rocks occur in the single outcrop of Jefferson west of Slough Creek, and this a-typical Jefferson appears to be widespread in the north-central and northeastern parts of Yellowstone National Park.

On Three Rivers Peak, the Jefferson Formation is about 240 feet thick. On Meridian Peak, it is about 175 feet thick, but it is only 125 feet thick near Cooke City, Mont., a few miles farther east (Lovering, 1930, p. 27). Lovering attributed this westward thickening to the erosional unconformity at the base of the Jefferson.

Three Forks Formation

The distribution of the Three Forks Formation is similar to that of the Jefferson Formation, which it conformably overlies. Although the rocks in the formation differ across the park, the Logan Gulch Member (Sandberg, 1965, p. 10-14) can be tentatively identified in the Gallatin Range, and the Logan Gulch and Trident Members can be tentatively identified near the Northeast Entrance.

In the Gallatin Range, where all of the rocks included in the formation are considered equivalent to the Logan Gulch Member, the lower 30-35 feet is mainly light-olive mudstone that weathers pale orange to grayish orange. It includes a few interbeds, each about 0.5 foot thick, of light-olive-gray yellowish-weathering very fine grained dolomite. The upper 45-50 feet of the formation is light-olive-gray grayish-orange- to light-gray-weathering very fine grained dolomite and silty dolomite, in 0.1- to 0.2- foot-thick beds.

In the northeast corner of the park, the Three Forks includes three units — lower shale, middle limestone, and upper shale. The lower shale and middle limestone are probably equivalent to the Logan Gulch Member and to the rocks included in the formation in the Gallatin Range, and the upper shale is probably equivalent to the Trident Member. The lower shale, about 50 feet thick, is made up of interbedded grayish-green to yellowish-brown to dark-gray paper-thin shale and calcareous mudstone, and yellowish-brown fine-grained limestone, in beds 0.1-1 foot thick. The middle limestone, about 30 feet thick, is light-olive-gray to pale-yellowish-brown very fine grained, typically brecciated limestone and dolomite. It weathers light brown to pale reddish brown and locally forms a conspicuous red cliff. The upper shale, about 50 feet thick, is yellowish-brown to grayish-green mudstone and shale in its lower half, grading upward into grayish-red calcareous mudstone and interbedded grayish-green shale. The upper 10 feet of this unit includes thin interbeds of light-gray or pale-yellowish-brown fossiliferous limestone. The Three Forks Formation at Cooke City, Mont. (Lovering, 1930, p. 28), appears to be almost identical to the formation in the northeastern part of the park.

In the Gallatin Range, the formation is about 80 feet thick (stratigraphic section, see appendix). In the northeast corner of the park it is about 130 feet thick, and near Cooke City, Mont., it is about 100 feet thick (Lovering, 1930, p. 28). The upper shale unit in the northeast corner of the park apparently was eroded farther west before deposition of the Lodgepole Limestone.

MISSISSIPPIAN

The thick sequence of limestone and dolomite of Mississippian age in the northern part of Yellowstone National Park is divided into two formations, the Lodgepole and Mission Canyon Limestones that together form the Madison Group. The thickness of the group is about 1,300 feet. These thin-bedded to massive carbonate rocks underlie much of the central part of the Gallatin Range and crop out intermittently to the Northeast Entrance. The regional stratigraphy of the Madison Group, and the development of the nomenclature followed in mapping these rocks in the park, was recently reviewed by Roberts (1966), Robinson (1963, p. 38-44), and Sando (1967a); the characteristics of the Madison Group in the Madison Range, west of the park, were discussed by Witkind (1969).

Iddings and Weed (1899, p. 7, 32, 35) measured and described several sections of Madison Limestone in the

Gallatin Range and described the formation where it crops out near Soda Butte Creek and on Abiathar Peak. Their main sections were on the crest of Crowfoot Ridge, now known to be faulted, and from Panther Creek through the cirque on the east side of Bannock Peak. Both sections are useful, but they are not so well exposed as a section measured on the south face of Bannock Peak by W.J. Mapel, A.E. Roberts, and E.K. Maughan (written commun., 1967; this report, stratigraphic section, see appendix), which is designated here as a reference section both for the Mississippian Madison Group and for its individual formations, the Mission Canyon and Lodgepole Limestones.

Lodgepole Limestone

The characteristic thin-bedded limestones of the Lodgepole overlie the Three Forks Formation with seeming conformity, although the absence of both the Sappington Member of the Three Forks (Sandberg, 1965, p. N14-N17) and the upper shale unit of the Three Forks in the Gallatin Range suggests that some of the Three Forks sediments were eroded before the Lodgepole was deposited. The Lodgepole commonly forms a series of low cliffs and ledges separated by grassy slopes, and it is not so well exposed as the cliff-forming massive rocks of the overlying Mission Canyon Limestone.

The limestone is light brownish gray or pale yellowish brown to light olive gray and medium gray and fine to medium grained; it occurs in beds 0.1-2 feet thick that are separated by partings, 0.1-0.2 foot thick, of yellowish-gray to pale-reddish-brown calcareous siltstone and silty limestone. In places in the lower part of the formation, the limestone beds may be as much as 5 feet thick and generally form units as much as 60 feet thick of mixed thin and thick beds. The limestones typically weather light gray or yellowish gray with a distinctive bluish tint. The lower half of the formation contains locally abundant yellowish-brown to dark-gray chert, both as irregular nodules, 0.2-0.3 foot thick and as much as 2 feet long, and as irregular beds, 0.1-0.2 foot thick.

The Lodgepole Limestone is about 485 feet thick on the south face of Bannock Peak (stratigraphic section, see appendix), and it maintains this thickness throughout the Gallatin Range. Near the Northeast Entrance it is at least 550-600 feet thick, but at that locality it is the youngest sedimentary rock exposed beneath the blanketing volcanic rocks, so its total thickness is uncertain.

The formation contains abundant fossils, mainly corals, brachiopods, and crinoid fragments; coarse-grained beds made up almost entirely of fossils or fossil fragments are common.

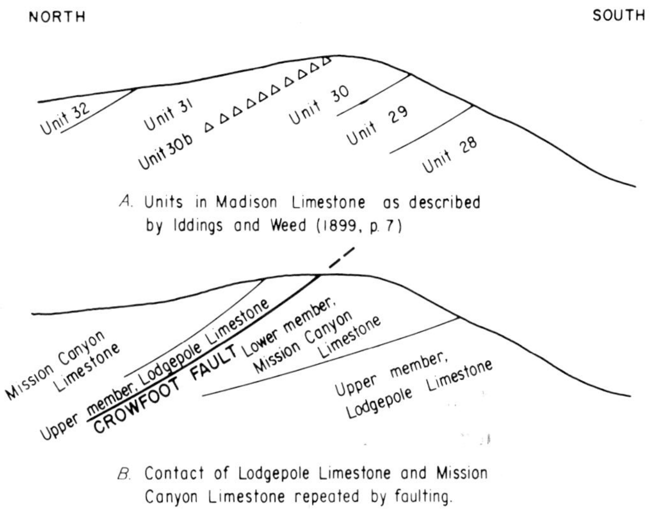

Figure 3: Schematic cross section, viewed toward the east, of part of Crowfoot Ridge, showing interpretation of units of Madison Group.

The Lodgepole probably includes units 24 through 28 of Iddings and Weed's Crowfoot Ridge section (1899, p. 7; Girty, 1899, p. 483) and units 1 through 10 of their Bannock Peak section (Iddings and Weed, 1899, p. 32). The stratigraphic positions of fossils described and illustrated by Girty (1899, p. 483-599) were located by reference to the numbered beds of Iddings and Weed's Crowfoot Ridge section.

Mission Canyon Limestone

The dominantly thick to massive limestone and dolomite beds of the Mission Canyon Limestone form cliffs above the less resistant thinner beds of the Lodgepole in the Gallatin Range and crop out in a few places farther east in the park, partly in isolated outcrops surrounded by volcanic rocks.

Most of the Mission Canyon is limestone, but dolomite and dolomitic limestone form some of the most distinctive units and compose about one-fourth of the formation. The limestone is medium gray to medium light gray, light olive gray, and pale yellowish brown and is fine to coarse grained; it generally occurs in beds 1-5 feet thick or massive, although some are as thin as 0.2 foot, and there are some zones of thinly laminated beds. Brownish-gray chert in nodules, as much as 0.2 foot thick and 2 feet long, and siliceous encrustations are common in the limestone. The coarser grained limestones are bioclastic, some beds are coquinas made up of fossil trash, and many beds are fossiliferous, containing abundant brachiopods and corals.

Dolomite, dolomite breccia, and dolomitic limestone are very light gray to pale orange and are finer grained and more massive than the limestone. Dolomite and dolomitic limestone form a massive unit 80-100 feet thick at the base of the formation. The upper half of the formation includes two conspicuous units of dolomite breccia that persist throughout the Gallatin Range. The lower dolomite breccia unit, about 300 feet below the top of the formation, is 60 feet thick and forms a prominent cliff that is pocked with small caverns. The upper dolomite breccia unit, about 100 feet thick, forms a light-colored massive cliff at the top of the formation.

The Mission Canyon Limestone is 814 feet thick on the south face of Bannock Peak (W. J. Mapel, A.E. Roberts, and E.K. Maughan, written commun., 1967; this report, stratigraphic section, see appendix). Farther east on Bannock Peak, Iddings and Weed (1899, p. 32) measured 810 feet of rocks — units 11 through 22 of their Madison Limestone — which is equivalent to the Mission Canyon of this report. The Crowfoot Ridge section of Iddings and Weed (1899, p. 7) includes about 1,200 feet of rocks — units 29 through 32 — that are about equivalent to the Mission Canyon, but these rocks are cut by a strand of the flat Crowfoot fault — the "red limestone conglomerate" of their unit 30b is actually fault gouge and breccia — and by steeply dipping fracture zones that probably are faults. Unfaulted sections of the Mission Canyon nearest the Crowfoot Ridge section of Iddings and Weed are on the east valley wall of the Gallatin River, where they are about 800 feet thick, and the Crowfoot Ridge section is about 400 feet too thick as a result of duplication by faulting. Units 29 and 30 of Iddings and Weed's section are probably the lower part of the Mission Canyon, but part of unit 31 is Lodgepole repeated above the Crawford fault. Hence, unit 31 probably includes part, or all, of unit 28, as well as all of units 29 and 30 and younger rocks (fig. 3). W. J. Sando (written commun., 1967) recognized the anomalous thickness of the Crowfoot Ridge section, and has pointed out the difficulties in stratigraphic interpretation resulting from Crowfoot Ridge having been used as the type locality for many of the Mississippian species described by Girty (1899, p. 483-599).

PENNSYLVANIAN

Pennsylvanian rocks include the Amsden Formation and the overlying Quadrant Sandstone, which together have a maximum thickness of almost 400 feet. These rocks are in the Gallatin Range, on Horseshoe Hill, and on the ridge to the north, and in slices of the Gardiner fault zone where the zone is crossed by the Grand Loop Road. Farther east, the Pennsylvanian rocks were removed either by erosion or by faulting before eruption of the volcanic rocks that now blanket most of the eastern part of the park. No rocks younger than Mississippian are known in the north-central or northeastern parts of the park.

Quadrant Mountain, in the Gallatin Range, is the type locality of the Quadrant Sandstone (Weed, 1896, p. 5; Iddings and Weed, 1899, p. 33-34), although the name was first used by Peale (1893, p. 39) for rocks near Three Forks, Mont. As originally defined, the Quadrant Sandstone of Quadrant Mountain included all rocks between the top of the Madison Limestone and the base of the cherty sandstone (Shedhorn Sandstone of this report) that was the basal part of the Teton Formation. A section of the formation on the southeast spur of Quadrant Mountain measured by Iddings and Weed (1899, p. 34) was about 400 feet thick. The basal part of the formation, 100 feet thick, was shown in the original measured section as concealed, and it is clear that Iddings and Weed did not recognize that at least part of the concealed beds are not sandstone, but, rather, are red shale and limestone - the Amsden Formation of this report. Condit (1919, p. 112, 117-118) recognized the presence of red beds, but he included them in the basal Quadrant when he redefined the overlying Teton Formation. Scott (1935) remeasured the type Quadrant section and recognized the Amsden in northern Yellowstone National Park for the first time. His section (Scott, 1935, p. 1017) includes about 230 feet of Quadrant and about 100 feet of Amsden. Scott's Quadrant Mountain section was later revised (Thompson and Scott, 1941, p. 350; Scott, 1945), and 43 feet of the beds that he had originally placed in the Amsden were assigned to Sacajawea (?) Formation of Mississippian age. The name Amsden was thus dropped completely from the Quadrant Mountain section. The rocks added to the Quadrant belonged there because of their lithologies; the name Sacajawea (?) was applied by Thompson and Scott following the usage of Branson (1937), to rocks in which unspecified megafossils thought to be of Mississippian age were found. The name Sacajawea Formation, however, has not been widely or consistently used (Sando, 1967b, p. 546-550; 1968), and because these rocks are lithologically most like the Amsden exposed in adjacent areas, I assign them to the Amsden Formation.

The regional stratigraphy of the Amsden and Quadrant in southwestern Montana was discussed by Sloss and Moritz (1951, p. 2158-2160, 2164-2165), Robinson (1963, p. 38-39, 47-52), and more recently by Maughan and Roberts (1967); Maughan and Roberts also considered the long-standing problem of placement of the Mississippian-Pennsylvanian boundary in this region. Because specific Mississippian fossils were not found in the Amsden of the northern part of Yellowstone National Park, I follow the usage of Maughan and Roberts and consider the Amsden to be entirely of Pennsylvanian age, rather than partly Mississippian and partly Pennsylvanian, as in most earlier reports.

Neither the Amsden nor the Quadrant Formations were remeasured during the fieldwork leading to this report. The formations are well exposed at many places on Quadrant Mountain and Bannock Peak, but the best exposures are on the arete that connects Bannock Peak to Quadrant Mountain. The Amsden is also well exposed on the eastern summit of Bannock Peak.

Amsden Formation

The dominantly red rocks of the Amsden Formation, of Early and Middle Pennsylvanian age (Maughan and Roberts, 1967, p. B6, B20-B23), were deposited on an irregular karst surface developed on the Mission Canyon Limestone, a surface that is well exposed on Bannock Peak.

In its thickest sections, on Bannock Peak, the Amsden consists of a lower and an upper unit. The lower unit is pale-red or grayish-red to yellowish-orange chippy mudstone, siltstone, and shale containing interbeds of similarly colored to light-gray fine-grained to very fine grained calcareous sandstone. The upper unit is grayish- and pale-red calcareous or dolomitic siltstone and contains a few thin beds of dolomite and shale. The basal rock of the lower unit typically is grayish-red conglomeratic sandstone, 0.5-1 foot thick. In other parts of the Gallatin Range, and north of Horseshoe Hill, the formation consists of only the lower unit; the upper unit apparently was removed by erosion before deposition of the Quadrant Sandstone.

The thickness of the formation ranges from 0 to about 60 feet. Because the changes in thickness are erratic, they probably partly reflect irregularities on the underlying karst surface. The greatest thickness is on Bannock Peak; the formation is absent or only a few feet thick on the east face of Quadrant Mountain. Elsewhere, it is generally 20-50 feet thick.

The two units of the Amsden Formation on Bannock Peak are lithologically similar to formations included in the Amsden Group in central Montana, as summarized by Maughan and Roberts (1967); the lower, most widespread unit is tentatively correlated with the Lower Pennsylvanian Tyler Formation, and probably with the Stonehouse Canyon Member of that formation. The upper, dolomitic unit resembles the Middle Pennsylvanian Devils Pocket Formation. The middle formation, the Lower and Middle Pennsylvanian Alaska Bench Limestone of the Amsden Group, does not seem to be represented in the Amsden of the northern part of Yellowstone National Park, and its absence may reflect the regional erosion that Maughan and Roberts (1967, p. B15) suggested took place before deposition of the Devils Pocket Formation.

Quadrant Sandstone

The red rocks of the Amsden Formation are unconformably overlain by the light-brown sandstones and carbonate rocks of the Quadrant Sandstone of Middle Pennsylvanian age. These resistant rocks and the overlying Shedhorn Sandstone form the largest of the cliff-rimmed dipslopes of the Gallatin Range: Quadrant Mountain, the slope dipping north from Bannock Peak, and many smaller, similar slopes and bedrock-controlled benches.

The Quadrant Sandstone is composed mainly of yellowish-brown to very light gray fine-grained clean quartz sandstone and quartzitic sandstone in beds that commonly are 0.5-3 feet thick but in places are as much as 8 feet thick. The thicker beds are mainly in the middle part of the formation. Much of the sandstone is conspicuously cross-laminated and consists of well-sorted subrounded to subangular clear glassy quartz grains about 0.2 mm in diameter. Most of the sandstone in the lower half of the unit is calcareous. Interbeds of medium-light-gray to light gray fine-grained dolomite, mostly less than 1 foot thick but in places as much as 3 feet thick, are common in the basal part and upper half of the formation, but are not common in the rest of the formation. On the east side of Quadrant Mountain, the basal part of the formation contains several lenses, from 20 to 40 feet thick and a few hundred feet long, of light-gray very fine grained dolomite. The upper-most bed included in the formation is a very light gray very fine grained dolomite, 3-10 feet thick; this bed is underlain by the clean fine-grained sandstones of the Quadrant and is overlain by shale and the dirty sandstones of the Shedhorn Sandstone of Permian age.

The thickness of the Quadrant appears to be fairly uniform in the Gallatin Range — about 300-325 feet. This thickness is about the same as that measured by Iddings and Weed (1899, p. 34) if the Amsden rocks are excluded from their section, but it is substantially less than that given by Condit (1919, p. 117-118), which was almost 400 feet if the Amsden rocks are excluded. It is a little thicker than the 279 feet measured by Thompson and Scott (1941, p. 350).

PERMIAN

Rocks of Permian age, a little more than 100 feet thick, were originally included as the basal unit of the Teton Formation (Weed, 1896, p. 5; Iddings and Weed, 1899, p. 34; Stanton, 1899, p. 600) of "supposed Triassic" age, forming the "cherty beds" of limestone and sandstone in the lower half of that formation. Condit (1919, p. 113-114, 117; Condit and others, 1928, p. 188) examined these rocks on Quadrant Mountain and Bannock Peak, recognized them as Permian, and included them in the Phosphoria Formation. He measured and described a section in The Pocket, a small cirque on the north side of Quadrant Mountain, which is one of the best exposed sections of Permian rocks in the northern part of the park. The rocks were later included in the Shedhorn Sandstone (Cressman, 1955) in the revised nomenclature of Permian rocks in the western phosphate field (McKelvey and others, 1956, p. 2852). In accordance with this revised nomenclature, the name Shedhorn Sandstone is used in this report for the sandy, phosphatic and cherty rocks of Permian age that disconformably overlie the Quadrant Sandstone and that underlie rocks of Triassic age.

Shedhorn Sandstone

Even though the Shedhorn Sandstone is a relatively thin formation, about 115 feet thick, it is widely exposed in the Gallatin Range, where its outcrop pattern conforms to that of the underlying Quadrant Sandstone.

The Shedhorn Sandstone is especially distinguished by a few beds of phosphate rock, by phosphatic oolites, pellets, and glauconite scattered through many of the rocks, by abundant chert in beds and nodules and in tubular concretions normal to bedding, and by phosphatized fishbones common in some beds. The lithologic details of the formation differ somewhat from outcrop to outcrop because of lateral changes in the rocks, but in a general way the formation is in three parts: (1) the lower half is dominantly sandstone, but includes a few beds of phosphate rock and is the main phosphatic zone; (2) the middle part is a conspicuously cherty group of beds that make up about one-fourth of the formation; and (3) the rest of the formation is interbedded sandstone and dolomite. The sandstone beds in the lower half of the formation are yellowish gray, pale yellowish brown, or medium gray to light olive gray; they are fine grained, are commonly at least slightly calcareous, and contain sparse grains of glauconite and from a few percent to 50 percent phosphatic oolites and pellets. Beds are 0.1-3 feet thick. Some beds contain irregular nodules, commonly about 0.2 foot thick and 1 foot long, of brownish-black phosphatic chert that weathers yellowish gray. The basal part of the formation, above a thin basal bed of sandstone, includes a 10-foot-thick bed of yellowish-gray shale capped by a 1- to 2-foot-thick bed of yellowish-gray fine-grained dolomite. The thickest bed of phosphate rock or of very phosphatic sandstone, about 3 feet thick, is a few feet above the shale. The beds of phosphate rock grade laterally into phosphatic sandstone, and the more phosphatic sandstone beds commonly contain abundant phosphatized fishbones. The top of the lower sandstone unit is marked by a 5- to 10-foot-thick sandstone bed that contains tubular concretions of brownish-gray chert, normal to bedding, as much as 0.3 foot in diameter and 4 feet long. The overlying cherty beds, about 30 feet thick, are brownish-gray to grayish-yellow phosphatic chert in beds 0.1-1 foot thick; they include a few light-gray to pale yellowish-gray phosphatic and glauconitic quartz sandstone beds that contain abundant chert nodules.

The upper part of the formation is mostly medium-to light-gray fine-grained to very fine grained partly calcareous quartz sandstone that contains only sparse grains of chert, phosphate, and glauconite and, as a result, is much cleaner than the lower sandstones. The sandstone is in beds 0.1-0.5 foot thick and includes some equally thin interbeds of dolomite, which is medium light gray to brownish gray, fine grained to very fine grained, siliceous, locally cherty, partly fossiliferous, and vuggy.

MESOZOIC SEDIMENTARY ROCKS

Mesozoic sedimentary rocks, more than 4,000 feet thick, crop out only in the Gallatin Range and near Mammoth Hot Springs. Although most of them are not as resistant to erosion as the older rocks, they are well exposed in many places, perhaps because they have been intruded by many sills that form resistant layers and control the steep, cliff-circled slopes common from Little Quadrant Mountain north to Electric Peak and the Yellowstone National Park boundary.

The oldest Mesozoic rocks, marine deposits of Triassic and Jurassic age, are overlain by nonmarine rocks of later Jurassic and Early Cretaceous age and a thick sequence of marine and nonmarine clastic rocks of later Cretaceous age. Triassic and Jurassic rocks constitute about 10 percent of the total Mesozoic section, and although these rocks are dominantly clastic, they include marine limestones of Jurassic age. The remaining 90 percent of the Mesozoic section, of Cretaceous age, is a monotonous sequence of shale, mudstone, and sandstone, which includes a few conglomerate beds and, in the upper part of the section, a few beds of coal and carbonaceous sandstone and shale.

The Mesozoic rocks in the Gallatin Range were originally divided into the Teton Formation of "supposed Triassic" age, the Ellis Formation of Jurassic age, and the Dakota, Colorado, Montana, and Laramie Formations of Cretaceous age (Iddings and Weed, 1899, p. 36-55; Stanton, 1899, p. 600-607; this report, fig 2). Extensive collections of fossils from these formations were described and illustrated by Stanton (1899) and by Knowlton (1899). In the present report, the rocks are separated into units that include one unit of Triassic age, two of Jurassic age, and 10 of Cretaceous age. Some of these units include several formations that in other areas have been mapped separately.

TRIASSIC

Triassic rocks, in outcrops made conspicuous by the red color of some of the rocks, are present in the Gallatin Range, and the best exposed sections are between Bannock Peak and Fawn Pass. These rocks were first mapped as the upper part of the Teton Formation (Weed, 1896, p. 5). Condit (1919, p. 113-114, 117-118), who recognized that both Permian and Triassic rocks were included in the "Teton Formation," suggested that the name be discontinued, and it was subsequently abandoned (McKelvey and others, 1956, p. 2835). The upper 145 feet of the rocks originally included in the Teton Formation (Iddings and Weed, 1899, p. 34, units 34-38), is here treated as an undivided unit of rocks that includes the Dinwoody, Woodside, and Thaynes (?) Formations of Triassic age.

Dinwoody Formation

The Dinwoody Formation of Early Triassic age (Kummel, 1954, p. 167-168; Condit and others, 1928, p. 187) is exposed south of Fawn Pass but apparently is absent farther east. The formation is light-gray to yellowish-gray limestone that is very fine grained and thin bedded and characteristically weathers pale brown or pale yellowish brown. It is 0-50 feet thick. Condit, Finch, and Pardee (1928, p. 187) described about 30 feet of thin-bedded white limestone of Triassic age on top of Quadrant Mountain, rocks that would be equivalent to the Dinwoody Formation, but I could not identify this white limestone in my mapping, and I believe that on Quadrant Mountain the younger Woodside Formation unconformably overlies the Shedhorn Sandstone, although erosional remnants of the Dinwoody may be present in some places.

Woodside Formation

The red rocks of the Woodside Formation of Early Triassic age (Condit and others, 1928, p. 187; Kummel, 1954, p. 181, pl. 37) are conspicuous and form the red hills that crown Quadrant Mountain and a bright-red band that extends from the upper valley of Fawn Creek into the Gallatin River valley. The lower 10-20 feet of the formation is greenish-gray shale that grades upward into 1- to 5-foot beds of grayish-red to moderate-reddish-brown mudstone, siltstone, and very fine grained micaceous sandstone, and a few interbeds as much as 3 feet thick of greenish-gray mudstone and siltstone. The red rocks form most of the unit, and the greenish-gray rocks in the lower part are commonly not exposed. Iddings and Weed (1899, p. 34) described a 10-foot-thick calcite cemented conglomerate at the base of the formation near the southeast spur of Quadrant Mountain, but the conglomerate does not appear to be present any place else. The Woodside is 75-100 feet thick.

The Woodside overlies the Dinwoody Formation south of Fawn Pass, but a short distance to the east on Quadrant Mountain, it unconformably overlies the Shedhorn Sandstone. The disappearance of the Dinwoody and the local presence of conglomerate, as noted by Iddings and Weed (1899, p. 34), suggest that part or all of the Dinwoody in this area was eroded before deposition of the Woodside and that the Woodside is unconformable on the Dinwoody, as well as on the Shedhorn Sandstone.

Thaynes(?) Formation

The Thaynes(?) Formation forms a ledge above the Woodside Formation near Fawn Creek and south of Fawn Pass. The unit commonly is 15-50 feet thick but in some places it is absent, and the overlying Jurassic rocks rest unconformably on the Woodside. The sandstone is light gray, fine grained, clean, and calcareous, is made up of well-sorted well-rounded quartz grains, and occurs in beds 0.2-5 feet thick. The rock weathers pale yellowish brown, and the weathered beds yield a blocky talus that fringes the cliffy outcrops.

The age of the sandstone is uncertain, as no fossils were found. The Lower Triassic Thaynes Formation, however, is predominantly calcareous sandstone in adjacent, southwestern Montana (Kummel, 1954, p. 177), and Witkind (1969) recognized in the Madison Range a unit of calcareous siltstone about 10 feet thick that he thought might be part of the Thaynes. Furthermore, Kummel (1954, p. 177) pointed out the widespread effects of pre-Jurassic erosion, perhaps reflected by the erratic changes in thickness of this sandstone. These regional considerations suggest to me that the sandstone is most probably a thin, eastward erosional edge of the Thaynes Formation, and, accordingly, I have mapped it with Triassic rocks. Fraser, Waldrop, and Hayden (1969, p. 20), however, studied the same unit near Gardiner, Mont., the only other area where it has been recognized, and considered it to be a basal sandstone representing marine onlap in Jurassic time. Crickmay (1936, p. 549-552) also assigned the sandstone to the Jurassic and stated that its regional continuity and lithologic contrast with the underlying Triassic red beds indicate that it was a basal Jurassic unit. However, the regional relations of the sandstone are unknown, and its lithology differs considerably from that of the overlying rocks, as well as from that of the underlying ones.

JURASSIC

Jurassic rocks crop out in the Gallatin Range and near Mammoth Hot Springs and in slices along the Gardiner fault zone. These rocks are separable into four formations — the marine Sawtooth, Rierdon, and Swift Formations, all in the Ellis Group, and the nonmarine Morrison Formation — which have a total thickness of about 400 feet. Iddings and Weed (1899, p. 37-38, 51) included most of the Jurassic rocks in the "Ellis limestone" and "Ellis sandstone," but their descriptions of these units differ from place to place, and at some localities they seemingly included part of the Morrison Formation in the Cretaceous Dakota Formation. The Jurassic rocks in Fawn Pass and at the head of Fan Creek were measured and described by Crickmay (1936, p. 549-552), who also reported on fossils collected from them. He assigned these rocks to different formations — Morrison(?), Sundance (formerly the upper part of the Ellis), and Ellis (formerly the lower part of the Ellis) — from those used in this report, but his generalized lithologic descriptions are clear, and the lists of fossils contained in the rocks are extensive. Crickmay's section (1939, p. 549-550) at the head of East Fork Fan Creek is the best exposed section of Jurassic rocks in the northern part of Yellowstone National Park and is a reference sequence for Jurassic rocks, even through the Morrison Formation is somewhat thinner there than in most other places in the Gallatin Range. Fossils collected from Jurassic rocks in various places in the Gallatin Range were also described by Stanton (1899).

Sawtooth Formation (Ellis Group)

The Sawtooth Formation in Yellowstone National Park consists of the three lithologic units that characterize the formation in south-central Montana (Imlay and others, 1948; Imlay, 1952, p. 968; Peterson 1957, p. 404-408): a lower red-bed member, a middle limestone member, and an upper red-bed member. The lower red-bed member, 20-30 feet thick and consisting of grayish-red and grayish-green siltstone, mudstone, and shale, unconformably overlies Triassic rocks; the member commonly is concealed by soil and float. The middle limestone member, 60-75 feet thick, is medium light gray to medium dark gray, very fine grained, and platy to chippy; it weathers light gray to olive gray. The limestone commonly is in beds 0.1-1 foot thick, many of which contain abundant fossils. Interbeds of medium-dark-gray calcareous shale, less than 1 foot thick, are common, and Crickmay (1936, p. 549) showed a 14-foot-thick shale bed in the upper part of the member. The upper red-bed member, 25-40 feet thick, is pale-red to grayish-red siltstone, mudstone, and shale that almost everywhere is concealed by red soil.

The beds that Crickmay (1936, p. 549-552) labeled units Yb through Yj and included in the Ellis Formation and the "Lower Sundance formation," constitute the Middle Jurassic Sawtooth Formation of this report. The section measured by Crickmay (1936, p. 549-550) is about 140 feet thick.

Rierdon Formation (Ellis Group)

The upper red-bed member of the Sawtooth Formation is conformably overlain by the basal oolitic limestone of the Rierdon Formation, a resistant rock that almost everywhere forms a low ledge or is represented by abundant float. The Rierdon Formation, 40-60 feet thick, consists mostly of mudstone, siltstone, and shale. The basal limestone is 1-6 feet thick, medium gray, fine grained, and very oolitic and weathers medium light gray. It is overlain by pale-red to grayish-red calcareous mudstone, siltstone, and shale, about 20-25 feet thick, which in turn, is overlain by 20-30 feet of medium-gray calcareous mudstone, the uppermost unit of the Rierdon. A few medium- to light-gray partly oolitic limestone beds, generally less than 0.5 foot thick, are interbedded in the middle and upper parts of the formation, but they are generally concealed beneath the soil-covered smooth slope developed on the formation.

The Rierdon Formation is not as richly fossiliferous as the underlying Sawtooth Formation, but is does contain many fossils (Crickmay, 1936, p. 549- 551, units Yk and Yl) that date it as Late Jurassic (Crickmay, 1936, p. 550; Imlay, 1952, p. 968).

Swift Formation (Ellis Group)

The uppermost formation in the Ellis Group is the Swift Formation, which consists of 15-60 feet of pale-yellowish-brown to light-olive-gray fine- to medium-grained calcareous sandstone and very sandy limestone. These distinctive rocks are a mixture of subangular to subrounded quartz grains, subrounded grains of brownish chert, glauconite, oolites, and abundant fragments of fossils, all cemented by calcite. Locally, the rocks are sandy coquinas. Beds are generally 0.1-1 foot thick and are cross-laminated. Almost the entire thickness of the Swift is commonly exposed in a cliff that is made by prominent by the adjacent smooth slopes developed on the underlying and overlying mudstones.

Some of the fossils in the Swift Formation in Yellowstone National Park were described by Crickmay (1936, p. 549-564, unit Ym), who included the unit in the Sundance Formation, and these fossils established the formation as Late Jurassic.

Morrison Formation

The Morrison Formation, of Late Jurassic age, is poorly exposed in northwestern Yellowstone National

Park. In most places it forms a gently slope between ledges of Swift sandstone below and the basal conglomerate and sandstone of the Kootenai Formation above. The formation is about 200 feet thick and consists mainly of grayish-red, dark-gray, and grayish-green partly calcareous shale, mudstone, and siltstone. The basal part of the Morrison is grayish-red shale and mudstone that overlies the Swift Formation, and there are at least a few beds, 1-2 feet thick, of olive-gray fine-grained limestone in the lower part of the formation. The upper half includes lenticular interbeds, as much as 3 feet thick, of yellowish-brown to light-gray siltstone and fine-grained to very fine grained calcareous cross-laminated quartz sandstone. The uppermost part of the formation is mainly dark-gray shale and platy mudstone, but contains many sandstone interbeds and a few thin interbeds of medium-light-gray chippy limestone. Northwest of Fawn Pass at the head of Stellaria Creek, these upper beds yielded fragments from the end of a leg bone of a moderately large dinosaur, but the bone pieces were too fragmentary to permit more positive identification (G. Edward Lewis, written commun., 1967). The same beds contain abundant fossil wood.

CRETACEOUS

About half the sedimentary rocks exposed in the northern part of Yellowstone National Park are of Cretaceous age. This sequence, almost 4,000 feet thick, consists, in ascending order, the Kootenai Formation, Thermopolis Shale, Mowry Shale, Frontier Sandstone, Cody Shale, Telegraph Creek Formation, Eagle Sandstone, Everts Formation, and Landslide Creek Formation (fig. 2). The rocks between the base of the Kootenai and the upper part of the Eagle are best exposed on the ridge southwest of Electric Peak; the Cody Shale and all the younger Cretaceous formations are exposed on Mount Everts. These formations were studied and described in great detail in the adjacent Gardiner, Mont., area by Fraser, Waldrop, and Hyden (1969). The Cretaceous rocks in the two areas do not differ in any significant respect, so only summary descriptions of them are given in this report; complete descriptions and measured sections are given in the report on the Gardiner area by Fraser, Waldrop, and Hyden (1969).

The regional relations, stratigraphy, and age of Cretaceous rocks in other nearby areas were discussed by Roberts (1965).

The Cretaceous rocks in the northern part of the park were divided by Iddings and Weed (1899, p. 37-38, 50-55) into the Dakota, Colorado, Montana, and Laramie Formations. The descriptions and measured sections by Iddings and Weed do not permit specific correlations with the present mapped units. Moreover, their assignments of the rocks are not consistent. The Cinnabar Mountain measured section (Iddings and Weed, 1899, p. 53-54), which was given as typical for the Gallatin region, does, however, suggest that equivalent units are about as shown in figure 2.

LOWER CRETACEOUS FORMATIONS

Lower Cretaceous formations crop out in the Gallatin Range and in fault slivers along the Gardiner fault zone. They are the Kootenai Formation, Thermopolis Shale, and Mowry Shale (fig. 2) and have an aggregate thickness of about 1,000 feet. Units of comparable age mapped by Fraser, Waldrop, and Hyden (1969, p. 22-25) near Gardiner, Mont., are the Cloverly Formation, which includes equivalents of the Kootenai and the lower sandstone member of the Thermopolis Shale, which includes equivalents of the middle shale and upper sandstone members of the Thermopolis of this report, and the Mowry Shale, which is the same as the Mowry of this report.

The base of the Kootenai — oldest unit of the Cretaceous sequence — is a gray massive cross-laminated partly calcareous chert-pebble conglomerate and quartz-chert sandstone and conglomerate sandstone, 20-50 feet thick. This basal unit, assigned to the Pryor Conglomerate Member of the Kootenai near Livingston, Mont., by Roberts (1965, p. B55), unconformably overlies the Morrison Formation. The basal conglomerate is overlain by medium-dark-gray and grayish-red shale and mudstone containing a few thin interbeds of light-gray limestone and sandstone. The upper part of the formation contains the distinctive "gastropod limestone," a medium- to light- gray fine-grained limestone, 20-30 feet thick, that in many places is made up almost wholly of poorly preserved molds of gastropod shells. This limestone is overlain by 10-20 feet of medium- to light-gray shale that is the uppermost part of the Kootenai Formation. The formation is from 250 to 300 feet thick.

The Kootenai is overlain with apparent conformity by the Thermopolis Shale, which includes a lower sandstone member, 75-100 feet thick; and an upper sandstone member, about 100 feet thick. The lower sandstone member is medium light gray to yellowish gray, clean, medium to thick bedded, and fine grained and includes subordinate interbeds of medium-dark-gray shale and siltstone. The member forms a cliff above the Kootenai. The middle shale member is dark-gray very fissile shale containing a few interbeds, 1-3 feet thick, of olive-gray fine-grained sandstone. The upper sandstone member is mainly yellowish-gray to grayish-brown fine-grained partly glauconitic calcareous feldspathic cross-laminated sandstone. Some of the sandstone contains chips of mudstone, and the upper member also contains a few interbeds of dark-gray shale and mudstone. These dirty sandstones and interbedded finer grained rocks contain *Ostrea anomioides* Meek (W. A. Cobban, oral commun., 1967), which is restricted to the Taft Hill Member of the Blackleaf Formation or its equivalents.

The youngest of the Lower Cretaceous units, the Mowry Shale, conformably overlies the upper sandstone of the Thermopolis, and consists of about 300 feet of medium- to dark-gray hard chippy to pencilly mudstone and shale. Some beds in the upper part of the formation contain abundant fish scales. The formation includes a few interbeds of olive-gray fine- to medium-grained sandstone, generally less than 1 foot thick, and a few distinctive light-gray bentonite beds, as much as 3 feet thick. The bentonite beds are made especially conspicuous by the abundant ground squirrel holes dug in the soft rock.

UPPER CRETACEOUS FORMATIONS

Upper Cretaceous rocks consist, in ascending order, of the Frontier Sandstone, Cody Shale, Telegraph Creek Formation, Eagle Sandstone, Everts Formation, and part of the Landslide Creek Formation (fig. 2). The Everts Formation and the Landslide Creek Foramtion that overlies it (Fraser and others, 1969, p. 32-34) are equivalent to rocks that elsewhere in Montana have been included in the Livingston Group. (See Roberts, 1965, p. B59-B60, for summary.)

The Frontier Sandstone, lowest of the Upper Cretaceous formations, conformably overlies the Mowry Shale and is mainly yellowish-gray to medium-gray fine-grained sandstone speckled with abundant grains of heavy minerals; the Frontier is partly calcareous, partly tuffaceous, ripple marked and cross-laminated. It is in 1- to 5-foot-thick beds that are marked by worm trails and burrows. The sandstone beds are separated by 0.1- to 0.2-foot-thick interbeds of dark-gray fissile shale. Fish teeth are in some of the sandstone beds, and small iron concretions and glauconite are in the uppermost beds. The beds at the top of the formation typically are stained reddish brown or yellowish brown. The formation is a little more than 100 feet thick on the southwest spur of Electric Peak, but is only 40-60 feet thick on the southeast spur of Electric Peak and on Mount Everts, where it forms a low cliff. The Frontier Sandstone mapped in Yellowstone National Park is correlative with the lower sandstone member of the Frontier in the Gardiner area.

The Frontier is conformably overlain by the Cody Shale, which is dominantly dark-gray shale and lenticular brownish-gray dirty sandstone. It is separable into three units: a lower shale member, 350-400 feet thick; a middle sandstone member, about 50 feet thick, equivalent to the Eldridge Creek Member near Livingston, Mont. (W.A. Cobban, oral commun., 1967; Roberts, 1965, p. B58); and an upper shale member, about 800 feet thick.

The lower shale member is medium- to dark-gray and yellowish-gray shale and mudstone, interbedded with minor amounts of gray and yellowish-gray siltstone and fine-grained sandstone (stratigraphic sections, see appendix.)

The middle sandstone member — dominantly glauconitic sandstone — is yellowish gray to brownish gray, very fine grained, and rusty weathering and is speckled with heavy minerals and glauconite. It occurs in beds, 1-2 feet thick, that are partly cross-laminated. The member includes some interbeds of dark-gray shale, generally only about 1 foot thick. It contains a distinct *Inoceramus* fauna (W. A. Cobban, oral commun., 1967) and is much more fossiliferous than the rest of the Cody Shale.

The upper shale member is mainly medium-to dark-gray platy to chippy mudstone and shale, but it includes some interbeds of dark-greenish-gray to brownish-gray very fine grained sandstone and siltstone. The sandstone beds become more abundant in the upper 150 feet of the member, as the Cody Shale grades into the overlying Telegraph Creek Formation.

The Cody Shale on Mount Everts was studied in detail by Fox (1939), who described the faunal zones and discussed some of their regional relations. The formation is about 1,200 feet thick (Fox, 1939) on Mount Everts and about the same on Electric Peak. On Electric Peak, however, its apparent thickness is almost doubled by the many intruded sills.

The lower shale member and middle sandstone member were included in the Frontier Formation in the Gardiner area by Fraser, Waldrop, and Hyden (1969). The lower shale member of the Cody as mapped in Yellowstone National Park is the same as the middle shale member of the Frontier in the Gardiner area, and the middle sandstone member in Yellowstone National Park, or the Eldridge Creek Member of the Cody near Livingston, Mont., is the same as the upper sandstone member of the Frontier in the Gardiner area.

The Telegraph Creek Formation, about 300 feet thick, is composed mainly of sandstone, and is a gradational unit between the underlying Cody Shale and the overlying Eagle Sandstone. The Telegraph Creek is dominantly brownish-gray to medium-light-gray fine-grained to very fine grained calcareous sandstone, in beds 0.2-0.5 foot thick, with interbedded medium-gray platy mudstone and shale. Medium-light-gray beds of sandstone are predominant in the upper part of the formation. The formation grades into the overlying Eagle Sandstone, by decrease in mudstone and shale beds and increase in clean thick-bedded to massive sandstone.

The Eagle Sandstone, about 800 feet thick, is the youngest Cretaceous formation present on Electric Peak, where it caps the peak and forms the long dipslope that reaches northward to the park boundary. The Eagle also forms the lowest massive cliffs on Mount Everts, where it was measured and described by Fraser, Waldrop, and Hyden (1969). The base of the formation, the Virgelle Sandstone Member, about 160 feet thick, consists of medium-light-gray thick-bedded to massive cross-laminated fine-grained arkosic sandstone beds that form a prominent cliff on Mount Everts. Above the Virgelle Member, the Eagle is mainly sandstone, but it is less massive and includes many beds, as much as a few feet thick, of dark-gray carbonaceous shale and thin lenticular beds of coal. The sandstone beds are light gray to medium gray, thin to thick bedded, fine to medium grained, calcareous, and cross-laminated and are speckled with abundant grains of heavy minerals. The top of the formation on Mount Everts consists of a massive cliff-forming sandstone, about 20 feet thick, that lies above the highest coal bed (Fraser and others, 1969).

An erosional unconformity separates the Eagle from the overlying Everts Formation, the type section of which is on Mount Everts (Fraser and others, 1969). The Everts Formation, about 1,200 feet thick, forms the upper cliffs on Mount Everts and is mainly light- to medium-gray thin-bedded cross-laminated fine- to medium-grained calcareous and tuffaceous sandstone speckled with grains of heavy minerals. It contains interbeds, 1-2 feet thick, of dark-gray to olive-gray shale, mudstone, and siltstone. The upper half of the formation is dominantly sandstone, and the finer grained interbeds are most abundant in the lower half.

The Everts Formation is overlain, probably unconformably, by the Landslide Creek Formation (Fraser and others, 1969, p. 34). In the report area, only remnants of this formation are preserved on the northeast slope of Mount Everts, where the rocks are more or less concealed by glacial debris. All these limited exposures are of drab greenish-gray poorly sorted sandstone and conglomerate, made up principally of sand, pebbles, and cobbles of andesitic volcanic rocks.

APPENDIX

STRATIGRAPHIC SECTIONS

Section measured from top of Three Forks Formation to base of Pilgrim Limestone on northeast spur of Three Rivers Peak, Gallatin Range, Yellowstone National Park, Wyo.

	Thickness (feet)
Lodgepole Limestone:	
62. Limestone, medium-gray, medium-light-gray-weathering, fine-grained, fossiliferous; contains abundant medium-dark-gray chert in nodules as much as 0.2 ft thick and 0.1–3 ft long; in beds 0.2–0.5 ft thick	Not measured
Three Forks Formation (reference section):	
61. Dolomite, light-olive-gray, grayish-orange- or light-gray-weathering, very fine grained; in beds 0.1–0.2 ft thick. Lower part of unit strongly sheared and crumpled	48.2
60. Mudstone, light-olive, pale-orange- to grayish-orange-weathering	4.0
59. Dolomite; similar to unit 61	2.0
58. Mudstone; similar to unit 60; about 5 percent of unit is dolomite similar to unit 61; in interbeds 0.1–0.5 ft thick	22.0
57. Dolomite; similar to unit 61; in beds 0.3–0.5 ft thick	4.0
Total thickness, Three Forks Formation	80.2
Jefferson Formation (reference section):	
56. Dolomite, pale-yellowish-brown to brownish-gray, yellowish-gray- to light-olive-gray-weathering, very fine grained, massive	31.0
55. Shale, greenish-gray, fissile, papery	3.0
54. Dolomite, pale-yellowish-brown, grayish-orange-weathering, extremely fine grained, and dolomitic siltstone that is very pale yellowish brown, grayish orange weathering, and thinly platy	3.0
53. Sill; not measured	.....
52. Dolomite, medium-dark-gray, medium-gray-weathering, fine-grained, sugary, fetid; faintly mottled medium light gray; laminated in some places	5.0
51. Dolomite and interbedded silty dolomite; dolomite is medium gray; weathers pale grayish orange to light-olive-gray; is extremely fine grained; has brittle blocky fracture; in crinkly beds 0.5–2.0 ft thick; some beds thinly laminated (0.1 mm); forms cliffy outcrop. Silty dolomite is very pale yellowish brown; weathers grayish orange; is thinly platy; grades into dolomitic siltstone	11.0
50. Sill; not measured	.....
49. Dolomite, brownish-gray, dark-yellowish-brown- to pale-yellowish-brown-weathering, fine-grained (0.1 mm), sugary, fetid; in beds 0.1–3.0 ft thick; laminated in places. Uppermost 0.8 ft of unit bleached by sill	5.8
48. Dolomite; similar to dolomite in unit 51	17.5
47. Dolomite and interbedded dolomitic siltstone; dolomite is brownish gray; weathers grayish orange; is extremely fine grained; blocky weathering. Dolomitic siltstone is very pale yellowish brown; weathers grayish orange; is thinly platy	6.3
46. Dolomite; similar to unit 49	6.6
45. Dolomite and dolomitic siltstone; similar to unit 47; middle of unit includes 1.0-ft-thick bed of dolomite similar to unit 49	7.0
44. Dolomite; similar to unit 49; thinly laminated	1.5
43. Dolomite and dolomitic siltstone; similar to unit 47	6.5
42. Dolomite; similar to unit 49; massive; contains abundant stromatoporoids	13.0
41. Dolomite, brownish-gray, grayish-orange-weathering, extremely fine grained, blocky weathering	2.9

Jefferson Formation—Continued — *Thickness (feet)*

40. Siltstone, dolomitic, very pale yellowish brown, grayish-orange-weathering, thinly platy 3.5
39. Dolomite; similar to unit 49; becomes massive in upper half of unit; includes a few lighter colored beds of intraformational breccia 63.0
38. Dolomite, yellowish-brown and grayish-brown, pale-yellowish-brown- to light-gray-weathering, extremely fine grained; in beds 0.2–0.5 ft thick; laminated, blocky weathering; forms ledge 20.9
37. Dolomite, medium-dark-gray, brownish-gray-weathering, fetid; in beds 2.0–3.0 ft thick; laminated. Interbedded dolomite similar to unit 38. Unit includes a few thin beds of solution breccia 10.0
36. Limestone, dark-gray, medium-gray-weathering, extremely fine grained; contains fossil fragments 2.5
35. Dolomite, medium-gray; weathers light gray to very pale yellowish brown; very fine grained; in beds 0.5–1.0 ft thick; laminated 9.0
34. Limestone; similar to unit 36 .... 1.0
33. Dolomite; similar to unit 35 .... 1.0
32. Dolomite, pale-brown to brownish-gray; weathers same color; fine grained, sugary; in beds 1.5–3.0 ft thick; partly laminated. Unit is massive appearing; contains small horn corals and stromatoporoids 9.9
31. Dolomite, light-gray and medium-gray; weathers same color; in beds 0.1–0.5 ft thick; thinly laminated 2.9

Total measured thickness, Jefferson Formation 243.8

Bighorn Dolomite (reference section):

30. Dolomite, light-brownish-gray; weathers very pale yellowish brown to light gray; extremely fine grained ($<$ 0.1 mm); in beds 3.0 ft thick to massive; hackly weathering; contains large receptaculitids typical of Bighorn Dolomite (R. J. Ross, written commun., 1969) 28.0
29. Dolomite; similar to unit 30, but beds are 0.1–0.3 ft thick and laminated (0.1 mm); platy weathering; contains medium-light-gray to yellowish-orange chert in lenses 0.05–0.1 ft thick and 1.0–5.0 ft long. Dolomite in middle 20 ft of unit is medium gray; weathers medium light gray to light gray, but is otherwise similar to rest of unit 75.0

Total thickness, Bighorn Dolomite 103.0

Snowy Range Formation (reference section):

Grove Creek Member:

28. Dolomite, light-brownish-gray, fine-grained, fetid; upper part of unit contains lenses, 0.1 ft thick and as much as 1.0 ft long, of sandstone that contains fragments of brachiopods and trilobites not identifiable, but suggesting Cambrian age (A. R. Palmer, written commun., 1969) 10.0
27. Dolomite, light-brownish-gray, very fine grained; in beds 0.1–0.3 ft thick; irregularly bedded; upper half of unit is dolomitic limestone; includes 5-ft-thick unit of massive dolomite composed of vertical algal columns 17.0
26. Limestone, light-gray; weathers same color; very fine grained, ribboned; similar to unit 24; contains *Billingsella* fragments (A. R. Palmer, written commun., 1969) 1.0
25. Dolomite, light-brownish-gray, very fine grained; in beds 0.1–0.3 ft thick; irregularly bedded. Unit forms base of cliff 11.6

Total thickness, Grove Creek Member 39.6

Sage Limestone Member:

24. Limestone, medium-gray, medium-light-gray-weathering, very fine grained, laminated; irregularly ribboned with grayish-orange to dark-yellowish-orange very fine grained sandy limestone and calcareous sandstone; in beds 1.0–1.5 ft thick. And alternating interbeds, 1.5–3.0 ft thick, of similar rocks that are mottled, rather than ribboned. Upper 9.0 ft of unit is massive mottled limestone. Locally coquinoid; contains silicified *Billingsella,* dolomoldic siliceous steinkerns of a moderately large *Pelagiella,* and a moderately high spired sinistral gastropod that could be *Scaevogyra* (A. R. Palmer, written commun., 1967) 31.0
23. Limestone; similar to ribboned rocks of unit 24; platy weathering; forms ledge. Bedding surfaces of upper beds contain abundant chitinous brachiopods. Unit contains assemblage of fossils from *Taenicephalus* zone, including *T. shumardi* (Hall), *Maustonia nasuta* (Hall), *Angulotreta* sp., *Pelagiella* and *Billingsella* (A. R. Palmer, written commun., 1967) 13.1

Total thickness, Sage Limestone Member 44.1

Dry Creek Shale Member:

22. Shale, greenish-gray to light-olive-gray, papery 0.8
21. Limestone, medium-gray, medium-light-gray-weathering, very fine grained to aphanitic, sandy; in irregular beds 0.1–0.2 ft thick; includes thin beds of flat-pebble conglomerate 2.0
20. Shale; similar to unit 22 .... 4.0
19. Limestone, medium-light-gray, light-gray-weathering, oolitic; coquinoid (contains trilobite fragments); glauconitic; contains moderately abundant limestone chips 2.6
18. Concealed; underlain by interbedded greenish-gray to light-olive-gray paper-thin shale and less common thin beds of light-olive-gray very fine grained calcareous sandstone. Base of unit is marked by 1.0-ft-thick bed of yellowish-orange to yellowish-brown very fine grained thinly platy calcareous sandstone 31.2

Total thickness, Dry Creek Shale Member 40.6

Total thickness, Snowy Range Formation 124.3

	Thickness (feet)
Pilgrim Limestone (reference section):	
17. Limestone, medium-gray; weathers mottled medium gray to medium light gray and yellowish gray, pale yellowish brown, and light olive gray; fine to medium grained, oolitic, massive; forms prominent ledge	44.0
16. Concealed; underlain mainly by limestone flat-pebble conglomerate that is medium gray; weathers medium light gray; medium to coarse grained and glauconitic; contains abundant fossil fragments and a few beds marked by silty ribbons. Center of unit includes 10.0-ft-thick bed of medium-gray light-gray-weathering very fine grained to aphanitic limestone. Top 6 ft of unit is light-gray very fine grained limestone, irregularly mottled with pale-yellowish-orange silty limestone	60.0
15. Limestone, medium-gray, medium-light-gray-weathering, very fine grained; ribboned with medium-gray limestone that weathers moderate yellowish brown to grayish orange and contains abundant very fine sand and silt in irregular layers and seams 0.01–0.1 ft thick. Beds 0.2–0.5 ft thick. Includes interbedded oolitic limestone that is medium gray; weathers same color; is glauconitic; contains abundant trilobite fragments; in beds 0.3–1.0 ft thick	14.7
14. Limestone, oolitic; similar to oolitic rocks in unit 15; in beds 0.3–3.0 ft thick	10.6
13. Limestone flat-pebble conglomerate; similar to that in unit 16	.3
12. Shale, light-olive-gray, fissile	.2
11. Limestone, oolitic; similar to unit 14	1.3
10. Limestone and interbedded oolitic flat-pebble conglomerate and ribboned rocks; similar to rocks in units 15 and 16	1.7
9. Limestone, ribboned; similar to ribboned limestone in unit 15	3.9
8. Limestone flat-pebble conglomerate; similar to unit 16	1.7
7. Limestone, oolitic, medium-gray; weathers same color; contains moderately abundant layers of flat-pebble conglomerate as much as 0.1 ft thick; glauconitic, fossiliferous	4.4
6. Concealed; underlain by ribboned limestone similar to that of unit 15	6.0
5. Hornfels, greenish-gray to grayish-green; weathers same color; contains abundant lenses, as much as 0.1 ft thick and 0.1–0.3 ft long, of greenish-gray relict limestone that weathers very light gray	18.0
4. Sill; not measured	....
3. Hornfels; similar to that of unit 5. Upper part of unit includes a few thin interbeds of chippy medium-dark-gray argillite. Lower part of unit contains cordierite	40.0
Total measured thickness, Pilgrim Limestone	206.8
Park Shale:	
2. Shale, medium-dark-gray, chippy metamorphosed	5.0
Sill:	
1. Indian Creek laccolith; not measured	....
Base of section.	

Measured section of Madison Group on south face of Bannock Peak, southern Gallatin Range, Yellowstone National Park, Wyo.

[Measured by W. J. Mapel, A. E. Roberts, and E. K. Maughan in 1964]

	Thickness (feet)
Amsden Formation.	
Madison Group (reference section):	
Mission Canyon Limestone (reference section):	
Upper member:	
24. Dolomite, brecciated and very well recemented, very light gray, finely crystalline, cliff-forming; massively bedded. Light color is very conspicuous. Scattered lenses of quartzose sandstone. A few feet of pale-green to yellow dolomitic siltstone near base, mostly covered	97
23. Limestone, cherty, fine to coarsely crystalline; abundant veins and vugs of very granular white calcite. Unit forms rubbly slope	210
22. Breccia, dolomite; angular to subangular yellowish-brown fragments, as much as 1 ft, set in matrix of medium-gray limestone. Unit massive to poorly bedded; forms very cavernous cliff	60
Total thickness, upper member	367
Lower member:	
21. Dolomite, light-olive-brown; medium to thick bedded in lower part; thinly laminated in upper part and containing algal structures. Generally forms rubbly slope	23
20. Dolomite, dark-olive-brown; forms small massive cliff	5
19. Limestone, laminated, cherty, yellowish-gray to brownish-gray; some low-angle cross-laminae and yellowish chert stringers. Generally forms slope	41
18. Limestone, brownish-gray, light-medium-gray-weathering, coarsely crystalline, abundantly fossiliferous; numerous vugs, as much as 4 in. in diameter, filled with calcite crystals. Generally forms massive cliff	26
17. Limestone, brownish- to yellowish-gray, fine to coarsely crystalline; commonly laminated with light- and dark-gray bands; cherty; less cherty and less laminated and more thin and medium beds upward. Forms rubbly slope	20
16. Limestone, brownish- to medium-gray, very fossiliferous; corals about 40 feet from base; thick bedded to massive; generally forms cliff	80
15. Limestone, massive to medium-bedded, light-gray, arenaceous, laminated, fine to coarsely crystalline	19
14. Limestone, laminated, cherty, light-gray	10
13. Limestone, massive to medium-bedded, grayish-white, coarsely to finely crystalline; highly porous with abundant pinpoint vugs. Upper two-thirds of unit is arenaceous limestone; yellowish- to olive-gray alternate laminae, locally penecontemporaneously deformed. Forms slope	35

	Thickness (feet)
Madison Group—Continued	
Mission Canyon Limestone—Continued	
Lower member—Continued	
12. Limestone; dolomitic in lower 5 ft; medium dark gray; weathers medium-gray; medium to thin bedded; moderately to abundantly fossiliferous, mostly crinoid trash, colonial and horn corals; forms prominent cliff with ledges about 10–12 ft apart	82
11. Dolomite, medium-olive-gray, dense to very fine grained; thin bedded and shattered by vertical fractures; includes pods of crinoid coquina. Forms prominent reentrant on face of cliff	4
10. Limestone, slightly dolomitic, abundantly fossiliferous; composed mostly of crinoid trash, similar to coquina of unit 11; olive brown; thick and massive bedded. Forms part of cliff	20
9. Dolomite, massive, very light gray, greenish-white-weathering, medium to coarsely crystalline; limy in upper part. Intermixed porous and nonporous zones; some low-angle cross-stratification; in a few places porosity seems to be bedding controlled; some stylolites; forms prominent massive cliff; only slightly fossiliferous; a few thin moderately fossiliferous zones	82
Total thickness, lower member	447
Total thickness, Mission Canyon Limestone (reference section)	814
Lodgepole Limestone (reference section):	
8. Limestone, thin-bedded, grayish-brown, silty; fossiliferous throughout; generally forms rubble slope. Sample collected 25 ft from top of unit contains *Zaphrentites* sp., *Homalophyllites* sp., and *Vesiculophyllum* sp. (W. J. Sando, written commun., 1968)	46
7. Limestone and siltstone interbedded; thin beds less than 1 ft thick. About two-thirds limestone, one-third siltstone; cross-laminated; fine to coarsely crystalline; coarse material is mostly crinoid trash. Limestone is blue gray; siltstone is yellowish gray. Limestone is arenaceous. Basal 3–5 ft generally poorly exposed and forms slope, but most beds can be seen. Most of unit forms series of thin ledges. Silt gradually decreases upward	110
6. Limestone; almost coquina of crinoids and other fossiliferous trash; abundant brachiopods; scattered lenses of siltstone	2
5. Limestone, thin-bedded; fine to coarsely crystalline, coarse is mostly crinoid trash; fossiliferous, brownish gray; weathers yellowish gray. Slope poorly exposed	9
4. Limestone, fine to coarsely crystalline, fossiliferous; abundant crinoid trash; brownish gray; light gray weathering; some thin and some massive beds, but mostly medium to thick beds. Mostly cliff forming, but with a talus slope at base	60
3. Limestone, coarse to finely crystalline, grayish-brown. Abundant crinoid trash in many beds. Thin to medium bedded, with very thin siltstone partings	22
2. Limestone, cherty, medium-gray, medium-light-gray-weathering, very finely crystalline. Chert is medium gray; forms irregular nodules and lenses; weathers brownish gray; mostly thin, some medium bedded. Very thin siltstone partings. Forms cliff	125
1. Limestone, mostly covered, sandy, coarsely crystalline, pale-yellowish-brown to medium-light-gray; abundant crinoid columnals; thinly bedded to laminated; weathers to slope, generally talus covered	112
Total thickness, Lodgepole Limestone (reference section)	486
Total thickness, Madison Group (reference section)	1,300

Three Forks Formation.

Measured section of lower member of Cody Shale and Frontier Sandstone, Mount Everts, opposite Sheepeater Canyon Bridge, Yellowstone National Park, Wyo.

	Approximate thickness (feet)
Cody Shale:	
Lower member:	
17. Largely concealed; forms smooth slope underlain by interbedded shale, mudstone, siltstone, and sandstone. Shale is medium dark gray, papery. Mudstone and siltstone are medium dark gray to light olive gray. Sandstone is light olive gray to medium dark gray, very fine grained, and speckled with abundant heavy mineral grains; partly calcareous; platy weathering; in beds 1–4 ft thick. Unit is fossiliferous and contains numerous dark-yellowish-brown calcareous nodules	200
16. Shale, medium-dark-gray, papery	15
15. Siltstone, light-brown to dark-yellowish-brown, nodular, ocherous	3
14. Shale, medium-gray, papery; contains abundant veinlets of white calcite and gypsum	15
13. Sandstone, medium-gray to yellowish-gray, fine-grained, biotitic, glauconitic, platy, and interbedded dark-gray shale	35
12. Shale, dark-gray, papery to pencilly	20
11. Sandstone, yellowish-gray to light-olive-gray, very fine grained; contains abundant heavy minerals; glauconitic; contains abundant brown calcareous nodules; in beds 1.0 ft thick; platy weathering. Beds separated by partings of dark-gray shale	15
10. Shale, yellowish-gray to very light gray, bentonitic, papery to chippy	5

	Approximate thickness (feet)
Cody Shale—Continued	
Lower member—Continued	
9. Sandstone, medium-light-gray, very fine grained; speckled with heavy minerals; in beds 1–2 ft thick	5
8. Siltstone and mudstone, pale-yellowish-gray, bentonitic. Unit includes a few thin beds of very fine grained quartzite	12
7. Siltstone and shale, grayish-red to very dusky purple; nodular-weathering beds 0.1–0.2 ft thick	10
6. Siltstone and mudstone; similar to unit 8	20
5. Siltstone, mudstone, and shale; similar to unit 7. Upper 4 ft of unit is blocky and siliceous	16
Approximate thickness, lower member of Cody Shale	371
Frontier Sandstone:	
4. Sandstone, medium-gray to grayish-purple to pale-yellowish-brown, fine-grained to very fine grained; in beds 1–5 ft thick; partly glauconitic; and interbedded dark-gray shale	20
3. Sandstone, yellowish-gray to light-olive-gray, very fine grained to fine-grained; speckled with abundant heavy minerals; cross-laminated; partly calcareous; in beds 2–5 ft thick; beds commonly marked by worm trails and borings; beds separated by 0.1- to 0.2-ft-thick partings of dark-gray papery shale. Top of unit platy weathering	25
Approximate thickness, Frontier Sandstone	45
Mowry Shale:	
2. Sandstone, medium-dark- to medium-light-gray, very fine grained; speckled with heavy minerals; in beds 0.1–0.2 ft thick; and medium-dark-gray fissile shale, in beds 0.1–0.2 ft thick	10
1. Mudstone and shale, medium- to medium-dark-gray, fissile to platy, and interbeds, 0.2–2.0 ft thick, of medium-dark-gray siltstone and very fine grained sandstone; underlies smooth dark-colored slope ....about	200
Base of section.	

REFERENCES CITED

Benson, A. L., 1966, Devonian stratigraphy of western Wyoming and adjacent areas: Am. Assoc. Petroleum Geologists Bull., v. 50, no. 12, p. 2566–2603.

Boyd, F. R., 1961, Welded tuffs and flows in the rhyolite plateau of Yellowstone Park, Wyoming: Geol. Soc. America Bull., v. 72, no. 3, p. 387–426.

Branson, C. C., 1937, Stratigraphy and fauna of the Sacajawea Formation, Mississippian, of Wyoming: Jour. Paleontology, v. 11, no. 8, p. 650–660.

Brookins, G. D., 1968, Rb-Sr and K-Ar age determinations from the Precambrian rocks of the Jardine-Crevice Mountain area, southwestern Montana: Earth Sci. Bull., v. 1, no. 2, p. 5–9.

Brookins, G. D., and Brown, L. E., 1966, Sulfur isotopic analyses from Park County, Montana: Econ. Geology, v. 61, no. 3, p. 613–617.

Brown, C. W., 1961, Cenozoic stratigraphy and structural geology, northeast Yellowstone National Park, Wyoming and Montana: Geol. Soc. America Bull., v. 72, no. 8, p. 1173–1193.

Brown, L. E., 1965, Genesis of the ores of the Jardine–Crevasse Mountain area, Park County, Montana: Manhattan, Kans., Kansas State Univ. unpub. M.S. thesis.

Calvert, W. R., 1912, The Electric coal field, Park County, Montana: U.S. Geol. Survey Bull. 471–E, p. 406–422.

Christiansen, R. L., and Blank, H. R., Jr., 1969, Volcanic evolution of the Yellowstone rhyolite plateau and eastern Snake River Plain, U.S.A. [abs.], *in* Symposium on volcanoes and their roots: Oxford, England, Internat. Assoc. Volcanol. Chemistry of Earth's Interior, V. Abs., p. 220–221.

Condit, D. D., 1919, Relations of late Paleozoic and early Mesozoic formations of southwestern Montana and adjacent parts of Wyoming: U.S. Geol. Survey Prof. Paper 120–F, p. 111–121.

Condit, D. D., Finch, E. H., and Pardee, J. T., 1928, Phosphate rock in the Three Forks–Yellowstone Park region, Montana: U.S. Geol. Survey Bull. 795–G, p. 147–209.

Cressman, E. R., 1955, Physical stratigraphy of the Phosphoria formation in part of southwestern Montana: U.S. Geol. Survey Bull. 1027–A, p. 1–31.

Crickmay, C. H., 1936, Study in the Jurassic of Wyoming: Geol. Soc. America Bull., v. 47, no. 4, p. 541–564.

Deiss, C. F., 1936, Revision of type Cambrian formations and sections of Montana and Yellowstone National Park: Geol. Soc. America Bull., v. 47, no. 8, p. 1257–1342.

Dorf, Erling, 1960, Tertiary fossil forests of Yellowstone National Park, Wyoming *in* West Yellowstone — Earthquake area, Billings Geol. Soc. Guidebook 11th Ann. Field Conf., Sept. 1960: p. 253–260.

Dorf, Erling, and Lochman, Christina, 1940, Upper Cambrian formations in southern Montana: Geol. Soc. America Bull., v. 51, no. 4, p. 541–556.

Foose, R. M., Wise, D. U., and Garbarini, G. S., 1961, Structural geology of the Beartooth Mountains, Montana and Wyoming: Geol. Soc. America Bull., v. 72, no. 8, p. 1143–1172.

Fox, S. K., Jr., 1939, Stratigraphy and micropaleontology of the Cody shale in southern Montana and northern Wyoming: Princeton, N.J., Princeton Univ. unpub. Ph. D. dissert.

Fraser, G. D., Waldrop, H. A., and Hyden, H. J., 1969, Geology of the Gardiner area, Park County, Montana: U.S. Geol. Survey Bull. 1277, 118 p.

Gilbert, G. K., 1877, Report on the geology of the Henry Mountains [Utah]: Washington, U.S. Geog. and Geol. Survey Rocky Mountain region (Powell), 160 p.

Girty, G. H., 1899, Devonian and Carboniferous fossils [of Yellowstone National Park]: U.S. Geol. Survey Mon. 32, pt. 2, p. 479–599.

Goddard, E. N., chm., and others, 1948, Rock-color chart: Washington, Natl. Research Council (repub. by Geol. Soc. America, 1951), 6 p.

Grant, R. E., 1965, Faunas and stratigraphy of the Snowy Range Formation (Upper Cambrian) in southwestern Montana and northwestern Wyoming: Geol. Soc. America Mem. 96, 171 p.

Hague, Arnold, 1896, Yellowstone National Park sheets; general description: U.S. Geol. Survey Geol. Atlas, Folio 30.

Hague, Arnold, and others, 1899, Geology of the Yellowstone National Park: U.S. Geol. Survey Mon. 32, pt. 2, 893 p.

Hall, W. B., 1961, Geology of part of the Upper Gallatin Valley of southwestern Montana: Laramie, Wyo., Wyoming Univ. unpub. M.S. thesis.

Hamilton, Warren, 1960, Late Cenozoic tectonics and volcanism of the Yellowstone region, Wyoming, Montana, and Idaho, *in* West Yellowstone — Earthquake area, Billings Geol. Soc. Guidebook 11th Ann. Field Conf., Sept. 1960: p. 92–105.

———1963, Overlapping of late Mesozoic orogens in western Idaho: Geol. Soc. America Bull., v. 74, no. 6, p. 779–787.

Hanson, A. M., 1952, Cambrian stratigraphy in southwestern Montana: Montana Bur. Mines and Geology Mem. 33, 46 p.

Hayden, F. V., 1872, Preliminary report of the U.S. Geological Survey of Montana and portions of adjacent Territories, being a fifth annual report of progress: Washington, U.S. Govt. Printing Office, 538 p.

———1873, Sixth annual report of the U.S. Geological Survey of the Territories * * * for the year 1872: Washington, U.S. Govt. Printing Office, 844 p.

Holmes, W. H., 1883, Report on the geology of Yellowstone National Park: U.S. Geol. and Geog. Survey of the Territories (Hayden) 12th Ann. Rept., pt. 2, p. 1–57.

Horberg, C. L., 1940, Geomorphic problems and glacial geology of the Yellowstone Valley, Park County, Montana: Jour. Geology, v. 48, no. 3, p. 275–303.

Howard, A. D., 1937, History of the Grand Canyon of the Yellowstone: Geol. Soc. America Spec. Paper 6, 159 p.

Iddings, J. P., 1898, Bysmaliths: Jour. Geology, v. 6, p. 704–710.

———1899, The intrusive rocks of the Gallatin Mountains, Bunsen Peak, and Mount Everts *and* The igneous rocks of Electric Peak and Sepulchre Mountain: U.S. Geol. Survey Mon. 32, pt. 2, p. 60–148.

———1904, A fracture valley system [Montana]: Jour. Geology, v. 12, p. 94–105.

Iddings, J. P., and Weed, W. H., 1894, Livingston atlas sheet [Montana]: U.S. Geol. Survey Geol. Atlas, Folio 1, 4 p.

———1899, Descriptive geology of the Gallatin Mountains: U.S. Geol. Survey Mon. 32, pt. 2, p. 1–59.

Imlay, R. W., 1952, Correlation of the Jurassic formations of North America, exclusive of Canada: Geol. Soc. America Bull., v. 63, no. 9, p. 953–992.

———1967, Twin Creek Limestone (Jurassic) in the Western Interior of the United States: U.S. Geol. Survey Prof. Paper 540, 105 p.

Imlay, R. W., Gardner, L. S., Rogers, C. P., Jr., and Hadley, H. D., 1948, Marine Jurassic formations of Montana: U.S. Geol. Survey Oil and Gas Inv. Prelim. Chart 32.

Kirkham, V. R. D., 1927, A geologic reconnaissance of Clark and Jefferson and parts of Butte, Custer, Fremont, Lemhi, and Madison Counties, Idaho: Idaho Bur. Mines Geology Pamph. 19, 47 p.

———1931, Snake River downwarp: Jour. Geology, v. 39, no. 5, p. 456–482.

Knowlton, F. H., 1899, Fossil flora [of Yellowstone National Park]: U.S. Geol. Survey Mon. 32, pt. 2, p. 651–882.

Kummel, Bernhard, 1954, Triassic stratigraphy of southeastern Idaho and adjacent areas [Wyoming-Montana]: U.S. Geol. Survey Prof. Paper 254-H, p. 165–194.

Lochman-Balk, Christina, 1950, Status of Dry Creek shale of central Montana: Am. Assoc. Petroleum Geologists Bull., v. 34, no. 11, p. 2200–2222.

———1956, The Cambrian of the Rocky Mountains and southwest deserts of the United States and adjoining Sonora Province, Mexico, *in* Rodgers, John, ed., El Sistema Cambrico, su paleogeografia y el problema de su base — Symposium, Pt. 2: p. 529–661.

Love, J. D., 1956, Summary of geologic history of Teton County, Wyoming, during Late Cretaceous, Tertiary, and Quaternary times, *in* Wyoming Geol. Assoc. Guidebook 11th Ann. Field Conf., 1956: p. 140–150.

———1961, Reconnaissance study of Quaternary faults in and south of Yellowstone National Park, Wyoming: Geol. Soc. America Bull., v. 72, no. 12, p. 1749–1764.

———1968, Stratigraphy and structure — A summary, *in* Behrendt, J. C., Tibbetts, B. L., Bonini, W. E., and Lavin, P. M., A geophysical study in Grand Teton National Park and vicinity, Teton County, Wyoming: U.S. Geol. Survey Prof. Paper 516–E, p. E3–E12.

Lovering, T. S., 1930, The New World or Cooke City mining district, Park County, Montana: U.S. Geol. Survey Bull. 811–A, p. 1–87.

Maughan, E. K., and Roberts, A. E., 1967, Big Snowy and Amsden Groups and the Mississippian-Pennsylvanian boundary in Montana: U.S. Geol. Survey Prof. Paper 554–B, 27 p.

McKelvey, V. E., Williams, J. S., Sheldon, R. P., Cressman, E. R., Cheney, T. M., and Swanson, R. W., 1956, Summary description of Phosphoria, Park City, and Shedhorn formations in western phosphate field [advance summ.]: Am. Assoc. Petroleum Geologists Bull., v. 40, no. 12, p. 2826–2863.

McMannis, W. J., 1962, Devonian stratigraphy between Three Forks, Montana, and Yellowstone Park, *in* Three Forks–Belt Mountains area, and symposium — The Devonian system of Montana and adjacent areas, Billings Geol. Soc. Guidebook 13th Ann. Field Conf., 1962: p. 4–12.

McMannis, W. J., and Chadwick, R. A., 1964, Geology of the Garnet Mountain quadrangle, Gallatin County, Montana: Montana Bur. Mines and Geology Bull. 43, 47 p.

Pampeyan, E. H., Schroeder, M. L., Schell, E. M., and Cressman, E. R., 1967, Geologic map of the Driggs quadrangle, Bonneville and Teton Counties, Idaho, and Teton County, Wyoming: U.S. Geol. Survey Mineral Inv. Field Studies Map MF-300.

Pardee, J. T., 1950, Late Cenozoic block faulting in western Montana: Geol. Soc. America Bull., v. 61, no. 4, p. 359–406.

Peale, A. C., 1893, The Paleozoic section in the vicinity of Three Forks, Montana, with petrographic notes, by G. P. Merrill: U.S. Geol. Survey Bull. 110, 56 p.

Peterson, J. A., 1957, Marine Jurassic of Northern Rocky Mountains and Williston Basin: Am. Assoc. Petroleum Geologists Bull., v. 41, no. 3, p. 399–440.

Pierce, W. G., 1957, Heart Mountain and South Fork detachment thrusts of Wyoming: Am. Assoc. Petroleum Geologists Bull., v. 41, no. 4, p. 591–626.

———1960, The "break-away" point of the Heart Mountain detachment fault in northwestern Wyoming, *in* Short papers in the geological sciences: U.S. Geol. Survey Prof. Paper 400–B, p. B236–B237.

Reid, R. R., 1957, Suggested modification of pre-Beltian history in southwestern Montana [abs.]: Geol. Soc. America Bull., v. 68, no. 12, pt. 2, p. 1871–1872.

Richards, P. W., and Nieschmidt, C. L., 1961, The Bighorn dolomite and correlative formations in southern Montana and northern Wyoming: U.S. Geol. Survey Oil and Gas Inv. Map OM-202.

Rittmann, Alfred, 1952, Nomenclature of volcanic rocks: Bull. Volcanol., ser. 2, v. 12, p. 75–102.

Roberts, A. E., 1965, Correlation of Cretaceous and lower Tertiary rocks near Livingston, Montana, with those in other areas of Montana and Wyoming, *in* Geological Survey research 1965: U.S. Geol. Survey Prof. Paper 525–B, p. B54–B63.

———1966, Stratigraphy of Madison Group near Livingston, Montana, and discussion of karst and solution-breccia features: U.S. Geol. Survey Prof. Paper 526–B, 23 p.

Robinson, G. D., 1963, Geology of the Three Forks quadrangle, Montana, *with sections on* Petrography of igneous rocks, by H. F. Barnett: U.S. Geol. Survey Prof. Paper 370, 143 p.

Rubel, D. N., 1964, Geology of the Independence area, Sweet Grass and Park Counties, Montana [abs.]: Ann Arbor, Mich., Michigan Univ., Ph. D. thesis, Dissert. Abs., v. 25, no. 6, p. 3515–3516.

Ruppel, E. T., 1966, Preliminary evaluation of radar imagery, southern Gallatin Range and vicinity, *in* Preliminary evaluation of radar imagery of Yellowstone Park, Wyoming: Natl. Aeronautics and Space Adm. Tech. Letter 30.

Sandberg, C. A., 1965, Nomenclature and correlation of lithologic subdivisions of the Jefferson and Three Forks Formations of southern Montana and northern Wyoming: U.S. Geol. Survey Bull. 1194–N, 18 p.

Sando, W. J., 1967a, Mississippian depositional provinces in the northern Cordilleran region, *in* Geological Survey research 1967: U.S. Geol. Survey Prof. Paper 575–D, p. D29–D38.

———1967b, Madison Limestone (Mississippian), Wind River, Washakie, and Owl Creek Mountains, Wyoming: Am. Assoc. Petroleum Geologists Bull., v. 51, no. 4, p. 529–557.

———1968, A new member of the Madison Limestone (Mississippian) in Wyoming: Geol. Soc. America Bull., v. 79, no. 12, p. 1855–1857.

Scott, H. W., 1935, Some Carboniferous stratigraphy in Montana and northwestern Wyoming: Jour. Geology, v. 43, no. 8, pt. 2, p. 1011–1032.

———1945, Age of Amsden formation [Wyoming and Montana] [abs.]: Geol. Soc. America Bull., v. 56, no. 12, pt. 2, p. 1195.

Seager, G. F., 1944, Gold, arsenic, and tungsten deposits of the Jardine-Crevasse Mountain district, Park County, Montana: Montana Bur. Mines and Geology Mem. 23, 111 p.

Sloss, L. L., and Laird, W. M., 1947, Devonian system in central and northwestern Montana: Am. Assoc. Petroleum Geologists Bull., v. 31, no. 8, p. 1404–1430.

Sloss, L. L., and Moritz, C. A., 1951, Paleozoic stratigraphy of southwestern Montana: Am. Assoc. Petroleum Geologists Bull., v. 35, no. 10, p. 2135–2169.

Stanton, T. W., 1899, Mesozoic fossils [of Yellowstone National Park]: U.S. Geol. Survey Mon. 32, pt. 2, p. 600–650.

Tansley, Wilfred, Schafer, P. A., and Hart, L. H., 1933, A geological reconnaissance of the Tobacco Root Mountains, Madison County, Montana: Montana Bur. Mines and Geology Mem. 9, 57 p.

Thompson, M. L., and Scott, H. W., 1941, Fusulinids from the type sections of the lower Pennsylvanian Quadrant formation [Wyoming]: Jour. Paleontology, v. 15, no. 4, p. 349–353.

Weed, W. H., 1896, Yellowstone National Park; sedimentary rocks: U.S. Geol. Survey Atlas, Geol. Folio 30, p. 4–5.

———1900, Geology of the Little Belt Mountains, Montana: U.S. Geol. Survey 20th Ann. Rept., pt. 3, p. 257–461.

Wilson, C. W., Jr., 1934a, Geology of the thrust fault near Gardiner, Montana: Jour. Geology, v. 42, no. 6, p. 649–663.

———1934b, Section of Paleozoic and Mesozoic rocks measured at Cinnabar Mountain, Park County, Montana, and at Mount Everts, Yellowstone National Park, Wyoming: Am. Assoc. Petroleum Geologists Bull., v. 18, no. 3, p. 368–379.

Witkind, I. J., 1964, Preliminary geologic map of the Tepee Creek quadrangle, Montana-Wyoming: U.S. Geol. Survey Misc. Geol. Inv. Map I–417.

———1969, Geology of the Tepee Creek quadrangle, Montana-Wyoming: U.S. Geol. Survey Prof. Paper 609, 101 p.

Thirty-Third Annual Field Conference — 1982
Wyoming Geological Association Guidebook

HYDROTHERMAL ALTERATION IN RESEARCH DRILL HOLE Y-11 FROM A VAPOR-DOMINATED GEOTHERMAL SYSTEM AT MUD VOLCANO, YELLOWSTONE NATIONAL PARK, WYOMING

KEITH E. BARGAR AND L.J. PATRICK MUFFLER[1]

ABSTRACT

U.S. Geological Survey research diamond-drill hole Y-11 in the Mud Volcano area of Yellowstone National Park was drilled to a depth of 105.7 m. The hole penetrated about 2.1 m of Holocene fluvial sedimentary deposits, 13.1 m of glacio-sedimentary deposits associated with the Pinedale Glaciation (≥45,000-14,000 yr B.P.), approximately 4.3 m of tuff beds of the Upper Basin Member of the Plateau Rhyolite (~265,000 yr B.P.), and 86.2 m of a 600,000 year-old ash-flow tuff (the Lava Creek Tuff of the Yellowstone Group). Cavities in the welded ash-flow tuff contain abundant sanidine and tridymite from vapor-phase crystallization during cooling of the flow. The rhyolitic glassy groundmass had undergone extensive devitrification to form alkali feldspar and alpha-cristobalite before hydrothermal alteration.

Hydrothermal alteration in drill core Y-11 was produced by an early chloride-rich near-neutral hot-water system and a later vapor-dominated system. The hot-water system precipitated minerals such as chalcedony, quartz, septechlorite, calcite, rhodochrosite, bastnaesite, mordenite, fluorite, and some beta-cristobalite, pyrite, and montmorillonite. These minerals cannot have been deposited from the steam and bicarbonate-sulfate condensate found in the present geothermal system. Later minerals related to the current vapor-dominated regime include opal, kaolinite, halloysite, alunite, and some beta-cristobalite, pyrite, and montomorillionite. These younger minerals are, in part, superimposed on the earlier formed minerals. Relations between the two generations of hydrothermal minerals are best observed in the drill core within the later cross-cutting fractures at 74.6-77.6 m, where later fracture fillings of kaolinite and pyrite cut across earlier chalcedony fracture fillings.

INTRODUCTION

Y-11 is one of 13 research diamond-drill holes completed by the U.S. Geological Survey in 1967 and 1968 in selected hot-spring and geyser areas of Yellowstone National Park. These holes were drilled to obtain detailed physical and chemical data from the shallow part of a high-temperature geothermal system (White and others, 1975). Drill hole Y-11 is located in the Mud Volcano area (~145 m northwest of and 5 m higher than the Yellowstone River), about 8 km north-northwest of Yellowstone Lake (Fig. 1). The drill collar, at an elevation of 2350 m, is approximately 75 m northeast of a large vigorously boiling mudpot known as Old Sulphur Caldron. Other mudpots, acid-sulfate springs, and steam vents in the area are concentrated along north-northeast-trending lineations (White and others, 1971) and have equally ominous sounding names, such as Dragon's Mouth, Churning Caldron, Sizzling Basin, Grizzly Fumarole, and Black Dragon's Caldron. According to Allen and Day 1935), the cumulative discharge from all the thermal features in the approximately 2.5 km^2 geothermal area is only about 80 L/min.

Drill hole Y-11 was spudded May 14, 1968, and reached a maximum depth of 105.7 m on May 22, 1968. Water

[1]U.S. Geological Survey, 345 Middlefield Road, Menlo Park, CA 94025

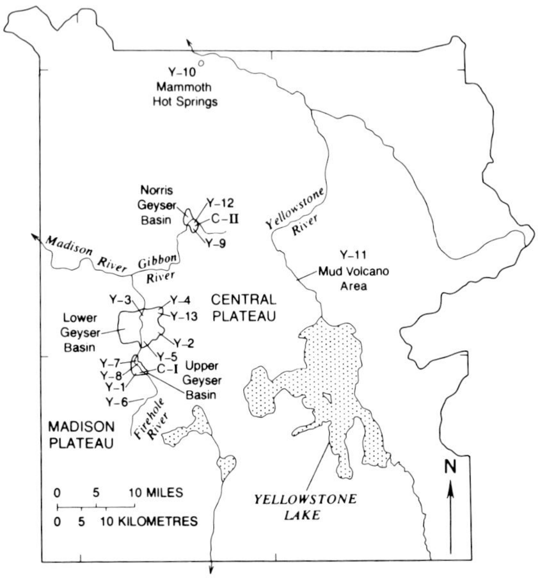

Figure 1: Index map showing location of drill hole Y-11 and other research drill holes in Yellowstone National Park.

was used as the circulating fluid. Core diameter was 11.4 cm from 0 to 6.1 m, 6.0 cm from 6.1 to 27.7 m, and 4.4 cm from 27.7 to 105.7 m (the hole bottom). Core recovery by wireline method averaged 39.6% from 0 to 19.8 m and 98.2% from 19.8 to 105.7 m.

The drill core, generally taken at intervals of 1.5 or 3.0 m, was logged at the drill site by L.J.P. Muffler, and A.H. Truesdell, and approximately 25% of the core was selected for detailed laboratory study. Selected samples either contain contained unusual features (e.g., veins, vugs, contacts, textural relations) or were more nearly representative of a given interval. Most of the analytical data reported here were determined from this "skeleton" core; the bulk core was used only for detailed study of the interval from 80-85 m. The "skeleton" core was systematically studied by X-ray-diffraction methods (>400 X-ray diffractograms) and petrographic microscope (62 thin sections). Whole-rock chemical analyses, semiquantitative spectrographic analyses, and bulk- and powder-density measurements were obtained for nine representative samples; three additional semiquantitative spectrographic analyses were obtained for selected minerals.

The physical measurements in drill hole Y-11 made during and after drilling were described in detail by White and others (1971, 1975). The maximum temperature measured in the drill hole was 191.4°C at the hole bottom (105.7 m); actual rock temperatures were interpreted to be slightly higher at this depth, as shown by the dashed temperature curve in Figure 2, which approximates the pre-drilling ground temperatures. The maximum pressure was 12.7 kg/cm^2 at the bottom of the drill hole.

According to White and others (1971), the physical data and eruptive characteristics indicate that at about 73 m drill hole Y-11 penetrated a vapor-dominated reservoir which contained less than 10 wt. % liquid water. At shallower depths, the rocks are saturated with a dilute near-neutral low-chloride bicarbonate-sulfate water (28 ppm Ca, 0.47 ppm Mg, 105 ppm Na, 12.6 ppm K, 0.18 ppm Li, 3.2 ppm NH_4, 258 ppm HCO_3, 74 ppm SO_4, 9.6 ppm Cl, 0.2 ppm NO_3, and 0.1 ppm B; temperature, 131.7°C; White and others, 1971; analysis 7, Table 1). This water was interpreted to be essentially the condensate from rising geothermal steam and gases (98.4% CO_2, 0.01% H_2, Trace CH_4, 0.0% C_2H_6, 0.8% N_2, 0.013%A, 0% H_2S, and 0.2% O_2; total, 99.42 vol %, excluding H_2O; White and others, 1971; analysis 5, Table 2), modified somewhat by reaction with the wallrock. At depths greater than 73 m, White and others (1971) concluded that both vapor (steam plus CO_2 and H_2S) and liquid are present in the rocks, and that the vapor is concentrated in the larger pores and fractures.

Vertical electrical soundings, resistivity horizontal profiles, and self-potential and induced-polarization measurements were made in the Mud Volcano area after the drilling of hole Y-11 (Zohdy and others, 1973). These geophysical studies show a shallow vapor-dominated reservoir, with a high resistivity of about 75-130Ω-m, that extends for about a mile along a northwest-southeast profile. The data of Zohdy and others also indicate that this vapor-dominated reservoir is overlain by a low-resistivity (~2-6.5 Ω-m) layer and has borders of moderate-resistivity (~30 Ω-m).

STRATIGRAPHY

Gravel and Conglomerate

From 0 to 2.1 m, drill hole Y-11 penetrated weakly cemented poorly sorted polymictic gravel, probably deposited by the Yellowstone River when its level was higher than at present (Richmond, 1977). Beneath this gravel layer (to a depth of 15.2 m) is a uniform moderately well cemented conglomerate, deposited during the Pinedale Glaciation about 45,000-14,000 yr B.P. (Pierce and others, 1976), that is composed almost entirely of granules and pebbles of black obsidian and white rhyolitic pumice. These two units are scarcely affected by hydrothermal alteration (Fig. 2), even though drill hole Y-11 is located near the outer edge of a large area of surficial acid-sulfate alteration (see White and others, 1971 Fig. 4). Both units are weakly cemented by opal and contain traces of beta-cristobalite, kaolinite, montmorillonite, and gypsum(?). Core recovery of these units was poor, and the minerals

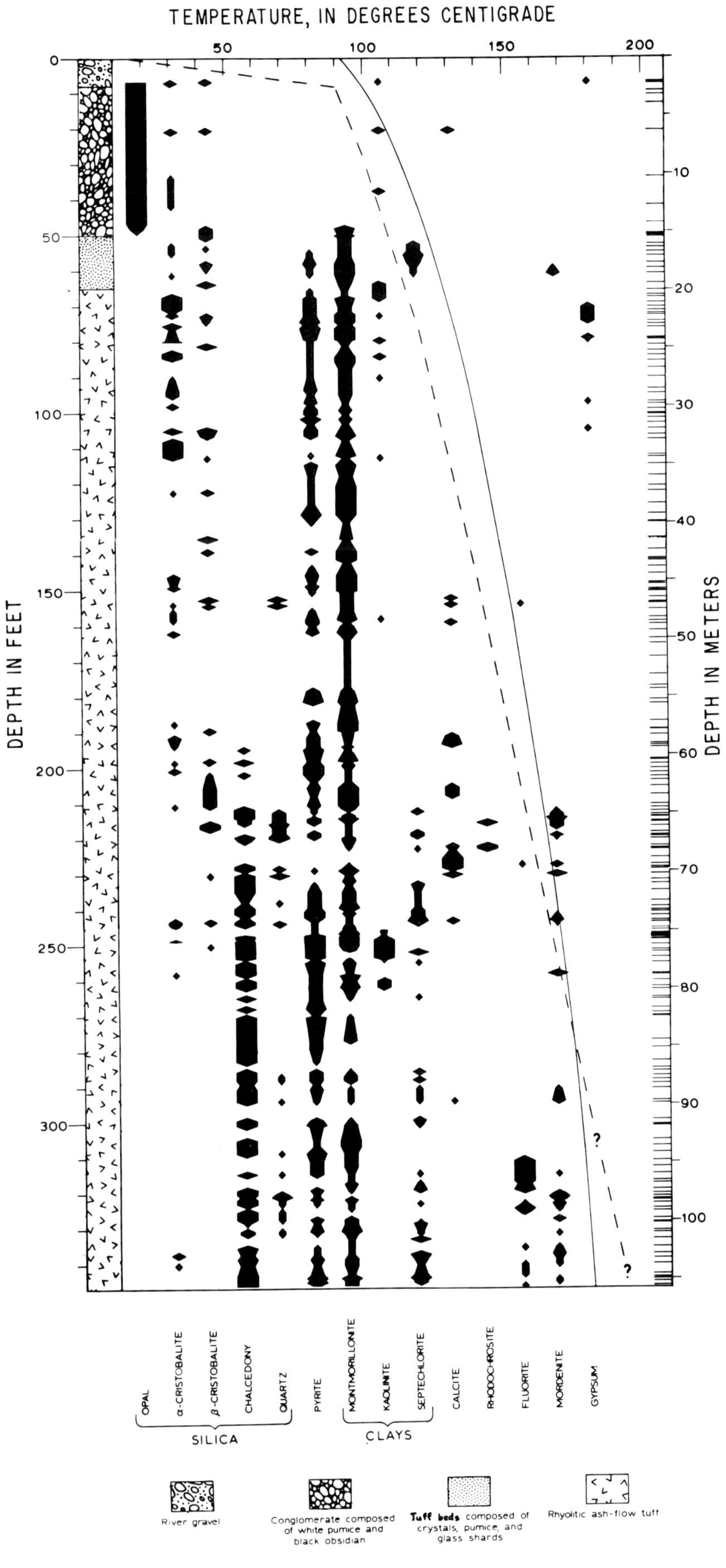

Figure 2: Distribution of devitrification alpha-cristobalite and hydrothermal-alteration minerals in drill core from hole Y-11. Left column shows a generalized stratigraphic section of rock units penetrated by the drill hole. Horizontal lines at right indicate distribution of samples studied. Width of mineral columns provides an estimate of relative abundance based on microscopic observation and X-ray-diffraction analyses. Vertical continuity in mineral presence between samples is assumed, except where a mineral abundance drops to zero, in which case the zero point is arbitrarily placed 0.3 m from last occurrence of the mineral. Dashed curve is an approximation to predrilling ground temperatures. Solid curve shows reference boiling temperatures for pure water, assuming a surface altitude of about 2350 m and a near-surface water table. Temperature data from White and others (1971).

listed are undoubtedly more abundant than shown in Figure 2. Obsidian and rhyolite pumice clasts in the conglomerate are unaltered.

Tuff of Uncle Tom's Trail

From 15.2 to 19.8 m, drill hole Y-11 penetrated a greenish-brown nonwelded lithic to vitric ash-flow tuff that was correlated with the Tuff of Uncle Tom's Trail of the Upper Basin Member of the Plateau Rhyolite (Christiansen and Blank, 1975). One rhyolite flow of the Upper Basin Member (the Scaup Lake flow) has a K-Ar age of about 265,000 years (J.D. Obradovich and R.L. Christiansen, unpub. data, 1973). In drill hole Y-11, the Tuff of Uncle Tom's Trail is extremely friable and contains abundant silt- to sand-sized crystals of quartz and sanidine.

In contrast to the overlying gravel and conglomerate, the tuffaceous interval is conspicuously altered. The typical hydrothermal-mineral assemblage consists of montmorillonite, beta-cristobalite, pyrite, and septechlorite; these secondary minerals are pervasive throughout the tuff beds. Mordenite fibers are intergrown with the clay minerals in several samples near the bottom of the unit; one sample at 19.5 m also contained minor kaolinite.

Lava Creek Tuff

From 19.8 m to the hole bottom at 105.7 m, the drill core from hole Y-11 consists of rhyolitic ash-flow tuff that was correlated by Christiansen and Blank (1975) with members B (19.8-36.3 m) and A (36.3-105.7 m) of the Lava Creek Tuff of the Yellowstone Group. A sanidine K-Ar age of 600,000 yr was reported for the Lava Creek Tuff (Christiansen and Blank, 1972). Stratigraphic relations and generalized lithology of the Lava Creek Tuff were discussed by Christiansen and Blank (1972) and Christiansen (1979). A detailed petrographic description of the Lava Creek Tuff encountered in drill hole Y-5 was given by Keith and Muffler (1978).

In drill hole Y-11, the Lava Creek Tuff contains phenocrysts of quartz, sanidine, and sporadic plagioclase in a devitrified, partially welded groundmass, consisting primarily of alkali feldspar and alpha-cristobalite. Feldspar phenocrysts are partially to completely altered to montmorillonite. Sporadic clots of clay and hematite appear to represent original pyroxene phenocrysts. Open spaces in partly collapsed pumice fragments, in lithophysae, and along partings are lined or filled with tridymite, alkali feldspar (cf. Keith and Muffler, 1978, Figs. 2-5), and magnetite. These three minerals were deposited from super critical vapor during cooling of the ash-flow tuff from magmatic temperatures (Ross and Smith, 1961). The high temperature vapor-phase minerals are unrelated to the lower temperature hydrothermal alteration later superimposed on the ash-flow tuff.

At the time of hydrothermal alteration, the ash-flow tuff in drill hole Y-11 had already undergone devitrification and vapor-phase crystallization, and thus contained little or no glass. Accordingly, in contrast to most other Yellowstone drill holes (C-I, Fenner, 1936; Y-1, Honda and Muffler, 1970; Y-2, Bargar and Beeson, 1981; Y-3, Bargar and others, 1973; Y-7 and Y-8, Keith and others, 1978b; Y-13, Keith and others, 1978a), which contained abundant highly reactive glass just before hydrothermal alteration, the ash-flow tuff in drill hole Y-11 displays less conspicuous alteration of the groundmass. At depths less than about 75 m, the groundmass consists of alkali feldspar and alpha-cristobalite; below 75 m, some of the alpha-cristobalite has been transformed to the more stable quartz phase (Fig. 2.) Montmorillonite is disseminated in the groundmass in all but the least altered samples. The more highly altered specimens are rich in montmorillonite, pyrite, and (at depths >75 m) septechlorite, and have a conspicuous green cast.

The drill core from hole Y-11 is highly fractured, similar to that from hole Y-5, for which Keith and Muffler (1978) associated the fracturing with magmatic resurgence of the Mallard Lake dome (~150,000 yr B.P.; Eaton and others, 1975). The Mud Volcano area lies just inside the boundary of the Sour Creek dome (Christiansen, 1979); resurgent doming and fracturing(?) in this area took place shortly following collapse of the Yellowstone caldera (~600,000 yr B.P.; Eaton and others, 1975). The onset of hydrothermal alteration within the Yellowstone caldera occurred about 150,000 yr B.P. (Fournier and others, 1976).

HYDROTHERMAL ALTERATION

Figure 2 plots the distribution of devitrification alpha-cristobalite and hydrothermal-alteration minerals with depth in drill hole Y-11. Other devitrification, magmatic, vapor-phase or hydrothermal minerals that have only a minor distribution are not plotted — for example, earthy ferric oxides and hydroxides, which occur throughout the drill core. Hydrothermal-alteration minerals in drill hole Y-11 occur primarily as linings or fillings of void space, as coatings on fractures, and as matrix alteration, and can be divided into early- and late-stage minerals (Muffler and Barger, 1974). The early stage is characterized by chalcedony, septechlorite, calcite, quartz, mordenite, fluorite, iron oxides and hydroxides, rhodochrostite, bastnaesite, and some beta-cristobalite, pyrite, and montmorillonite; the late stage is charcacterized by fracture fillings of kaolinite along with halloysite, alunite, opal and some beta-cristobalite, montmorillonite, and pyrite. Relations between the two stages are clear only in the interval 74.4-77.6, m where later veins of kaolinite and pyrite cut across early breccia veins of chalcedony.

Silica Minerals

Opal occurs mainly as a white cementing material in the glacial sedimentary deposits between 2.1 and 15.2 m along with associated beta-cristobalite, montmorillonite, and halloysite. One sample at 6.4 m also displayed a subhorizontal 6-mm-thick band of opal that was bounded above and below by a 0.35-mm calcite coating; the opal and calcite were cut by a later fracture lined by a mixture of opal, kaolinite, and alunite. Opal, though probably a hot-water deposit, appears to be closely associated with vapor-dominated systems in drill hole Y-11 and in other areas of surficial alteration, as discussed by Schoen and others (1974), that contain the same suite of minerals.

Figure 3: Vapor-phase cavity filling at 66.1 m. Tabular vapor-phase alkali feldspar crystals (bottom center) are overgrown by botryoidal chalcedony (bottom right), radiating prismatic quartz, fibrous mordenite, and a pervasive thin coating of septechlorite and iron oxide or hydroxide.

Beta-cristobalite makes up some of the cementing material in the glacial sedimentary deposits and the matrix of the underlying tuffaceous beds. In the ash-flow tuff, scattered occurrences of beta-cristobalite line the bottoms of a few cavities and several fractures above 76.2 m, in association with montmorillonite with or without pyrite and chalcedony. Beta-cristobalite also occurs at 24.8, 46.5-47-1, and 62.7 m as white botryoidal cavity fillings along with montmorillonite, calcite, prismatic quartz, or chalcedony. Colorless, vitreous, very low birefringent beta-cristobalite (refractive index, 1.46) at 76.2 m coats a late vertical planar fracture containing pyrite and kaolinite.

Open-space chalcedony deposits, which are abundant in the drill core below 57.2 m, vary considerably in morphology and mineral associations; these deposits are interpreted as having precipitated during more than one generation. Vapor-phase cavities between 65.1 and 67.0 m are lined by botryoidal chalcedony associated with prismatic quartz and fibrous mordenite, all three of which are coated by iron oxide or hydroxide, gray or green septechlorite, and some pyrite (Fig. 3). A similar botryoidal chalcedony deposit in a vapor-phase cavity at 95 m is coated by septechlorite, hematite and pyrite, later flourite, still later mordenite and prismatic quartz, and, finally, montmorillonite. Other cavity fillings down to 99.7 m contain orange-or-red-stained botryoidal chalcedony associated with later hydrothermal quartz and mordenite. Scattered cavity fillings of white botryoidal chalcedony between 62.7 and 78.7 m are younger than calcite and septechlorite and partly younger than mordenite (see Fig. 7). A few other cavities at 58.3-64.5 and 80.6 m consist of layered, clear white, gray, or purple-brown chalcedony geopetal fillings, with associated pyrite and montmorillonite.

Fractures at 57.8-65.3, 84.3, and 91.6 m are coated by white or grayish-tan, greasy or clay-like chalcedony deposits, with associated montorillonite and pyrite. Other fractures at 57.2-105.6 m are commonly lined by purple to brown, in some places banded (parallel to the fracture surfaces) chalcedony associated with septechlorite, montmorillonite, and iron oxide, and later pyrite(?), fluorite, mordenite, and kaolinite.

Cloudy to clear prismatic quartz crystals or sprays of radiating quartz prisms (Fig. 3) are deposited in vapor-phase cavities and a few fractures at 46.5-47.1, 64.8-74.3, and 87.4-100.9 m. The quartz crystals, generally about 0.5 to 1.0 mm long, were deposited later than chalcedony with or without pyrite or calcite and earlier than mordenite, flourite, montmorillonite, septechlorite, iron oxide or hydroxide, and pyrite. Hydrothermal quartz also replaces vapor-phase sanidine in thin sections from 72.5 and 100.7 m (cf. Keith and Muffler, 1978, Fig. 8). A vapor-phase cavity at 97.7 m (Fig. 4) showed red-stained ragged tabular vapor-phase alkali fedspar (Or_{80-85}) crystals that probably were at lease partly replaced by SiO_2.

Iron Oxide, Hydroxide, and Sulfide

No iron oxide or hydroxide minerals are plotted in Figure 2, although yellow, orange, and red iron oxide or hydroxide staining of rocks and open-space fillings is fairly common throughout most of the drill core from hole Y-11. X-ray diffraction analysis of numerous samples suggests that the most of these iron-stained deposits consist of amorphous iron oxide or hydroxide, although only a few samples show sufficient crystallinity for identification of hematite or goethite. The iron content (especially FE_2O_3) in several chemical analyses of the drill core from hole Y-11 (Table 1) is only slightly greater than that in analyses of unaltered Lava Creek tuff (total Fe, ~1.1-1.7 wt %; R.L. Christiansen, unpub. data, 1974) except for samples Y11-16.5 and Y11-104.7. In sample Y11-16.5 (actually a tuff bed overlying the Lava Creek Tuff in drill hole Y-11), the whole rock is stained red, and sample Y11-104.5 is cut by a large red-stained fracture. The large FeO (6.0 wt %) and loss-on-ignition (1.4%) values of the deeper sample (Y-11-104.7) probably reflect pyrite, which is fairly abundant in some fractures (S was not looked for).

Pyrite occurs in most samples below 16.8 m. Euhedral cubic pyrite crystals may be diseminated in the groundmass of the ash-flow tuff but more commonly are deposited as fracture or cavity fillings. The pyrite generally is associated with earlier hydrothermal deposits, but sometimes is also closely associated with late kaolinite fracture fillings.

Clay Minerals

Varicolored montmorillonite occurs in most vugs and fractures, replaces plagioclase, sanidine, and mafic phenocrysts, and is disseminated in the groundmass throughout the volcanic section of the drill core from hole

Figure 4: Vapor-phase cavity at 97.7 m lined by ragged tabular vapor-phase alkali feldspar. Red-stained chalcedony coats and partly replaces(?) the feldspar crystals. Large (max. length 7mm), clear to cloudy quartz crystals line bottom of cavity along with later fibrous mordenite.

Table 1. Chemical analyses drill core from hole Y-11[1]

	Glacial conglomerate	Tuff of Uncle Tom's Trail		Lava Creek Tuff (member B)			Lava Creek Tuff (member A)		
Sample[2]	Y11-6.4	Y11-16.5	Y11-19.5	Y11-23.0	Y11-29.9	Y11-36.0	Y11-47.1	Y11-75.7	Y11-104.7
				Major oxides (wt%)					
SiO_2	73.9	58.5	84.4	76.8	75.0	74.8	73.2	66.8	69.0
Al_2O_3	10.9	21.1	7.6	11.1	11.8	12.1	12.9	17.1	9.2
Fe_2O_3	0.62	4.2	0.73	1.1	2.1	1.7	2.0	1.8	6.0
FeO	0.52	0.12	0.02	0.04	0.12	0.16	0.12	0.04	6.0
MgO	0.06	0.53	0.14	0.27	0.10	0.06	0.07	0.15	0.74
CaO	0.49	0.93	0.35	0.48	0.38	0.48	0.46	0.57	0.31
Na_2O	0.45	1.2	0.67	2.1	3.1	3.0	3.4	3.7	0.16
K_2O	3.3	2.4	1.9	4.8	5.3	5.1	5.4	6.5	1.1
H_2O^+	3.5	4.3	1.4	0.82	0.41	0.47	0.27	0.55	3.0
H_2O^-	4.2	4.8	1.2	0.86	0.44	0.41	0.43	1.1	1.2
TiO_2	0.15	0.31	0.08	0.16	0.16	0.14	0.21	0.26	0.16
P_2O_5	0.05	0.04	0.01	0.03	0.04	0.03	0.03	0.04	0.05
MnO	0.03	0.04	0.02	0.01	0.01	0.01	0.03	0.02	0.23
CO_2	0.02	0.06	0.01	0.02	0.02	0.02	0.02	0.01	0.01
Loss on ignition	0.23	--	--	--	--	--	--	--	1.4
Total	98.42	98.52	98.53	98.59	98.98	98.48	98.54	98.38	98.56
Bulk density	1.18	-	1.66	1.57	2.17	2.28	2.14 2.30	1.72	2.14 2.24
Powder density	2.15	1.92	2.17	2.27	2.30	2.38	2.39	2.40	2.51
Calculated porosity %	45	-	24	31	6	4	10 4	28	15 11
				Semiquantitative spectrographic analyses (ppm)[3]					
B	15	<2	70	5	7	7	5	3	<2
Ba	1000	700	500	1000	500	500	1500	1500	150
Be	5	7	7	3	7	7	5	5	10
Ce	300	200	<50	300	200	150	200	300	<50
Co	<1	3	2	<1	<1	<1	<1	5	5
Cr	10	10	3	7	1	<0.7	2	2	5
Cs	20	200	20	20	20	20	20	20	<20
Cu	5	10	7	7	5	5	5	5	15
Eu	<1	<1	<1	<1	<1	<1	5	5	<1
Ga	20	30	15	20	20	30	30	30	15
La	150	100	50	150	150	70	100	150	70
Li	<5	20	10	10	<5	<5	10	10	50
Mo	7	<2	3	5	5	7	5	5	<2
Nb	50	50	50	50	50	50	20	30	20
Ni	5	15	5	7	3	5	3	7	10
Pb	100	70	30	50	50	50	50	50	70
Rb	50	50	20	150	150	150	150	200	50
Sc	3	5	2	3	2	3	3	5	5
Sn	10	20	15	7	7	10	7	7	150
Sr	200	70	50	50	20	20	70	70	30
V	5	20	5	5	<1	<1	<1	3	5
Y	70	100	20	70	70	50	20	70	70
Yb	7	10	3	10	10	7	7	15	15
Zn	50	100	50	70	150	100	100	100	150
Zr	300	300	200	300	300	300	300	300	200

[1] Y-11 chemical analyses by rapid rock methods; N. Skinner, analyst.

Semiquantitative spectrographic method; J. Kent and M. Retzloff, analysts.

[2] Sample numbers correspond to depth (in meters).

[3] Looked for but not found: Ag, As, Au, Bi, Cd, Ge, Hf, In, Nd, Pd, Pr, Pt, Re, Sb, Sm, Ta, Te, Th, Tl, U, and W.

Y-11. Montmorillonite deposited in open spaces (Figs. 8, and 9) appears to have formed during both the early hot-water stage of hydrothermal alteration (commonly intergrown with septechlorite) and the later vapor-dominated stage in association with kaolinite. Most montmorillonite samples analyzed by X-ray diffraction have a basal (001) spacing between 14.5 and 15.5 Å, and, using the criteria of Grim (1968), Ca^{2+} is the major cation; only a few samples have a smaller basal spacing characteristic of Na^{+} montmorillonite. In all the samples tested, the (001) peak expands to a 17 Å with glycolation.

Kaolinite is scattered throughout the upper 79.8 m of the drill core and is present in some abundance at 19.5-20.7 and 75.7-79.8 m. Although kaolinite was detected on X-ray -diffraction patterns of altered whole-rock samples in the glacial sedimentary deposits and near the top of the ash-flow tuff, the mineral occurs primarily as a fracture filling, commonly in association with montmorillonite and pyrite. Other mineral associations include: opal at 2.2 and 6.4 m, alunite at 6.4 m, and beta-cristobalite in several samples scattered throughout the kaolinite deposits. The kaolinite in drill hole Y-11 is generally white, although other colors occur owing to associated minerals. Kaolinite has an (001) X-ray-diffraction peak at about 7.13 Å and a (002) peak at about 3.57 Å; only rarely was there sufficient sample material for other diagnostic peaks to be present on X-ray traces. Infrared spectrographic analyses of two samples from 19.5 and 75.7 m show that the mineral is kaolinite rather than dickite (P.R.L. Browne, written commun., 1973). Halloysite, a more hydrous form of kaolinite, was identified by X-ray diffraction at 2.3 m.

Gray to green septechlorite (serpentine-kaolin group) occurs at 16.2-18.7 m and in scattered places between 64.8 m and the bottom of the drill hole. Septechlorite in the upper zone is restricted to the Tuff of Uncle Tom's Trail. The mineral, which was identified on whole-rock X-ray diffractograms along with montmorillonite, pyrite, and beta-cristobalite, may be an alteration product of the glass. In the deeper zone, septechlorite fills fractures and vapor-phase cavities in association with montmorillonite, pyrite, and a few deposits of rhodochrosite (Figs. 5, 6), quartz, chalcedony, mordenite, calcite, iron oxide or hydroxide and flourite.

X-ray-diffraction analysis shows a 7-Å (001) spacing for septechlorite in the drill core from hole Y-11, a value similar to the kaolinite basal spacing. The refractive indices and color of kaolinite and septechlorite differ but are not completely reliable for discrimination because both minerals are commonly intergrown with montmorillonite. The two minerals could be distinguished only by an intersalation technique (Andrew and others, 1960), by which the unknown clay mineral was first treated with potassium acetate and then with ammonium nitrate; X-ray-diffraction patterns are obtained after both treatments. No change occurs in the spacing of septechlorite peaks, whereas the kaolinite lattice expands to 14 Å with the potassium acetate treatment (in the drill core from hole Y-11, this peak is commonly masked by montmorillonite) and contracts to about 11.6 Å after treatment with 10 N ammonium nitrate.

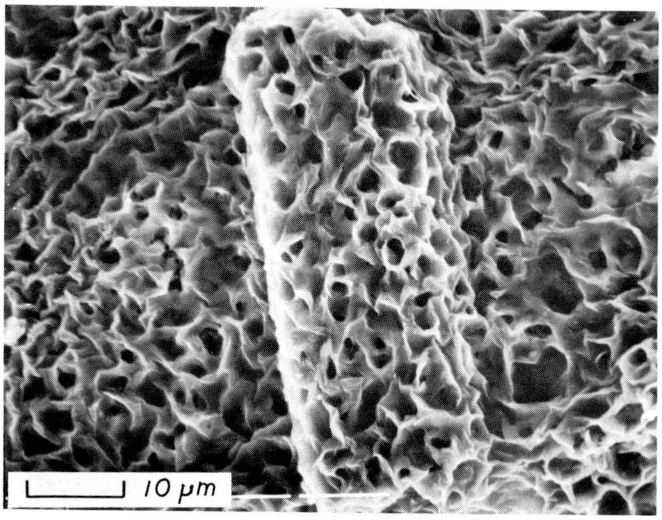

Figure 5: Scanning electron micrograph showing boxwork of septechlorite coating rhodochrosite at 68.0 m.

Carbonate Minerals

Foliate vapor-phase cavities at 64.8, 65.8, 67.8-68.0 m contain a boxwork of rhodochrosite crystals (Fig. 6) in association with chalcedony, mordenite, pyrite, septechlorite, montmorillonite(?), hematite, and goethite(?). The rhodochrosite in drill hole Y-11 appears to have formed earlier than the septechlorite at 64.8 m but is coated by a thin layer of septechlorite at 65.8 and 68.0 m (Figs. 5, 6). Hydrothermal rhodochrosite also occurs in drill hole Y-6 at a temperature measured during drilling of about 140° to 194°C (K.E. Bargar and M. H. Beeson, unpub. data, 1982). A reconstructed ground-temperature profile for drill hole Y-11 (Fig. 2) shows a temperature between about 150° and 160°C for the depth at which rhodochrosite was found in this drill core.

In the drill core from hole Y-6, the position of the d(104) X-ray-diffraction peak of rhodochrosite varies somewhat with chemical composition and is about 2.84 Å for nearly pure $MnCO_3$ and nearly the same for rhodochrosite containing about 12 to 14 wt % $FeCO_3$; the d(104) peak increases to about 2.88 Å with increase in calcium content. No chemical analyses were obtained for the rhodochrosite from drill hole Y-11. A d(104) X-ray peak at approximately 2.84 Å indicates that the trigonal crystals may be nearly pure rhodochrosite; however, the close association with hematite and goethite(?) suggests the presence of a significant iron component.

A small amount of calcite was observed in a thin section of glacial sedimentary material at 6.4 m. Hydrothermal calcite at 46.5-48.6 m and 58.3-74.2 m in drill hole Y-11 typically lines vapor-phase cavities on top of iron-oxide-or-hydroxide-stained alkali feldspar and tridymite. The tiny white-bladed crystals are also intergrown with white botryoidal silica (beta-cristobalite above 64.9 m and chalcedony below that depth) and coated by radiating clusters of clear prismatic quartz crystals. Later, white fibrous mordenite coats several of the calcite-silica cavity fillings, and flourite, pyrite, montmorillonite(?), and earlier dark-green septechlorite (Fig. 7) are each associated with one of the calcite-silica samples. A fracture filling at 89.6 m contained a thick sequence of purple, brown, and clear chalcedony covered by creamy bladed calcite and, still later, white fibrous mordenite.

Figure 6: Scanning electron micrograph of tabular rhodocrosite crystals mostly coated by later septechlorite at 68.0 m.

Flourite

X-ray diffraction data indicate that flourite is present at depths of 47.1 and 69.4 m. Clear to white octahedral flourite crystals line several fractures and vapor-phase cavities from 94.7 m to the bottom of the drill hole at 105.7 m, where predrilling ground temperatures are probably about 180° to 190+°C. These temperatures are a little hotter than the temperature range at which flourite was found in the drill cores from holes Y-5 (160° - 170°C) (Keith and Muffler, 1978) and Y-6 (140° - 180°C) (K.E. Bargar and

Figure 7: Vapor-phase cavity at 74.2 m coated by dark (green) septechlorite, and intergrown white bladed calcite and clear to white botryoidal chalcedony. Cavity also contains late prismatic quartz and mordenite.

M.H. Beeson, unpub. data, 1982). However, the single occurrence of flourite in the drill core from hole Y-2 was at a temperature of about 200°C (Bargar and Beeson, 1981), and the measured temperature at which flourite occurs in drill hole Y-3 ranges from about 132.7°C (Bargar and others, 1973) to approximately 196°C at the bottom of the drill hole (White and others, 1975). In the drill core from hole Y-11, fluorite is a late hydrothermal mineral, as in the other Yellowstone drill cores, and it is commonly associated with earlier chalcedony, iron oxide or hydroxide(?), pyrite, septechlorite, and later montmorillonite, quartz, and mordenite.

Bastnaesite

Bastnaesite, the rare-earth-element carbonate-flouride mineral $(Ce,La)(CO_3)F$, (not shown in Fig. 2), lines a few fractures in the drill core from hole Y-11 at 35.0-36.0, 48.2, and 54.4 m. A sample of the rare-earth-element mineral from 36.0 m consists of tiny white soft clayey spherical deposits (Fig. 8). Bastnaesite commonly has a lamellar structure consisting of aggregates of tabular hexagonal crystals (Glass and Smalley, 1945). In Figure 8, the bastnaesite appears to be composed of spherical clusters of platy crystals. Closer views of the mineral (Fig. 9), however, suggest that the platiness of the deposit may be due to a coating of montmorillonite. In drill hole Y-11, bastnaesite is commonly associated with later red or gray montmorillonite, iron oxide or hydroxide that appears to be amorphous to X-rays, and pyrite (although no pyrite was seen in the scanning electron micrographs of Figs. 8, 9).

Bastnaesite, which is a fairly rare mineral, has been

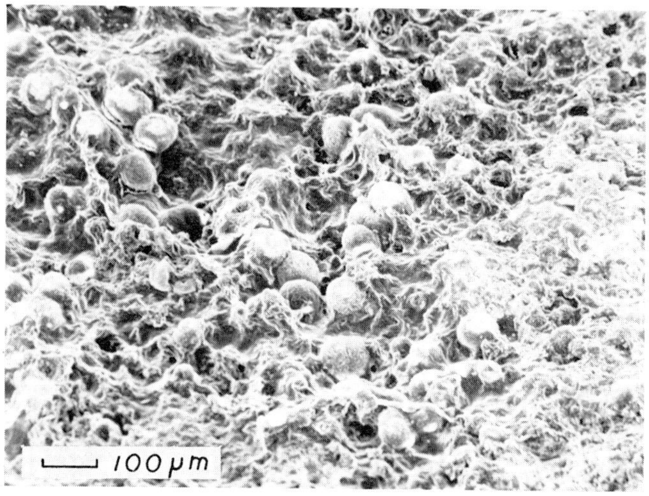

Figure 8: Scanning electron micrograph of montomorillonite-coated spherical bastnaesite crystal clusters(?) lining a fracture at 36.0 m. Some spheres appear to consist of aggregates of bladed crystals.

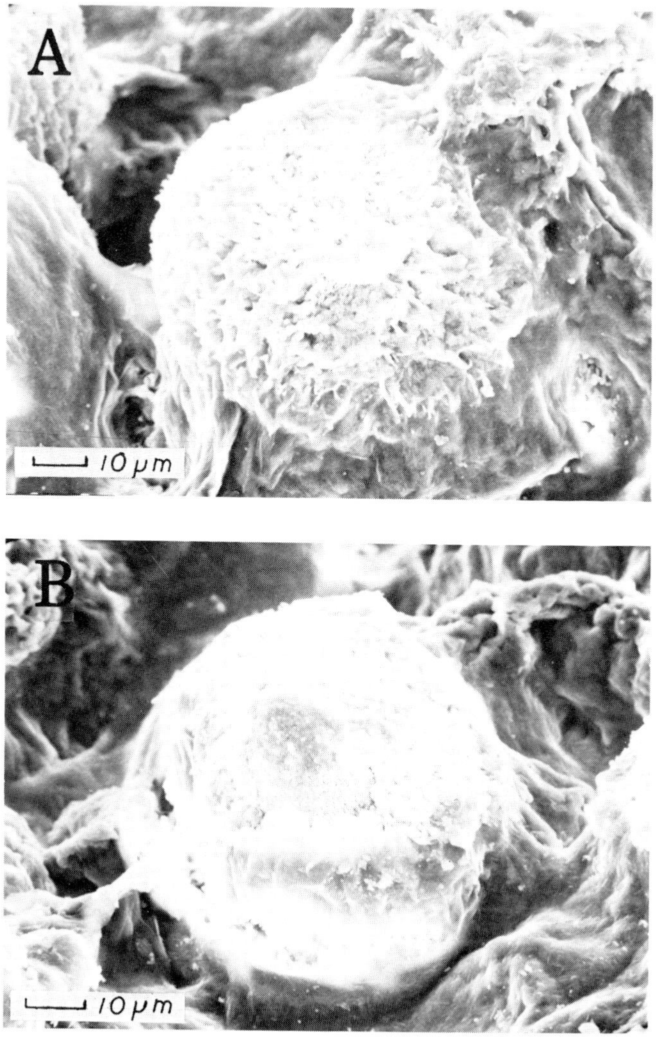

Figure 9: Scanning electron micrographs of bastnaesite spheres at 36. m. *A*, Platey montmorillonite coating of bastnaesite that probably accounts for bladed appearance of several bastnaesite spheres in Figure 8. *B*, Montmorillonite coating bastnaesite sphere. Break in montmorillonite coating reveals bastnaesite(?) with minute granular appearance.

reported from contact-metamorphic zones, carbonatite or fluorite deposits, and pegmatites (Glass and Smalley, 1945; Olson and others, 1954; Adams and Young, 1961; Mineyev and others, 1974). Apparently, this rare-earth-element mineral is a common alteration product of allanite (a rare-earth-element member of the epidote group).

Table 2 lists a semiquantitative spectrographic analysis of the bastnaesite from drill hole Y-11. Cerium and lanthanum are very abundant, as would be expected from previously reported analyses (Glass and Smalley, 1945; Olson and others, 1954). Most other rare-earth elements in the analysis are also fairly abundant, particularly dysprosium neodymium, praseodymium, and yttrium. More than 150 samples of unaltered and hydrothermally altered rocks and hydrothermal minerals from the 13 Yellowstone drill holes were analyzed by the semiquantitative method; nearly all the analyzed samples have low concentrations of rare-earth elements, generally near or below the detection limits.

The concentration of trace elements in the bastnaesite sample (Table 2) in comparison with that in the whole-rock analyses (Table 1) (note particularly the sample at 30.0 m, the same sample from which the bastnaesite analysis was obtained) suggests that the rare-earth elements in the bastnaesite were not concentrated from immediately adjacent volcanic rocks. Instead, the rare-earth elements appear to have been concentrated from the thermal waters that flow through the rocks. However, analyses of hot-spring and steam-water samples from Yellowstone Park (Rowe and others, 1973; Thompson and others, 1975; Thompson and Yadav, 1979; Stauffer and others, 1980) do not include rare-earth elements.

Previous studies of bastnaesite (Glass and Smalley, 1945; Olson and others, 1954) have not reported the presence of cadmium, which is fairly abundant in the bastnaesite analysis from drill hole Y-11 along with zinc and lead (Table 2). Cadmium is commonly associated with such sulfide minerals as sphalerite. Pyrite was the only sulfide mineral detected in the sample analyzed from a scraping of a fracture surface (Table 2); other minerals include bastnaesite, montmorillonite, and goethite. Cadmium has a very low concentration (generally below the detection limit) in all other analyses of altered Yellowstone drill core, and so its presence here is puzzling.

Table 2. Semiquantitative six-step spectrographic analysis of a fracture filling containing bastnaesite from 36.0 m[1-2]

Element[3]	Concentration (wt)
Si	>10
Al	2
Fe	>10
Mg	0.05
Ca	1
K	3
Ti	0.05

Element[3]	Concentration (ppm)
B	70
Ba	100
Be	70
Cd	3,000
Ce	50,000
Cr	30
Cu	50
Dy	500
Er	150
Eu	70

Element[3]	Concentration (ppm)
Ga	<15
Gd	300
Ho	70
La	7,000
Mn	2,000
Nb	300
Nd	3,000
Ni	7
Pb	300
Pr	2,000
Sc	500
Sm	300
Sr	30
Tb	100
Tm	30
W	150
Y	1,500
Yb	150
Zn	500
Zr	200

1/ Analyst: C. Heropoulos, U.S. Geological Survey.

2/ Sample contaminated by vapor-phase tridymite, amorphous iron oxide or hydroxide, pyrite, and montmorillonite.

3/ Looked for but not found: Ag, As, Au, Bi, Co, Ge, Hf, In, Li, Lu, Mo, Na, P, Pd, Pt, Re, Sb, Sn, Ta, Te, Th, Tl, U, and V.

Mordenite

Mordenite was the only zeolite mineral found in the drill core from drill hole Y-11. Zeolitization is more extensive in Yellowstone drill cores that contain abundant glass at the onset of hydrothermal alteration (C-I, Fenner, 1936; Y-1, Honda and Muffler, 1970; Y-2, Bargar and Beeson, 1981; Y-3, M. H. Beeson and K. E. Bargar, unpub. data, 1982; Y-7 and Y-8, Keith and others, 1978b; Y-13, Keith and others, 1978a). In cores containing older rhyolite flows in which the glass was mostly devitrified before hydrothermal alteration (Y-5, Keith and Muffler, 1978; and Y-6, K. E. Bargar and M. H. Beeson, unpub. data, 1982), the zeolite minerals are less profuse and generally consist of open-space deposits of mordenite.

In drill hole Y-11, mordenite occurs at 15.5-18.7 m and at scattered places in the drill core below 64.8 m. The upper mordenite zone is restricted to the groundmass of the Tuff of Uncle Tom's Trail and probably formed by alteration of glass. In the lower zone, mordenite occurs as sprays of radiating needlelike crystals or masses of cottony in-

Figure 10: White fibrous mordenite coating botryoidal chalcedony in cavity at 78.7 m.

tergrown fibrous crystals (Fig. 10) that line open-space vapor-phase cavities and fractures. At 65.3-67.0 m, the depositional sequence of open-space hydrothermal minerals is: (1) chalcedony, (2) prismatic quartz, and (3) mordenite, all coated by gray-green, yellow-orange, red iron oxide or hydroxide, associated septechlorite, and later pyrite at 66.9 m. Below 67.0 m, the principal paragenetic sequence remains: (1) chalcedony, (2) prismatic quartz, and (3) mordenite; however, one or more of the minerals hematite, pyrite, septechlorite, flourite, montmorillonite, and calcite may also be interspersed at either end or within the primary sequence.

Sulfate Minerals

Gypsum was identified by X-ray diffraction in the cementing material of the conglomerate at 2.2 m, along with kaolinite, opal, and beta-cristobalite(?). Tiny white gypsum needles also line a few fractures in crystal-poor tuff between 21.5 and 29.7 m; associated minerals are montmorillionite and pyrite. The temperature at which gypsum was found exceeds 57°C. At such temperatures, anhydrite is the stable calcium sulfate phase, and so the gypsum in the drill core from hole Y-11 is probably not a product of hydrothermal alteration (Honda and Muffler, 1970; Holland and Malinin, 1979). Instead, gypsum very likely formed after the core was removed from the drill hole, owing to oxidation of pyrite; however, no pyrite was identified at 2.2 m.

X-ray-diffraction traces show the presence of a small amount of alunite associated with kaolinite and opal in a soft white fracture filling at 6.4 m. Alunite, kaolinite, and opal are commonly found in close association in areas of surficial acid (H_2SO_4) alteration in geothermal areas throughout the world (Browne, 1978); details of the formation of this potassium aluminum sulfate mineral were discussed by Schoen and others (1974) and Steiner (1977). Alunite also occurs near the surface in drill holes Y-3 (M. H. Beeson and K. E. Bargar, unpub. data, 1982) and Y-13 (Keith and others, 1978a) and in the upper parts of drill holes from Norris Geyser Basin (C-II, Y-9, and Y-12; T. E. C. Keith, written commun., 1982).

DISCUSSION

The hydrothermal-alteration mineralogy in the drill core from hole Y-11 consists of: (1) an early group characterized by chalcedony, septechlorite, quartz, calcite, rhodochrosite, bastnaesite, fluorite, iron oxide and hydroxide, mordenite, and some beta-cristobalite, pyrite, and montmorillonite; and (2) a later group consisting of opal, kaolinite, halloysite, alunite, and some beta-cristobalite, pyrite, and montmorillonite.

The early alteration minerals (except for bastnaesite) are typical of hydrothermal minerals that have been identified from drill cores in Upper and Lower Geyser Basins (Fenner, 1936; Honda and Muffler, 1970; Keith and Muffler, 1978; Keith and others, 1978a,b; Bargar and Beeson, 1981), where the waters are near-neutral and contain high Cl, low SO_4, and a great dominance of Na over K. Alteration-mineral suites in the hot-water geothermal areas of Upper and Lower Geyser Basins are represented by silica minerals, zeolites, clays, carbonates, sulfides, fluorite (similar to the early-formed alteration minerals in drill core Y-11), and little, if any, kaolinite except in acid-sulfate environments at or near the ground surface.

The fluids encountered in drill hole Y-11 and in the surrounding Mud Volcano geothermal area differ strikingly from those in the hot-water areas of Upper and Lower Geyser Basins. Pores at depth are filled either with steam or with a bicarbonate-sulfate or acid-sulfate water interpreted as a steam condensate that is modified somewhat by rock-water interaction (White and others, 1971). The dilute near-neutral water in drill hole Y-11 typically is low in Cl and Na and relatively high in Ca, Mg, and SO_4. This water is responsible for the latest formed alteration minerals in the drill core, which are present above 19.5 m, overprint earlier alteration between 19.5 and 56.7 m, and occur in conspicuous late crosscutting fractures at 74.6-77.6 m.

According to the model of White and others (1971), the waters at Mud Volcano differed in the past from the acid-sulfate or near-neutral bicarbonate-sulfate fluids present today. White and others suggest that at one time the Mud Volcano area was a typical hot-water geothermal system, as evidenced by old sinter deposits found in the area, and thus had a near-neutral silica-sodium-bicarbonate-chloride water from which the deposition of zeolites, calcite, silica minerals, fluorite, etc. would be expected. At some time in the history of the Mud Volcano system, more hot water boiled off than was replaced by recharge, and the system became vapor dominated. At shallow levels (estimated at >73 m by White and others, 1971) fluids in the pores consist of steam condensate with a composition similar to that of the bicarbonate-sulfate water reported by White and others (1971, analysis 6, Table 1), whereas at deeper levels the voids were filled with either steam or steam condensate. This model seems to fit well with the hydrothermal alteration observed in drill hole Y-11 because it is unlikely that the same hydrothermal fluids could be responsible for the two mineral assemblages present in the drill core.

ACKNOWLEDGEMENTS

The authors were assisted in this study by T.E.C. Keith, P.R.L. Browne, R.L. Oscarson, C.N. Bargar, and N. Prime in sample selection, infrared spectroscopy, scanning electron microscope, X-ray diffraction, and photographic studies, respectively. The manuscript benefited from the reviews and constructive criticisms of M.H. Beeson, T.E.C. Keith, and D.E. White.

REFERENCES

Adams, J.W., and Young, E.J., 1961, Accessory bastnaesite in the Pikes Peak granite, Colorado: art. 254 *of* Short papers in the geologic and hydrologic sciences, U.S. Geological Survey Professional Paper 424-C, p. C292-C294.

Allen, E.T., and Day, A.L., 1935, Hot springs of the Yellowstone National Park: Carnegie Institution of Washington Publication 446, 525 p.

Andrew, R.W., Jackson, M.L., and Wada, K., 1960, Intersalation as a technique for differentiation of kaolinite and chloritic minerals by X-ray diffraction: Soil Science Society of America Proceedings, v. 24, p. 422-424.

Bargar, K.E., and Beeson, M.H., 1981, Hydrothermal alteration in research drill hole Y-2, Lower Geyser Basin, Yellowstone National Park, Wyoming: American Mineralogist, v. 66, p. 473-490.

Bargar, K.E., Beeson, M.H., Fournier, R.O., and Muffler, L.J.P., 1973, Present-day deposition of lepidolite from thermal waters in Yellowstone National Park: American Mineralogist, v. 58, p. 901-904.

Browne, P.R.L., 1978, Hydrothermal alteration in active geothermal fields: Annual Review of Earth and Planetary Sciences, v. 6, p. 229-250.

Christiansen, R.L., 1979, Cooling units and composite sheets in relation to caldera structure: in Chapin, C.E., and Elston, W.E. (eds.), Ash-flow tuffs, Geological Society of America Special Paper 180, p. 29-42.

Christiansen, R.L., and Blank, H.R., Jr., 1972, Volcanic stratigraphy of the Quaternary rhyolite plateau in Yellowstone National Park: U.S. Geological Survey Professional Paper 729-B, p. Bl-B18.

Christiansen, R.L., and Blank, H.R., Jr., 1975, Geologic map of the Canyon Village Quadrangle, Yellowstone National Park, Wyoming: U.S. Geological Survey Geologic Quadrangle Map GQ-1192.

Eaton, G.P., Christiansen, R.L., Iyer, H.M., Pitt, A.M., Mabey, D.R., Blank, H.R., Jr., Zietz, I., and Gettings, M.E., 1975, Magma beneath Yellowstone National Park: Science, v. 188, p. 787-796.

Fenner, C.N., 1936, Bore-hole investigations in Yellowstone Park: Journal of Geology, v. 44, 225-315.

Fournier, R.O., White, D.E., and Truesdell, A.H., 1976, Convective heat flow in Yellowstone National Park: Proceedings Second United Nations Symposium on Development and Use of Geothermal Resources, San Francisco May 20-29, 1975, v. 1, U.S. Government Printing Office, Washington D.C., p. 731-739.

Glass, J.J., and Smalley, R.G., 1945, Bastnaesite: American Mineralogist, v. 30, p. 601-615.

Grim, R.E., 1928, Clay mineralogy: McGraw-Hill, San Francisco, 595 p.

Holland, H.D., and Malinin, S.D., 1979, The solubility and occurrence of non-ore minerals: in Barnes, H.L. (ed.), Geochemistry of hydrothermal ore deposits, John Wiley & Sons, New York, p. 461-508.

Honda, S., and Muffler, L.J.P., 1970, Hydrothermal alteration in core from research drill hole Y-1, Upper Geyser Basin, Yellowstone National Park, Wyoming: American Mineralogist, v. 55, p. 1714-1737.

Keith, T.E.C., and Muffler, L.J.P., 1978, Minerals produced during cooling and hydrothermal alteration of ash flow tuff from Yellowstone drill hole Y-5: Journal of Volcanology and Geothermal Research, v. 3, p. 373-402.

Keith, T.E.C., Beeson, M.H., and White, D.E., 1978a, Hydrothermal minerals in U.S. Geological Survey research drill hole Y-13, Yellowstone National Park, Wyoming: Geological Society of America Abstracts with Programs, v. 10, p. 432-433.

Keith, T.E.C., White, D.E., and Beeson, M.H., 1978b, Hydrothermal alteration and self-sealing in Y-7 and Y-8 drill holes in northern part of Upper Geyser Basin, Yellowstone National Park, Wyoming: U.S. Geological Survey Professional Paper 1054-A, p. A1-A26.

Mineyev, D.A., Rozanov, K.I., Smirnova, N.V., and Matrosova, T.I., 1974, Bastnaesitization products of accessory orthite: Doklady Akad. Nauk. SSSR, v. 210, p. 149-152.

Muffler, L.J.P., and Bargar, K.E., 1974, Hydrothermal alteration of rhyolitic ash-flow tuff in the vapor-dominated geothermal system at Mud Volcano, Yellowstone National Park, U.S.A. (abs.): International Symposium on Water-Rock Interaction, Prague, Czechoslovakia, Abstract Volume, p. 52.

Olson, J.C., Shawe, D.R., Pray, L.C., and Sharp, W.N., 1954, Rare-earth mineral deposits of the Mountain Pass District, San Bernardino County, California: U.S. Geological Survey Professional Paper 261, 75 p.

Pierce, K.L., Obradovich, J.D., and Friedman, I., 1976, Obsidian hydration dating and correlation of Bull Lake and Pinedale Glaciations near West Yellowstone Montana: Geological Society of America Bulletin, v. 87, p. 703-710.

Richmond, G.M., 1977, Surficial geologic map of the Canyon Village quadrangle, Yellowstone National Park, Wyoming: U.S. Geological Survey Miscellaneous Geologic Investigations Map I-652.

Ross, C.S., and Smith, R.L., 1961, Ash-flow tuffs: their origin, geologic relations, and identification: U.S. Geological Survey Professional Paper 366, 81 p.

Rowe, J.J., Fournier, R.O., and Morey, G.W., 1973, Chemical analysis of thermal waters in Yellowstone National Park, Wyoming, 1960-65: U.S. Geological Survey Bulletin 1303, 31 p.

Schoen, R., White, D.E., Hemley, J.J., 1974, Argillization by descending acid at Steamboat Springs, Nevada: Clays and Clay Minerals, v. 22, p. 1-22.

Stauffer, R.E., Jenne, E.A., and Ball, J.W., 1980, Chemical studies of selected trace elements in hot-spring drainages of Yellowstone National Park: U.S. Geological Survey Professional Paper 1044-F, p. F1-F20.

Steiner, A., 1977, The Wairakei geothermal area, North Island, New Zealand: Its subsurface geology and hydrothermal rock alteration: New Zealand Geological Survey Bulletin 90, 136 p.

Thompson, J.M., and Yadav, S., 1979, Chemical anaylses of waters from geysers, hot springs and pools in Yellowstone National Park, Wyoming from 1974 to 1978: U.S. Geological Survey Open-File Report 79-704, 49 p.

Thompson, J.M., Presser, T.S., Barnes, R.B., and Bird, D.B., 1975, Chemical analyses of the waters of Yellowstone National Park, Wyoming from 1965-1973: U.S. Geological Survey Open-File Report 75-25, 59 p.

White, D.E., Muffler, L.J.P., and Truesdell, A.H., 1971, Vapor-dominated hydrothermal systems compared with hot-water systems: Economic Geology, v. 66, p. 75-97.

White, D.E., Fournier, R.O., Muffler, L.J.P., and Truesdell, A.H., 1975, Physical results of research drilling in thermal areas of Yellowstone National Park, Wyoming: U.S. Geological Survey Professional Paper 892, 70 p.

Zohdy, A.A.R., Anderson, L.A., and Muffler, L.J.P., 1973, Resistivity, self-potential, and induced-polarization surveys of a vapor-dominated geothermal system: Geophysics, v. 38, p. 1130-1144.

Thirty-Third Annual Field Conference — 1982
Wyoming Geological Association Guidebook

THE GEOCHEMISTRY OF SHOSHONE GEYSER BASIN, YELLOWSTONE NATIONAL PARK

A.H. TRUESDELL[1] AND J.M. THOMPSON[1]

INTRODUCTION

The chemistry of surface springs issuing from a geothermal system can yield information on subsurface temperatures and flow patterns. A relatively small, high-temperature geothermal system was chosen to test the utility of extensive water and gas sampling in geothermal exploration.

LOCATION OF SPRINGS AND FUMAROLES

Shoshone geyser basin is situated at the west end of Shoshone Lake, 8 miles south-southeast of Upper Geyser Basin in Yellowstone National Park, Wyoming, USA. The geography and thermal activity of Shoshone geyser basin were described by Hayden (1883) and by Allen and Day (1935). The western part of the basin is characterized by alkaline boiling springs that have built up sinter terraces on both sides of Shoshone Creek (Fig. 1). To the east of Shoshone Basin in the low hills between Shoshone Creek and Shoshone Lake and along the lake shore, the activity consists entirely of fumaroles and acid springs. The shallow bays of the lake adjacent to the area of fumaroles and acid springs have numerous gas vents.

Almost all of the activity occurs along a series of northeast-trending lineaments, the two to the northwest trending N 50° E and those to the southeast trending N 40° E. The change in character of activity from alkaline springs in the west to fumaroles, acid springs and gas vents in the east cuts across the lineaments.

WATER AND GAS COLLECTION AND ANALYSIS

Water samples were collected from major springs and gas samples were collected from springs with visible gas bubbles and from fumaroles. Samples of water and gas were collected in only a few places from the same spring because: (1) the springs of greatest flow in the western part of the basin, although favourable for water sampling, were generally superheated and contained little gas and (2) the water compositions of acid springs and drowned fumaroles in the eastern part of the basin were not considered to reflect deep conditions (see later discussion).

Water samples were analyzed by standard methods and have been published (Thompson et al., 1974). Gas samples were collected in gas-tight syringes with CO_2 and H_2S absorbed in 3 N NaOH solution. H_2S was fixed with Cd acetate and analyzed by thiosulfate-iodine titration. CO_2 was analyzed gravimetrically by precipitation of $BaCO_3$, (Truesdell and Pering, 1974).

CHEMISTRY OF WATERS AND GASES

Regular variations in water and gas chemistry were observed. Ratios of CO_2/H_2S increased (Fig. 2), and CO_2 + H_2S)/ other gases decreased (Fig. 3) across the basin from east to west. The ratio of total gas/water also decreased toward the west as shown by dominance of fumeroles in the east and superheated springs in the west. Superheated springs, with temperatures 1-3°C above the normal boiling point, must have relatively low dissolved gas contents (below saturation). If this were not so exsolving bubbles would nucleate boiling and the springs would no longer be superheated.

The alkaline springs generally show a decrease in Cl(1) and SiO_2 and an increase in HCO_3 from east to west (Fig. 4-6). The ratios of HCO_3/Cl increased (Fig. 6) and Cl/SO_4 decreased(Fig. 7)to the west, but the B/Cl ratio of analyzed springs was constant (90% within ± 20%, all within ± 10%). This variation is due largely to error in the analyses with boron contents ranging from 0.9 to 3.6 mg/1. Subsurface temperatures estimated from SiO_2 contents (Fouriner and Rowe, 1966) and Na-K-Ca relations (Fouriner and Truesdell, 1973) are 203 ± °C for the easternmost alkaline springs, 185 ± °C for numerous springs in the center of the basin, and 160°C for the spring at the western extremity. The distribution of oxygen isotopes between the dissolved sulfate and water of springs throughout the basin suggests temperatures of last sulfate-water isotopic equilibration of 270 ± 15°C using fractionation factors from Mizutani and Rafter (1969) and Mizutani

[1]U.S. Geological Survey, Menlo Park, California 94025 U.S.A.

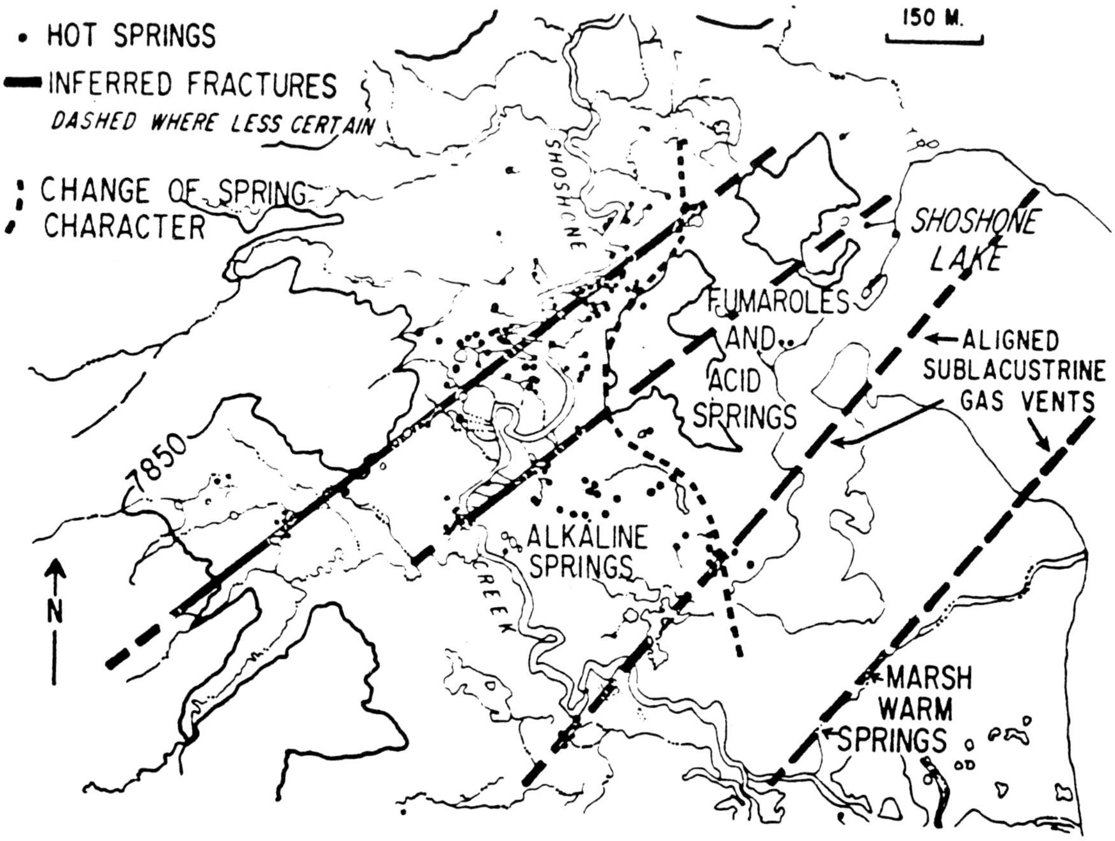

Figure 1: Shoshone geyser basin showing inferred fractures and springs.

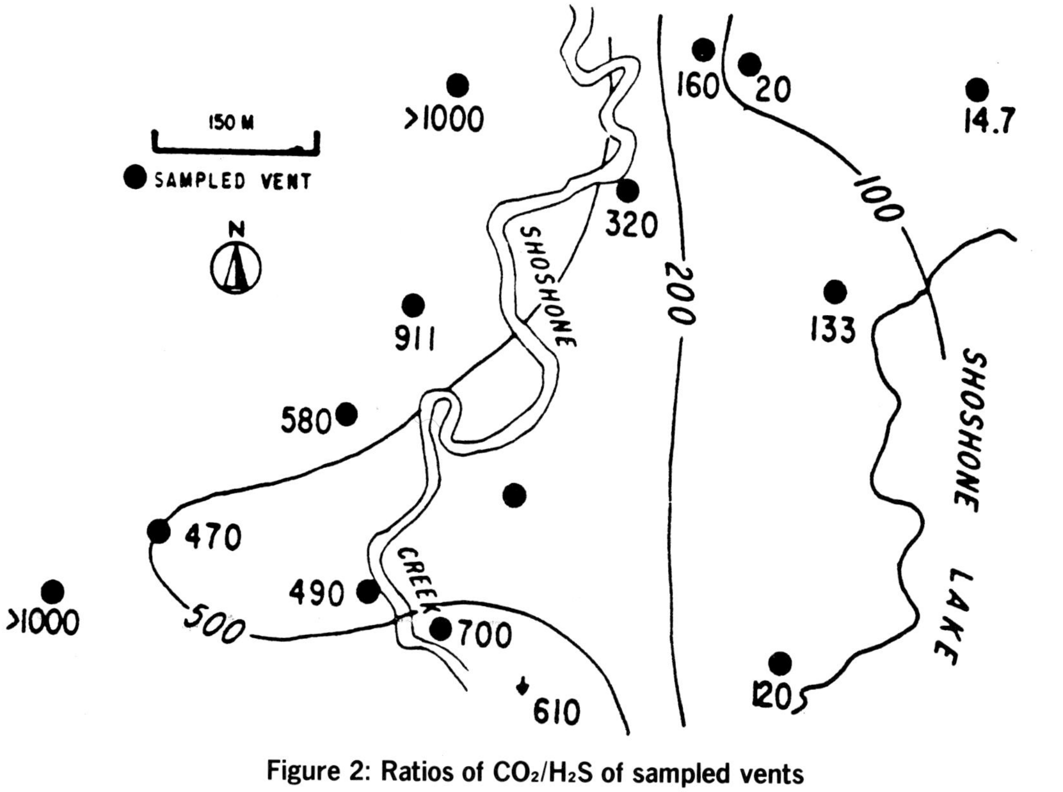

Figure 2: Ratios of CO_2/H_2S of sampled vents

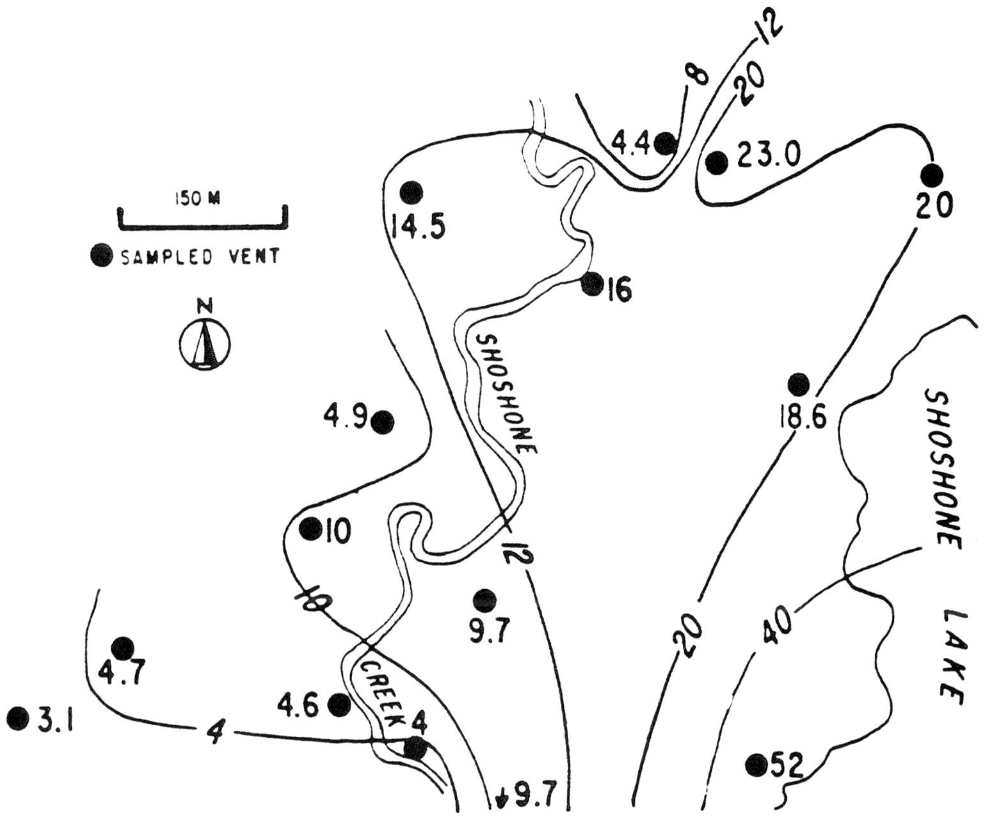

Figure 3: Ratios of $(CO_2 + H_2S)$/other gases of sampled vents

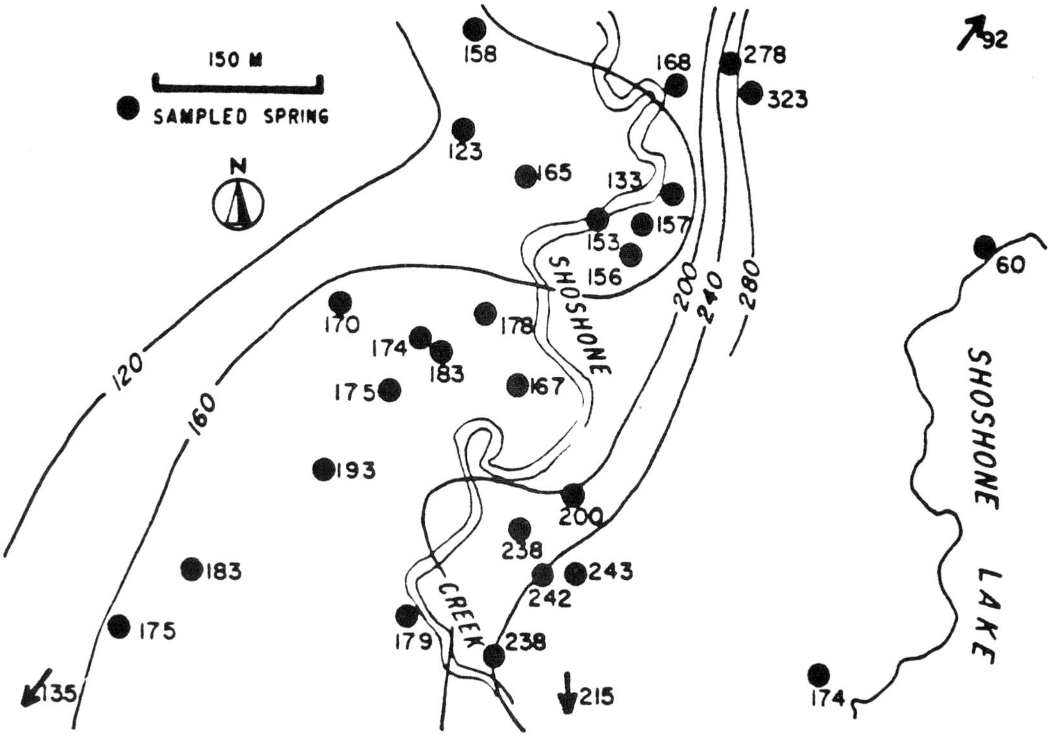

Figure 4: Concentration of Cl, in parts per million, in sampled springs

(1972) and correcting for SO_4 of lower temperature origin (see later discussion). This same temperature was also obtained by a new mixing calculation method.

INTERPRETATION

These chemical relations are interpreted in Fig. 8. Thermal water at 270 ± °C and with a chemical composition similar to (but less concentrated than) that of the eastern most alkaline springs moves up a major fracture zone striking N 50° E and dipping steeply to the southeast. This water first boils at about 600 meters depth and a large part of the dissolved CO_2, H_2S and other gases partitions into steam (Ellis and Golding, 1963; Kozintseva, 1964). Part of the steam ascends to the surface along small, probably near vertical fractures oriented N 40° E and gives rise to the eastern fumaroles, acid springs and gas vents. At the surface some of the H_2S contained in the steam reacts with air to produce sulfur and sulfuric acid. The acid leaches the rock in the immediate area to produce characteristic amphitheatres in the hillside. The resulting springs are acid ($pH<2$) with high SO_4, low Cl, and metal ions inherited from the rock. This origin of acid springs in hot water systems was first described by White (1957). Minor amounts of chloride-bearing water rise along the fractures with the steam to produce low-Cl flowing springs along the lake shore with the same B/Cl ratio as the alkaline springs.

At about 185 meters depth a large fracture is intersected that leads part of the thermal water to the surface to form the easternmost alkaline springs. These springs are highest in Cl and SiO_2 and are interpreted to be representative of the deep thermal water that has not undergone near-surface dilution but has been concentrated by about 36% owing to steam separation as the temperature dropped from 270 to 92°C. The ratio of gas to water in these springs is lower than that in the fumaroles to the east because of the strong partitioning of the gases into steam relative to liquid water.

If these three eastern fractures are vertical as drawn in Figure 8, then the major fracture zone dips southeast 80°. If, on the other hand, their angle of dip is not vertical, then they must also dip to the southeast and the dip of the major fracture zone will be less.

With decreasing temperature the more reactive gases, CO_2 and H_2S, are removed from the fluids by water-rock interactions and by reactions caused by mixture of fluids so that the ratio of $CO_2 + H_2S)$/other gases decreases. The ratio of CO_2/H_2S increases with decreasing temperature even though H_2S is more soluble than CO_2 in liquid water (Kozinseva, 1964) because of water-rock interactions to form pyrite (Fe S_2).

$$2\,H_2S + \text{Fe oxides} = FeS_2, + H_2O. \quad (1)$$

CO_2 in turn reacts with Na silicates to form HCO_3 ions, alkaline ions, and clay minerals,

$$CO_2 + H_2O + \text{Na silicate} = HCO_3^- + N_a^+ + \text{H silicate} \quad (2)$$

The proportion of HCO_3 in the water depends on the temperature, partial pressure of CO_2, and the pH of the water, which is buffered by aluminosilicate mineral pairs in the rock (Ellis, 1967, Truesdell and Singers, 1974). As the temperature decreases with the activity of alkali ions remaining nearly constant, the pH increases (Hemley, 1959) and more CO_2 is converted to HCO_3 with coupled introduction of Na into the water. The increase in HCO_3 with decreasing temperatures causes the observed increase in HCO_3/Cl ratios to the west (Fig. 6).

Neither Cl nor B enter into near surface reactions with gases or rocks, and thus the Cl/B ratio is nearly constant.

Most of the thermal water continues upward along the major fracture zone and enters a shallow aquifer at about 117 meters (Fig. 8). This aquifer is assumed to lie along the contact of rhyolite flow and glacial sediments. Similar aquifers were found at the base of glacial sediments in research drill holes in the upper and lower geyser basins (White et al., 1974).

In this aquifer the thermal water is diluted with varying proportions of dilute cold ground water of meteoric origin. The shallow ground water contributes negligible solutes directly to the mixed thermal water, however oxygen dissolved in the cold water reacts with an excess of H_2S dissolved in the thermal water by the reaction

$$2\,O_2 + H_2S = SO_4^{2-} + 2\,H^+ \quad (3)$$

to produce additional SO_4 and the Cl/SO_4 ratios of the mixed waters decrease (Fig. 7) with dilution of the thermal water. Cold water at 10°C can dissolve about 10 ppm of oxygen from the atmosphere and can produce a maximum of 15 ppm of SO_4 by oxidation of H_2S with complete removal of dissolved O_2.

The ratio of thermal to cold water in each mixed spring may be calculated from its Cl and SiO_2 contents and the Cl contents of an undiluted spring. For example, the westernmost spring is a mixture of dilute water containing about 2 ppm Cl and 0.6 ppm SO_4 and thermal water containing 208 ppm Cl and 45 ppm SO_4. The concentration of Cl and SO_4 in the thermal water is calculated from the concentrations in the most concentrated spring of the basin allowing for 36% concentration by loss of steam on cooling from 270°C to 92°C. The concentrations in the dilute water are typical of western Yellowstone cold springs. The fraction of thermal water in the mixture is 0.567 calculated from the Cl

[1]Except in the equations, chemical species are shown without valance.

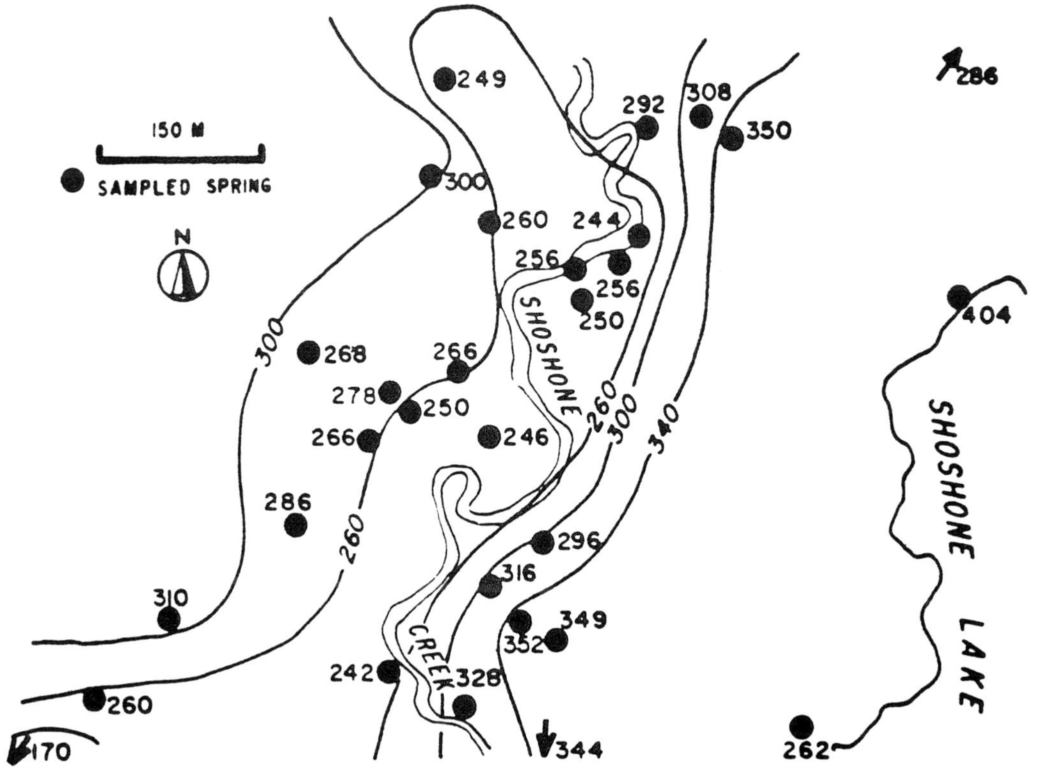

Figure 5: Concentration of SiO_2, in parts per million, in sampled springs.

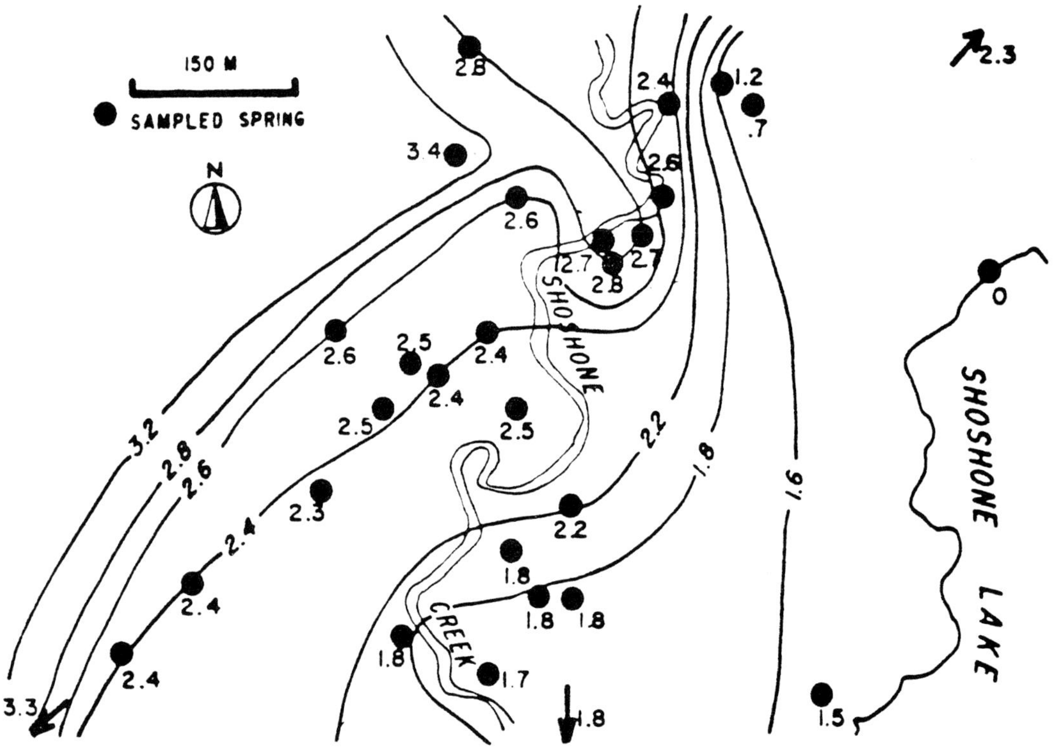

Figure 6: Ratio of HCO_3/Cl, by weight, in sampled springs

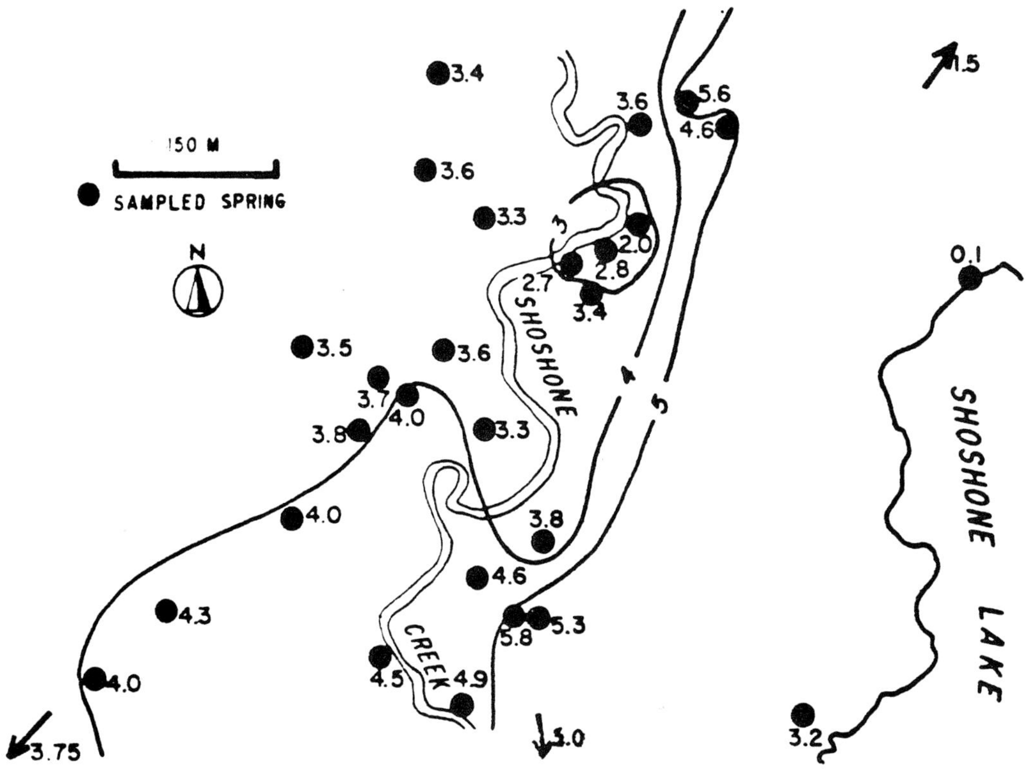

Figure 7: Ratio of Cl/SO_4, by weight, in sampled springs

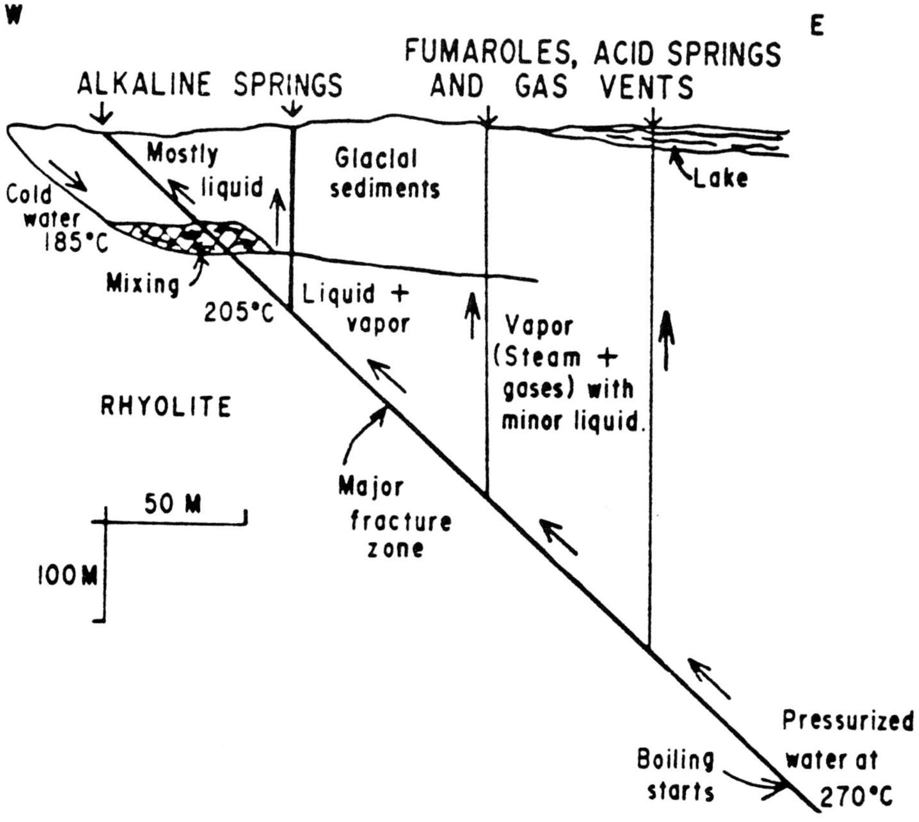

Figure 8: Hypothetical east-west cross section of part of the Shoshone geothermal system.

contents of the springs again with allowance for cooling from 160°C (from SiO_2 contents) to 94°C by loss of steam after mixing. The 10 ppm oxygen dissolved in the dilute water reacting (by equation 3) with an excess of H_2S introduced with thermal water produces 6.5 ppm additional SO_4 ion, which changes the CL/SO_4 ratio from 5 ± of the thermal water (or the easternmost alkaline springs) to 3.7. These calculations show most of the variation in Cl/SO_4 ratios is due oxidation of H_2S by this process.

The mixing and therefore the reaction occurs at 160-185°C and thus is almost certainly abiogenic. This is supported by the finding that water in similar rocks originally at 203°C from drillhole Y-13 in Lower Geyser Basin, Yellowstone was completely sterile (Brock, pers. commun., 1968). The oxidation of H_2S to SO_4 in thermal areas has been assumed to require sulfuroxidizing bacteria (Schoen and Ehrlich, 1968; Schoen and Rye, 1970), but the chemical relations at Shoshone suggest that this is not always true.

After dilution and reaction of dissolved gases, the thermal waters rise to the surface along the trace of the westernmost N 50° E fracture.

SUMMARY

Dilution, boiling, and temperature dependent reactions of water with rock and of dissolved gases with water and rock produce marked chemical changes in the thermal waters of Shoshone geyser basin as they ascend along a steeply dipping fracture system. The changes in the concentrations of certain gases observed in fumaroles and springs, tapping water and steam from increasingly shallow parts of the fracture system include decreasing ratios of total gas/water, $(CO_2 + H_2S)$/ other gases and H_2S/CO_2. Changes in alkaline springs include decreases in SiO_2, Cl, and other constituents and in the ratios of Cl/HCO_3 and Cl/SO_4. These chemical changes may be expected to occur to a greater or lesser degree in other geothermal systems and may be of value in geothermal exploration.

REFERENCES

Allen E.T. and Day A.L. (1935): Hot springs of Yellowstone National Park. - Carnegie Inst. of Wash. pub. no. 466.

Ellis A.J. (1967): The chemistry of some explored geothermal systems: Chapt. 2 in Geochemistry of Hydrothermal Ore Deposits. H.L. Barnes, ed., 670 p., Holt, Rinehart, Winston, New York.

Ellis A.J. and Golding R.M. (1963): The solubility of carbon dioxide above 100°C in water and in sodium chloride solutions. - Am. Jour. Sci. 261, 47-60.

Fournier R.O. and Rowe J.J. (1966): Estimation of underground temperatures from the silica content of water from hot springs and wet-steam wells. - Am. Jour. Sci. 264, 685-697.

Fournier R.O. and Truesdell A.H. (1973): An empirical Na-K-Ca geothermometer for natural waters. - Geochim. Cosmochim. Acta 37, 1255-1275.

Hayden F.V. (1883): Twelfth annual report of the U.S. Geological an Geographical Survey of the Territories for 1878, p. II, 488 p.

Hemley J.J. (1959): Some mineralogical equilibria in the system K_2O-Al_2O_3-SiO_2-H_2O. - Am. Jour. Sci. 257, 241-270.

Kozintseva T.N. (1964): Solubility of hydrogen sulfide in water at elevated temperatures. - Geochem. Internat. 4, 750-756.

Mizutani Y. (1972): Isotopic composition and underground temperature of the Otaki geothermal water Kyushu, Japan. - Geochem. Jour. 6, 67-73.

Mizutani Y. and Rafter T.A. (1969): Oxygen isotopic composition of sulfates. Part 3, Oxygen isotope fractionation in the bisulfate ion-water system. - New Zeal. Jour. Sci. 12, 54-59.

Schoen R. and Ehrlich G.G. (1968): Bacterial origin of sulfuric acid in sulfurous hot springs. - Proc. 23rd Int. Geol. Cong. Prague, Czech., v. 17, p. 171-178.

Schoen R. and Rye R.O. (1970): Sulfur isotope distribution in solfataras, Yellowstone National Park. - Science 1970, 1082-1084.

Thompson J.M., Willey L., Presser T.S., Barnes R., and Byrd D.B. (1974): Chemical analyses of waters of Yellowstone National Park. Wyoming 1965 to 1973. - U.S. Geol. Survey open-file rept.

Truesdell A.H. and Pering K.L. (1974): Geothermal gas sampling methods. - U.S. Geological Survey open-file rept.

Truesdell A.H. and Singers W. (1974): The calculation of aquifer chemistry in hot-water geothermal systems. - U.S. Geol. Survey Jour. Research 2, 271-278.

White D.E. (1957): Thermal waters of volcanic origin. - Bull. Geol. Soc. Am. 68, 1637-1658.

White D.E., Fournier R.O., Muffler L.J.P., and Truesdell A.H. (1974): Physical data from research drilling in thermal areas of Yellowstone Park, Wyoming. - U.S. Geol. Survey Prof. Paper (in press).

Thirty-Third Annual Field Conference — 1982
Wyoming Geological Association Guidebook

CONTINUOUS MONITORING OF THERMAL ACTIVITY IN YELLOWSTONE NATIONAL PARK

BY IRVING FRIEDMAN[1] AND DANIEL R. NORTON[1]

Monitoring of the thermal activity in Yellowstone National Park is a cooperative effort of the U.S. Geological Survey and National Park Service and is being carried out for two reasons: 1) to secure background information as to the natural variations in activity in order to assess the influence of proposed geothermal and gas and oil development adjacent to the park, and 2) to determine the long term trends in activity and to relate such trends to seismic activity and movement of magma in and adjacent to the park.

Due to financial limitations we have been restricted to the activation of a single station that monitors stream flow, temperature and electrical conductivity of a thermal spring on the west side of Boundary Creek, about 2½ miles from the west boundary of the Park, and 2½ miles southeast of Buffalo Lake. The data is collected twice daily, at noon and midnight, by a Campbell Scientific Co. model CR21 micrologger[2] and stored on a magnetic tape cassette. Although the instrument is capable of running for about 1 year on internal dry cells, the site has been visited 2 or 3 times a year for servicing and to retrieve the data cassette.

The spring is being monitored by recording the depth of water in a weir located about 100 meters downstream from the source which discharges at a temperature of about 80°C. For the first 9 months temperatures were being recorded at the weir. At present temperatures are being recorded at the source as well as at the weir. An electrical conductivity cell is mounted in the weir.

Figure 1 shows the data from 1 October, 1980 to 15 June, 1981. Note the cooling due to snow melt in April and early May 1981. The electricity conductivity and water flow also reflect the input of "pure" snow melt during this period. During late May and June 1981 the water flow remained high, but the temperature increased, as did the electrical conductivity. This is attributed to increased spring flow and a decreasing contribution of snow melt.

The technique of recording data at remote sites using a micrologger has proven successful. If funding can be secured we plan to instrument other thermal areas near the Park boundaries as well as the major thermal features.

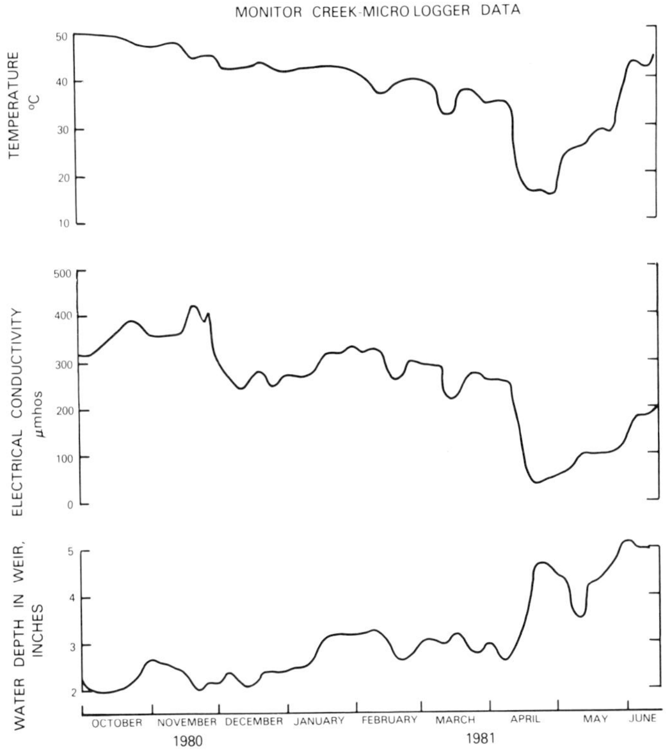

Figure 1.

[1]U.S. Geological Survey, Box 25046, Federal Center, Denver, CO 80225.
[2]Use of trade names in this report is for descriptive purposes only and does not constitute endorsement by the U.S. Geological Survey.

Thirty-Third Annual Field Conference — 1982
Wyoming Geological Association Guidebook

THE CODY HYDROTHERMAL SYSTEM

HENRY P. HEASLER[1]

INTRODUCTION

The hot springs and associated travertine and sulfur deposits near Cody, Wyoming were first discovered by John Colter in 1807 or 1808 (Mattes, 1962). Because of Colter's description of the thermal areas along the Stinkingwater River (now called the Shoshone River), fur trappers called the area "Colter's Hell" (Mattes, 1962), a name which has sometimes mistakenly been applied to Yellowstone National Park.

The hot springs of Colter's Hell are the surface manifestations of a much larger hydrothermal system. That system has been studied to define its extent, maximum temperature, and mechanism of operation. The study area covers 2,700 km² (1,040 mi²) in northwest Wyoming (see Figures 1 and 2). Research and field work included locating and sampling the hot springs, geologic mapping, thermal logging of available wells, measuring thermal conductivities, analyzing over 200 oil and gas well bottom-hole temperatures, and compiling and analyzing hydrologic data. These data were used to generate a model for the hydrothermal system.

General Geologic Setting

The Cody study area covers portions of the western Bighorn Basin and eastern Absaroka Mountains. The uplifts shown in Figure 2 are Laramide in age (Fanshawe, 1952). Uplifts which are discussed include the Rattlesnake anticline and the Horse Center anticline. The Rattlesnake anticline trends northwest, has an exposed core of Precambrian granite and granitic gneiss, and is located in T.53N., R. 103W., the southwest corner of T.53N., R.102W., and the north half of T.52N., R.102W. (see Figure 2). The Horse Center anticline trends north to north-northwest, has an exposed core of Triassic Chugwater Formation, and is located in the southeast corner of T.52N., R.102W., the southwest quarter of T.52N., R.101W., and the northwest quarter of T.51N., R.101W. (see Figure 2). Both structures exert control on the areal extent and magnitude of the hydrothermal system.

Absaroka volcaniclastic material of Eocene age (Smedes and Prostka, 1972) covers the western portion of the study area. Any heat generated by the 47-million-year-old volcanism has dissipated and no longer effects the thermal regime of the area. The volcanics are important because they cover and obscure the underlying sedimentary structure and may also influence local groundwater flow and chemistry.

Sedimentary rocks in the area represent the Cambrian through Quarternary systems, with only Silurian deposits absent. Thicknesses of the sedimentary rocks used in the thermal modeling were those of Pierce (1966, 1970, 1978), who gives a total sediment thickness of over 7 km in the eastern portion of the study area.

The thrust faults shown on Figure 2 — the Southfork thrust fault and Heart Mountain Detachment fault of Pierce (1975, 1978) - are not believed to exert significant control on the hydrothermal system.

Evidence for the Cody hydrothermal system

The most obvious evidence for the Cody hydrothermal system is the four sets of hot springs in the study area. All are located along the Shoshone River as shown in Figure 2. Chemical, temperature, and flow data for the springs are listed in Table 1.

The easternmost springs are the DeMaris Hot Springs, a group of at least seven major vents and more than ten minor vents along 220 meters of the northern shore of the Shoshone River. Measured temperatures for the vents range from 24 to 34°C, but Breckenridge and Hinckley (1978) report temperatures as high as 37°C (see Table 1). The springs occur on the southeast flank of the Rattlesnake anticline and flow from the Park City and Dinwoody Formations. A 41-meter-deep well located on a terrace 150 meters southwest of the DeMaris Hot Springs, which supplies 790 liters per minute (208gpm) of 37°C water to the Cody Health Spa, is believed to have the same source (see Table 1).

1.2 km west, along the northern shore of the Shoshone River, is a group of at least five vents called Shoshone Hot Springs (see Figure 2). Measured temperatures at three accessible vents were 28 to 34°C, with flow from the Tensleep Sandstone. Further west, the next set of hot springs is a series of 23°C seeps flowing from road fill that covers the Pilgrim Limestone. These are probably the Needle Hot Springs described but not located by Breckenridge

[1]Department of Geology and Geophysics, University of Wyoming, Laramie, Wyoming 82071

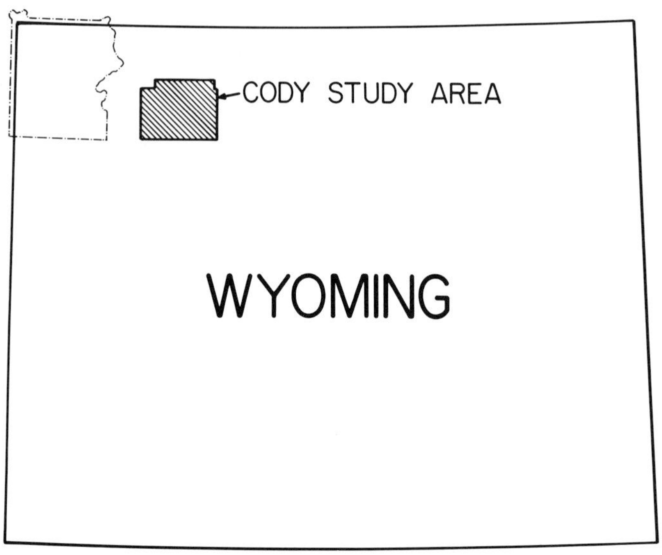

Figure 1: Location of the Cody Study Area.

and Hinckley (1978). The fourth set of springs is that reported by Love and Good (1970) to have been covered in 1907 by Buffalo Bill Reservoir. Love and Good locate four possible thermal springs beneath the reservoir. Examination of pre-1907 photographs, field mapping of travertine deposits, and observation of ice-free areas on the reservoir indicate that the largest of the four springs is located beneath the reservoir as shown on figure 2. This is probably the site of John Colter's "boiling tar spring" (see Love and Good, 1970, pages B19-B21), believed to have been an active oil seep in historic times.

Additional evidence for the hydrothermal system is travertine and sulfur deposits. These deposits cover an .8-to 1.2-km-wide strip from DeMaris Hot Springs southeast 4 km along the eastern flank of Rattlesnake anticline near the contact of the Park City, Dinwoody, and Chugwater Formations (Woodruff, 1908; Wideman, 1957). Production of sulfur from these deposits was unsuccessfully attempted in 1906, 1926, and 1951 (Wideman, 1957). Travertine deposits are also located above present reservoir level near the site of the fourth set of springs (Pierce, 1970; unpublished mapping by Heasler, 1980).

Subsurface evidence for the hydrothermal system consists of reported oil well bottom-hole temperatures (BHT's) which are higher than normal. Thermal gradients were calculated as

$$\text{Thermal Gradient} = \frac{\text{BHT-(Mean Annual Air Temperature)}}{\text{Depth of well}}$$

in °F/1,000 ft (which equals 1.823°C/km); these are shown in Figure 2 for each well. Most oil-well-BHT-derived gradients for wells in Figure 2 are in the range 18 to 37°C/km (10 to 20°F/1000 ft). However, along the crest of Horse Center anticline and near the southern termination of Rattlesnake anticline, thermal gradients are high: eleven are in the range 49 to 205 °C/km (27 to 112°F/1000 ft).

Normal and high thermal gradients were verified by the precision thermal logging of available wells by the Wyoming Geothermal Resource Assessment Group, Department of Geology and Geophysics, University of Wyoming. Thermistor probes precise to ±.005°C were lowered at 10-meter intervals to give temperature-depth plots (see Decker, 1973 for a complete description of this thermal technique). The temperature-depth plots were analyzed to define zones of hydrologic flow and of different thermal gradients through various rock types.

Table 2 summarizes all the thermally logged well data for the Cody area. The Husky and BLM wells listed in Table 2 are east of the thermal area and have normal gradients of 26 and 20°C/km (14 and 11°F/1000 ft). The measured thermal gradient values from Table 2 for the Horse Center anticline are equal to or greater than the oil well BHT gradients of 49 to 205°C/km (27 to 112°F/1000 ft). This substantiates the existence of the hydrothermal system in the area of the Horse Center anticline.

Of particular interest are thermal logs of three wells on the Horse Center anticline and five wells on the southeastern flank of Rattlesnake anticline. The location of these wells is shown in figure 3, the temperature-depth plots in Figures 4 and 5.

Figure 4 shows the temperature-depth plots of three oil exploration holes along the axis of the Horse Center anticline. Wells Letha C-4 and Gains Govt C-2, completed in the Tensleep Sandstone, show a primarily linear increase in temperature with depth. Well Rose Govt 1 was completed in Cambrian sediments with a maximum temperature of 47.54°C at 500 meters in the Madison Limestone. The Rose Govt 1 log indicates that temperatures continue to increase below the Tensleep Sandstone. The nature of this temperature increase will be discussed in the modeling section; it is believed it to be due to warm water flow in the Tensleep Sandstone and Madison Limestone. The temperature decrease and isothermal character below 500 meters in this well are not characteristic of the thermal regime of the area. Rather, they probably represent water flow down a fault system to the drill hole and then down the outside of the well casing. Note that all three of these wells have high thermal gradients (greater than 94°C/km, 52°F1000 ft) and bottom-hole temperatures (39-48°C) higher than those of the hot springs (maximum 37°C).

Figure 5 shows temperature-depth plots down five temperature gradient holes drilled by the Wyoming Geothermal Resource Assessment Group. (A complete description of the drilling project can be found in Heasler and Decker, 1980). Well UWH-1 was completed in the

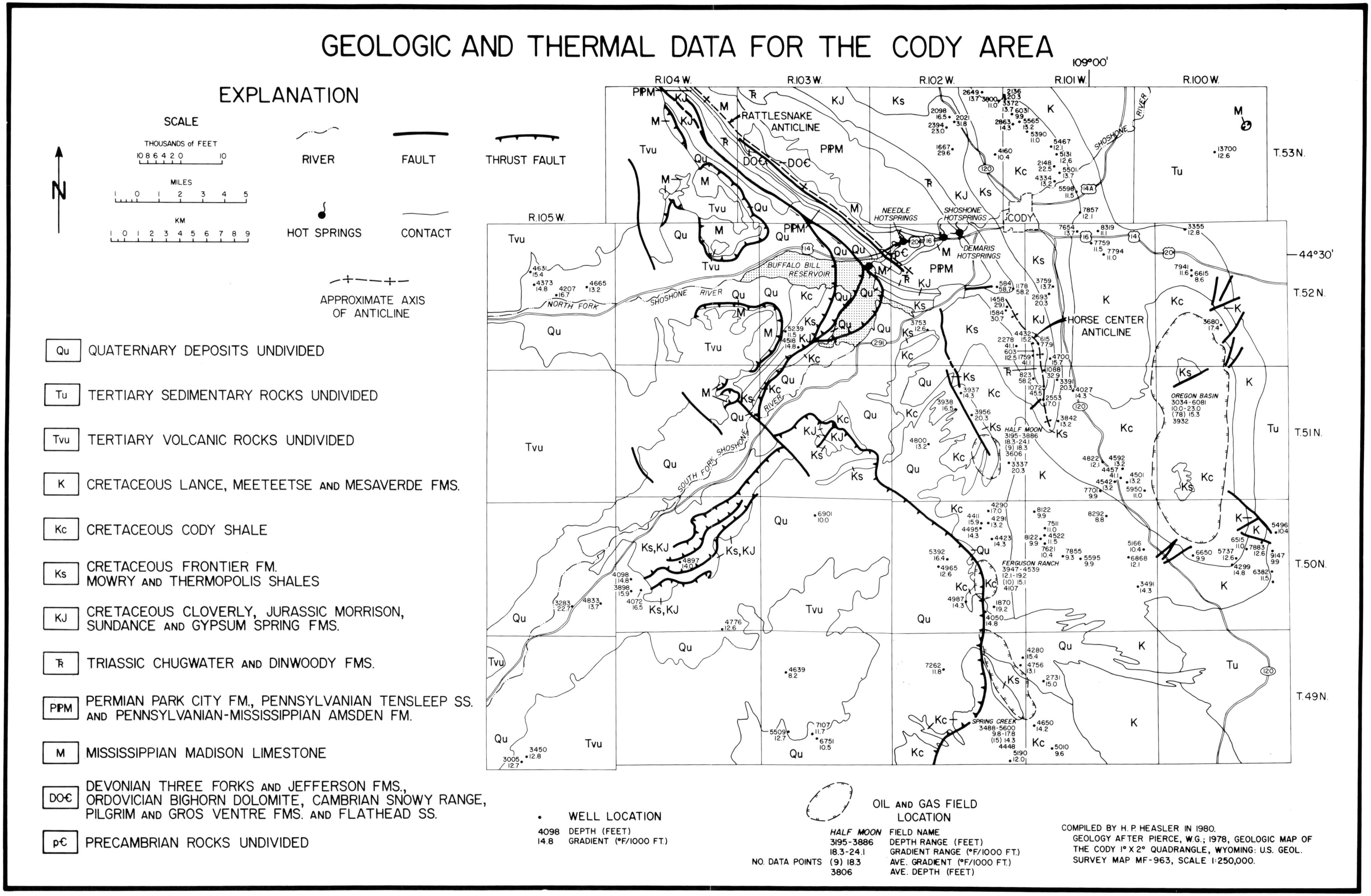
GEOLOGIC AND THERMAL DATA FOR THE CODY AREA
EXPLANATION
SCALE
THOUSANDS of FEET
10 8 6 4 2 0 10
MILES
1 0 1 2 3 4 5
KM
1 0 1 2 3 4 5 6 7 8 9
N
RIVER
FAULT
THRUST FAULT
HOT SPRINGS
CONTACT
APPROXIMATE AXIS OF ANTICLINE
Qu QUATERNARY DEPOSITS UNDIVIDED
Tu TERTIARY SEDIMENTARY ROCKS UNDIVIDED
Tvu TERTIARY VOLCANIC ROCKS UNDIVIDED
K CRETACEOUS LANCE, MEETEETSE AND MESAVERDE FMS.
Kc CRETACEOUS CODY SHALE
Ks CRETACEOUS FRONTIER FM. MOWRY AND THERMOPOLIS SHALES
KJ CRETACEOUS CLOVERLY, JURASSIC MORRISON, SUNDANCE AND GYPSUM SPRING FMS.
TR TRIASSIC CHUGWATER AND DINWOODY FMS.
PPM PERMIAN PARK CITY FM., PENNSYLVANIAN TENSLEEP SS. AND PENNSYLVANIAN-MISSISSIPPIAN AMSDEN FM.
M MISSISSIPPIAN MADISON LIMESTONE
DO€ DEVONIAN THREE FORKS AND JEFFERSON FMS., ORDOVICIAN BIGHORN DOLOMITE, CAMBRIAN SNOWY RANGE, PILGRIM AND GROS VENTRE FMS. AND FLATHEAD SS.
p€ PRECAMBRIAN ROCKS UNDIVIDED
WELL LOCATION
4098 DEPTH (FEET)
14.8 GRADIENT (°F/1000 FT.)
OIL AND GAS FIELD LOCATION
HALF MOON FIELD NAME
3195-3886 DEPTH RANGE (FEET)
18.3-24.1 GRADIENT RANGE (°F/1000 FT.)
NO. DATA POINTS (9) 18.3 AVE. GRADIENT (°F/1000 FT.)
3806 AVE. DEPTH (FEET)
COMPILED BY H. P. HEASLER IN 1980.
GEOLOGY AFTER PIERCE, W.G.; 1978, GEOLOGIC MAP OF THE CODY 1° X 2° QUADRANGLE, WYOMING: U.S. GEOL. SURVEY MAP MF-963, SCALE 1:250,000.
109°00'
R.105 W.
R.104 W.
R.103 W.
R.102 W.
R.101 W.
R.100 W.
T.53 N.
T.52 N.
T.51 N.
T.50 N.
T.49 N.
44°30'
RATTLESNAKE ANTICLINE
NEEDLE HOTSPRINGS
SHOSHONE HOTSPRINGS
DEMARIS HOTSPRINGS
CODY
BUFFALO BILL RESERVOIR
SHOSHONE RIVER
NORTH FORK
SOUTH FORK SHOSHONE RIVER
HORSE CENTER ANTICLINE
OREGON BASIN
3034-6081
10.0-23.0
(78) 15.3
3932
HALF MOON
3195-3886
18.3-24.1
(9) 18.3
3606
FERGUSON RANCH
3947-4539
12.1-19.2
(10) 15.1
4107
SPRING CREEK
3488-5600
9.8-17.8
(15) 14.3
4448

Figure 2:

TABLE 1

Chemical Analyses[1]

No.	Name	Location	Flow[3] (1/m)	Temperature (°C)	Total Dissolved Solids	pH	Cond.	Ca.	Mg	Na	K	HCO_3	SO_4	Cl	NO_3
1	DeMaris Hot Springs[2]	NW¼ sec. 3, T.52N. R.102W.		24				217	173	258		681	39	17	
2				37				188	367	256		872	641	29	
3				27		7.1	1960	369	63	33	16	993	418	20	.1
4				28		6.9	1970	354	72	33	16	952	422	21	.1
5				27	1320	7.0	1880	322	66	30	18	917	402	16	0
6				27.5	1340	7.0	1790	324	70	30	17	900	418	17	0
7				27	1340	6.5	1570	300	74	28	13	880	400	17	0
8				36	1730	7.0	1800	380	86	47	20	1200	500	25	0
9				34	1740	6.7	2200	380	78	43	22	1100	470	21	0.0
10				34	1690	6.5	2260	410	82	45	21	1200	480	28	0.0
11	Cody Spa Well	NW¼ sec. 3, T.52N., R.102W.	790	34	1700	6.6	2150	420	78	42	22	1100	470	22	0.0
12	Shoshone Hot Springs	NW¼ sec. 4, T.52N., R.102W.	190	34	1530	6.5	2020	390	68	38	18	1100	360	18	1.0
13	Needle Hot Springs	SW¼ Sec. 5, T.52N., R.102W.	190	28	876	6.9	1320	210	34	24	10	800	118	6	.1

No.	F	S	SiO_2	B	Fe	CU	Mn	Zn	Ni	As	Se	Hg	Cd	Cr	Pb	Ag	Ba	Date	Reference
1		205	19															1926	Breckenridge and Hinckley, 1978
2		104	38															1926	Breckenridge and Hinckley, 1978
3	2.0		18		.2													2/15/68	Lowry and Lines, 1972
4	2.0		18		.06													2/15/68	Lowry and Lines, 1972
5	1.9		19	.55	.05													8/14/70	Lowry and Lines, 1972
6	1.9		18	.10	.08													8/14/70	Lowry and Lines, 1972
7	2.7	<.001	17	.33	0	<.01	<.05	<.02	<.1	<.05	<.001	.003	<.01	<.1	<.1	<.5	<.5	8/17/76	Breckenridge and Hinckley, 1978
8	2.9	<.001	21	.49	0	<.01	<.05	<.02	<.1	<.05	<.001	.001	<.01	<.1	<.1	<.5	<.5	8/18/76	Breckenridge and Hinckley, 1978
9	1.6	8.7	19	.6	.02	<.05	.06	<.02	<.1	<.005	<.005		<.01	<.1	<.2	<.05	<.1	10/6/79	State Lab. No. 0-1393[4]
10	2.8	1.3	24	.55	.2	<.04	.06	<.02	<.1	<.005	<.005	<.001	<.03	<.05	<.2	<.05	<.1	2/5/81	State Lab. No. 1-2476[4]
11	1.6	9.2	19	.5	.2	<.05	<.05	<.02	<.1	<.005	<.005	<.001	<.01	<.1	<.2	<.05	<.1	10/6/79	State Lab. No. 0-1386[4]
12	1.5	13	17	.5	.1	<.05	.06	<.02	<.1	<.005	<.005	<.001	<.01	<.1	<.2	<.05	<.1	10/6/79	State Lab. No. 0-1388[4]
13	1.3	.6	17	.3	.1	<.05	.09	<.02	<.1	<.005	<.005	<.001	<.01	<.1	<.2	<.05	<.1	10/6/79	State Lab. No. 0-1387[4]

1 Values for major cations, anions, and trace elements are given in milligrams/liter which is nearly equivalent to parts per million by weight.
2 Total flow for the DeMaris Hot Springs is approximatley 6400 liters per minute.
3 Flow amounts are estimates given in liters per minute for the sampled vent. To convert to gallons per minute, multiply by .264.
4 Analyses by the Wyoming Department of Agriculture, Division of Laboratories, Laramie, Wyoming.

TABLE 2

Measured Thermal Gradient Data

Well Name	Location	Depth (meters)	Bottom-hole Temp. (°C)	Interval for calculation of Thermal Gradient (meters)	Thermal Gradient[1] (°C/km)
Husky 16-69 7D	NW¼ sec. 4, T.52N., R.101W.	1916.4	60.31	26-1916	25.8
UWH-1	SW¼ sec. 2, T.52N., R.102W.	99.0	26.72	3-99	157.5
UWH-2	NW¼ sec. 11, T.52N., R.102W.	117.0	35.70	10-117	216.5
UWH-3	SW¼ Sec. 11, T.52N., R.102W.	72.0	25.72	5-70	189.9
UWH-4	NW¼ sec. 14, T.52N., R.102W.	97.0	12.38	5-97	3.6
UWH-5	SW¼ sec. 14, T.52N., R.102W.	109.0	18.14	5-105	76.9
UWH-6	NW¼ sec. 15, T.52N., R.102W.	85.1	20.43	18-85	108.1
BLM	NE¼ sec. 19, T.52N., R.101W.	61.4	12.41	7-61	20.4
Gaines Govt C-2	NE¼ sec. 31, T.52N., R.102W.	185.0	45.11	0-184	177.2
Rose Govt 1	NE¼ sec. 6, T.51N., R.102W.	740.0	45.44	20-220	101.3
Letha C-4	SE¼ sec. 6, T.51N., R.102W.	323.0	39.62	5-280	94.9

[1] Thermal gradient values were calculated using a linear least squares statistical technique for all temperature-depth values in the listed depth interval. To convert form °C/km to °F/1000ft multiply by .549.

Chugwater Formation, wells UWH-2, UWH-3, and UWH-4 in the Tensleep Formation, and well UWH-5 in the Sundance Formation. All wells except UWH-4 have high thermal gradients of 77 to 216°C/km. None of the holes encountered water except UWH-2, which found water at 85 meters, below which the temperature increases very rapidly. This confirms the Tensleep Sandstone as an aquifer containing warm water.

Hydrothermal system model

Using geological and geothermal data, a model of the Cody Hydrothermal system was devised. In the model, waters in the Tensleep Sandstone and Madison Limestone are heated at depth in a syncline, due to the regional thermal gradient. The heated waters gain access to shallower depths and the surface by moving upward along the steep western limb of the Horse Center anticline. This is shown diagrammatically in Figure 6.

The direction of water flow in such a system is important because it defines the structure through which the water flows. Pierce (in Love and Good, 1970) suggests water flow from the southeast which divides and flows around Rattlesnake Mountain. However, analysis of over 80 oil well drill stem tests and other hydrologic data show that the primary water flow direction is from the west with a small component from the southwest around the southern edge of the Rattlesnake anticline (see Figure 7). This direction of water flow is through a syncline that occurs to the west of Rattlesnake and Horse Center anticlines. The water flow direction as shown in Figure 7 would explain the existence of the thermal springs as located on Figure 2.

Conductive thermal modeling techniques were then used as described by Heasler (1981) to estimate subsurface temperatures in the strata to the west of Rattlesnake anticline. Temperatures at a given depth in the syncline are modeled using

$$T_S = T_0 + (Q/K_1)\, dx_1 + (Q/K_2) dx_2 + \ldots$$

where T_S is the unknown temperature in the syncline, T_0 is the mean annual surface temperature, Q is the regional heat flow from the earth, K_1 and dx_1 are the thermal conductivity and thickness of lithologic unit closest to the ground surface, K_2 and dx_2 are the thermal conductivity and thickness of the lithologic unit below unit 1, and so on. This method reflects temperature increase with depth in the earth as a function of the amount of heat flowing through the earth and the thermal conductivity of the rocks covering the earth. (Thermal conductivity differences among formations is the reason for the different linear thermal gradient segments above 500 meters in the Rose Govt 1 well as plotted in Figure 4).

Results of the conductive thermal model are shown in Table 3. The range of heat flow values for the Beartooth Mountains, Absaroka Mountains, and Bighorn Basin is 54 to 75 milliwatts per square meter (Heasler et al., 1982; Buelow, 1980; Heasler, 1978; Baker, 1976; Blackwell, 1969). An upper limit of 67 milliwatts per square meter for the Cody study area was based on the work of Heasler (1978) and Heasler et al. (1982).

The data in Table 3 show Tensleep Sandstone temperatures in the syncline to the west of Rattlesnake anticline to be higher than those of the hot springs, but these data generally agree with the measured temperatures in Horse Center anticline. The model's predicted temperatures for the top of the Tensleep Sandstone are in the range 46 to 54°C. This information, coupled with the direction of water flow, confirms the earlier suggestion by Fisher (1906) that the hot water in the Cody area is the result of normal thermal gradients within the earth. Woodruff's (1908) explanation of a slowly cooling magmatic heat source is not considered viable for many reasons. These reasons include lack of data for any recent igneous activity in the Cody area, the areal distribution of the thermal anomaly as shown in Figure 2, the low magnitude of the thermal anomaly, the flow direction of the water that is transporting heat, and the success of conductive thermal modeling in predicting the measured temperatures.

TABLE 3

Results of Conductive Thermal Modeling

Formation	Thickness[1] (meters)	Thermal Conductivity[2] (W/meter °K)	Temperature (°C) at base of formation[3] Using 54mW/m²	Using 67 mW/m²
Cody Shale	500	1.7	23	27
Frontier Fm.	137	3.1	25	30
Mowry Shale	122	1.6	29	35
Thermopolis Shale	168	2.5	33	40
Cloverly and Morrison Fms.	183	3.6	36	43
Sundance and Gypsum Spring Fms.	152	3.0	39	46
Chugwater and Dinwoody Fms.	244	2.9	44	52
Park City Fm.	21	4.0	44	52
Tensleep and Amsden Fms.	122	3.3	46	54
Madison Limestone	274	4.0	50	59
Three Forks, Jefferson Fms., and Bighorn Dolomite	180	3.2	53	63
Grove Creek, Snowy Range Fms., Pilgrim Limestone, Gros Ventre Fm., and Flathead Sandstone	396	2.1	63	76

[1]Thicknesses are generalized from Pierce (1966, 1970, 1978)
[2]Thermal conductivities in units of watts per meter °K are from Heasler (1978) and unpublished data.
[3]Assuming a mean annual surface temperature of 7°C. Results are expressed to the nearest °C. Heat flow values are expressed in milliwatts per square meter.

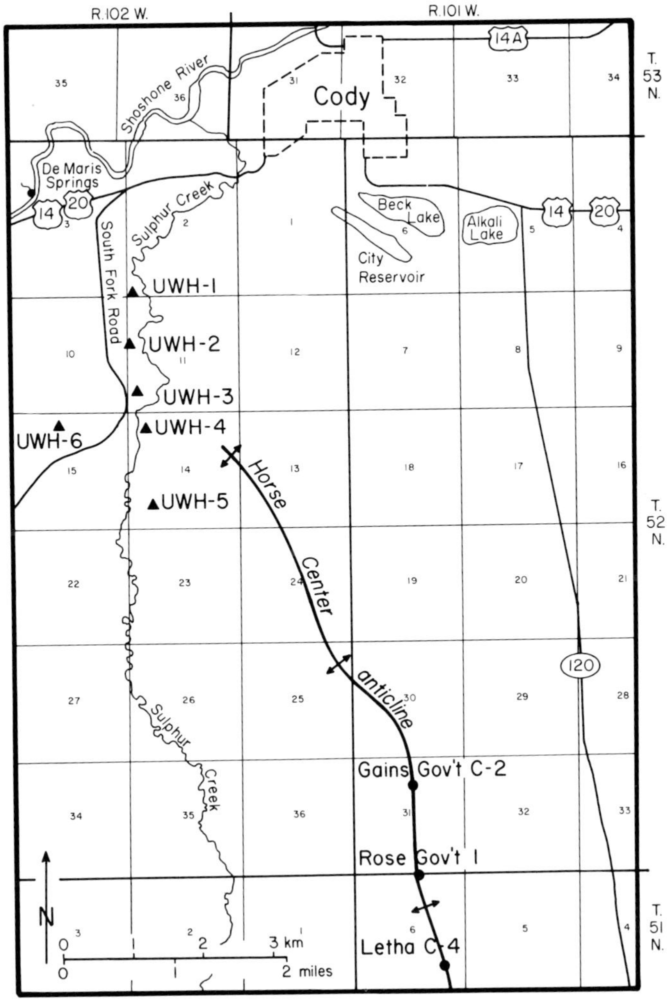

Figure 3: Location of drilled temperature gradient holes (triangles) and three thermally logged oil wells (circles).

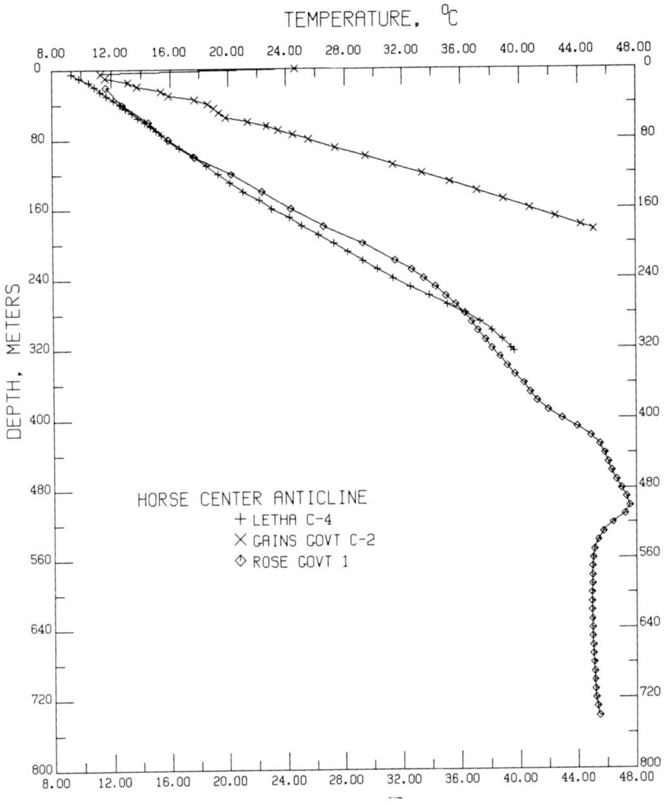

Figure 4: Temperature plots for the three oil wells shown on Figure 3.

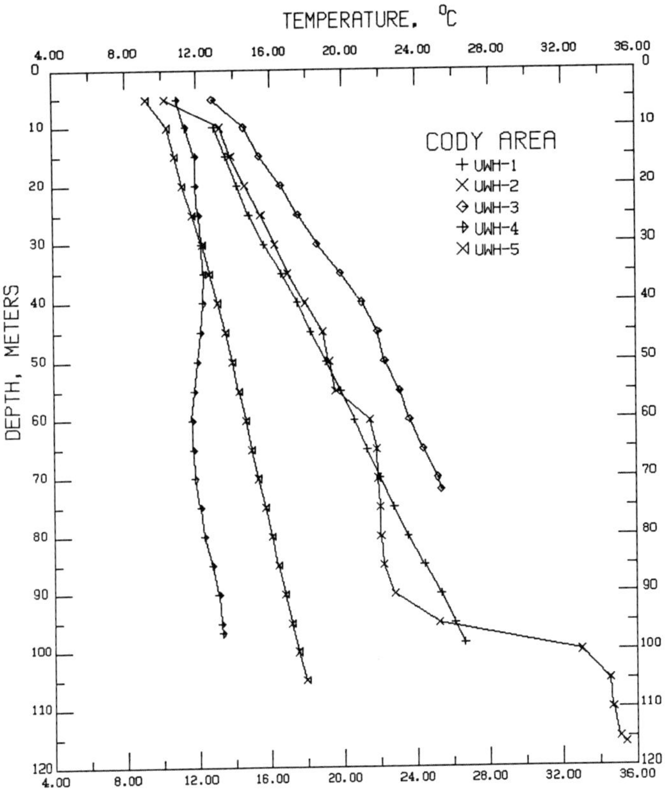

Figure 5: Temperature plots for five of the drilled temperature gradient holes shown on Figure 3.

Conclusion and Summary

The Cody hydrothermal system covers an area 1 to 2 km in width that extends from the DeMaris Hot Springs (west of Cody) south 11 km along the Horse Center anticline. To the east of this area the important aquifers (the Tensleep Sandstone and Madison Limestome) slope into the Bighorn Basin and soon become too deep to be economically drilled for hot water use. In the north portion of this area, warm water of over 34°C in the Tensleep Sandstone can generally be reached at shallow depths of 50 to 300 meters. Higher temperatures of over 40°C are expected in the Madison Limestone at depths of 250 to 600 meters.

The heat for the hydrothermal system results from the earth's normal temperature increase with depth. Water flows through the syncline west of Rattlesnake anticline absorbing heat. This heated water then flows up over Horse Center anticline and a portion flows around the southern

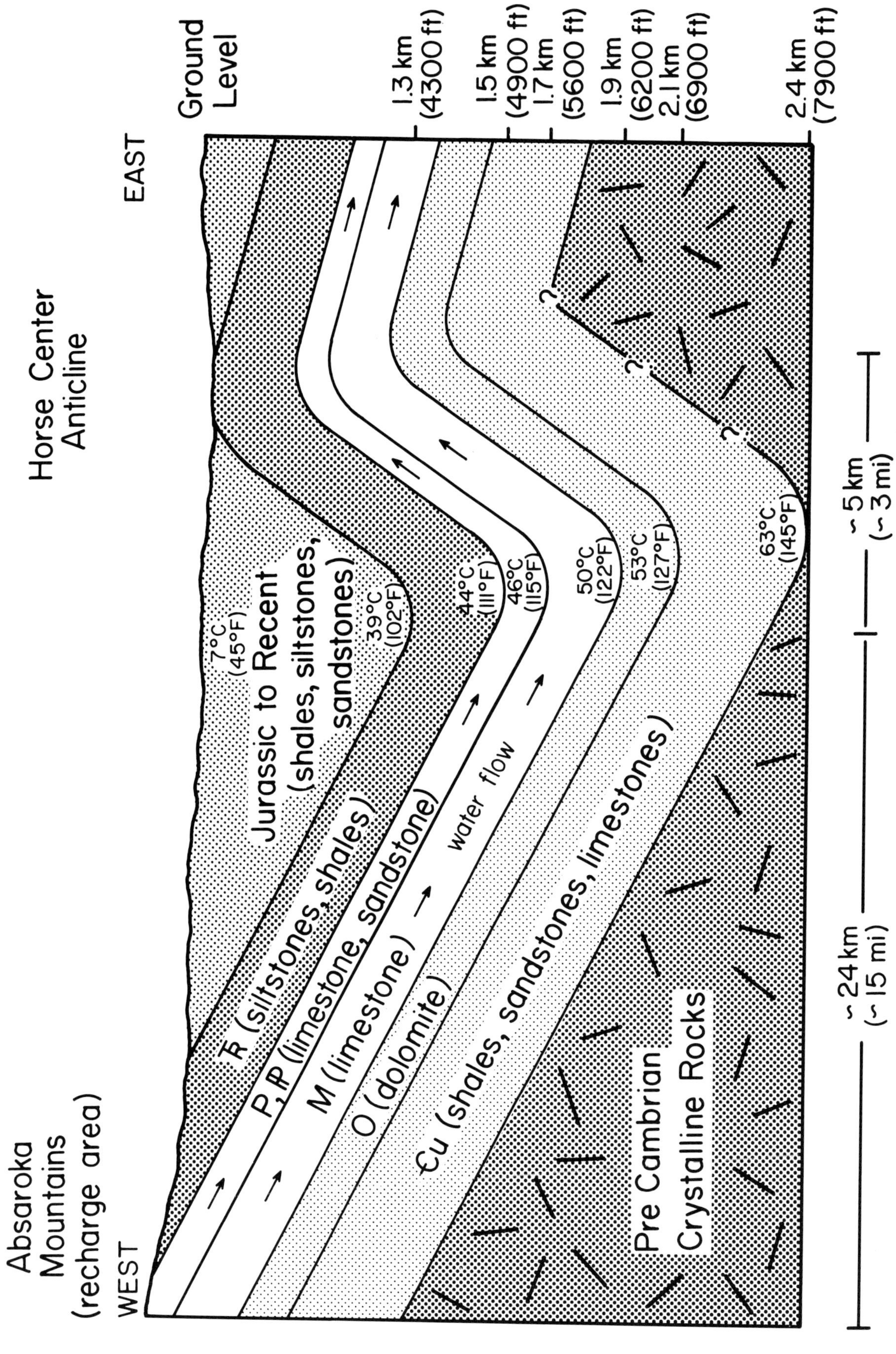

Figure 6: Generalized model of the Cody hydrothermal system showing simplified structure and approximate temperatures and depths.

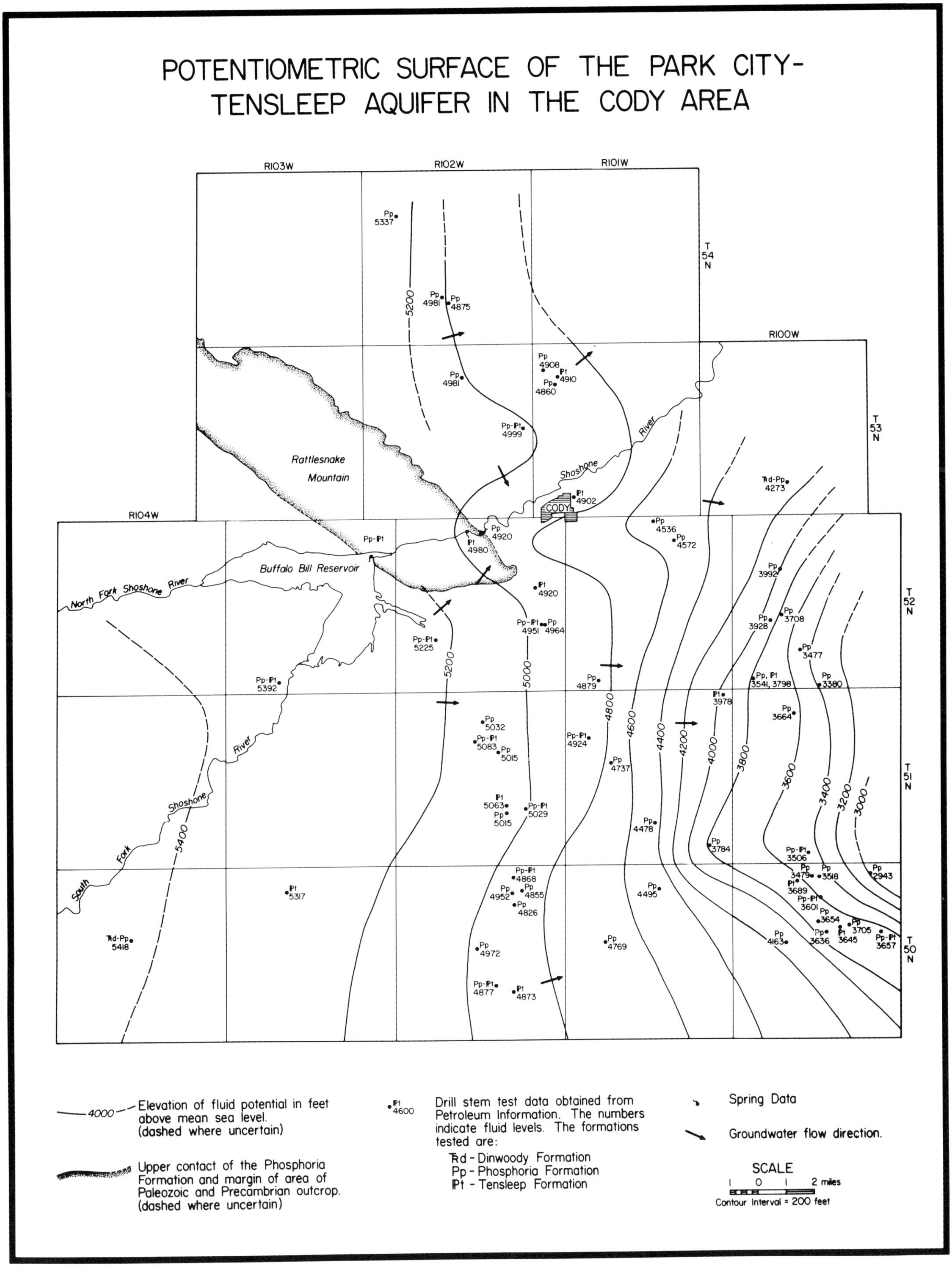

Figure 7: Potentiometric map showing water flow directions.

end of Rattlesnake anticline. Maximum temperatures in the hydrothermal system for the Tensleep Sandstone are predicted to be 44 to 52°C and for the Madison Limestone 46 to 59°C. These occur in the south portion of the hydrothermal area.

Although the Cody hydrothermal system is a low-temperature geothermal resource, temperatures are high enough for space heating and low-temperature industrial uses (Brown et al., 1982). Given its favorable location near Cody and nearby active subdivision areas, the Cody hydrothermal system represents a valuable energy resource.

Acknowledgements

The author wishes to thank Coronado Oil Company of Denver and Husky Oil of Cody for granting permission to thermally log selected wells. Also thanked are C. Gail, J. Housel, C. Koster, G. Nielson, and D. Pike, who allowed drilling of the temperature gradient wells. Dale M. Doremus was responsible for the analysis and contouring of the drill stem test date for the potentiometric map.

Funding for the project was from Cooperative Agreement No. DED-FC07-79ID12026 between the U.S. Department of Energy and the Department of Geology and Geophysics, University of Wyoming.

REFERENCES CITED

Baker, K.H., 1976, Heat Flow Studies in Colorado and Wyoming: M.S. Thesis, Univ. Wyo., Laramie, 142p.

Blackwell, D.D., 1969, Heat Flow Determinations in the Northwestern United States: J. Geophys. Res., vol. 74, p. 992-1007

Breckenridge, R.M., and Hinckley, B.S., 1978, Thermal Springs of Wyoming: Wyo. Geol. Survey, Bull. 60, 104p.

Brown, K., Burgess-Lyon, P., and Marcotte, K., 1982, Cody, Wyoming, Site Specific Development Analysis: Wyoming Geothermal Commercilization Office, Univ. Wyo., Report to the U.S. Dept. of Energy, 139 p.

Buelow, K.L., 1980, Geothermal Studies in Wyoming and Northern Colorado, with a Geophysical Model of the Southern Rocky Mountains near the Colorado-Wyoming Border: M.S. Thesis, Univ. Wyo., Laramie, 150p.

Decker, E.R., 1973, Geothermal Measurements by the University of Wyoming: Wyoming Contributions to Geology, vol. 12, p. 21-24.

Fanshawe, J.R., 1952, Big Horn Basin Tectonics: Wyo. Geol. Assoc. Guidebook, 7th Ann. Field Conf., p. 19-21.

Fisher, C.A., 1906, Geology and Water Resources of the Bighorn Basin, Wyoming: U.S. Geol. Survey Prof. Paper 53, 72p.

Heasler, H.P., 1978, Heat Flow in the Elk Basin Oil Field: M.S. Thesis, Univ. Wyoming, Laramie, 168p.

Heasler, H.P., 1981, Conductive Thermal Modeling of Wyoming Geothermal Systems: U.S. Dept. of Energy, State Coupled Resource Assessment Meeting, May 4-7, 1981, Glenwood Springs, Colorado, Proceedings p. 301-313.

Heasler, H.P., and Decker, E.R., 1980, Preliminary Data from Six Temperature Gradient Holes near Cody, Wyoming: Department of Geology and Geophysics, University of Wyoming, Laramie, 35p.

Heasler, H.P., Decker, E.R., and Buelow, K.L., 1982, Preliminary Results of Heat Flow Studies in Wyoming, 1979 to 1981: U.S. Dept. of Energy, State Coupled Resource Assessment Meeting, April 5-7 1982, Salt Lake City, Utah, Proceeding in press.

Love, J.D., and Good, J.M., 1970, Hydrocarbons in Thermal Areas, Northwestern Wyoming: U.S. Geol. Survey, Prof., Paper 644-B, 23p.

Lowry, M.E., and Lines, G.C., 1972, Chemical Analysis of Groundwater in the Bighorn Basin: Wyo. Dept. Economic Planning and Development Report, 15p.

Mattes, M.J., 1962, *Colter's Hell and Jackson's Hole:* Yellowstone Library and Museum Association and the Grand Teton Natural History Association, 87p.

Pierce, W.G., 1966, Geologic Map of the Cody Quardangle, Park County, Wyoming: U.S. Geol. Survey, Geol. Quadrangle Map GQ-542.

Pierce, W.G., 1970, Geologic Map of the Devils Tooth Quadrangle, Park County, Wyoming: U.S. Geol. Survey, Geol. Quadrangle Map GQ-817.

Pierce, W.G., 1975, Principal Features of the Heart Mountain Fault and the Mechanism Problem: Wyo. Geol. Assoc. Guidebook, 27th Ann. Field Conf., p. 139-148.

Pierce, W.G., 1978, Geologic Map of the Cody 1° by 2° Quadrangle, Wyoming: U.S. Geol. Survey, Miscellaneous Field Studies Map MF-963.

Smedes, H.W., and Prostka, H.J., 1972, Straitigraphic Framework of the Absaroka Volcanic Supergroup in the Yellowstone National Park Region: U.S. Geol. Survey, Prof. Paper 729-C, 33p.

Wideman, F.L., 1957, A Reconnaissance of Sulphur Resources in Wyoming, Colorado, Utah, New Mexico, and Arizona: U.S. Bur. Mines, Information Circular 7770, 61p.

Woodruff, E.G., 1908, Sulphur Deposits at Cody, Wyoming: U.S. Geol. Survey, Bul. 340-L, p. 451-456.

Thirty-Third Annual Field Conference — 1982
Wyoming Geological Association Guidebook

THE HEART MOUNTAIN DETACHMENT FAULT, NORTHWEST WYOMING: INVOLVEMENT OF ABSAROKA VOLCANIC ROCK

THOMAS A. HAUGE[1]

INTRODUCTION

The Heart Mountain detachment fault of northwestern Wyoming and adjacent Montana is well known as a gravity slide largely as a result of the work of Pierce (publications 1938-1980). Pierce's work demonstrates that a once-continuous slab of Paleozoic sedimentary rocks became detached along a subhorizontal bedding-plane fault and now occurs as several tens of allochthonous masses distributed over an area more than two times larger than the 1300 km.^{2} area of the bedding-plane detachment (Fig. 1). Extension of the Paleozoic slab involved displacement of allochthonous rocks to the southeast across progressively younger sedimentary strata, eventually onto the Eocene land surface. An estimated 50% (Pierce, 1968) of the bedding-plane detachment is presently overlain by Eocene volcanic rock of the Absaroka Supergroup. According to previous workers (primarily Nelson, Pierce, and Prostka of the USGS — see references), these volcanic rocks are in part allochthonous and in part in depositional contact with the autochthon. The occurrence of volcanic rocks in depositional contact with the fault surface was viewed as the result of: (1) tectonic denudation of a portion of the autochthon during Heart Mountain faulting, followed immediately by; (2) widespread deposition of volcanic rocks upon the tectonically denuded surface.

The most recently published maps of the detachment terrane (Prostka, 1978; Pierce, 1978, 1980; Nelson et al, 1980) reflect continuing disagreement among the previous workers regarding how much of the volcanic rock in contact with the bedding-plane detachment is allochtonous and how much is *in situ*. All agree that the Wapiti Formation is *in situ* where it overlies the bedding-plane detachment and that the Cathedral Cliffs Formation within the detachment terrane is allochthonous. Rocks of the Lamar River Formation "in part slid onto and in part may have been deposited upon [the bedding-plane detachment]" (Nelson et al, 1980, Plate 1). Pierce (1978, 1980) maps the great majority of volcanic rocks in contact with the bedding-detachment as Wapiti Formation. Prostka (1978) and Nelson et al (1980) map many of these same rocks as the older Lamar River and Cathedral Cliffs Formation.

[1]Dept. of Geol. Sciences, University of Southern California.
Los Angeles, CA 90007.

The author's recent (1977-1981) detailed field study of the Heart Mountain detachment fault terrane included examination of many areas where volcanic rocks overlie the bedding-plane detachment. These studies emphasized analysis of tectonic deformation of the volcanic rocks rather than resolution of stratigraphic problems. The resulting assessment of the involvement of volcanic rocks in Heart Mountain faulting was therefore independent of the problematic stratigraphy.

FIELD RELATIONSHIPS

Volcanic rock overlying the Heart Mountain bedding-plane detachment was studied throughout its area of exposure, from the western margin of the detachment terrane (the break-away fault of Pierce, 1960 and 1980) to exposures 40 km. southeast near Russel Peak and White Mountain. Detailed descriptions of the observed relationships are in preparation for publication at a later date, and only a partial summary of observed relationships is presented here.

The volcanic rocks vary widely in texture and include well-bedded tabular flows and breccias, well-bedded epiclastic sequences, and massive breccias of irregular form. Some of the breccias are intrusive (Prostka, 1978; Nelson et al, 1980), and subvertical porphyritic dikes are locally abundant. Where a primary planar fabric is recognized, the extent of tectonic deformation is readily ascertained, but where primary fabrics are obscure or nontabular, assessment of tectonic deformation is difficult.

Truncation of Stratified Volcanic Rocks Among Faults

In several areas (e.g., north of Pilot Creek, east of Painter Gulch, west of Cathedral Cliffs, east of Reef

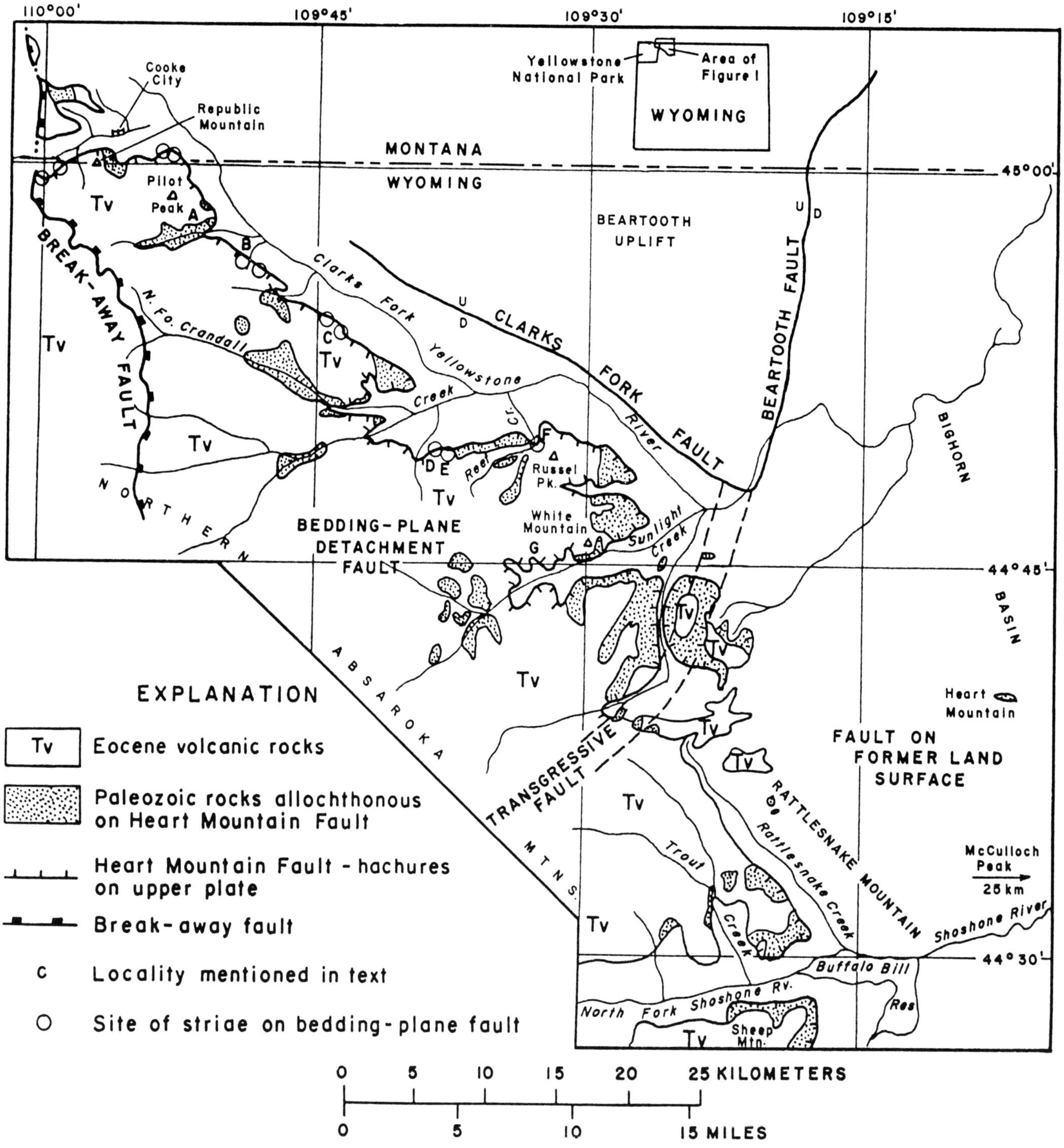

Figure 1: Map showing location of four phases of Heart Mountain fault and localities mentioned in text. Modified from Pierce (1979). A, north of Pilot Creek; B, Jim Smith Creek; C, Squaw Creek; D, west of Oliver Gulch; E, west of Cathedral Cliffs; F, Reef Creek; G, east of Reef Creek; H, east of Painter Gulch.

Creek), inclined sequences of volcanic flows and breccias occur in the hanging walls of high-angle faults that extend down to, but not below, the bedding-plane detachment fault. The hanging-wall strata dip toward and are truncated along the high-angle faults, and in some areas stratigraphically lower units are demonstrably truncated by the bedding-plane detachment. Footwall rocks may be volcanic or Paleozoic sedimentary rocks. Throw across these faults ranges from 10's to 100's of meters. Truncation of subvertical porphyry dikes along faults in the Heart Mountain fault allochthon is observed west of Oliver Gulch, where the dikes occur within massive volcanic breccias.

In all these areas, structural relationships require that the faulting of volcanic rocks occurred as the rocks moved along the bedding-plane detachment. The volcanic rocks must be allochthonous.

Sheared, Brecciated, and Faulted Volcanic Rocks Along the Bedding-Plane Detachment

In many areas where the bedding-plane fault and superjacent rocks are well exposed, a thin (1 to 2 mm) veneer of dark-colored microbreccia adheres to the buff-colored dolomite forming the top of the autochthon. The upper surface of the microbreccia layer is commonly slickensided and striated. Thin-section examination reveals clasts of volcanic rock and carbonate rock within the microbreccia. Striated microbreccia along the bedding-plane detachment was observed at many localities extending from within one km of the break-away fault to the Reef Creek area 40 km to the southeast (see Fig. 1).

The degree of tectonic deformation of volcanic rock along the bedding-plane detachment is highly variable, including: brecciation of a basal zone up to 15 m thick; penetrative shearing of the basal few centimeters of volcanic rock; anastomosing faults subparallel to and within a few meters of the bedding-plane detachment; and no apparent deformation. Highly deformed zones of volcanic rock along the bedding-plane detachment are commonly bounded at the top by shear surfaces, often gently undulose, above which the volcanic rock is undeformed or only slightly deformed. At Jim Smith Creek, steeply dipping volcanic strata are truncated downward along such a shear surface.

The occurrence of brecciation, faulting, or pervasive shearing of volcanic rocks along the bedding-plane detachment indicates that the overlying volcanic rocks are allochthonous on the Heart Mountain fault. Where the basal tectonized zone is very thin or poorly exposed, such as along Squaw Creek, the allochthonous nature of otherwise little-deformed volcanic rocks may go unrecognized. Similarly, many allochtonous masses of Paleozoic sedimentary rocks are very little deformed except along their bases.

Subvertical Igneous Dikes

Many hundreds of subvertical igneous dikes occur in rocks overlying the Heart Mountain bedding-plane detachment. Igneous dikes in the lower plate are very rare. No dikes extending across the bedding-plane detachment were observed. The upper-plate dikes usually strike N$\pm$30°, are commonly 2-5 m thick, and collectively represent extension of the upper plate in excess of 1 km. Cross-cutting relationships indicate that some dikes predate local upper-plate faulting, some were intruded between episodes of upper-plate faulting, and others postdate local upper-plate faulting. At one locality west of Oliver Gulch, several tens of subvertical dikes occur within a 200 m long exposure of volcanic rock, indicating roughly 50% extension of the allochthon in this area. The dikes are probably laterally intruded (Pierce, 1963), and the upper-plate extension they represent must have been accommodated by eastward translation along the bedding-plane detachment.

Summary

Examination of rocks along the Heart Mountain bedding-plane detachment fault, at numerous localities from throughout the extent of its exposure, indicates that the contact between the autochthon and the overlying volcanic rocks is everywhere tectonic. The volcanic rocks superjacent to the bedding-plane detachment are therefore allochthonous, including those thought by previous workers to be *in situ* on a surface of tectonic denudation.

DISCUSSION

In contrast with previous work, in which volcanic rocks were seen as in depositional contact with tectonically denuded portions of the bedding-plane detachment fault, the present study indicates that the contact between volcanic rock and the lower plate is everywhere tectonic. As a result, tectonic denudation is not a necessary part of the geologic history indicated by the present study.

If tectonic denudation of the bedding-plane detachment occurred during Heart Mountain faulting, the duration of exposure of the tectonically denuded surface to subaerial processes must have been brief. Pierce (1968) pointed out the lack of erosion of the surface of tectonic denudation, indicating that "the length of time involved in tectonic denudation . . . was extremely brief" (p 197) and "the area of tectonic denudation . . . was buried . . . by volcanic rock almost immediately after it was formed" (p 191). Pierce (1973) believed that movement on the Heart Mountain fault was "very rapid, probably cataclysmic" (p 492). Voight (1974) was impressed by the "conspicuous absence of erosion on the tectonically-denuded bedding-plane fault surface" (p 32) indicating "very rapid" (p 32) fault movements and immediately subsequent burial by younger volcanic rocks. The present study agrees with previous studies regarding the lack of post-tectonic erosion of the bedding-plane detachment horizon but suggests that tectonic denudation of the detachment fault did not occur. If there was no tectonic denudation, then the extraordinary velocities of faulting and rates of igneous extrusion suggested by previous workers are not required.

Furthermore, if tectonic denudation did not occur, gravity sliding of numerous detached blocks along the bedding-plane detachment is not indicated. Instead, field relationships suggest spreading of a continuous mass of upper-plate Paleozoic sedimentary and Eocene volcanic rocks. The volcanic rocks overlying the bedding-plane detachment are seen as comprising a graben that formed as the slab of Paleozoic rock underwent discontinuous brittle extension.

Work to date has not produced a complete tectonic map spanning any of the allochthonous graben of volcanic rocks that separate the large allochthonous horsts of Paleozoic sedimentary rocks. However, the infilling of the graben is believed to have been accommodated by the following structural features, all of which have been observed in various places within the allochthonous volcanic terrane:

1) high-angle, sometimes listric normal faults at contacts with allochthonous Paleozoic sedimentary rocks and within the volcanic terrane;
2) low-angle normal faults observed at several structural levels within the volcanic terrane, including the Reef Creek fault of Pierce (1963);
3) drag-faulting of volcanic strata along high-angle faults;
4) swarms of subvertical dikes;
5) zones of chaotic deformation, including brecciation, faulting, and pervasive shearing, primarily along the bedding-plane detachment.

Intrusion of irregular bodies of autobrecciated andesite (Prostka et al, 1975; Prostka, 1978) and deposition of extrusive rocks probably contributed to the continuing infilling of the graben as they extended. At Pilot Peak, allochthonous volcanic rock is at least 600 m thick (see Nelson et al, 1980, Fig. 9). In most other areas the total thickness of allochthonous volcanic rock is not known.

The present study focused upon the bedding-plane detachment fault and superjacent rocks. Detailed field study must be extended to the Shoshone River drainage in order to assess the involvement of volcanic rock in the former land-surface phase of Heart Mountain faulting. Prostka (1978) and Nelson et al (1980) mapped allochthonous volcanic rocks along Trout Creek, more that 15 km. southeast of the transgressive fault. Reconaissance during the present study suggests the presence of allochthonous volcanic rock on Sheep Mountain 10 km farther southeast. In the Bighorn Basin, where the most easterly remnants of allochthonous Paleozoic rocks occur, the basal fault horizon is widely dissected. Volcanic rock is presently not abundant in this area, although small remnants at McCulloch Peaks caused Rouse (1937) to suggest that Absaroka volcanic rocks had once extended that far east. At the present time, it is not known whether all Heart Mountain fault klippe were emplaced as part of the continuous allochthonous mass observed above the bedding-plane detachment. Heart Mountain, McCulloch Peaks, or other klippe on the former land surface may have been shed as gravity slide blocks from the toe of the spreading allochthon as it extended beyond the transgressive fault.

CONCLUSIONS

Field evidence indicates that all volcanic rock in contact with the Heart Mountain bedding-plane detachment fault was fault-emplaced. Relationships suggest that tectonic denudation of the bedding-plane fault did not occur during Heart Mountain faulting. Instead, the Heart Mountain fault allochthon, at least where it overlies the bedding-plane detachment, moved as a continuous extending mass rather than as numerous separate gravity slide blocks. The extraordinary high velocities of faulting indicated by previous interpretations of field relationships are no longer indicated. Hypotheses explaining the mechanism of detachment and displacement of the Heart Mountain fault allochthon require re-evaluation in light of this new information.

Mapping of the volcanic rocks involved in Heart Mountain faulting is incomplete. However, structural relationships observed to date permit creation of the following model of the history of fault-emplacement of volcanic rock onto the bedding-plane detachment.

During Heart Mountain faulting, extension of the once-continuous slab of Paleozoic sedimentary rock was accompanied by the formation of ten or more graben now composed dominantly of Absaroka volcanic rocks. As extension progressed, the graben, which were initially very narrow, increased in width as rocks within the graben were structurally lowered. A zone of complex deformation developed at the bases of the graben along the bedding-plane detachment fault. Low- and high-angle faulting (normal, normal-oblique, and strike-slip), intrusion of volcanic breccias, and deposition of younger volcanic rocks periodically replenished the distending graben as extension continued. Some extension was accommodated by lateral intrusion of porphyritic dikes. Extension of the upper-plate occurred gradually over a protracted period of time (probably 10^6 years) coeval with on-going Absaroka volcanism. When Heart Mountain faulting ceased, volcanic rock had been structurally lowered in this manner to cover roughly 50% of the area of the bedding-plane detachment fault.

ACKNOWLEDGEMENTS

The content of this manuscript has improved greatly from my discussions with many colleagues over the past several years. Particular thanks for recent discussions and criticism of the manuscript go to P. H. Cashman, G. A. Davis, C. G. Sammis, and L. C. Thurn. Meggan Mosher kindly typed the manuscript and Gay Havens adapted the figure.

REFERENCES

Nelson, W. H., Prostka, H. J., and Williams, F. E., 1980, Geology and mineral resources of the North Absaroka Wilderness and vicinity, Park County, Wyoming; U.S. Geol. Survey Bull., 1447 with sections on Mineralization of the Cooke City mining district by James E. Elliott and Aeromagnetic survey by Donald L. Peterson.

Pierce, W. G., 1960, The "break-away" point of the Heart Mountain detachment fault in northwestern Wyoming: *in* Geol. Survey Research 1960, U.S. Geol. Survey Prof. Paper, 400-B, p. B236-237.

Pierce, W. G., 1963, Reef Creek detachment fault, northwestern Wyoming: Geol. Survey of America Bull., v. 74, no. 10, p. 1225-1236.

Pierce, W. G., 1968, Tectonic denudation as exemplified by the Heart Mountain fault, Wyoming: *in* Orogenic Belts, 23rd International Geological Congress, Prague, Czechoslovakia, 1968, Report, Section 3, Proceedings: p. 191-197.

Pierce, W. G., 1973, Principal features of the Heart Mountain fault, and the mechanism problem: *in* DeJong, K. A., and Scholten, R. (eds.), Gravity and Tectonics, New York, John Wiley and Sons, p. 457-471.

Pierce, W. G., 1978, Geologic map of the Cody 1° x 2° quadrangle, northwestern Wyoming: U.S. Geological Survey Miscellaneous Field Studies Map MF-963.

Pierce, W. G., 1979, Clastic dikes of Heart Mountain fault breccia, northwestern Wyoming, and their significance: U.S. Geological Survey Prof. Paper 1133, p. 1-25.

Pierce, W. G., 1980, The Heart Mountain break-away fault, northwestern Wyoming: Geol. Soc. America Bull., Part 1, v. 91, p. 272-281.

Prostka, H. J., 1978, Heart Mountain fault and Absaroka volcanism, Wyoming and Montana, U.S.A.: *in* Voight, B. (ed.) Rockslides and Avalanches, I, Natural Phenomena, Amsterdam, Oxford, New York, Elsevier Scientific Pub. Co.

Prostka, H. J., Ruppel, E. T., and Christiansen, R. L., 1975, Geologic map of the Abiathar Peak quadrangle, Yellowstone National Park, Wyoming and Montana: U.S. Geological Survey Quadrangle Map GQ-1244.

Rouse, J. T., 1937, Genesis and structural relationships of the Absaroka volcanic rocks, Wyoming: Geol. Soc. America Bull., v. 48, p. 1257-1296.

Voight, B., 1974, Architecture and mechanics of the Heart Mountain and South Fork rockslides: *in* Voight, B. and Voight, M. A., (eds.) Rock Mechanics: The American Northwest, 3rd Congress International Society of Rock Mechanics Expedition Guidebook, Spec. Pub. Experiment Station, College of Earth and Mineral Sciences, The Pennsylvania State University, p. 26-36.

Thirty-Third Annual Field Conference — 1982
Wyoming Geological Association Guidebook

RELATION OF VOLCANIC ROCKS TO THE HEART MOUNTAIN FAULT

BY WILLIAM G. PIERCE[1]

ABSTRACT

In a paper in this volume, Hauge concludes that the Heart Mountain fault movement did not produce tectonic denudation. He also believes that the volcanic rocks associated with it — some heretofore regarded as pre-faulting and others as post-faulting — are still all part of the fault movement. His studies emphasize analysis of tectonic deformation of the volcanic rocks rather than stratigraphic data for structural interpretation.

The stratigraphic controls which are used to determine the relations of the volcanic rocks to the Heart Mountain fault are the Cathedral Cliffs Formation which is pre-faulting and the Wapiti Formation which is post-faulting. There is general agreement that the Cathedral Cliffs Formation is pre-faulting, for it is part of the upper plate. Some of the relations indicating the Wapiti Formation is post-faulting in age are: (1) The break-away fault, which is part of the Heart Mountain fault, cuts the Cathedral Cliffs Formation but is overlain by the Wapiti Formation which is several thousand feet thick and is not cut by the fault. The Wapiti Formation is clearly younger than the Heart Mountain fault along the 22 mile extent of the break-away fault. (2) The upper plate of the Heart Mountain fault broke up into numerous blocks which separated as movement continued. Large open spaces formed between the blocks which are now filled with Wapiti Formation. The volume or amount of Wapiti Formation filling the spaces between blocks is much too great to have been derived from the upper plate. However, if the Wapiti was deposited in the open spaces following the faulting, there is no need for a volume relationship to the upper plate. (3) The fault breccia at the base of many of the upper plate blocks has been examined and found to be composed entirely of carbonate fault breccia. If the open spaces between the upper plate blocks were filled with volcanic rocks which were also moving as part of the upper plate, there should be a fair proportion of volcanic material in the fault breccia, but it is lacking. (4) Near Pilot Peak, steeply dipping, tectonically emplaced Cathedral Cliffs Formation is overlain by flat-lying Wapiti which in turn is overlain by flat-lying Trout Peak trachyandesite. If the Wapiti was fault emplaced but the overlying Trout Peak was not, can it be assumed that it is just a coincidence that the stratification is parallel in the two formations here? Such an assumption is questionable, for there are similar examples elsewhere in the area of the Heart Mountain fault. (5) It seems to me that a rapidly accumulating body of volcanic flows, pouring out over an irregular layer of fault breccia is likely to produce features similar in some respects to those produced by sliding mass of the same volcanic material. There is a need to look for diagnostic differences between them.

[1]U.S. Geological Survey, Menlo Park, CA

In his paper in this volume, Hauge concludes that tectonic denudation did not occur as part of the Heart Mountain fault movement. He also concludes that the volcanic rock formations (which other investigators believe can be categorized as pre- and post-Heart Mountain faulting in age) are all part of the fault movement. He does not, however, use the stratigraphic controls that are available and used by others in identifying and mapping the volcanic rock formations. Those stratigraphic controls are the Cathedral Cliffs Formation, which is pre-faulting in age and by definition a part of the upper plate (Pierce, 1963) and the Wapiti Formation (Nelson, W. H., and Pierce, W. G., 1968) which is post-faulting. Although he does not indicate the formations involved at the localities he interprets as being in fault contact with the Heart Mountain fault surface, some are Cathedral Cliffs, and others are Wapiti, according to my observations.

The specific observational data on which Hauge bases his conclusions are not yet available for discussion except for his paper in this volume. Some comments on a few well known and generally accepted features related to the fault that bear on questions that have been raised at this time are as follows:

1. The break-away fault (Pierce, 1960, 1980), more than 22 miles long, is one of the 4 phases of the Heart Mountain fault. It cuts the Cathedral Cliffs Formation but is overlain by the Wapiti Formation, which is several thousand feet thick and is not cut by the fault. These relations seem to clearly indicate that the Wapiti Formation is younger than the Heart Mountain fault. If so, then how is this to be reconciled with Hauge's conclusion that the Wapiti Formation and the Cathedral Cliffs Formation are both involved in the same manner in the fault movement?

2. The areal extent of the upper plate of the Heart Mountain fault at the time movement began was about 500 square miles. After movement ceased the upper plate blocks were scattered over an area of about 1300 square miles (Nelson and others, 1972). Large voids or spaces must have been formed between the upper plate blocks to produce this distribution as known today from detailed mapping (Pierce, 1973) and partially shown in figure 1 of Hauge (this volume). These open spaces comprised about half the original upper plate area. Most of the larger upper plate blocks have as their uppermost stratigraphic unit the Cathedral Cliffs Formation. There are only a few places where large masses of Cathedral Cliffs Formation are known to have been dislodged from the upper plate and come to rest on the Heart Mountain fault surface. Most of the spaces between the upper plate blocks are now occupied by Wapiti Formation resting on the fault surface. If this Wapiti was fault emplaced, as Hauge believes, there is a volume problem to be taken into account. The area covered by Wapiti Formation is much larger than the area covered by upper plate rocks; if the Wapiti was fault emplaced as part of the upper plate the increased amount of a part of it must be explained. There is no volume problem, however, if the Wapiti was depositionally emplaced.

3. The composition of the fault breccia that occurs at the base of upper plate blocks of Paleozoic carbonate rock should be an indicator of the kind of rocks that have moved along the Heart Mountain fault. As I understand Hauge's interpretation of the fault, the volcanic rocks now filling the spaces between the upper plate blocks also moved on the fault surface, so the fault breccia beneath the upper plate blocks should presumably contain a sizable proportion of volcanic fragments to carbonate fragments. I have not found such a relationship, or anything near it. I have examined the fault breccia beneath many of the upper plate blocks of carbonate rock and found it to consist entirely of carbonate breccia. These blocks have moved several to many miles along the fault. If there was at least as much if not more volcanic rock moving right along with and between the carbonate rocks, shouldn't there be a mixture of volcanic and carbonate components in the fault breccia beneath the upper plate? This lack of volcanic material indicates to me that there was very little movement of volcanic rocks on the Heart Mountain fault, but there would be some presently found in a few places where upper plate Cathedral Cliffs volcanics toppled down and moved on the fault surface.

4. The volcanic rock Hauge (this volume) mentions at Pilot Peak is an interesting illustration of the desirability, if not necessity, of identifying the volcanic formation involved. His reference to Figure 9 in Nelson et al, 1980, is to a photograph which I took a few years ago, showing volcanic rocks dipping 50 degrees overlain by flat-lying Wapiti that in turn is overlain by flat-lying Trout Peak trachyandesite. We agree that the steeply dipping rocks were tectonically emplaced for they are a tilted mass of Cathedral Cliffs volcanic rocks that are part of the upper plate. But what about the large area of flat-lying Wapiti and Trout Peak volcanic rock overlying it? Those of us who mapped that area think that the Wapiti is in depositional contact with the underlying rocks and the Trout Peak depositional on the Wapiti. If, as Hauge believes, the Wapiti was fault emplaced on the rocks on which it rests, was the Trout Peak trachyandesite, which here and elsewhere rests on the Wapiti with a depositional contact, carried along with it as part of the Heart Mountain fault? Or was it deposited on a faulted block of Wapiti and is it just a coincidence that their stratification is parallel? Other areas of Wapiti within the Heart Mountain fault area that are overlain by trout Peak similarly show no angular discordance between the two formations, and no suggestion

that the Wapiti is fault emplaced.

5. My field work, observations, and mapping conducted over a period of 40 years led me to the conclusion based on several lines of evidence that tectonic denudation was produced by the Heart Mountain fault movement, and very soon thereafter deposition of several thousand feet of Wapiti Formation rapidly blanketed a large area, including the Heart Mountain fault. The data supporting those conclusions is scattered through a number of publications as the sequence of events unfolded, but at the moment there is not time to assemble them for publication here. Some of the field evidence to support each of the divergent points of view, i.e., tectonic or depositional, are similar in some respects. What are the diagnostic differences between them? Would a rapidly accumulating body of volcanic breccia flowing out over an irregular layer of fault breccia have markedly different features than similar breccia that had slid? I have observed and described a host of features associated with these volcanic rocks that cannot be reasonably explained as tectonic. It is not feasible to recount them here, except to mention as an example, some clastic carbonate dikes of Heart Mountain fault breccia that extend upward from the fault for tens of feet into the Wapiti Formation (Pierce, 1979). The country rock adjoining the dikes appears to have flowed upward similar to the upward flow of the carbonate breccia in the dike, which suggests that the volcanic rock of the Wapiti Formation was in a somewhat mobile or semi-fluid state at the time the dike was injected. If true, this interpretation does not lend support to Hauge's concept that these volcanic rocks were fault emplaced.

REFERENCES CITED

Nelson, W. H., and Pierce, W. G., 1968, Wapiti Formation and Trout Peak trachyandesite, northwestern Wyoming: U. S. Geological Survey Bulletin 1254-H, p. H1-H11.

Nelson, W. H., Pierce, W. G., Parsons, W. H., and Brophy, G. P., 1972, Igneous activity, metamorphism, and Heart Mountain faulting at White Mountain, northwestern Wyoming: Geological Society of America Bulletin, v. 83, p. 2607-2620.

Nelson, W. H., Prostka, H. J., and Williams, F. E., 1980, Geology and mineral resources of the North Absaroka Wilderness and vicinity, Park County, Wyoming: U.S. Geological Survey Bulletin 1447, with sections on Mineralization of the Sunlight Mining region and geology and mineralization of the Cooke City mining district by James E. Elliott and Aeromagnetic survey by Donald L. Peterson (in press).

Pierce, W. G., 1960, The "break-away" point of the Heart Mountain detachment fault in northwestern Wyoming, in Geological Survey research 1960: U.S. Geological Survey Professional Paper 400-B, p. B236-237.

______, 1963, Cathedral Cliffs Formation, the Early Acid Breccia unit of northwestern Wyoming: Geological Society of America Bulletin, v. 74, p. 9-22.

______, 1973, Principle feature of the Heart Mountain fault and the mechanism problem, *in* DeJong, K. A., and Scholten, Robert, eds., Gravity and tectonics: New York, John Wiley & Sons, p. 457-471.

______, 1979, Clastic dikes of Heart Mountain fault breccia, northwestern Wyoming, and their significance: U.S. Geological Survey Professional Paper 1133, p. 1-25.

______, 1980, The Heart Mountain break-away fault, northwestern Wyoming: Geological Society of America Bulletin, v. 91, p. 272-281.

Thirty-Third Annual Field Conference — 1982
Wyoming Geological Association Guidebook

CATASTROPHIC LARGE-SCALE LATE CENOZOIC DETACHMENT FAULTING OF EOCENE VOLCANIC ROCKS, SOUTHEAST ABSAROKA RANGE, NORTHWEST WYOMING

THOMAS M. BOWN[1]

INTRODUCTION

Recent geologic mapping in the Owl Creek — Grass Creek area of the southeast Absaroka Range documented the presence of approximately 120 detachment masses (here called klippen) that are remnants of a formerly much larger body (or bodies) of Eocene volcanic rocks probably displaced during Quaternary time (Brown, 1982). Similar masses in the same region occur in bedded Eocene volcanic rocks. These masses were probably displaced during Eocene time. About 30 other klippen of similar lithology and structural character were mapped by earlier workers (Rohrer, 1966; Wilson, 1970, 1975) farther north in the adjacent Soapy Dale Peak and Adam Weiss Peak Quadrangles. Recognition of the wide distribution, stratigraphic sources, and structural and geomorphic relations of these allochthonous volcanic rocks in the southeast Absaroka Range and southwest Bighorn Basin (Fig. 1) provides evidence of an episode of detachment faulting of hitherto unrecognized young age and great magnitude. A summary of new evidence pertaining to the scale, mechanisms, and age of this faulting is presented here in advance of a more detailed report (T.M. Bown and J. D. Love, unpublished data).

Several important oil and gas structures have been drilled along the southeast margin of the Absaroka Range (for example, the Wagonhound Creek, Prospect Creek, Grass Creek, Aspen Creek, and Fourbear fields), and it is likely that other important structures remain to be found beneath volcanic cover in the central and southeast Absaroka Range. In many areas, detachment faulting of Eocene and Pliocene-Pleistocene age has resulted in disharmonic structural deformation of the surficial volcanic sequence that can be misleading in oil and gas exploration mapping studies. It is emphasized that these aberrant structural relations are the result of gravity displacement of Teritary volcanic rocks and neither affect nor reflect subsurface trends.

[1]/U.S. Geological Survey, Denver, Colorado 80225

STRATIGRAPHY

In the southeast Absaroka Range, Tertiary rocks consist of the non-volcanic lower Eocene Willwood Formation (20-100 m) and, in ascending stratigraphic order, the volcaniclastic and volcanic igneous Aycross (290 m), Tepee Trail (400-600 m), and Wiggins (500-900 m) Formations (Love, 1939; Brown, 1982). The contracts of these units with one another and with all older rocks are unconformable. In many places, Tepee Trail and Wiggins strata were displaced by detachment faulting and now comprise masses that lie on the eroded Aycross and Willwood Formations along the southeast margin of the Absaroka Range, and on the eroded lower and middle Eocene Tatman Formation at Squaw Teats in the south central Bighorn Basin (Rohrer, 1966; Pierce, 1978).

PREVIOUS WORK

Rouse (1940) originally described anomalous structural relations of volcanic rocks in the southeast Absaroka Range, where intensely deformed rocks locally overlie less deformed strata in the valley of the Wood River (Fig. 1). He believed that this deformation was local and attributed it to adjustment of volcanic rocks along the mountain front, subsequent to emplacement of the Kirwin intrusives (for example, Wilson, 1963, 1964A, 1964B), but before lithification of the volcanics.

Rohrer (1966) was the first to recognize the allochthonous nature of these as well as similar rocks that cap Squaw Teats in the south-central Bighorn Basin (fig. 1). He mapped them as detachment fault remnants in Adam Weiss Peak Quadrangle and believed that the contact of volcanic strata with underlying non-volcanic rocks delimits the detachment fault contact in that area.

Wilson (1970) mapped additional klippen in the adjacent Soapy Dale Peak Quadrangle, and followed Rohrer in drawing the fault contact between volcanic and non-

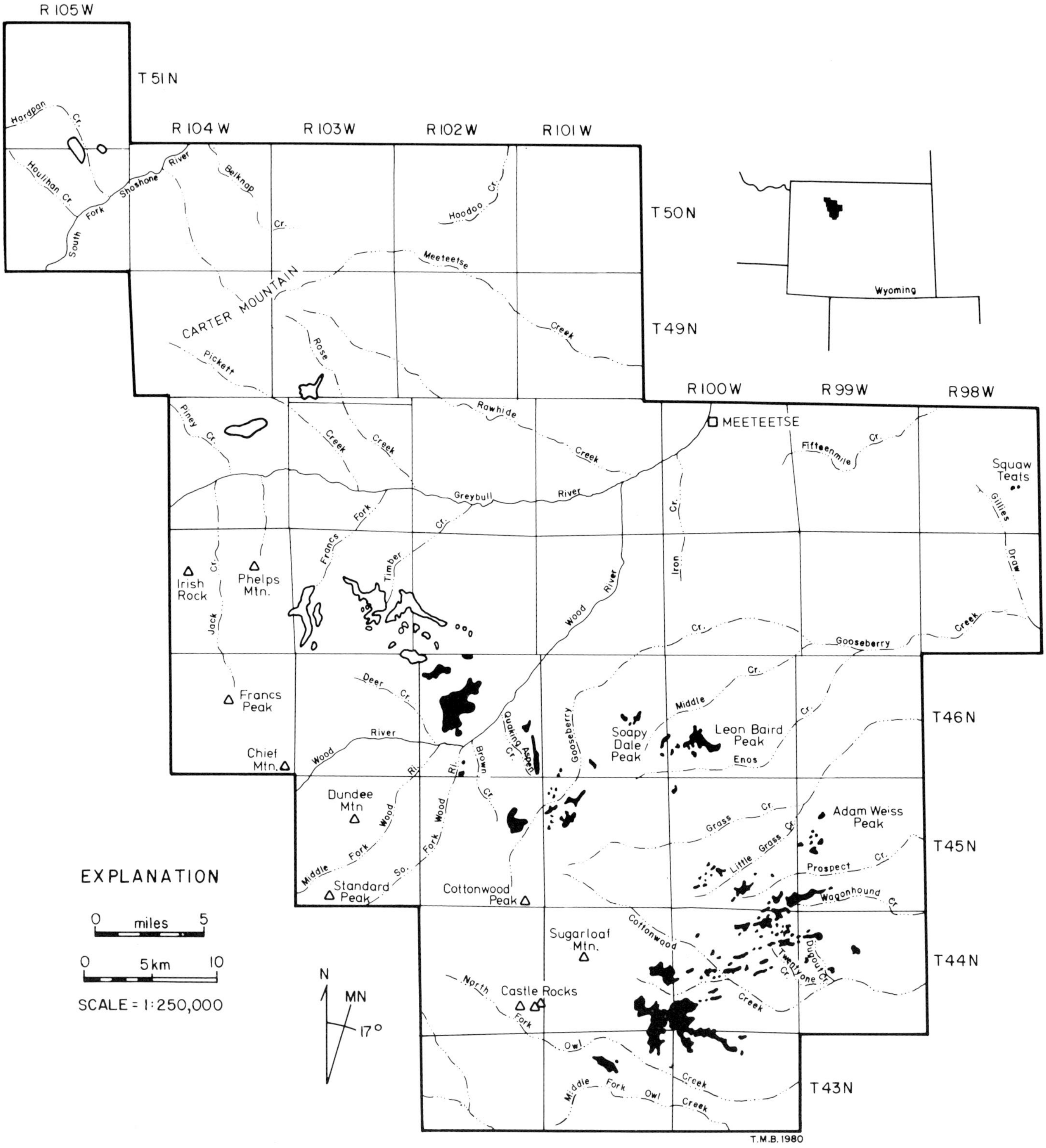

Figure 1: Map showing distribution of displaced rocks of the Enos Creek-Owl Creek detachment fault (black areas), and other displaced masses (open areas) in a part of the southeast Absaroka Range and southwestern Bighorn Basin.

volcanic rocks. In 1975, Wilson described what he believed are "break-away" (décollement) zones for the detached masses in the upper drainage of the Wood River. He believed that displaced rocks in that area were derived from the "Pitchfork" and Wiggins Formations (= Aycross and Tepee Trail Formations of this paper; see Bown, 1982: A15, A20-24), and that the "break-away" area is defined and bounded on the west by north-south trending normal faults that are downthrown to the east.

Bown (1982) recognized stratigraphic sources of these and other allochthonous rocks in the Tepee Trail and Wiggins Formations (sensu Brown, 1982), related all displaced masses in the Wood River-Owl Creek region to a single episode of detachment faulting, and postulated a late Tertiary or Quaternary age for the faulting. Roher (1966) believed that klippen in the Adam Weiss Peak Quadrangle resulted from post-middle Eocene to pre-Pleistocene detachment faulting and named this structure the Enos Creek detachment thrust. That name is here emended to Enos Creek-Owl Creek detachment fault to reflect its large scale. The mapping of Bown (1982, Pl. 1) provides detailed field evidence for this faulting in the Owl Creek-Grass Creek region, and it is recommended that map be used to supplement this text.

RECOGNITION OF DISPLACED ROCKS

In this study, allochthonous rocks of the Enos Creek-Owl Creek detachment fault are recognized solely by their structural discontinuity with, and occurrence on, all less deformed underlying rocks. Rocks beneath remnants of the displaced masses are invariably flat-lying to gently folded, whereas the allochthonous rocks are intensely folded and faulted; faults in them do not continue to the underlying strata, and S-shaped folds, recumbent folds, plastic folding, disconnected strewn out segments of interbeds, shingle-like fragments of interbeds, and breccia are common. This folding and faulting is rootless and is wholly unlike deformation seen in any autochthonous rocks in the Absaroka Range. Displaced rocks of the Enos Creek-Owl Creek detachment fault are nowhere overlain by other rocks. However, older detachment faulting, also involving volcanic rocks, is recognized elsewhere in the east-central part of the Range (see later section describing other displaced masses). In some of these instances, strongly deformed strata occur as parts of stratigraphic sequences.

Figure 2: Southeast margin of large type 1 klippe on the divide separating North Fork of Owl Creek and Lake Creek; Secs. 3 and 4, T. 43 N., R. 100 W., Hot Springs County, Wyoming. View to north. Displaced Tepee Trail rocks lie on eroded white mudstones and volcanic sandstones of the Aycross Formation.

DISTRIBUTION AND VOLUME OF DISPLACED ROCKS

Klippen of the Enos Creek-Owl Creek detachment fault are distributed from the Wood River on the north to the divide between the North and South Forks of Owl Creek on the south, and from a line connecting Lake Creek and Deer Creek on the west to Squaw Teats in the south-central Bighorn Basin on the east. A line connecting klippen on the periphery of this region (Fig. 1) circumscribes about 900 km²/, representing the minimum area affected by the fault. No detached rocks are known north or south of a line connecting Adam Weiss Peak and Squaw Teats, or east of Squaw Teats, but it seems clear that the original area of the klippen field may well have been much larger.

Because rocks displaced by this faulting have been extensively dissected by erosion, it is difficult to arrive at a meaningful estimate of the original volume of displaced rock. It is also possible that some of the rock was not transported as a single mass, but rather as an individual large blocks. Therefore, the area circumscribed by klippen might never have been completely covered with allochthonous rock. This is especially true for the area between Adam Weiss Peak and Squaw Teats. Nevertheless, the mean thickness of displaced rock and the density of klippen lying at various elevations in the area between Owl Creek and Enos Creek (Fig. 1) strongly suggests that this area, as far east as Adam Weiss Peak, was once entirely covered by fault debris. To the extent that this view is correct, a crude minimum estimate of the displaced volume can be obtained by averaging the mean thickness of klippen at the western and eastern margins of the field (as far east as Adam Weiss Peak) and multiplying by the area of the field. This calculation shows that at least 185 km³/ of rock were displaced.

STRUCTURE OF DISPLACED ROCKS

Bown (1982) described four varieties of klippen of the Enos Creek-Owl Creek fault that were distinguished from one another by a combination of their lithologic character and internal structure. However, additional study shows that three structurally important and distinctive types of allochthonous rocks can be distinguished; these are critical to interpretations of the mechanism, age, and speed of faulting. Type 1 (Figs. 2-4) are large, juxtaposed blocks of gently dipping to steeply dipping and recumbent rocks, separated by normal faults, and in which internal stratigraphic relations of beds are preserved. Type 2 (Figs. 5-7) are masses of internally incoherent rock with disjunct folds and segments of folds (including plastic folding) and minor amounts of breccia, in which internal stratigraphic relations are not preserved. Type 3 (Fig. 8) are masses of loose, uncemented breccia.

LOWER CONTACT OF DISPLACED ROCKS

In the course of mapping in the Owl Creek-Soapy Dale Peak area, it became clear very early that no planar detachment fault contacts are recognizable there. Rather, klippen of allochthonous rocks are preserved at varying elevations; they lie on high-level erosion surfaces, they fill ancient valleys (Figs. 5 and 8), and occur as patches on steep valley walls. Masses of plastically folded rock with rotated S-shaped folds and fold segments and shingled segments of interbeds swimming in a matrix of loose breccia commonly abut flat-lying, structurally undisturbed rocks (Fig. 5). Because Rohrer (1966: A30) and Wilson (1975: Fig. 1, cross-section) both interpreted this faulting in the Adam Weiss Peak-Soapy Dale Peak-Noon Point Quadrangles to be genetically related to thrusting, and because thrust faults commonly have somewhat plannar to curvilinear contacts separating the upper and lower plates, both authors mapped the detachment fault contact at the relatively planar contact between the non-volcanic Willwood Formation and the lowest volcanic rocks (Aycross Formation, sensu Bown, 1982; "Pitchfork" Formation, sensu Rohrer, 1966; and Wilson, 1970, 1975). This interpretation certainly includes all displaced rocks above their detachment fault contact, but is also includes a large volume of autochthonous rocks in their "upper plate" and does not satisfactorily explain many anomalous structural relations in which intensely deformed rocks are juxtaposed against wholly undeformed strata.

My reconnaissance mapping (Fig. 1) demonstrates that the lower part of the volcanic sequence in the Adam Weiss Peak-Soapy Dale Peak-Noon Point areas is clearly autochthonous and is separated from the non-volcanic Willwood Formation by an erosional unconformity and not a fault. This is the same contact relation observed between the Willwood and Aycross Formations throughout the remainder of the southeast Absaroka Range (Bown, 1982). At different localities, in place volcaniclastic rocks beneath displaced rocks include 20-290 m of Aycross strata and 0-180 m of Tepee Trail strata. All of the rocks beneath the detachment fault contacts mapped in these areas by Bown (1982, Pl. 1; Figs. 5 and 8, this paper) are flat-laying to gently inclined, though minor gouging occurs locally. There is no structural discontinuity related to detachment faulting at the Willwood Formation-Aycross Formation contact anywhere in the southeast Absaroka Range. Though thrust faults locally displace Aycross and Tepee

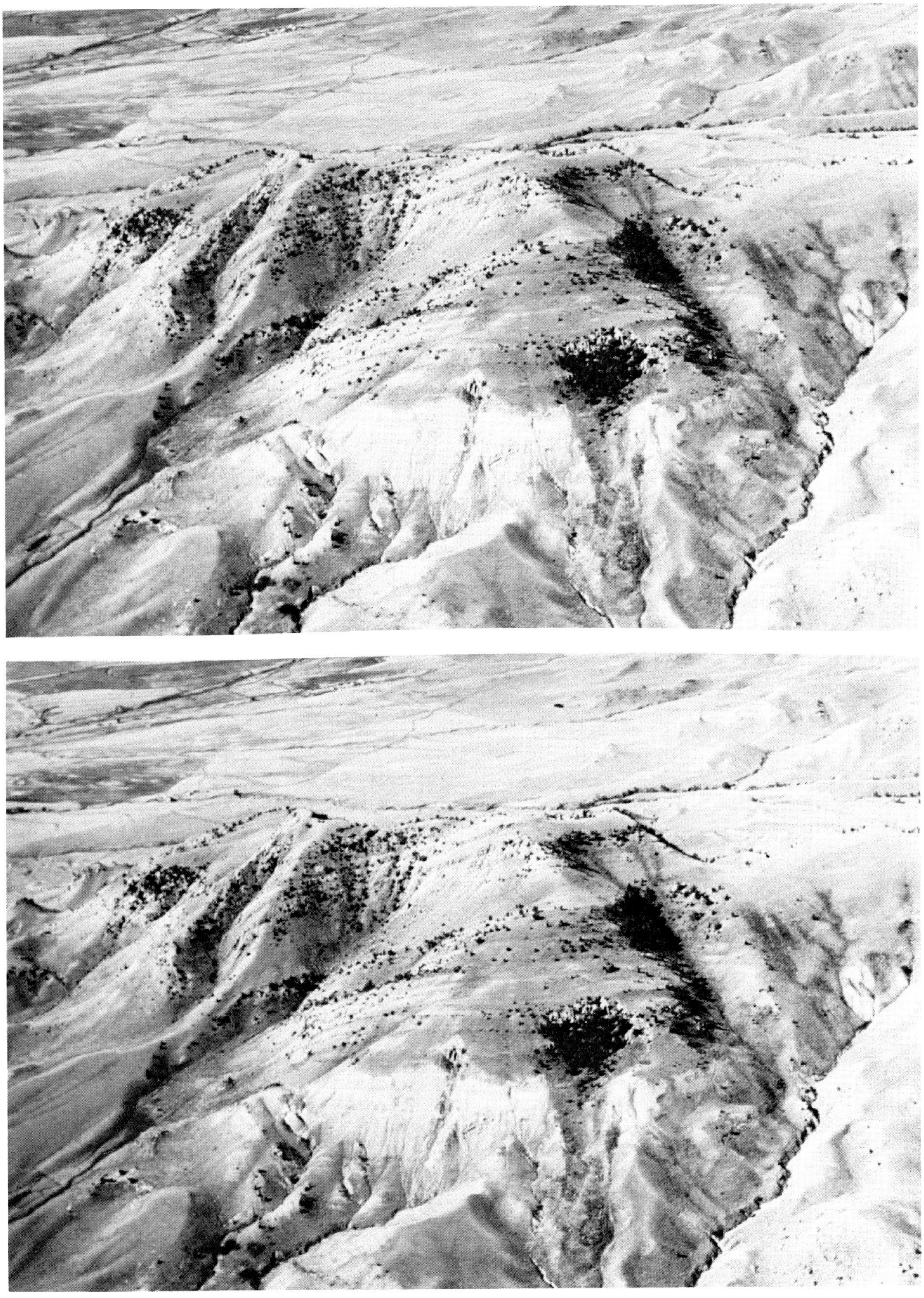

Figure 3: Oblique stereo aerial photographs of displaced Tepee Trail rocks on divide separating North Fork of Owl Creek and Lake Creek, showing juxtaposed blocks preserving internal stratigraphic relations and bounding of blocks by normal faults; Secs. 3 and 4, T. 43 N., R. 100 W. View to north. Top = left, bottom = right. Photographs by W.B. Hall, 1980.

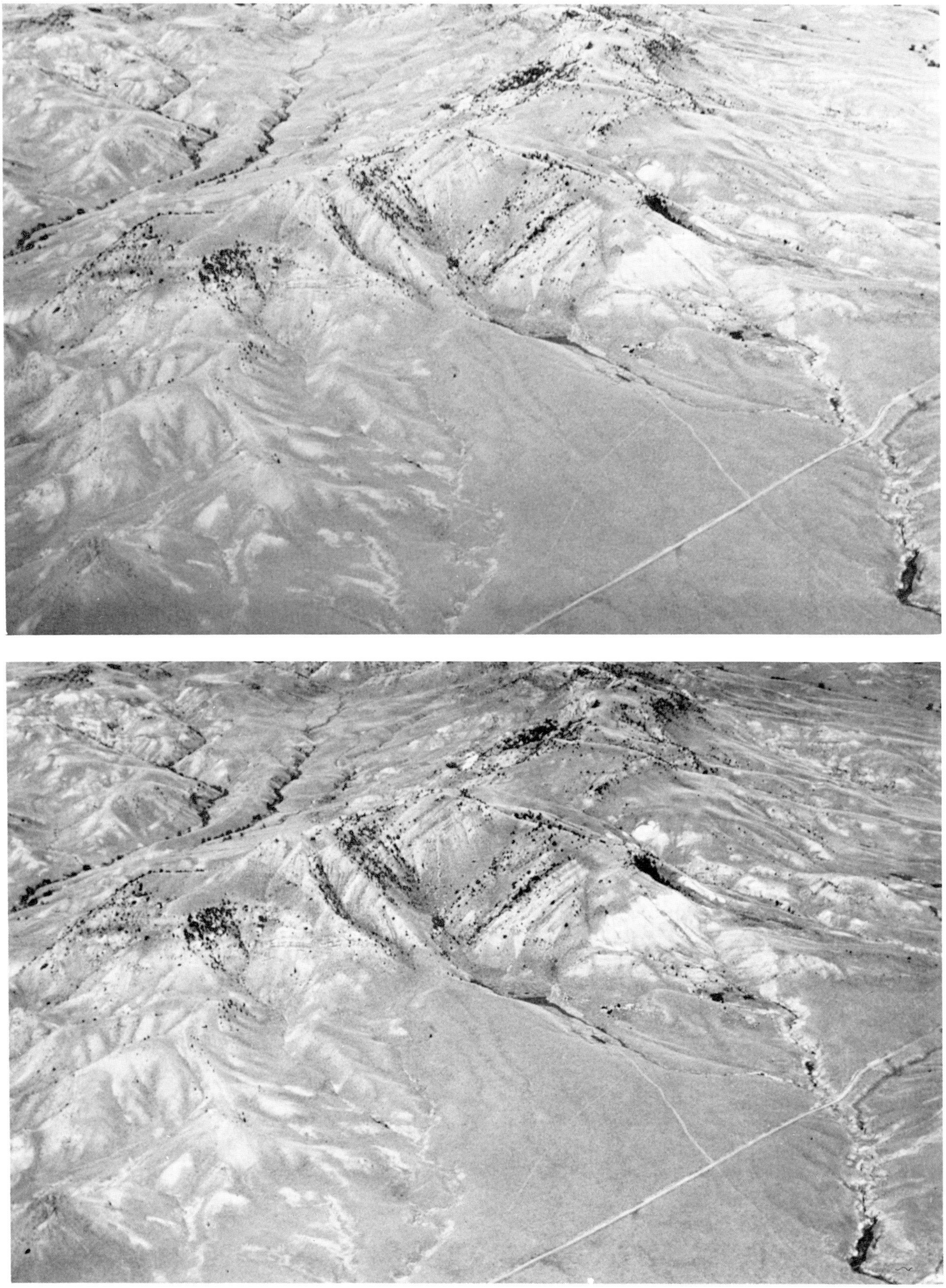

Figure 4: Oblique stereo aerial photographs of displaced masses of Tepee Trail rocks on divide separating North Fork of Owl Creek and Lake Creek, showing contact with eroded Aycross Formation and rise of pre-detachment transgression surface to west. View to west. Allochthonous rocks lie in Secs. 30-33, T. 44 N., R. 100 W., Secs. 3, 4, and 6, T. 43 N., R. 100 W., Sec. 1 and 23, T. 43 N., R. 101 W., and Secs. 25, 26, 35, and 36, T. 44 N., R. 101 W., Hot Springs County, Wyoming. Top = left, bottom = right. Photographs by W.B. Hall, 1980.

Figure 5: Type 2 klippe composed of undifferentiated rocks of Tepee Trail and Wiggins Formations on divide separating Wagonhound and Prospect Creeks, Secs. 31 and 32, T. 45 N., R. 99 W. View to north. Note relatively undistributed flat-lying Aycross strata beneath strongly folded and brecciated displaced rocks.

Figure 6: Detail of intensely deformed rocks composing type 2 klippe at head of Dugout Draw, NW¼ Sec. 7, T. 44 N., R. 99 W. View to northeast. Note vertical and overturned beds and disjunct segments of folded beds.

Figure 7: East side of type 2 klippe at Chimney Rocks, SW¼ Sec. 34, T. 44 N., R. 100 W. View to northwest. Note vertical and overturned beds of displaced Tepee Trail Formation.

Figure 8: Breccia of undifferentiated Tepee Trail and Wiggins origin composing type 3 klippe near head of Twentyone Creek, SE¼ Sec. 3, T. 44 N., R. 100 W. View to northwest. Note surrounding flat-lying light colored rocks of Aycross Formation and occurrence of fault debris as fill of an ancient valley (contact emphasized; Tac - Aycross Formation, Tud = displaced rocks).

Trail rocks (Bown, 1982, Pl. 1), these are very minor structural features. They do not affect allochthonous rocks which overlie the thrust fault traces, and the upper plates of the thrusts are made up of gently inclined strata that are nowhere as intensely deformed as rock displaced by the Enos Creek-Owl Creek detachment fault.

Utilizing more than 600 topographic datum points determined from the elevations of the contacts of allochthonous and autochthonous rocks in the Owl Creek-Grass Creek area, a topographic map of the surface transgressed by displaced masses was constructed (T.M. Bown and J.D. Love, unpublished data, 1980). This map and accompanying cross-sections show that allochthonous rocks crossed a region of ruggged topography with as much as 650 m of local relief. Low-lying areas and ancient stream courses were completely filled in some places with displaced rock (Fig. 8; Bown, 1982, Pl. 1, cross-section). Recognition of the lower contact of the displaced rocks as a rugged ancient land surface with valleys, hills, and tablelands, and not as a more planar or curvilinear, thrust-like sole clarifies complex discordant structural relations of detached Eocene volcanic rocks in the southeast Absaroka Range, and provides important evidence pertaining to the mechanics and age of faulting.

Figure 9: View to north of displaced rocks on Leon Baird Peak divide (in far distance) and their relation to remnant of high-level erosion surface (foreground) on north side of Grass Creek.

STRUCTURAL, AREAL, AND TOPOGRAPHIC RELATIONS OF DISPLACED ROCKS

As observed above, three types of klippen, distinguished by their differing degrees of internal structural deformation, make up all allochthonous rocks of the Enos Creek-Owl Creek detachment fault. The areal and topographic distributions of these different types of klippen are related,

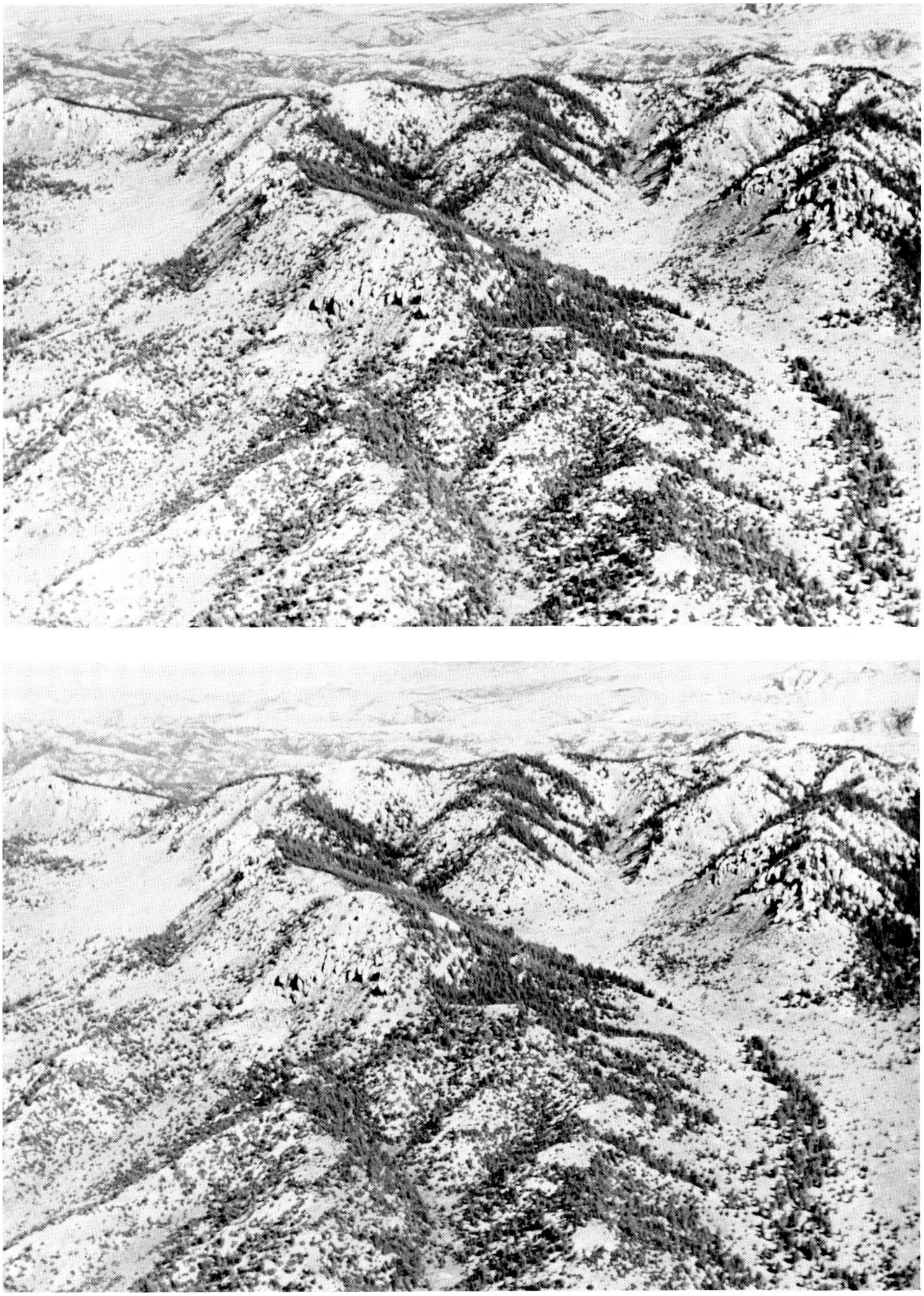

Figure 10: Oblique stereo aerial photographs of displaced rocks of undifferentiated Tepee Trail and Wiggins origin capping Leon Baird Peak divide between Enos Creek and Gooseberry Creek, Secs. 29 and 30, T. 46 N., R. 100 W., and Sec. 25, T. 46 N., R. 101 W. View to north. Note remnants of highlevel erosion surface on Willwood and Aycross Formations beneath displaced mass. Top = left, bottom = right. Photographs by W.B. Hall, 1980.

in a general way, to the different degrees of deformation in them. For example, type 1 klippen (the least deformed) occur only on relatively steep slopes at relatively high elevations at the western margin of the klippen field. They are composed wholly of Tepee Trail rocks which closely resemble adjacent autochthonous Tepee Trail exposures developed in the valleys of Owl Creek, Lake Creek, and the Wood River (Fig. 1). Type 2 klippen, on the other hand, invariably occur more basinward (eastward) and at lower elevations. Loss of internal stratigraphic relations, plastic folding, and minor brecciation are common. These klippen are composed of undifferentiated rocks of the Tepee Trail and Wiggins Formations, and of lithologies that are known in place only in areas several kilometers west of where the klippen occur. Uncemented breccias, characteristic of type 3 klippen, are from the same undifferentiated stratigraphic sources and are preserved only as ancient valley fill deposits at the lowest local topographic elevations in the klippen field. These relations are critical to interpretation of the displacement mechanisms of the Enos Creek-Owl Creek detachment fault.

MECHANISMS AND RECONSTRUCTION OF FAULTING

Multiple movement mechanisms of detachment faulting are indicated by the lower contact relations of displaced rocks and by the consistent associations of the three different types of deformation in displaced rocks with their areal and topographic distributions. None of these movement mechanisms bear any close relationship to thrust faulting. Large thrusts are unknown in the southeast Absaroka Range and all folding and faulting related to the Enos Creek-Owl Creek detachment fault are rootless. Moreover, the effect of this faulting has been to displace younger rocks on older rocks and to deform the younger rocks intensely and in a manner that bears no relation to the structure of the older rocks. A body force, i.e., gravity, is indicated as one of the principal mechanisms initiating and sustaining movement and, generally speaking, the Enos Creek-Owl Creek detachment fault is best regarded as an example of gravitationally induced mass-wasting on an enormous scale.

Type 1 klippen are best developed in the Owl Creek-Lake Creek area and in the upper drainage of the Wood River (Fig. 1). They are the least deformed, occur at relatively high elevations on 5-8° slopes, and are closest to the high Absaroka Range. They are made up of enormous juxtaposed blocks of stratified rock up to 400 m thick and up to 9 km² in area that are separated from one another by normal faults. Each of the blocks differs from adjacent blocks in structural grain, i.e., a mass of rock dipping vertically might abut another mass with dips of 5°, 50°, or 20° overturned, and with greatly contrasting strikes (Fig. 3; Bown, 1982, Pl. 1].

Preservation of internal stratigraphy, absence of plastic folding and brecciation, position on steep slopes at high elevations, and close proximity to source areas show that type 1 masses were moved the least distance. Minor backward (westward) rotation accompanied by strike and dip changes in some of these blocks, for example, at the near Noon Point in the Wood River drainage (Wilson, 1975, Fig. 1), suggests movement analogous to slumping. This is particularly evident at Noon Point where the rotated blocks lie adjacent to autochthonous Tepee Trail rocks in which equivalent strata can be identified. Rotation slumping movement was followed by separation of the blocks, rapid gravity-sliding, and further rotation and tumbling of blocks that piled up on one another on gentler slopes at lower elevations where friction became too great to support further movement. These relations are best seen on the high divide separating Lake Creek from the North Fork of Owl Creek (Figs. 1-3). Analogous movement, though on a much smaller scale, was termed "retrogressive block-gliding" by Voight (1973).

Because the volume of rock initially displaced was very large, most of the detached mass reached the much gentler (1-1.5°) slopes of a basinward-inclined high-level erosion surface and was propelled at least 45 km across it. This movement was accompanied by internal disordering due to frictional drag at the base of the mass and resulting velocity gradients developed within it. Confining pressure, coupled with velocity gradients, produced plastic folding and shingling of beds in the center of the mass, accompanied by loss of cohesion and separation of beds at its margins. The resulting structure is seen in type 2 klippen which occur at intermediate topographic elevations on the reconstructed surface of transgression, and at intermediate distances from source areas. These klippen are best developed in the Soapy Dale Peak, Leon Baird Peak, Adam Weiss Peak, and Twentyone Creek areas of the southeast Absaroka Range (Figs. 1, 5-7, 9 and 10; Bown, 1982, Pl. 1).

In the area between the Wood River and Grass Creek (Fig. 1), the high-level erosion surface transgressed by displaced rocks was relatively planar and undissected at the time of faulting. Remnants of this surface with type 2 klippen on them are well preserved there today (Figs. 9 and 10). However, in the area between Owl and Grass Creeks (Fig. 1), this surface was deeply dissected by basinward-flowing streams prior to faulting. The maximum depth of dissection was at least 300 m in the present Twentyone Creek area. As parts of the mobile mass plummeted into these erosional voids, they were brecciated and any remaining internal stratigraphy was obliterated. These remnants are preserved today as type 3 klippen (Figs. 5 and 8). In some areas, the base of the moving mass rumpled underlying autochthonous rocks and its leading edge collided with topographic highs in the valleys, or against valley walls, producing deep gouging. However, in other areas, brecciated rock fills ancient valleys with no gouge or other disruption of underlying rock whatsoever (figs. 5 and 8). These relations suggest that the dense, brecciated part of the mobile mass might well have locally trapped pockets of air in the relatively confined overridden valleys, and raced down them on a cushion of compressed air. Gradual dissipation of the air through the mass and from its periphery allowed settling of the rubble with little or no disturbance of underlying rock. Development of a cushion of compressed air beneath comminuted moving rock is a well known feature of many landslides (for example, Kent, 1966; Shreve, 1966, 1968). Brecciation, and hence fluidization of the mass is essential to confine the entrapped air completely (Kent, 1966), and large blocks of rock cannot be transported by this mechanism because air can easily escape around the edges of the blocks. Ancient landslide debris entombed in other rock is termed megabreccia, and megabreccias with both gouged and ungouged underlying strata are well known (e.g., Longwell, 1951; Kreiger, 1977).

The Enos Creek-Owl Creek detachment fault is a unique structure with multiple movement mechanisms closely akin to slumping, retrogressive block-gliding, gravity sliding, and landsliding. It is perhaps best termed a detachment fault because this name has fewer genetic constraints than the terms gravity glide fault or landslide fault, and allows inclusion of a variety of displacement mechanisms. These mechanisms prevailed differentially in different parts of the area affecting by faulting in response to physiographic features of the source area and the surface of transgression. The physiography of the pre-detachment land surface and the differential movement of the mass that is reflected by its structure are important in determining the age and speed of this displacement.

AGE OF FAULTING

It is tempting to relate the age of the Enos Creek-Owl Creek detachment fault to that of its three closest presumed genetic analogues in space; i.e., the Heart Mountain, Reef Creek, and South Fork detachment faults of the northeast Absaroka Range, some 35-40 km to the north. Pierce (for example, 1975) has shown that the Heart Mountain detachment fault is the youngest of these features, and a combination of stratigraphic, structural, and geomorphologic considerations demonstrates to me that the Enos Creek-Owl Creek detachment fault is considerably younger than the Heart Mountain detachment fault. The stratigraphic evidence alone is compelling. Displaced blocks of the Heart Mountain fault and the suface of tectonic denudation overridden by them are locally buried by hundreds of meters of volcaniclastic and volcanic igneous rocks of the middle Eocene Wapiti Formation, and the younger Trout Peak Trachyandesite and Langford Formation (Pierce, 1941, 1957, 1963). The upper part of the Wapiti Formation is a correlative of the Tepee Trail Formation (Bown, 1982), and both Tepee Trail and Wiggins rocks were displaced by the Enos Creek-Owl Creek detachment fault. Moreover, the Heart Mountain fault apparently originated as a bedding fault and principally displaces Paleozoic strata. The Enos Creek-Owl Creek fault has no bedding fault component, it displaces only middle Eocene volcanic rocks, and is rootless. Its mass-wasting type of displacement is unique in this region.

As mentioned above, several klippen of the Enos Creek-Owl Creek detachment fault rest upon remnants of a high-level erosion surface that is inclined basinward (eastward) about 1-1.5°. This surface can be dated relative to other physiographic features in the Bighorn Basin, thus providing a maximum age for faulting. The most easterly of the klippen, two masses capping Squaw Teats in the south central Bighorn Basin (Fig. 11), lie on a remnant of a high-level erosion surface at an elevation of about 6,020 feet (1,836 m). Twenty km north of Squaw Teats is Tatman Mountain, another high-level erosion surface developed at 6,230 feet (1,900 m) and capped by gravel and sandstone deposits of probable late Pliocene or Pleistocene age

Figure 11: Squaw Teats klippen in the south central Bighorn Basin, Secs. 23 and 26, T. 48 N., R. 98 W. View to northeast. The upper 60 m of the left (west) butte and the upper 32 m of the east butte are displaced Tepee Trail and Wiggins rocks that were transported at least 45 km.

(Rohrer and Leopold, 1963). Though its lower elevation suggests that the surface beneath Squaw Teats is slightly younger than the Tatman Mountain surface, they are believed to be approximately equivalent in age.

As observed by Mackin (1937), extending the profile of the Tatman Mountain surface westward at the same declivity as that of the adjacent Greybull River or its lower terraces causes this imaginary profile to enter the Absaroka Range in the area of the upper Greybull River valley at an elevation of 8,200-8,500 feet (2,500-2,560 m). This elevation almost exactly coincides with elevations at the bases of the highest type 1 detachment klippen which also lie on high-level surfaces in the Wood River area at elevations of 8,100-8,400 feet (2,470-2,560 m). Moreover, Mackin (1937) and Pierce and Andrews (1941) believed that several of the high-level surfaces at the east margin of the Absaroka Range were probably basin margin remnants of a formerly much more continuous surface that correlates with the Tatman Mountain surface in the central Bighorn Basin. In the southeast Absaroka Range, remnants of this surface with klippen of the Enos Creek-Owl Creek detachment fault perched atop them are well developed in the Adam Weiss Peak-Leon Baird Peak area (Figs. 9 and 10). The implication is clear that allochthonous rocks might well have been transported eastward across this essentially featureless surface before it was extensively dissected by streams.

Using stream downcutting rates for the Quaternary calculated by Palmquist (1978) from datable Pearlette-like volcanic ashes in stream terraces in the eastern Bighorn Basin, the age of the Tatman Mountain surface is 1.995 m.y. Calculations of age based on these downcutting rates as applied to thickness of allochthonous rock removed by erosion and rock removed from beneath the high-level surface in the Owl Creek-Grass Creek area and at Squaw Teats yield similar age estimates (1.745 and 2.060 m.y., respectively). These relations suggest that the age of the Enos Creek-Owl Creek detachment fault is latest Pliocene or earliest Pleistocene.

Because of the youthfulness of faulting, well into the latest Teritary or earliest Quaternary erosion cycle of the Bighorn Basin, it is extremely doubtful that exact "breakaway" areas of the fault will ever be recognized. This is especially true in view of the mass-wasting type of displacement which doubtlessly left a considerable scarp that has subsequently retreated westward due to erosion in the intervening 2.0-odd million years. The Heart Mountain detachment fault, on the other hand, was developed during or immediately preceding widespread deposition of large volumes of volcanic rock and break-away areas of that fault (Pierce, 1975) were engulfed by this rock and partly preserved. In the case of the Enos Creek-Owl Creek detachment fault, it is probable that most of the material was derived from a former salient of the Absaroka Range between Owl Creek and the Wood River, a salient that was largely destroyed during the displacement. Cottonwood Peak might well be a remnant of this salient (Fig. 1).

SPEED OF MOVEMENT AND TRIGGERING MECHANISM

The absence of field evidence for continued thrusting "push" of tectonic origin from the west (source area of displaced rocks), and the presence of landslide breccias underlain by rocks that are not strongly gouged suggest very rapid displacement. In geologic and even in historic terms, this movement must have been almost instantaneous, possibly on a scale of hours, if not minutes (see later discussions). This idea is reinforced by all aspects of the internal structure of the klippen which suggest movement by a body force (gravity). Due to friction, large-scale gravity movements cannot be sustained for geologically significant periods of time without periodic pushing. Moreover, the estimated breadth/thickness ratio of the displaced mass as a body was far too large to support emplacement by pushing from behind (see Hsü, 1969). The differential distribution of fault klippen as discrete structural entities is, as shown above, consistent with a large-scale, mass-wasting type of movement.

The Pliocene-Pleistocene age of displacement also argues for very rapid movement. There is no local or regional evidence for orogenic compressional forces being active at this time, and no intrusive volcanics were emplaced in or near this area in the time of faulting. Moreover, erosion rates at the time of displacement were relatively great, yet there is no evidence of the effects of erosion on a slowly or sporadically moving sheet or blocks of allochthonous rock as there is for parts of the Heart Mountain detachment fault (Pierce and Nelson, 1973). Even movement on a scale of months would, due to much greater friction, produce far more extensive and pervasive gouging of the incompetent underlying Aycross mudstones in deeply eroded pre-detachment valleys than is seen in the field.

The enormous size of the area known to be affected by this detachment faulting also suggests a single episode of displacement. Given slopes of 5-8° close to source areas and 1-1.5° farther basinward, an enormous volume of material must have been dislocated almost instantaneously to build the necessary momentum for transporting debris a minimum of 45 km to the east. The largest previously described mass-wasting feature is the Saidmarreh landslip in Iran (Harrison and Falcon, 1938). In that occurrence, the displacement produced a debris covered area of about 170 km^2, which is less than one-fifth the minimum area affected by the Enos Creek-Owl Creek detachment fault. Shreve (1968) suggested a speed of about 350 km/hr for the Saidmarreh landslip; the landslide phase of the Enos Creek-Owl Creek displacement probably attained a similar speed.

Bucher (1947) and Pierce (for example, 1963, 1975) believed that a combination of gravity and earthquake oscillations were responsible for triggering and sustaining movement on the Heart Mountain detachment fault. The same is proposed as a triggering mechanism for displacement of rocks associated with the Enos Creek-Owl Creek fault. Both areas of faulting lie near and partly within the Absaroka-Yellowstone volcanic field. Volcanic activity began there in the Paleocene (L.L. Love and others, 1976), and seismic activity is common today in parts of this and adjacent areas (Smith and Christiansen, 1980), though on a much diminished scale.

At about the time of the Pliocene-Pleistocene boundary, headward erosion of streams in the western Bighorn Basin had largely defined the Absaroka Range scarp somewhat east of its present position. Basinward, this scarp was met by one or more partly dissected high-level erosion surfaces formed by lateral planation of streams during pauses in the excavation of the Bighorn Basin sedimentary fill (Fig. 12A). The Absaroka Range scarp was developed essentially at the westward retreating margin of volcaniclastic fill, with harder, more competent pyroclastics of the Tepee Trail and Wiggins Formations resting on incompetent bentonitic strata of the Aycross Formation. This, then, was an ideal unstable setting for catastrophic mass-movement. It is postulated that a particularly violent episode of volcanic activity, possibly associated with emplacement of the Huckleberry Ridge Tuff in the Absaroka-Teton area about 2.05 m.y. ago (Love, 1977; J.D. Love and others, 1978; J.D. Love, personal communications, 1980-81), was accompanied by intense earthquakes in the eastern Absaroka Range. Violent and periodic oscillations of these earthquakes dislodged an

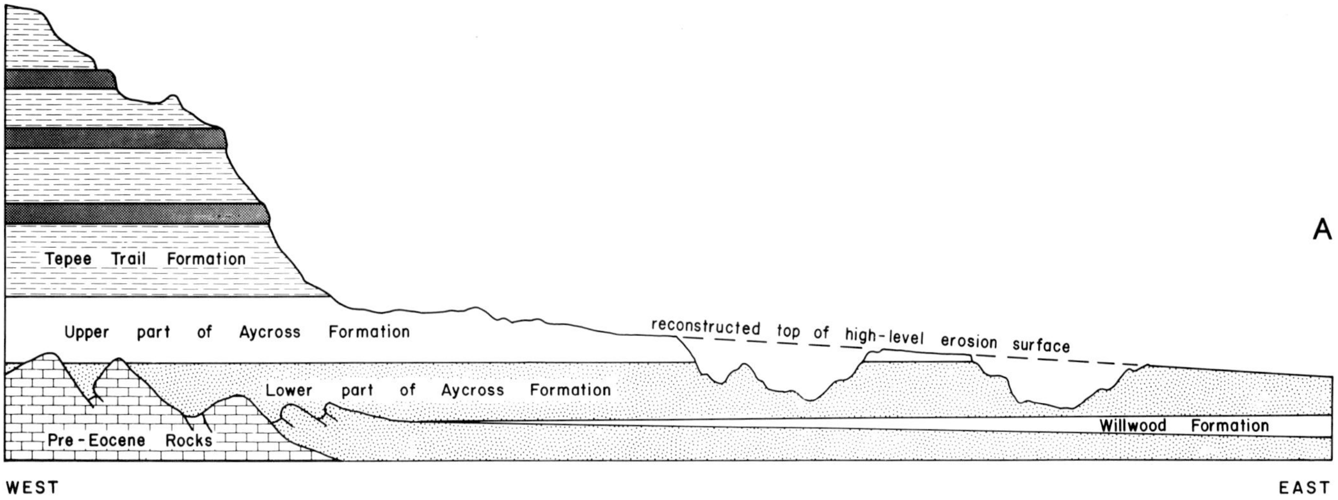

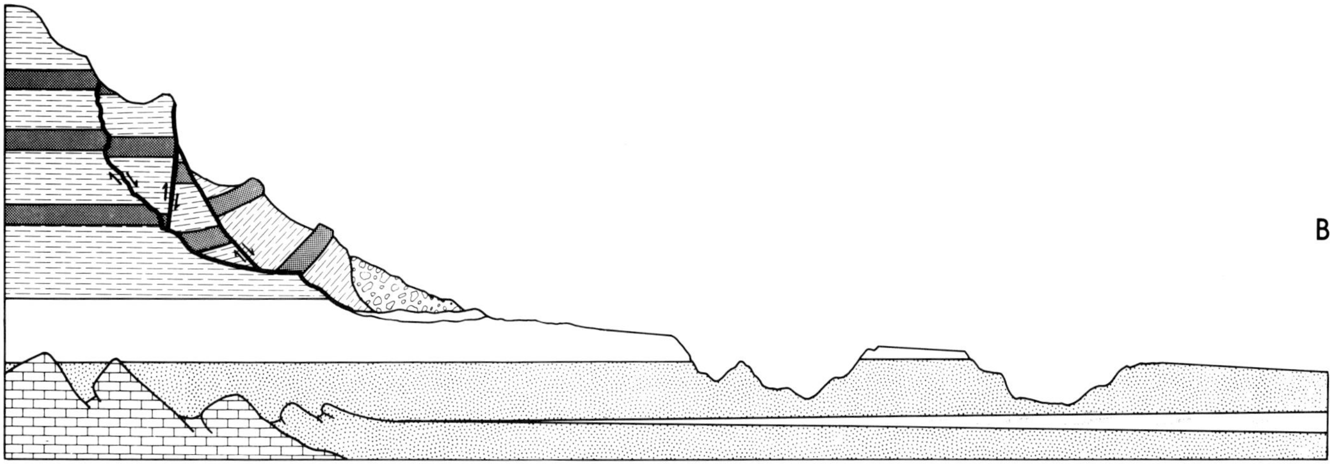

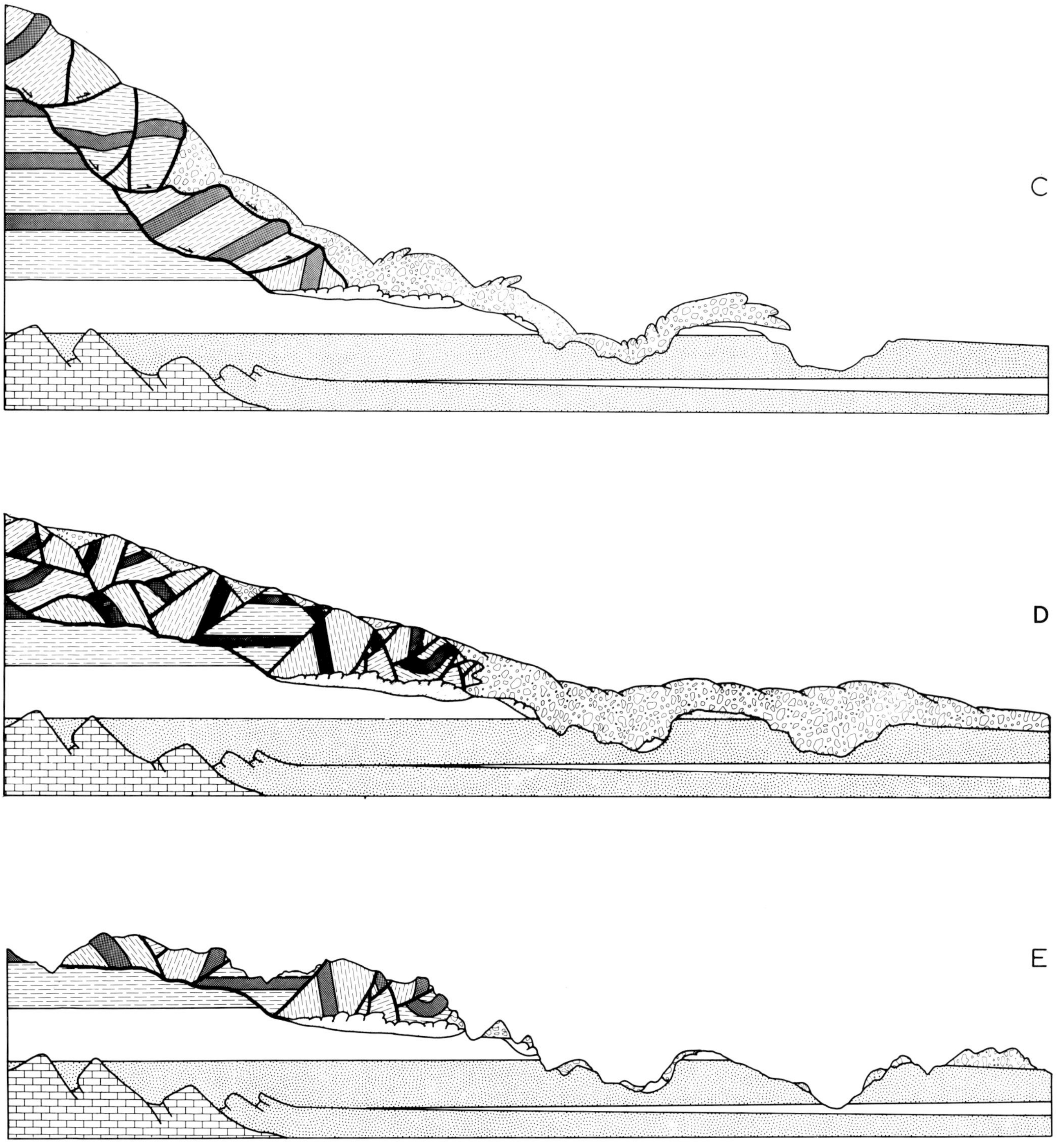

Figure 12: Hypothetical reconstruction of five successive stages in the displacement of Eocene volcanic rocks by the Enos Creek-Owl Creek detachment fault. A- Pre-detachment surface proximal to the Absaroka Range scarp in the Owl Creek area, showing east margin of Absaroka Range scarp made up of Aycross the Tepee Trail rocks, and reconstructed top of basinward-inclined high-level erosion of surface that has been dissected by streams in two places. Wiggins strata lie on the Tepee Trail Formation farther west, out of the area of the diagram. Surface profile is diagrammatic and highly exaggerated. B- Earthquake activity fractures the front of the Absaroka Range scarp, causing several large blocks of Tepee Trail rocks to become dislodged and slump basinward. A mass of breccia accumulates in front of the moving mass and rocks beneath its leading edge are deeply gouged. C- Continued earthquake oscillations cause tensional fractures to propagate farther to the west and more masses of Tepee Trail and Wiggins rocks are dislodged and move downslope, piling up at the break in slope where the high-level erosion surface meets the front of the Absaroka Range scarp. Masses of shattered stratified rock and breccia are propelled across the high-level erosion surface, filling ancient paleovalleys and, in some places, riding down the valleys on entrapped pockets of compressed

air. All internal stratigraphy is lost as the masses break apart due to basal friction and collision with underlying topographic highs. The leading edge of the mass of displaced rocks reaches the area of Squaw Teats in the south central Bighorn Basin, some 45 km east of source areas. D- Detached rocks in the Owl Creek area immediately after cessation of movement, showing jumbled gravity-glide blocks proximal to source areas in the Absaroka Range, breccia and shattered folded rock (both depicted here as breccia), and gouged areas in the autochthonous Aycross Formation beneath the detached rock. The maximum preserved thickness of displaced rock is 425 m. E- Diagrammatic view of structural and topographic relations of klippen of the Enos Creek-Owl Creek detachment fault in the Owl Creek area today, following removal, in some areas, of at least 350 m of displaced rock by Pleistocene and Holocene streams.

immense volume of stratified volcaniclastic rocks from the east-facing Absaroka Range scarp, along a north-south line nearly 50 km in length (Figs. 1, 12B and 12C). Following detachment, the mass was propelled rapidly by gravity at least 45 km eastward into the south-central Bighorn Basin (Fig. 12C and 12D). Post-detachment erosion removed most of the fault debris and resulted in the klippen field seen today in the southeast Absaroka Range and southwestern Bighorn Basin (Fig. 12E).

OTHER DISPLACED MASSES OF VOLCANIC ROCK IN THE EASTERN ABSAROKA RANGE

Because of the long history of volcanic activity in the Absaroka-Yellowstone volcanic field, there is no reason to suppose that large-scale gravity displacements of probable earthquake origin were limited in time to the Eocene or to the Pliocene-Pleistocene boundary, or confined to the area of the eastern Absaroka Range. Evidence for pre-Wiggins (late middle Eocene) faulting is obscured by as much as 1,200 m of Wiggins rocks and correlatives throughout most of the Absaroka Range. However, there is much field evidence for at least two other eposides of possible large-scale gravity faulting in this area, in addition to the South Fork, Reef Creek, Heart Mountain, and Enos Creek-Owl Creek detachment faults. It is very likely that many others remain to be identified.

Sundell (1980) described two kinds of large-scale gravity displaced masses that he attributed to detachment faulting in the middle part of the Wiggins Formation in the southern Absaroka Range. One of these, termed by him the "Castle Rocks Chaos", is the same as chaotic Wiggins rocks earlier described by Love (1934, 1939) from the southernmost Absaroka Range. The structural geology of the Castle Rocks Chaos and other nearby gravity displaced masses is currently under study by K.A. Sundell.

Hay (1954, 1956) described anomalous structural and stratigraphic relations of "tuff-breccias" in the "Pitchfork" and Wapitti Formations in the valleys of the Greybull River and the South Fork of the Shoshone River (Fig. 1). Hay believed that these masses were emplaced by postdepositional interstratal movement, perhaps by fluid injection. My reconnaissance study of this area in 1979 and 1980 indicates that at least some of this deformation, especially the intense disharmonic folding of volcanic mudstones and sandstones beneath the Trout Peak Trachyandesite on the north side of the Greybull River valley (Fig. 13), and steeply dipping to vertical rocks on flat-lying rocks in the Rose Creek area (Fig. 14), might well be the result of Eocene gravity faulting. Similar relations are seen locally on the north side of Carter Mountain and on the south side of the Greybull River valley, between Dick Creek and Francs Fork of the Greybull River. The top of the type section of Hay's (1956) "Pitchfork" Formation is clearly displaced by faulting, and contorted Eocene volcanics are well exposed in cliff faces stratigraphically beneath the Trout Peak Trachyandesite in the upper drainage of Gwinn Fork of Dick Creek (Fig. 15).

Large blocks of allochthonous rock preserving internal stratification are distributed across eroded rocks of the Wapiti Formation in the drainages of Houlihan and Hardpan Creeks, on the divide separating the North and South Forks of the Shoshone River (Fig. 1). These blocks owe their present positions to gravity displacement, but the age of faulting and its relationship, if any, to adjacent structurally anomalies "tuff-breccias" described by Hay (1854: Pl. 2, Figs. 1 and 2) are uncertain. Nelson and others (1980) believed that these features resulted from gravity-sliding in Wapiti (middle Eocene) time. Those authors also record large displaced blocks of the Lamar River Formation underlying Wapiti strata and the Trout Peak Trachyandesite on the south side of the Pilot Peak, near the northeast boundary of Yellowstone National Park. Continued investigation should result in a more complete understanding of the mechanisms and age relations of the several episodes of gravity displacement in the Absaroka-Yellowstone volcanic field.

ACKNOWLEDGMENTS

I am especially grateful to J. David Love for suggesting the research area and for his sustained support and

Figure 13: Disharmonic folding of ?Aycross rocks beneath lens of Trout Peak Trachyandesite on north side of Greybull River valley, Secs. 2 and 3, T. 48 N., R. 104 W., Park County, Wyoming. Deformation may indicate Eocene eposode of gravity faulting.

Figure 14: Steeply tilted and vertical rocks overlying flat-lying ?Aycross or Wapiti strata between Rose and Rawhide Creeks on south flank of Carter Mountain, Secs. 28, 32, and 33, T. 49 N., R. 103 W., Park County, Wyoming. Deformation may indicate Eocene episode of gravity faulting.

Figure 15: Disharmonically folded and brecciated sequence lying stratigraphically beneath the Trout Peak Trachyandesite at head of Gwinn Fork of Dick Creek, Sec. 25, T. 47 N., R. 103 W., Park County, Wyoming. Deformation may indicate Eocene episode of gravity faulting.

generous assistance and consultation. I thank J. D. Love, W. G. Pierce, K. A. Sundell, and J. G. Eaton for consultation and useful debate in the field, and J. D. Love, W. R. Keefer, and M. H. Hait, Jr., for review of the manuscript. W. B. Hall took many superb color oblique stereo aerial (COSA) photographs in conjunction with this study, six of which are reproduced here as Figures 3, 4, and 10 (see Hall and Walsh, 1977, for procedure). These photographs were useful both as mapping aids and in clarifying field relations that are different to grasp on the ground. Unfortunately, some of the sharp detail of these photographs was lost in their reproduction as black-and-white illustrations. A. McKenna, M. C. Campbell, R. Kraus, and M. J. Kraus were able field assistants.

This research was supported in part by a National Research Council Postdoctoral Fellowship with the U.S. Geological Survey and continued in conjunction with the U.S. Geological Survey Wyoming State Geologic Map under the supervision of J. D. Love.

REFERENCES CITED

Bown, T.M., 1982. Geology, paleontology, and corelation of Eocene volcaniclastic rocks, southeast Absaroka Range, Hot Springs County, Wyoming. U.S. Geological Survey Professional Paper 1201-A: 75 pp.

Bucher, W.H., 1947. Heart Mountain problem. Wyoming Geological Association 2nd Annual Field Conference Guidebook: 189-197.

Hall, W.B., and Walsh, T.H., 1977. Color oblique stereo aerial (COSA) photographs for planning rapid geologic reconnaissance and other types of geologic programs. Wyoming Geological Association 29th Annual Field Conference Guidebook: 751-760.

Harrison, J.V., and Falcon, N.L., 1938. An ancient landslip at Saidmarreh in southwestern Iran. Journal of Geology 46: 296-309.

Hay, R.L., 1954. Structural relationships of tuff-breccia in Absaroka Range, Wyoming. Geological Society of America Bulletin 65: 605-620.

________, 1956. Pitchfork Formation, detrital facies of Early Basic Breccia, Absaroka Range, Wyoming. American Association of Petroleum Geologists Bulletin 40: 1863-1898.

Hsü, K.J., 1969. Role of cohesive strength in the mechanisms of overthrust faulting and landsliding. Geological Society of America Bulletin 80: 927-952.

Kent, P.E., 1966. The transport mechanism in catastrophic rock falls. Journal of Geology 74: 79-83.

Krieger, M.H., 1977. Large landslides, composed of megabreccia, interbedded in Miocene basin deposits, southeastern Arizona. U.S. Geological Survey Professional Paper 1008: 25 pp.

Longwell, C.R., 1951. Megabreccia developed downslope from large faults. American Journal of Science 249: 343-355.

Love, J.D., 1934. The geology of the western end of the Owl Creek Mountains, Wyoming. Wyoming Geological Survey Bulletin 24:

________, 1939. Geology along the southern margin of the Absaroka Range, Wyoming. Geological Society of America Special Paper 20: 136 pp.

________, 1977. Summary of Upper Cretaceous and Cenozoic stratigraphy, and of tectonic and glacial events in Jackson Hole, northwestern Wyoming. Wyoming Geological Association 29th Annual Field Conference Guidebook: 585-593.

________, Leopold, E.B., and Love, D.W., 1978. Eocene rocks, fossils, and geologic history, Teton Range, northwestern Wyoming. U.S. Geological Survey Professional Paper 932-B: 40 pp.

Love, L.L., Kudo, A.M., and Love, D.W., 1976. Dacites of Bunsen Peak, the Birch Hills, and the Washakie Needles, northwestern Wyoming, and their relationship to the Absaroka volcanic field, Wyoming and Montana. Geological Society of America Bulletin 87: 1455-1462.

Mackin, J.H., 1937. Erosional history of the Bighorn Basin, Wyoming. Geological Society of America Bulletin 48: 813-894.

Nelson, W.H., Prostka, H.J., and Williams, F.E., 1980. Geology and mineral resources of the North Absaroka Wilderness and vicinity, Park County, Wyoming. U.S. Geological Survey Bulletin 1447: 101 pp.

Palmquist, R.C., 1978. Ash dated terrace sequence in the eastern portion of the Bighorn Basin, Wyoming. Geological Society of America Abstracts with Programs 10: 467.

Pierce, W.G., 1941. Heart Mountain and South Fork thrusts, Park County, Wyoming. American Association of Petroleum Geologists Bulletin 25: 2021-2045.

________, 1957. Heart Mountain and South Fork detachment thrusts of Wyoming. American Association of Petroleum Geologists Bulletin 41: 591-626.

________, 1963. Reef Creek detachment fault, northwestern Wyoming. Geological Society of America Bulletin 74: 1225-1236.

________, 1975. Principal features of the Heart Mountain fault and the mechanism problem. Wyoming Geological Association 27th Annual Field Conference Guidebook: 139-148.

________, 1978. Geologic Map of the Cody 1 x 2 degrees Quadrangle, Wyoming. U.S. Geological Survey Miscellaneous Investigations Map MF-963.

________, and Andrews, D.A., 1941. Geology and oil and coal resources of the region south of Cody, Park County, Wyoming. U.S. Geological Survey Bulletin 921-B: 99-180.

________, and Nelson, W.H., 1973. Crandall Conglomerate, an unusual stream deposit, and its relation to Heart Mountain faulting. Geological Society of America Bulletin 84: 2631-2644.

Rohrer, W.L., 1966. Geology of the Adam Weiss Peak Quadrangle, Hot Springs and Park Counties, Wyoming. U.S. Geological Survey Bulletin 1241-A: 39 pp.

________, and Leopold, E.B., 1963. Fenton Pass Formation (Pleistocene?), Bighorn Basin, Wyoming. U.S. Geological Survey Professional Paper 475-C: C45-C48.

Rouse, J.T., 1940. Structural and volcanic problems in the southern Absaroka Mountains, Wyoming. Geological Society of America Bulletin 51: 1413-1428.

Shreve, R.L., 1966. Sherman Landslide, Alaska. Science 154: 1639-1643.

________, 1968. The Blackhawk Landslide. Geological Society of America Special Paper 108: 47 pp.

Smith, R.B., and Christiansen, R.L., 1980. Yellowstone Park as a window on the earth's interior. Scientific American 242: 104-117.

Sundell, K.A., 1980. The geology of the North Fork of Owl Creek area, Absaroka Range, Hot Springs County, Wyoming. University of Wyoming unpublished M. Sci. Thesis, Laramie: 150 pp.

Voight, B., 1973. The mechanics of retrogressive block-gliding, with emphasis on the the evolution of the Turnagain Heights Landslide, Anchorage, Alaska. In DeJong, K.A., and Scholten, R., (Eds.), Gravity and Tectonics, New York, John Wiley & Sons, Interscience: 97-121.

Wilson, W.H., 1963. Correlation of volcanic rock units in the southern Absaroka Mountains, Wyoming. Contributions to Geology 2: 13-20.

________, 1964A. Geologic reconnaissance of the southern Absaroka Mountains, northwest Wyoming — Part I, the Wood River-Greybull River area. Contributions to Geology 3: 60-77.

________, 1964B. The Kirwin mineralized area, Park County, Wyoming. Wyoming Geological Survey Preliminary Report 2: 12 pp.

________, 1970. Geologic Map of the Soapy Dale Peak Quadrangle, Hot Springs County, Wyoming. Wyoming Geological Survey Map: 1 sheet, 1:24000.

________, 1975. Detachment faulting in volcanic rocks, Wood River area, Park County, Wyoming. Wyoming Geological Association 27th Annual Field Conference Guidebook: 167-171.

Thirty-Third Annual Field Conference — 1982
Wyoming Geological Association Guidebook

GEOMORPHIC ANALYSIS OF THE DEEP CREEK FAULT, UPPER YELLOWSTONE VALLEY, SOUTH-CENTRAL MONTANA

STEPHEN F. PERSONIUS[1]

INTRODUCTION

The Deep Creek fault bounds the western edge of the Beartooth uplift and forms the eastern margin of the upper Yellowstone (Paradise) Valley in south-central Montana (Fig. 1). This high-angle normal fault trends northeast for approximately 55 km and structurally separates the Gallatin and Beartooth foreland uplifts. Over 1500 m. of topographic relief exists between the valley floor and the uplifted Beartooth Plateau. Bonini and others (1972) used gravity data to estimate at least 5500 m of displacement along the central portion of the Deep Creek fault near Chico. The fault trace is characterized by scattered near-recent fault scarps[1] and is elsewhere buried by surficial debris or inferred by the presence of faceted spurs and other tectonic landforms.

This fault is one of several high-angle normal faults that form range fronts in south-central and southwest Montana. These northwest-to-northeast-trending structures characterize the extensional Basin and Range deformation which has occurred in this region since Middle (and possibly earlier) Tertiary time. Many of these faults (including the Deep Creek) are considered active today (Reynolds, 1979). This paper is an attempt to summarize the tectonic geomorphology of the eastern margin of the upper Yellowstone Valley, and in particular, the geomorphic expression of the near-recent fault scarps found along the Deep Creek fault. This data will then be applied to an interpretation of the near-recent history of movement along the fault.

FAULT GEOMETRY

The trace of the Deep Creek fault is generally quite straight, but close examination reveals that the fault shows several small-scale changes in geometry along its trace. *En echelon* breaks, bends, and splits or steps are present (Fig. 1). These elements are commonly expressed by other normal faults in the Basin and Range Province (Anderson, 1977). Most other range front normal faults in southwest Montana have geometries similar to the Deep Creek fault, although the various styles vary widely in scale. For instance, the faults bounding the Bridger and Madison ranges in southwest Montana have much larger scale bends than those seen along the Deep Creek trace. These geometric variations may be related to structural anisotropies in the underlying Precambrian basement. The Deep Creek fault trace parallels the strike of most Precambrian foliation along the mountain front, as well as several northeast-trending strike-slip faults mapped in the Beartooth uplift to the east. Reid and others (1975) mapped the Luccock Park fault as a normal fault superimposed on an older strike-slip fault, also trending northeast (Fig. 1). The bend in the Mill Creek area is probably related to intersection with the Mill Creek fault zone.

The dip of the Deep Creek fault is open to some speculation. Pardee (1950) described an excavated portion of the fault near Yankee Jim Canyon, in which a 50° west dip was measured. Bonini and others (1972) used a vertical dip to arrive at a best fit for their gravity data. The author has measured an approximate 60° west dip from a fault scarp exposed in an incised canyon 4 km northeast of Sixmile Creek. Recent Basin and Range structural studies such as that of Profett (1977) have determined that many range front normal faults are actually listric in shape; these faults stand at high angles (60° or greater) at the surface, and flatten gradually with depth. Seismic or deep drilling data will be required to accurately determine the dip of the Deep Creek fault.

RANGEFRONT GEOMORPHOLOGY

The west flank of the Beartooth uplift exhibits the rugged topography, great relief, and other geomorphic characteristics common to fault-bounded ranges in the western United States. Streams draining from this highland

[1]Department of Earth Sciences Montana State University, Bozeman, Montana 59717

into the Yellowstone Valley are deeply incised and have steep gradients, even where modified by glaciation. Foothills are absent, except where step faulting has created intermediate blocks between the valley bottom and range front. Faceted spurs are present, and are restricted to those areas not affected by piedmont glaciation. Detailed studies of the development of faceted spurs, and associated remnant pediments have been conducted in other parts of the Basin and Range Province, in particular along the Wasatch fault zone in northern and central Utah (Anderson, 1977;

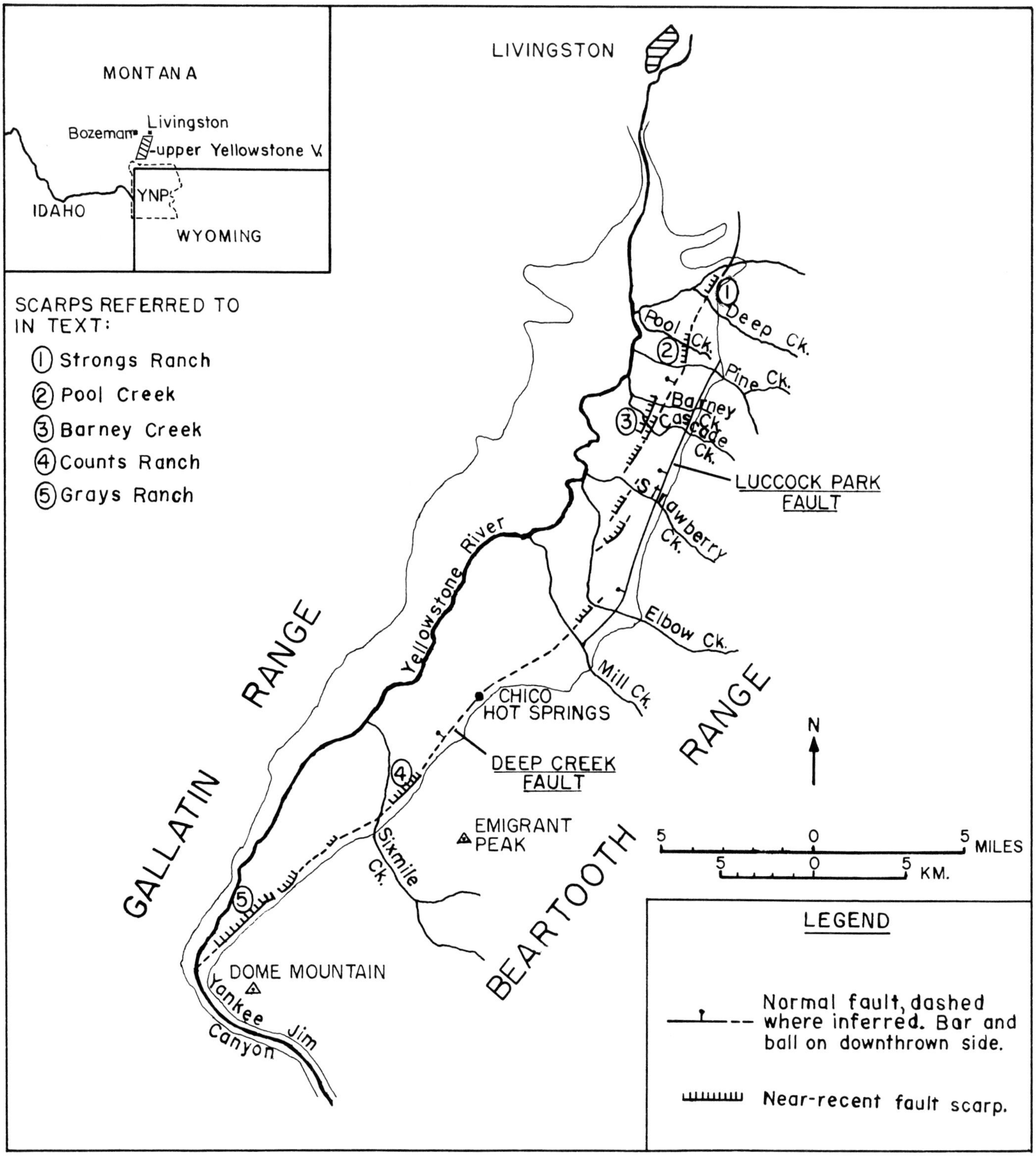

Figure 1: Map showing the outline of the upper Yellowstone Valley, trace of the Deep Creek and Luccock Park faults, and locations of near-recent fault scarps. Compiled from Montagne and Chadwick (1982, p. 4.), Reid and others (1975, fig. 1), and mapping by the author.

Hamblin, 1976; Osborne, 1978). These landforms are the geomorphic expression of recurrent uplift; it is widely believed that displacement on most normal faults occurs intermittently, with active periods of displacement separated by periods of quiescence of varying duration. Fault scarps and the faceted spurs developed from them are the result of active displacement and minor erosion between events. Remnant pediments are the result of backwasting of the mountain front during quiescent periods, with subsequent erosion and modification during the next phase of uplift. These relationships are not well expressed along the Deep Creek fault. Faceted spurs are preserved locally, but remnant pediments cannot be identified with any certainty along the mountain front. The upper 900 m of Emigrant Peak may exhibit remnants of these features above the glacial trim line.

There are several possible explanations for the absence of faceted spurs and pediments along the Deep Creek fault. 1) Glaciation has greatly modified the mountain front in many places. The probable existence of a huge, pre-Illinoisan ice sheet in the upper Yellowstone Valley (Montagne and Chadwick, 1982), may have severely altered any tectonic features along the mountain front. This disruption was compounded by the presence of extensive (though smaller) Pinedale piedmont ice. Glacial erosion affected the mountain front a minimum of 500 m above the valley floor; glacial erratics of piedmont origin have been found at this elevation (Horberg, 1940; Montagne and Chadwick, 1982). 2) The Precambrian crystalline rocks exposed in the footwall may be less susceptible to spur and pediment formation than the Paleozoic/Mesozoic sedimentary rocks exposed along other range fronts, such as the Wasatch Front. Osborne (1978) noticed that spur and pediment development was very subdued along areas where resistant units like the Tintic Quartzite were exposed along the Wasatch fault; this may be analogous to the resistant Precambrian rocks found in the Beartooth uplift. Lithologic resistance may also explain the lack of development along the fault-bound Teton Range in northwest Wyoming. 3) The Deep Creek fault may not be characterized by recurrent uplift. If fault displacement has been more or less continuous, the resultant range front geomorphology should be more uniform than that expected from recurrent uplift; and 4) a final possibility may be that the Deep Creek fault has been superimposed on an earlier structure, such as a pre-existing reverse or strike-slip fault, much like the Luccock Park fault (Reid and others, 1975). A combination of glacial erosion, lithologic resistance, and structural superposition may best explain the lack of spur and pediment development along the Deep Creek fault.

FAULT SCARP GEOMORPHOLOGY

Some of the most interesting geomorphic expressions of the Deep Creek fault are the near-recent fault scarps which occur near the base of the range front in several places along the fault trace. These scarps occur intermittently, and vary considerably in distance from the mountain front. Scarps of similar age appear from 600 m valleyward from the range front, to being coincident with the range front. Some range front "battering" has certainly occurred in those parts of the valley affected by piedmont ice (Montagne and Chadwick, 1982), but the magnitude of this erosion is unknown. The mountain front from the mouth of Yankee Jim Canyon to several kilometers north of Sixmile Creek has been severely battered by Pinedale and probably earlier piedmont ice sheets. The fault scarps located along this area post-date the last glaciation and are located from 50 to several hundred meters from the range front. The former figure may represent a minimum value of battering, if it is assumed that the fault zone and range front were coincident prior to glaciation. Regardless of the exact figure, glacial battering is surely a significant process in those valleys subject to piedmont glaciation.

Fault scarp lengths range from several hundred meters, to over 6 km. The shorter segments are found either where the trace breaks en echelon, or where the trace is curved. The longest segments are found where the trace is relatively straight.

Scarp Formation and Modification

The formation and modification of fault scarps in unconsolidated materials have been studied in detail by Nash (1980, 1981a, 1981b) and Wallace (1977, 1980) in other areas of the Basin and Range Province. Fault scarp morphology is initially a product of the mechanics of faulting, and the behavior of the faulted materials. Deep Creek fault scarps cut three general types of surficial materials: 1) glacial tills of various ages, 2) fanglomerates, and 3) rangefront colluvium. The texture of these materials is similar; they are very poorly sorted, with the coarse fractions dominated by angular cobbles and boulders of Precambrian crystalline rocks. For the purpose of simplification, the mechanical behavior of these materials appears to be similar enough to disregard material variability for scarp analysis. An initial scarp slope angle of approximately 60° is assumed; this figure appears to be a reasonable estimate based on theoretical and laboratory results, although higher figures are often seen in the field. These higher scarp angles usually result from tensional fracturing of surficial materials and not from direct movement on a faulted surface (Wallace, 1977).

Figure 2: Photograph of basal graben preserved along the Counts Ranch fault scarp. The prominent conifer in the right foreground is growing in the bottom of the graben, and the fault scarp trends off the photo to the left. Wooden staff (arrow) is 1.5 m long.

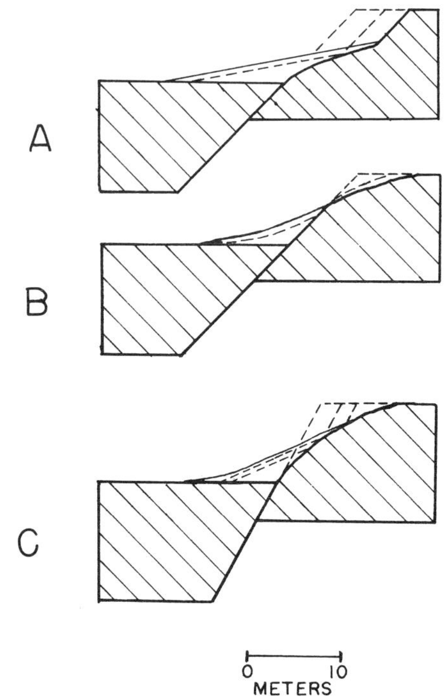

Figure 3: Diagram illustrating the erosional modification of a normal fault scarp. A, initial slope retreat phase; scarp retreats, forming a debris slope at the angle of repose (approximately 35° in Deep Creek fault materials). B, final slope rounding phase; crest and toe areas become progressively more rounded through time. C, summary diagram, showing slope retreat followed by slope rounding. After Nash (1981b, p. 8, 10, 11).

Basal moats or graben commonly form at the base of normal fault scarps. These features represent subsidence associated with secondary fracturing in the hanging wall block (Witkind, 1964) and may vary from a meter to tens of meters in width, and a meter or two in depth (Wallace, 1977). The graben will usually be rapidly filled in and covered (Nash, 1981b) but the larger ones may be quite persistent. Several of the scarps along the Deep Creek fault have basal graben (Fig. 2); these vary from 3 to 5 m in width, and .2 to .5 m in depth. These graben are most often found where scarps traverse upslope; they are apparently being modified as drainage channels, which may explain why they are still present. In relatively flat areas, they act as catch basins for water, and usually contain more luxuriant plant growth than their surroundings. Because of these modifications, profiles of these scarp areas should be avoided for geometric and analytical analysis. However, the graben may contain datable sediments, if drainage modification hasn't progressed too far.

The relationship between actual location of bedrock faults and fault scarps in overlying surficial materials is rarely well defined. With the possible exception of the excavated scarp previously described by Pardee (1950) (which is no longer exposed), the surface scarp/bedrock fault relationship is not well expressed along the Deep Creek fault. Witkind (1964), in his study of faults reactivated during the 1959 Hebgen Lake earthquake concluded that most surface scarps there coincided closely with bedrock faulting. Some scarps did occur upslope from the projected bedrock fault, although they did parallel the concealed fault trace. The relationship is best expressed on the Deep Creek fault just north of Elbow Creek, where an *en echelon* fault scarp coincides with the Precambrian/Quaternary contact and a steep break in slope. This information at least tentatively points to a close correlation between surface expression (fault scarps) and bedrock faulting. Shallow geophysical investigations would be required to confirm this hypothesis.

Erosional modification of normal fault scarps must be thoroughly understood before accurate dating can be attempted. The initial scarp slope angle is usually 60° or greater, which is generally higher than the angle of repose of most surficial materials. Debris sloughs off the scarp facing, causing the slope face to retreat back. The sloughed material collects at the base of the scarp and forms a debris slope at the angle of repose (Fig. 3A). This process was called "slope replacement" by Wallace (1977), and "slope retreat" by Nash (1981a). The retreat or replacement phase may vary from hundreds to several thousand years (Wallace, 1977) in arid climates, to several decades

(Wallace, 1980) in more humid regions. At the end of this phase, the slope face of the scarp will consist of a debris slope standing at the angle of repose of the underlying material (Nash, 1981b). The final stage of scarp modification was referred to as "slope decline" by Wallace (1977) and as "slope rounding" by Nash (1981a). After completion of the slope retreat phase, the gradient of the scarp slope decreases, and the crest and toe sections of the scarp become progressively more rounded (Fig. 3B). Figure 3C summarizes the complete modification of a normal fault scarp. All the scarps along the Deep Creek fault are believed to have completed the slope retreat phase, and are now undergoing slope rounding.

Methods

In order to quantify the description of fault scarps, accurate profiles must be measured across them. Although other methods are available (Bucknam and Anderson, 1979), in this study a Brunton compass, jacobs staff, and measuring tape were used to measure variations in slope across the scarps. The staff was placed flat on the slope, and segments with similar slope angle were measured by placing the compass directly on the staff. The profile of a typical fault scarp, and the parameters used to define it are shown in Figure 4.

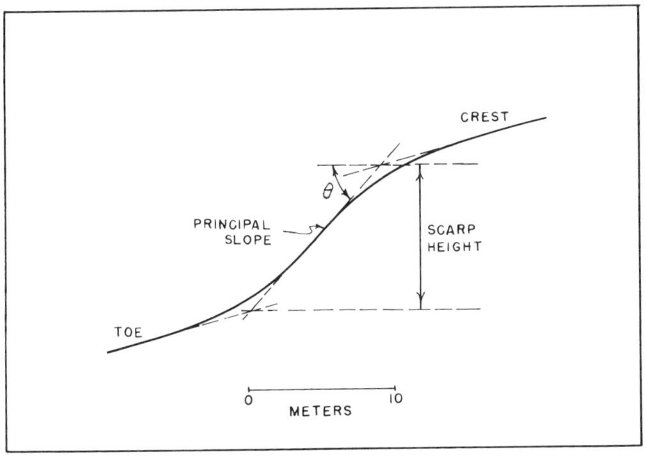

Figure 4: Diagram of a typical normal fault scarp, showing toe and crest segments, principal slope and principal slope angle (Θ), and scarp height. Modified from Bucknam and Anderson (1979, p. 13).

Field Analysis

The Deep Creek fault scarps were initially examined in the field with standard geomorphic techniques, including degree of dissection, freshness of morphology, and cross-cutting relationships.

The presence of several ages of glacial material in the upper Yellowstone Valley was particularly helpful in defining general ages for several fault scarps. The Pool Creek scarp (Fig. 5) apparently cuts Bull Lake lateral moraine (Montagne, 1982) but is buried by the adjacent Pinedale moraine. This relationship suggests a date of movement less than 150,000 years, and greater than 20,000 to 30,000 years. These dates are a bit speculative, because they are based on dated Bull Lake and Pinedale moraines in the West Yellowstone basin (Pierce, 1979; Pierce and others, 1976).

Figure 5: Photograph of the Pool Creek scarp, exposed across the central portion of the photo. View looking east; note the generally subdued expression of the scarp.

The Barney Creek, Counts Ranch, and Grays Ranch fault scarps all cut Pinedale age or younger surficial materials, and therefore are assumed to postdate the 13,000 to 14,000 year date obtained by Pierce (1979) for Pinedale deglaciation.

A third scarp age may be represented by the Strongs Ranch scarp (Fig. 6). This feature does not cut datable

Figure 6: Photographs of Strongs Ranch scarp, exposed in center of the photo. View looking east-southeast; again note the subdued expression on the scarp.

glacial material, but does show a subdued expression which probably indicates an age older than the Pool Creek scarp.

These types of observations are useful for relative-age dating, but additional information is required for establishment of absolute ages of faulting. The Grays Ranch scarp appears to offer the best chance of determining an absolute age from field relationships. This scarp is over 6 km long and represents the southernmost surface expression of the Deep Creek fault. The scarp cuts a large alluvial fan at the foot of Dome Mountain, 3.6 km north of Yankee Jim Canyon, but is buried by a smaller fan 750 m to the south (Fig. 7). This scarp/fan relationship is significant for several reasons: 1) the fans occur in an area subjected to severe glacial erosion by the Pinedale piedmont ice sheet, so they must be post-Pinedale in age (less than 14,000 yr); 2) from close examination of soil development, lichen growth, and general surface expression, these fans appear to be very similar in age; 3) the fans are steep (9-11° in the central portion of the fan), and have been deposited at the mouths of canyons which appear underfit with respect to the size of the fans; and 4) the fans appear to have been inactive for at least several thousand years. This evidence points to the conclusion that both fans were deposited by periglacial processes in late Quaternary/early Holocene time, immediately following Pinedale deglaciation. Ryder (1971) studied periglacial fans in southern British Columbia: in this study he described several unique characteristics of periglacial fans: 1) the fans were usually steep, and 2) fan size was not correlative with basin size. His data helps substantiate the association of the Dome Mountain fans with post-Pinedale periglacial processes. The duration of this post-glacial phase is open to some speculation, but these conditions probably didn't persist for more than a couple of thousand years after deglaciation.

Figure 7: Photograph of Grays Ranch scarp, view looking north. Scarp cuts upper portion of large fan in center of photo, and is then buried by smaller fan to south. Note the glacially battered mountain front upslope from fans.

Figure 8: Photograph of southernmost scarp in Barney Creek area, exposed in central portion of photo. View looking east; note relatively fresh expression of the scarp.

A late Quaternary/early Holocene age is also indicated for the Barney Creek scarps (Fig. 8), where association with periglacial landforms is also evident. In this case, the scarps cut a large alluvial fan at the mouth of Barney Creek, but are buried by debris flows along the mountain front and boulder torrent deposits (Montagne and Chadwick, 1982) along Cascade Creek. These landforms appear to be related to periglacial conditions. An age date of 12,000±1000 years is therefore thought to be a reasonable best-guess estimate for the Grays Ranch and Barney Creek scarps.

Closely spaced cross-cutting relationships of the Grays Ranch type are not associated with the Pool Creek, Strongs Ranch, and Counts Ranch scarps. The Counts Ranch scarp exhibits surface morphology very similar to the Grays Ranch and Barney Creek scarps, and is thought to be very similar in age. The Pool Creek and Strongs Ranch scarps are probably much older, perhaps as much as 100,000 years. It is probable that the hillslope processes (especially solifluction) active in the Pleistocene would have destroyed any scarps older than the Bull Lake Glaciation. Determination of recurrence intervals of major faulting must await absolute dating of these older features.

Analytical Techniques

Several recent studies have attempted to date fault scarps through analytical analysis of surface morphology. Wallace (1977) examined fault scarps in north-central Nevada, and devised several dating methods for dating scarp profiles. The two techniques that appeared most useful for analysis of the Deep Creek profiles were: 1) comparison of the width of break in slope at the crest of the scarp, and 2) comparison of the angle of the principal slope, or straight segment which dominates on the face of the scarp.

Bucknam and Anderson (1979) used statistical analysis to define a strong relationship between scarp slope (principal slope) angle, scarp height, and scarp age. This analysis indicated that the scarp slope/scope height relationship could be approximated by a logarithmic curve, and that regression lines and equations fit to this data can be used as relative-dating tools. Scarps of comparable age should have similar regression lines and equations; in addition, lines of older scarps should plot relatively below lines of younger scarps. They concluded (as did Wallace) that scarp angles decrease through time, and that for scarps with similar initial angle but dissimilar height, the lower scarp will degrade faster.

Nash (1980, 1981a, 1981b) has derived a quantitative model for morphologic dating of fault scarps, in which the erosional degradation of fault scarps was modeled with several computer programs. These models quantify the relationship between original and present scarp angles, scarp height, an empirically derived coefficient of diffusion (essentially an erosion factor), and the age of the scarp.

These techniques all have their own inherent limitations, especially when an attempt is made to apply them to other geographic areas where hillslope processes and climate may be significantly different.

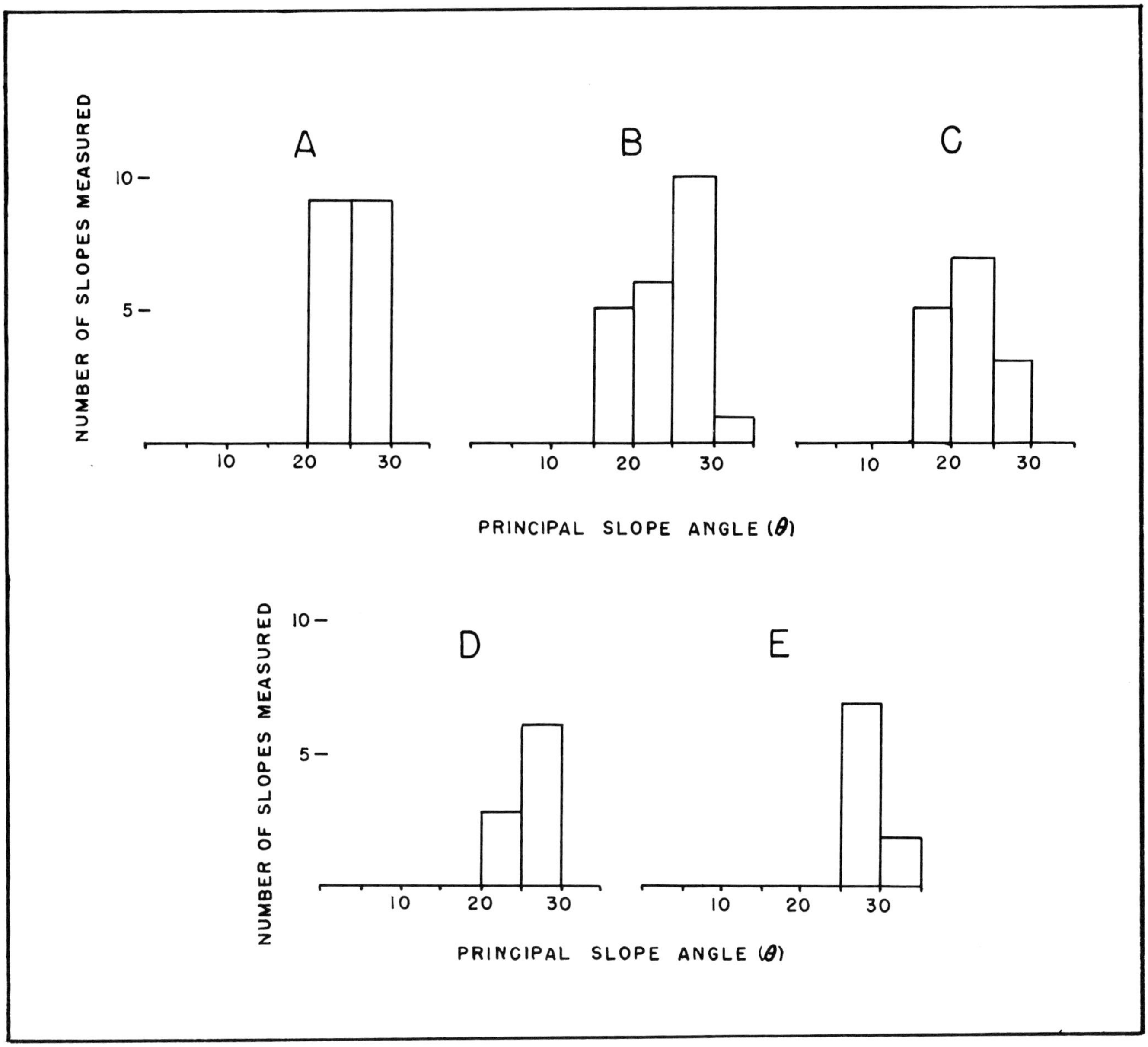

Figure 9: Histograms of principal slope angles of selected Deep Creek fault scarps. A, Grays Ranch; B, Barney Creek; C, Counts Ranch; D, Strongs Ranch; and E, Pool Creek.

Deep Creek Applications

An attempt was made to apply the dating methods described above to the profile data measured along the Deep Creek fault. This analysis met with only limited success, because of problems of limited scarp preservation, scarp modification by slumping, soil creep, and stream dissection, and variations in original slope.

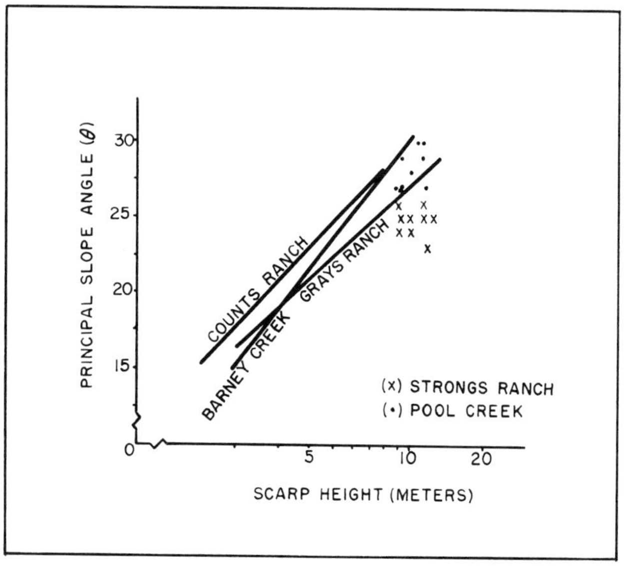

Figure 10: Regression lines fit to principal slope angle and scarp height measurements of Deep Creek fault scarps. Pool Creek and Strongs Ranch data insufficient to fit accurate regression lines.

The comparison of crestal scarp widths proved inconclusive; the wide variation in original slope and limited preservation of the older scarp invalidated this technique in the upper Yellowstone Valley. Comparison of principal slope angles was also inconclusive, because slope angles were clustered in the 25°-28° range for all scarps measured (Fig. 9). This may be a result of small increments of displacement (less than a meter) which would tend to maintain a relatively uniform slope (R. E. Wallace, pers. comm., 1981) or may represent some sort of quasi-stable slope for this environment.

Regression analysis of the scarp profile data proved to be more informative (Fig. 10, and Table 1). The Counts Ranch, Barney Creek, and Grays Ranch regression lines cluster quite closely and substantiate the conclusion that these scarps are of similar age. The dissection of the Strongs Ranch and Pool Creek scarps hindered acquisition of sufficient data to conclusively fit regression lines, but at least for the Strongs Ranch scarp, an older trend is indicated (Fig. 10). The position of the Pool Creek data may indicate that this scarp is more closely associated with the age of the Grays Ranch scarp, perhaps 20,000 to 50,000 years. Additional profiling may help to better define this relationship.

The application of Nash's morphological model is hampered by several problems: 1) the coefficent of diffusion is a constant, so that climatic variations during the Pleistocene probably restrict use of the model to the Holocene; 2) the model is designed for analysis of scarps

Table 1: Statistical analysis of Deep Creek fault scarp data. Note close correlation of Grays Ranch, Barney Creek, and Counts Ranch regression equations.

Scarp Location	*Regression Equation*	*R^2 (%)*	*SD (0)*	*Number of Measurements*
Grays Ranch	$\theta = 11.66 + 15.28 \log H$	23	2.3	18
Barney Creek	$\theta = 8.60 + 21.59 \log H$	70	4.6	22
Counts Ranch	$\theta = 10.55 + 21.26 \log H$	41	3.1	16
Strongs Ranch	———————	—	1.0	9
Pool Creek	———————	—	1.3	9

NOTE:
R^2 — coefficient of determination
SD — standard deviation of measured (θ)
H — scarp height in meters
θ — principal slope angle in degrees

with horizontal toe and crest segments (unfortunately the Deep Creek fault scarps all cut slopes ranging from 4° to over 15°); and 3) the model is designed for analysis of scarps in cohesionless materials, and may not adequately explain the behavior of surficial material along the Deep Creek fault. Preliminary results indicate that these problems will give a date substantially younger than that derived from other criteria (4000 years too young for the Grays Ranch scarp, using a coefficient of diffusion from the Hebgen Lake area). The model therefore needs to be modified to address these problems before accurate results can be obtained (D. B. Nash, pers. comm., 1982).

The history of displacement which produced these scarps will have an effect on the results of all these analytical techniques. If recurrent movement produces multifaceted scarps, erosional modification will not be constant; this will essentially "reset" the degradational sequence and modify the pre-existing scarp. All the scarps in the upper Yellowstone Valley are remarkably uniform in height along their lengths, with a notable decrease only at the preserved distal ends. Fault movements separated by long periods of quiescence should produce at least some multiple scarps, of which no clear-cut examples have been found. The Deep Creek fault scarps are therefore thought to be products of high intensity, short duration earthquake events, which are separated by essentially quiet periods of much longer duration. The heights of some scarps (10-12 m) possibly point to several periods of movement, although single-event displacements as great as 6.7 m were measured from historic faulting in the Hebgen Lake area (Witkind, 1964).

The scarp-forming events have been distributed intermittently along the trace of the Deep Creek fault, much like the distribution of recent seismic activity along the Wasatch fault zone (Smith and Sbar, 1974). This may imply that areas lacking fault scarp development are just as prone to earthquake occurrence as the areas with scarp development and perhaps even more susceptible.

SUMMARY AND CONCLUSIONS

The Deep Creek fault is a high-angle normal fault bounding the western flank of the Beartooth uplift. The fault trace is generally straight, with several small-scale bends and *en echelon* breaks. Near-recent fault scarps exposed along the trace express three major periods of recent faulting in the upper Yellowstone Valley. A fairly well documented period about 12,000 years BP was preceded by two poorly expressed periods of faulting; the Pool Creek scarp may have formed 20,000 to 50,000 years BP, and the Strongs Ranch scarp could be much older, perhaps more than 100,000 years BP.

The results of this study have shown that the near-recent paleoseismic history of an area may be reconstructed through fault-scarp analysis. In a general way, this reconstruction has documented the recurrence intervals of major faulting events in the near-recent history of the upper Yellowstone Valley. Readers should be cautioned against applying the results of this study to other geographic areas, because fault scarp morphology is strongly controlled by climate, aspect, and other local factors, as well as age. The author hopes the results of this study will be of use to others working with fault-scarp analysis.

ACKNOWLEDGEMENTS

The author wishes to thank Dr. John Montagne for sharing his knowledge of the upper Yellowstone, Dr. D. L. Smith for reviewing the manuscript, and Bruce Tremper for assistance with the statistical analysis.

REFERENCES CITED

Anderson, T. C., 1977, Compound faceted spurs and recurrent movement in the Wasatch fault zone, north central Utah: Brigham Young Univ. Geol. St., v. 24, pt. 2, p. 83-101.

Bonini, W. E., Kelley, W. N., Jr., and Hughes, D. W., 1972, Gravity studies of the Crazy Mountains and the west flank of the Beartooth Mountains, Montana: Guidebook, Montana Geological Society, 21st Annual Field Conference, Crazy Mountains Basin, p. 119-127.

Bucknam, R. C., and Anderson, R. E., 1979, Estimation of fault-scarp ages from a scarp-height-slope angle relationship: Geology, v. 7, p. 11-14.

Hamblin, W. K., 1976, Patterns of displacement along the Wasatch fault: Geology, v. 4, p. 619-622.

Horberg, Leland, 1940, Geomorphic problems and glacial geology of the Yellowstone Valley, Park County, Montana: Journal of Geology, v. 48, no. 3, p. 275-303.

Montagne, John, and Chadwick, R. A., 1982, Cenozoic history of the Yellowstone Valley south of Livingston, Montana: Geological Society of America, Rocky Mountain Section Guidebook, Field Trip no. 4, 67 p.

Nash, D. B., 1980, Morphologic dating of degraded normal fault scarps: Journal of Geology, v. 88, p. 353-360.

Nash, D. B., 1981a, FAULT: A FORTRAN program for modeling the degradation of active normal fault scarps: Computers and Geosciences, v. 7, p. 249-266.

Nash, D. B., 1981b, Fault scarp morphology: Indicator of paleoseismic chronology: Final Technical Report to U.S. Geological Survey, 132 p.

Osborne, S. D., 1978, Late Cenozoic movement on the central Wasatch fault, Utah: Brigham Young Univ. Geol. St., v. 25, pt. 2, p. 99-115.

Pardee, J. T., 1950, Late Cenozoic block faulting in western Montana: Geological Society of America Bulletin, v. 61, p. 359-406.

Pierce, K. L., 1979, History and dynamics of glaciation in the northern Yellowstone National Park area: U.S. Geological Survey Professional Paper 729-F, 90 p.

Pierce, K. L., Obradovich, J. D., and Friedman, Irving, 1976, Obsidian hydration dating and correlation of Bull Lake and Pinedale Glaciations near West Yellowstone, Montana: Geological Society of America Bulletin, v. 87, p. 703-710.

Proffett, J. M., Jr., 1977, Cenozoic geology of the Yerington district, Nevada, and implications for the nature and origin of Basin and Range faulting: Geological Society of America Bulletin, v. 88, p. 247-266.

Reid, R. R., McMannis, W. J., and Palmquist, J. C., 1975, Precambrian geology of the North Snowy block, Beartooth Mountains, Montana: Geological Society of America Special Paper 157, 135 p.

Reynolds, M. W., 1979, Character and extent of basin and range faulting, western Montana and east-central Idaho: RMAG-UGA 1979 Basin and Range Symposium, p. 185-193.

Ryder, J. M., 1971, Some aspects of the morphometry of periglacial alluvial fans in south-central British Columbia: Canadian Journal Earth Sciences, v. 8, p. 1252-1264.

Smith, R. B., and Sbar, M. L., 1974, Contemporary tectonics and seismicity of the western United States with emphasis on the intermountain seismic belt: Geological Society of America Bulletin v. 85, p. 1205-1218.

Wallace, R. E., 1977, Profiles and ages of young fault scarps, north-central Nevada: Geological Society of America Bulletin, v. 88, p. 1267-1281.

Wallace, R. E., 1980, Degradation of the Hebgen Lake fault scarps of 1959: Geology, v. 8, p. 225-229.

Witkind, I. J., 1964, Reactivated faults north of Hebgen Lake, *in* the Hebgen Lake, Montana, earthquake of August 17, 1959: U.S. Geological Survey Professional Paper 435, p. 37-50.

Thirty-Third Annual Field Conference — 1982
Wyoming Geological Association Guidebook

THE MADISON MYLONITE ZONE: A MAJOR SHEAR ZONE IN THE ARCHEAN BASEMENT OF SOUTHWESTERN MONTANA

ERIC A. ERSLEV[1]

ABSTRACT

The Madison Mylonite Zone is a major shear zone of probable middle Proterozoic age discontinuously exposed in the Archean basement of southwestern Montana. This northeast-trending, northwest-dipping zone of reverse faulting coincides with a profound topographic and geologic lineament extending from the Red Rock Lakes to Livingston.

The only full section through the Madison Mylonite Zone is in the southern Madison Range where the 3-km-thick shear zone is centered in a 10-km-thick aureole affected by retrograde metamorphism to epidote-amphibolite facies. The shear zone contains partially recrystallized ribbon quartz, plagioclase clasts, shredded micas, garnet, chloritoid, and blue-green hornblende. Garnet-biotite geothermometry yields temperatures of 500+50°C (+10°). Reverse faulting in the shear zone is indicated by sub-horizontal fold axes, down-dip mineral aggregate lineations, and large drag folds. The high temperature of the event combined with the lack of tectonic breccias or disrupted stratigraphy suggests entirely ductile deformation.

In the northwestern side of the Beartooth Mountains, a continuation of the Madison Mylonite Zone is offset by the Mill Creek-Stillwater fault zone. Normal faults utilizing the foliation of the shear zone have resulted in the burial of the northwestern side of the shear zone. Homogeneous muscovite-plagioclase-quartz mylonite and fine-grained amphibolite grade into porphyroclastic tonalitic gneiss at the southeastern contact of the shear zone.

The Madison Mylonite Zone parallels the Mullen Creek-Nash Fork Shear Zone of Wyoming and trends toward the boundary between the Churchill and Superior Provinces. However, it is not a major suture because Archean metasediments can be correlated across the shear zone. It may parallel a Hudsonian suture in the subsurface northwest of the exposures of Archean basement in Montana.

INTRODUCTION

The Archean basement of southwestern Montana and adjacent parts of Wyoming and Idaho is a complex metamorphic and igneous terrane which has puzzled geologists for nearly a century (James and Hedge, 1980). The discontinuous exposures of the basement complex are often in remote and precipitous mountain ranges resulting in an incomplete mapping coverage which in turn has hindered regional correlations. The intense metamorphic and accompanying deformational events undergone by these rocks has attentuated and recrystallized original depositional structures and unit contacts.

In the western part of the Archean terrane, the current controversy concerning the relative age of the "Cherry Creek," "Pon," and "pre-Cherry Creek" lithostratigraphic sequences (James and Hedge, 1980) started when Runner and Thomas (1928) recognized the ambiguity of the geologic relationships in the Gravelly Range where Peale (1896) subdivided the basement into older gnessic rocks and

[1]Lafayette College
Easton, PA 18042

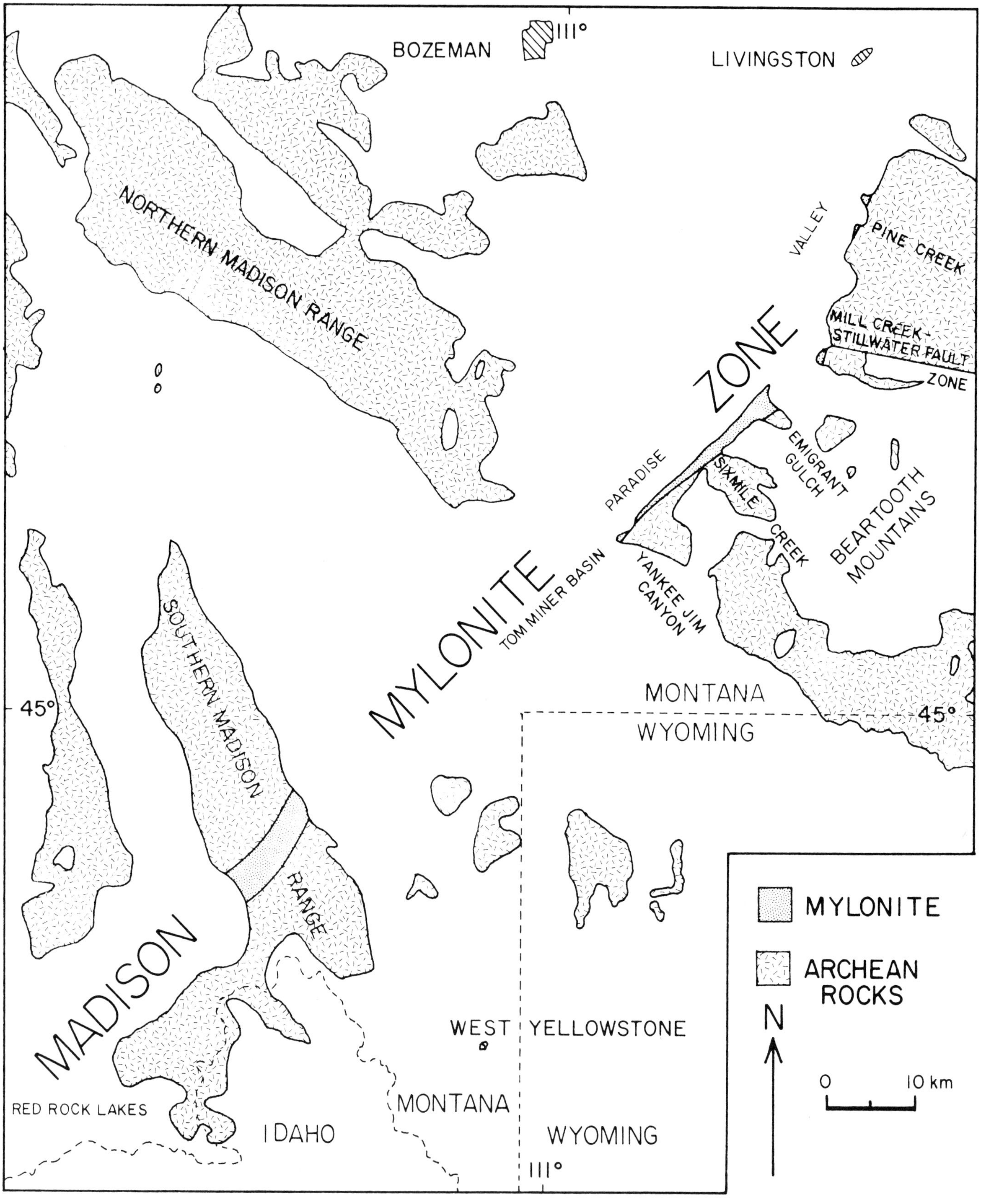

Figure 1: Exposures of the Madison Mylonite Zone in the Archean basement of southwestern Montana and adjoining parts of Wyoming and Idaho.

younger metasedimentary rocks. I have proposed a resolution to the controversy on the basis of metamorphic and structural relationships in the southern Madison Range (Erslev, 1980), but I cannot claim a general adherence to my hypothesis (cf., Wilson, 1981).

Even within well-mapped regions such as the Tobacco Root Mountains (Vitaliano et al., 1979) and the Ruby Range (Heinrich, 1960, Garihan, 1979), geologic relationships are ambiguous. No metamorphic isograds have been published to date despite the plethera of metamorphic petrologists working on the basement rocks. Units with a definite sequence and stratigraphy in the southern Madison Range and Gravelly Range (Erslev, 1981) are discontinuous in the Tobacco Root Mountains and Ruby Range. These relationships suggest an additional complication such as the disruption and offset of units on syntectonic and post-tectonic shear zones which may now be recrystallized beyond recognition.

Geological mapping in the southern Madison Range during the last five years has revealed a previously unknown, major shear zone, the Madison Mylonite Zone (Figure 1). The Madison Mylonite Zone is well exposed along the north side of Earthquake Lake where the Madison River was dammed by the Madison Landslide during the Hebgen Lake earthquake of August 17, 1959. Units within 5 kilometers of the 3-kilometer-thick zone of mylonites have been rotated and retrograded to epidote-amphibolite facies assemblages. Extensive petrography, mineral chemistry and mapping of the Madison Mylonite Zone in the southern Madison Range allowed the successful prediction of a continuation of this shear zone in the northwestern side of the Beartooth Mountains. Reconnaissance field studies and petrography have revealed that most of the northwestern margin of the Beartooth Mountains adjacent to the Yellowstone River consists of epidote-amphibolite facies mylonites.

The Madison Mylonite Zone is clearly a major structural feature of the pre-Beltian basement of southwestern Montana. This paper is a preliminary report on the structure, petrology, and possible tectonic consequences of the Madison Mylonite Zone. Similar shear zones of lesser width may be abundant throughout southwestern Montana and should provide additional constraints on the tectonic evolution of the Archean basement.

Madison Mylonite Zone in the Southern Madison Range

During the summer of 1977, geological mapping in the southern Madison Range revealed a gradational transition between granitic augen gneiss and dark-gray mylonite gneiss on the divide between the Rock Creek and Beaver Creek drainages. Further mapping delineated a northeasterly-striking, northwesterly-dipping slab of mylonite gneiss and phyllonite approximately three kilometers in thickness (Erslev, 1981). The best exposure of the mylonites is on the north wall of the Madison River canyon west of Hebgen Lake where the Madison River cuts through the narrow spine of the Madison Range. This area should be considered the type locality of the Madison Mylonite Zone for it is the only place where the full thickness of the shear zone is exposed.

The Madison Mylonite Zone contains the contact between a quartzofeldspathic gneiss terrane to the north and a metasedimentary terrane containing marbles and schists of the Cherry Creek Metamorphic Suite to the south (Erslev, 1981). The simplified block diagram in Figure 2 shows the generalized geologic patterns in the southern Madison Range. Petrologic and structural analysis of these dissimilar terranes has revealed that the northern domain has undergone an additional high-grade metamorphism and deformation not present in the lower-grade units of the Cherry Creek Metamorphic Suite (Erslev, 1981). The gneisses of the northern terrane contain relict granulite-facies assemblages including bronzite+hercynitic spinel+hornblende and salite+bronzite+garnet+hornblende+titaniferous biotite+feldspar, whereas the dolomitic marbles to the south contain dolomite+quartz — bearing assemblages. In addition, the highly variable fold orientations and styles in the gnessic terrane to the north reflect an early deformation absent in the metasedimentary terrane to the south. Thus, the Madison Mylonite Zone contains the contact between two distinct lithostratigraphic sequences of different ages. The shear zone may have been initiated at this contact due to the ductility contact between the two sequences.

Mylonite Petrography

The Madison Mylonite Zone is composed of fine-grained mylonite gneiss and phyllonite of the epidote-amphibolite facies. The northwestern side of the shear zone largely consists of dark-gray, quartz-muscovite-biotite-plagioclase mylonite gneiss and porphyroclastic epidote-amphibolite. The granitic, tonalitic, and amphibolitic gneisses to the north are the probable protolith of these high-temperature mylonites. The southeastern side of the Madison Mylonite Zone contains phyllonitic muscovite schists, chlorite-hornblende schists, finely-banded quartzite and a slabby dolomite-quartz marble.

The rocks within the Madison Mylonite Zone are slabby, brittle, and well-jointed. In marbles, quartz was recrystallized into grains an order of magnitude smaller than in marbles outside the shear zone. Ribbon quartz with a preferred orientation is common in feldspathic layers. Micas are shredded and wrapped around more equant grains of quartz, plagioclase, and epidote. Plagioclase and amphiboles form fractured porphroclasts rimmed by overgrowths and coronas. Garnets are uncommon; but where they exist, they often form ameboid porphyroblasts enclosing quartz of a finer grain size than the quartz outside the garnets.

Units within five kilometers of the Madison Mylonite Zone have undergone retrograde metamorphism to epidote-amphibolite facies due to fluids circulating in the shear zone. Mineral assemblages containing chloritoid, chlorite, hornblende and garnet are consistent with temperatures of equilibration of 500 ± 50°C (four specimens ± one standard deviation) (Erslev, 1981) calculated from Fe-Mg partitioning in garnet-biotite pairs using the calibration of Thompson (1976).

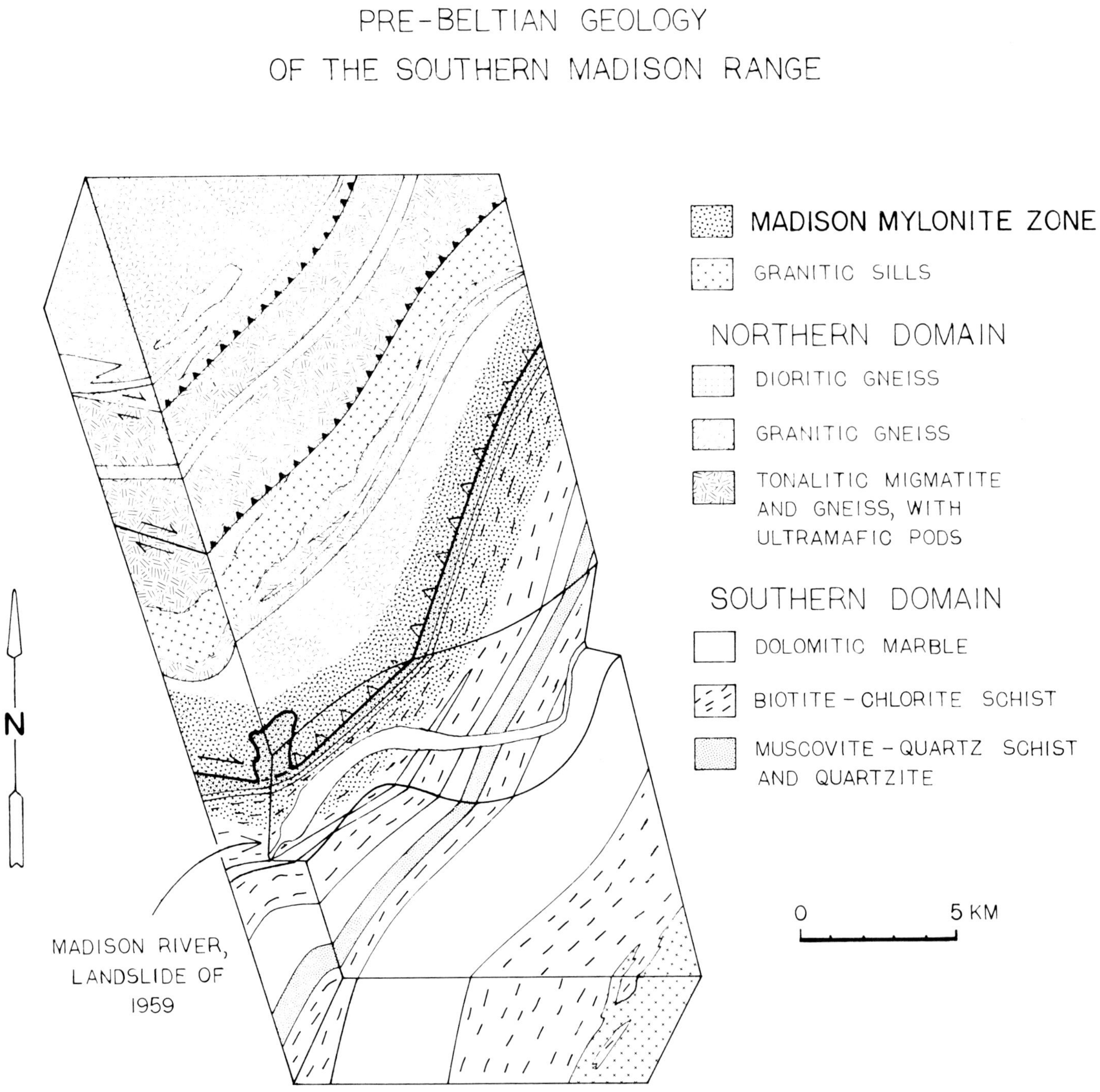

Figure 2: Simplified block diagram of the pre-Beltian (Archean) metamorphic rocks in the southern Madison Range.

STRUCTURAL GEOLOGY

The fabric diagrams in Figure 3 show that the foliation in the Madison Mylonite Zone dips moderately to the northwest. The sigmoidal map pattern of the shear zone shown schematically in Figue 2 may be the result of the later tectonism or an original, non-planar geometry. Fold axes are dominantly sub-horizontal and parallel the strike of the mylonite zone. The down-dip lineations are defined by elongate aggregates of minerals and may represent a streaking lineation parallel to tectonic transport.

The great thickness of this shear zone is the result of widely distribued strain during high temperature ductile faulting. A limited displacement on the Madison Mylonite Zone is indicated by the excellent lithostratigraphic correlation of the Cherry Creek Metamorphic Suite *south* of the Madison Mylonite Zone in the Madison Range with the type locality *north* of the Madison Mylonite Zone in the Gavelly Range (Erslev, 1981). The postulated dominance of ductile shearing without through-going faults is supported by the lack of disruption of thin marble and schist units within the shear zone and the apparent lack of tectonic breccias.

The Madison Mylonite Zone appears to be a zone of reverse faulting which thrust the gneissic terrane over the metasediments to the south. The mineral aggregate lineations and fold axes are consistent with this displacement. The steeply-dipping foliations of the northern and southern terranes are sharply flexed into concordance with the shear zone (Figure 4). The flexures appear to represent large-scale drag folds whose sense of shear supports the hypothesis of reverse faulting in the Madison Mylonite Zone.

Extension of the Madison Mylonite Zone

The thickness of the Madison Mylonite Zone in the southern Madison Range suggests that it is a major fault zone which should be exposed somewhere else in the Archean basement of southwestern Montana and adjoining areas. However, as Figure 1 shows, the exposure of Precambrian rocks in this area is discontinuous, with the trend of the Madison Mylonite Zone corresponding to a braod swath of Phanerozoic rocks.

Southwestern Extension

There is a 13-km-wide band of Tertiary volcanic rocks exposed along the extrapolated trend of the Madison Mylonite Zone to the southwest in the vicinity of Cliff and Hidden Lakes. Archean crystalline rocks of the Cherry Creek Metamorphic Suite are exposed to the southeast, but

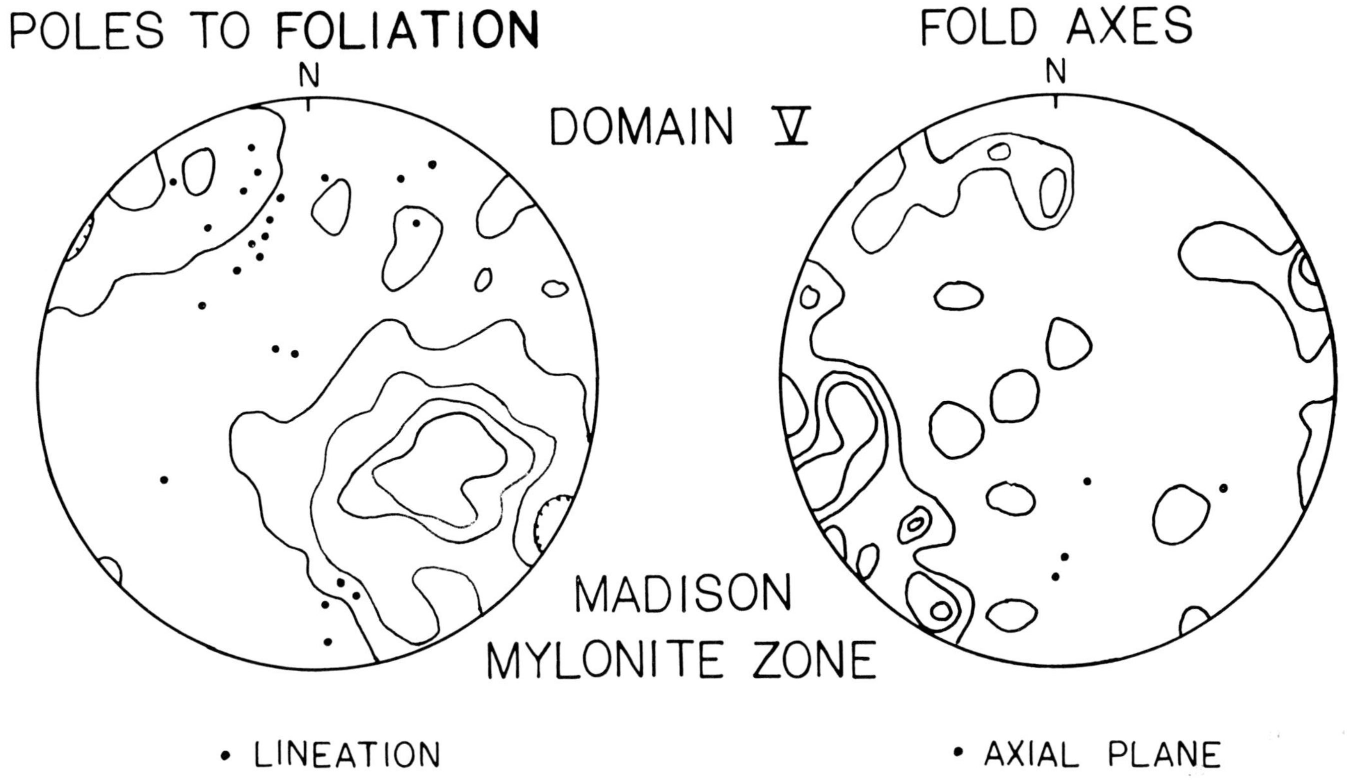

Figure 3: Fabric diagrams for the Madison Mylonite Zone in the southern Madison Range. 209 poles to foliation are contoured at intervals of .5, 2, 4, and 6% per 1% area. 48 poles to foliation are contoured at intervals of 1, 3, and 5% per 1% area. Mineral aggregate lineations and poles to axial planes of folds are plotted directly.

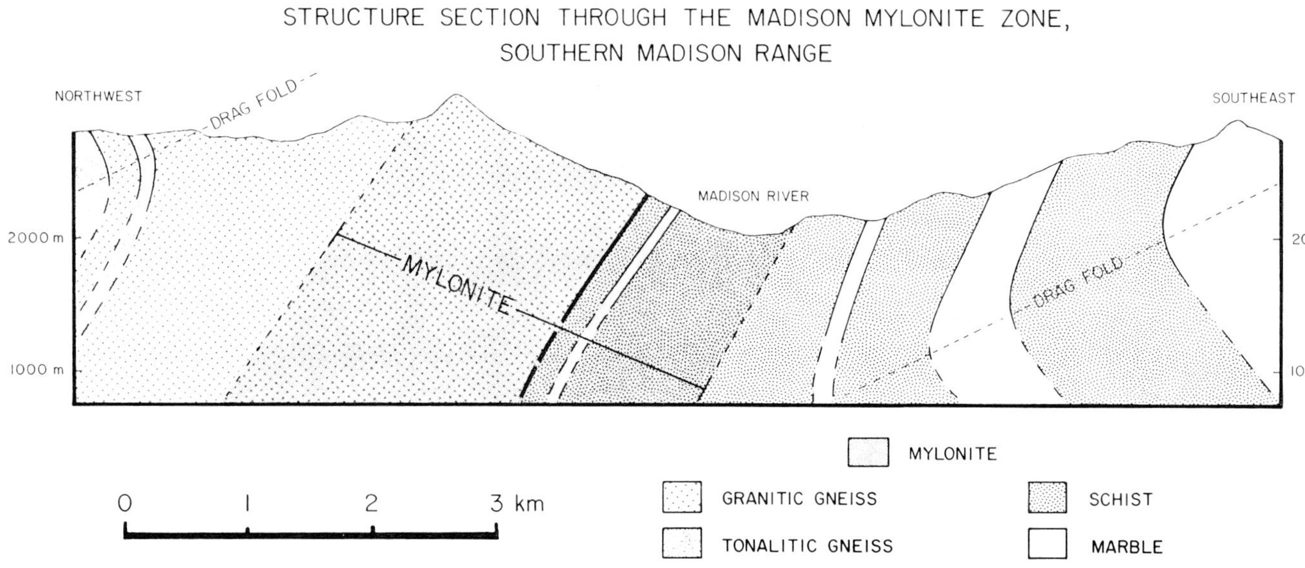

Figure 4: Structure section through the Madison Mylonite Zone illustrating the lack of tectonically disrupted units and the flexure of the adjoining rocks into concordance with the shear zone. The resulting drag folds are consistent with reverse faulting in the shear zone.

the lowermost units of the suite which are within the Madison Mylonite Zone in the southern Madison Range appear to be covered by the Tertiary volcanics to the southwest. On the western flank of the Centennial Mountains, the Phanerozoic strata are covered by volcanic rocks which presumably cover the subsurface extension of the Madison Mylonite Zone as it passes into the southwesterly-trending Snake River Plain of Idaho.

Within the swath of volcanic rocks to the southwest, there are a series of elongate lakes which parallel the trend of the Madison Mylonite Zone. From the southwestern arm of Cliff Lake, the Lost Mine Canyon leads to Hidden Lake and Elk Lake which drains to the southwest into Upper Red Rock Lake. It is certainly conceivable that this trough in the pre-Tertiary basement now filled with volcanic rocks and elongate lakes is the result of the reactivation of planes of weakness within the southwestern extension of the Madison Mylonite Zone.

Northeastern Extension

Northeast from the Madison Mylonite Zone in the southern Madison Range, the topographic and geologic lineament appears to die out under the Phanerozoic strata in the Monument Mountain area. However, the strong northeasterly lineament resurfaces in the elongate Tom Miner Basin which runs into the Paradise Valley of the Yellowstone River. Field reconnaissance during the summers of 1979 and 1981 revealed extensive areas of high-temperature mylonite gneiss and phyllonite on the southeastern side of these valleys. A representative sampling of foliation measurements was taken from five stream valleys cutting the mylonites in the northwestern flank of the Beartooth Mountains (Figure 5). A quick comparison with the fabric data from the southern Madison Range in Figure 3 shows a close correlation of both strikes and dips.

The recognition of the mylontic character of the rocks flanking the northwestern side of the Beartooth Mountains is difficult because the unmylontized protoliths are poorly exposed. The foliation planes of the mylonite zone have been utilized by en echelon normal faults whose hanging walls are largely covered with Tertiary volcanics and basin fill, effectively burying the protoliths to the northwest. The Absaroka volcanics capping the western Beartooth Mountains often cover the protoliths of the mylontic rocks on the southeastern side of the shear zone.

The southeastern margin of this mylonite zone is exposed one kilometer southwest of Miner in a road cut on the major road into Tom Miner Basin. This prophryoclastic gneiss contains large plagioclase clasts with clear rims and altered cores exhibiting highly-deformed twinning in a matrix of quartz, lozenges of muscovite, chlorite and biotite.

This zone of cataclasis is not exposed in Yankee Jim Canyon due to a southeasterly embayment of Quaternary alluvium along the Yellowstone River. However, to the northeast, the exposure of the Archean basement moves farther north, resulting in the exposure of the mylonite zone at Sixmile Creek. Burnham (1892) described a transition from quartzofeldspathic gneiss to the southeast to a two- to three-kilometer thick zone of mylonite bounding "lower grade fine-grained muscovite, biotite, and amphibolite schists which have undergone retrograde metamorphism". On the basis of a field reconnaissance and personal communications with Robert Burnham, it is clear to the author that these fine-grained schists are nearly identical to the rocks of the Madison Mylonite Zone in the southern Madison Range.

Farther to the northeast in Emigrant Gulch, there is a two- to three-kilometer-thick zone of phyllonitic mylonite whose unmylonitized protolith is covered by Tertiary volcanics. Porphyroclastic mylonite is exposed in the southeastern limit of the exposure. Mylonites farther to the northwest include finer-grained amphibolite with blue-green hornblended and mylonite gneiss containing quartz, plagioclase, biotite, and irregularly-shaped garnets. Garnet-biotite Fe-Mg exchange equilibria using the calibration of Thompson (1976) of one specimen yielded a temperature of 480°C, which is within the margin of error of temperature estimates from the Madison Mylonite Zone in the southern Madison Range.

The exposure of the Precambrian basement starts to pinch out at Mill Creek where a rapid transition from tonalitic gneiss to porphyroclastic mylonite containing highly deformed plagioclase clasts and abundant chlorite is exposed. The foliations from Mill Creek in Figure 4 are more northerly trending, an effect that can be ascribed to drag on the adjacent Mill Creek-Stillwater fault zone.

The seven kilometers of left-lateral strike separation on the Mill Creek-Stillwater fault zone (J. Robert Butler, personal communication) explains the apparent offset of the mylonite zone. Using the geologic maps of Reif, et al. (1975), two exposures of slabby mylonite were located in the foothills adjacent to Pine Creek. Reid et al. (1975) mapped these phyllonitic rocks as Davis Creek Schist, but they are much finer-grained and more homogeneous than units of Davis Creek Schist farther up Pine Creek. The late stage D_4 folds in the cross section of Reid et al. (1975) through the Pine Creek Lake block appear to be drag folds associated with reverse faulting in the mylonite zone.

The great lateral extent of the mylonitic units in the northwestern side of the Beartooth Mountains combined with the strong similarities in orientation, texture, and metamorphic facies indicates that they are correlative with the Madison Mylonite Zone of the southern Madison Range. If these zones are distinct, parallel shear zones, there should be two major shear zones exposed in the nearly continuous composite exposure from Henrys Lake in Idaho to the Tobacco Root Mountains. There are thin zones of cataclasis in the Gravelly Range (Millholland, 1976), but the Madison Mylonite Zone is the only major ductile shear zone reported from this area (Vitaliano, et al., 1979; Hadley, 1969; Erslev, 1981; Witkind, 1972).

DISCUSSION

The Precambrian age of the Madison Mylonite Zone is firmly established by its relatively high-temperature mineral assemblages and the absence of equivalent penetrative structures in the overlying sedimentary and volcanic rocks of Phanerozoic age. However, the current lack of isotopic ages from rocks within the Madison Mylonite Zone makes the time of shearing uncertain. The low-grade metamorphic assemblages in the Belt Supergroup in Montana appear to indicate that the Madison Mylonite Zone formed in pre-Beltian times, but the lack of

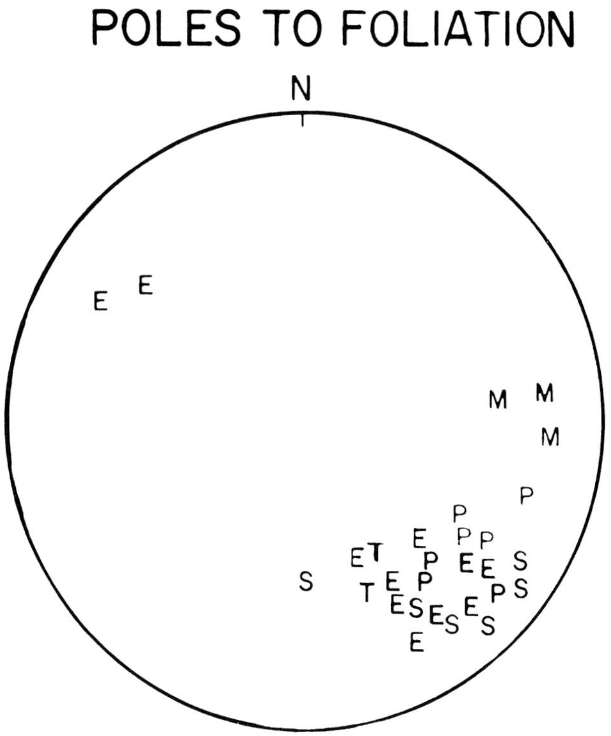

Figure 5: Representative foliation measurements of mylonite rocks exposed in the northwestern side of the Beartooth Mountains. T — Tom Miner Basin, S — Sixmile Creek, E — Emigrant Gulch, M — Mill Creek, and P — Pine Creek. See Figure 1 for the location of the sampled areas.

Belt sediments in the Madison Range itself does not allow absolute confidence in this inference.

There are several observations which seem to indicate an age in the range of 1.5 to 1.8 b.y. for the Madison Mylonite Zone. In the southern Madison Range, the ductile shearing is associated with a pulse of retrograde metamorphism which has been recognized throughout southwestern Montana. On the basis of K-Ar age measurements, Giletti (1966, 1971) deduced that the southeastern edge of a 1.6 b.y. regional metamorphism (Giletti's Line) was located parallel to and 20 kilometers north of the Madison Mylonite Zone. K-Ar ages in the 1.5 to 1.8 b.y. range have also been determined in the western Beartooth Mountains (Reid, et al., 1975; Peterman, 1979), so the re-equilibration may have had localized effects far from Giletti's Line. Montgomery (1981) reported that Rb/Sr model ages on biotites yielded a gradational transition from 2.5 b.y. to 1.9 b.y. as the Madison Mylonite Zone is approached in the southwestern Beartooth Mountains.

The Madison Mylonite Zone may represent a shear zone marginal to a middle Proterozoic retrograde metamorphic front which totally reset the K-Ar ages in the northwestern half of the pre-Beltian basement of southwestern Montana. It is conceivable that the shear zone had an Archean origin, but that possibility will have to await the results of isotopic analyses from the Madison Mylonite Zone itself.

The lack of exposed pre-Beltian basement southwest of Red Rock Lakes and northeast of the Beartooth Mountains makes any further extrapolations of the subsurface extent of the Madison Mylonite Zone quite speculative. However, there is a series of strong northeast-trending lineaments throughout the region. The Snake River Plain to the southeast appears to merge and follow the projected trend of the Madison Mylonite Zone. The shear zone may have provided a conduit for the magmas which formed the southwestern part of the Snake River Plain. A divergence from the shear zone trend may have been the result of a more easterly trajectory on the zone of melting presently located under Yellowstone National Park (Smith et al., 1977).

North of the Beartooth Mountains, a strong northeasterly-trending set of lineaments is apparent from magnetic data (Zietz, et al., 1971) and the preferred orientations of faults in the Phanerozoic cover. These lineaments trend in the direction of the Churchill-Superior boundary of the Canadian shield. The similarities between the northwestern edge of the Wyoming Province and the Superior-Churchill boundary in the Canadian Shield have been recognized by King (1976) and Peterman (1979). King (1976) summarized his thoughts in the following paragraph:

> "The mingling of Kenoran and Hudsonian dates in southwestern Montana and the Little Belt Mountains seems to represent a gradational boundary between two major Precambrian Provinces, analogous to the Superior and Churchill Provinces of the Canadian Shield. Here, however, in contrast to conditions in the Shield, there seems to be no sharply marked structural or stratigraphic boundary between the two provinces."

Fountain and Desmarais (1980) compared the rock types of the pre-Beltian basement of Montana with the Wabowden terrane of Manitoba and suggested a common geological evolution. Extrapolating from interpretations originating in the Canadian Shield, they suggested that a major continental suture should be evident in the pre-Beltian basement of Montana.

Initially, I hypothesized that the Madison Mylonite Zone was the missing structural boundary between the provinces because it trends toward the boundary between the Superior and Churchill Provinces. The higher pressure metamorphism of the northwestern part of the pre-Beltian terrane of Montana is similar to the conditions at the Superior-Churchill boundary in the Canadian Shield (Weber and Scoates, 1978). The low pressure metamorphism of the Beartooth Mountains is typical of the Superior Province (Fyfe, 1978). However, several lines of evidence have shown that the displacement on the mylonite zone was of limited magnitude and therefore the Madison Mylonite Zone cannot be a major continental suture.

There is no evidence for major units of Early Proterozoic age in the southwestern Montana corresponding to units of the Churchill Province (Peterman, 1979). If the boundary between the Churchill and Superior Province does cross through Montana, it must be limited to the subsurface northwest of the area of exposure. The Madison Mylonite Zone may parallel this postulated tectonic break between Archean and Proterozoic terranes. The retrograde metamorphism associated with the Madison Mylonite Zone may be the metamorphic overlap of a continuation of the Churchill Province in northwestern Montana.

ACKNOWLEDGMENTS

This study was initiated as part of doctoral research at Harvard University under the guidance of J. B. Thompson, Jr., J. Haller, and R. F. Dymek. The research was funded by the Department of Geological Sciences of Harvard University and the Committee on Advanced Study and Research at LaFayette College.

REFERENCES

Burnham, R. L., 1982, Mylonitic basement rocks in the Yankee Jim Canyon and Sixmile Creek areas, Park County, Montana (abs.): Geological Society of America Abstracts with Programs, v. 14, number 6, p. 305.

Erslev, E. A., 1980, Geology of the Precambrian metamorphic rocks of the southern Madison Range, southwestern Montana (abs.): Geological Society of America Abstracts with Programs, v. 12, number 6, p. 272.

Erslev, E. A., 1981, Petrology and Structure of the Precambrian metamorphic rocks of the southern Madison Range, southwestern Montana (Ph.D. thesis): Cambridge, Massachusetts, Harvard University, 133 pp.

Fountain, D. M. and Desmarais, N. R., 1980, The Wabowden terrane of Manitoba and the pre-Belt basement of southwestern Montana: A comparison, *in* Miller, M. R., ed., Guidebook of the Drummond-Elkhorn areas, west-central Montana: Montana Bureau of Mines and Geology Special Publication 82, pp. 35-46.

Fyfe, W. S., 1978, Crustal evolution and metamorphic petrology, *in* Metamorphism in the Canadian Shield: Canadian Geological Survey Paper 78-10, p. 1-4.

Garihan, J. M., 1979, Geology and structure of the central Ruby Range, Madison County, Montana: Geological Society of America Bulletin, Part II, v. 90, p. 695-788.

Giletti, B. J., 1966, Isotopic ages from southwestern Montana, Journal of Geophysical Research, v. 71, pp. 4029-4036.

Giletti, B. J., 1971, Discordant isotopic ages and excess argon in Biotites: Earth and Planetary Sciences Letters, v. 10, pp. 157-164.

Hadley, J. B., 1969, Geologic Map of the Cameron Quadrangle, Madison County, Montana: U.S. Geological Survey Map GQ 813.

Heinrich, E. W., 1960, Geology of the Ruby Mountains and nearby areas in southwestern Montana: Montana Bureau of Mines Memoir 38, pt. 2, pp. 23-40.

James, H. I. and Hedge, C. E., 1980, Age of basement rocks of southwest Montana: Geological Society of America Bulletin, Part I, v. 91, pp. 11-15.

King, P. B., 1976, Precambrian geology of the United States: An explanatory text to accompany the Geologic Map of the United States: U.S. Geological Survey Professional Paper 902, 84 pp.

Millholland, M. A., 1976, Mineralogy and petrology of Precambrian metamorphic rocks of the Gravelly Range, southwestern Montana (M.S. thesis): Bloomington, Indiana University, 134 pp.

Montgomery, C. W., 1981, Metamorphic history of Archean schists from the southwest Beartooth Mountains, Montana and Wyoming (abs.): Geological Society of America Abstracts with Programs, v. 13, number 4, p. 220.

Peale, A. C., 1896, Three Forks Folio, Montana: U.S. Geological Survey Folio 24.

Peterman, Z. E., 1979, Geochronology and the Archean of the United States: Economic Geology, v. 74, pp. 1544-1562.

Reid, R. R., McMannis, W. J., and Palmquist, J. C., 1975, Precambrian geology of the North Snowy block, Beartooth Mountains, Montana: Geological Society of America Special Paper 157, 135 pp.

Runner, J. J. and Thomas, L. C., 1928, Stratigraphic relations of the Cherry Creek Group in the Madison Valley, Montana (abs.): Geological Society of America Bulletin, v. 39, pp. 202-203.

Smith, R. B., Shuey, R. T., Pelton, J. R., and Bailey, J. P., 1977, Yellowstone hot spot: Contemporary tectonics and crustal properties from earthquake and aeromagnetic data: Journal of Geophysical Research, v. 82, pp. 3665-3676.

Thompson, A. B., 1976, Mineral reactions in pelitic rocks: II. Calculations of some P-T-X (Fe-Mg) phase relations: American Journal of Science, v. 276, pp. 425, 454.

Vitaliano, C. J., Burger, H. R., Cordua, W. S., Hanley, T. B., Hess, D. F., and Root, F. K., 1979, Geologic map of the southern Tobacco Root Mountains, Madison County, Montana: Geological Society of America Map and Chart Series MC-31.

Weber, W. and Scoates, R. F. J., 1978, Archean and Proterozoic metamorphism in the northwestern Superior Province and along the Churchill-Superior boundary, Manitoba, *in* Metamorphism in the Canadian Shield: Canadian Geological Survey Paper 78-10, pp. 5-17.

Wilson, M. L., 1981, The Montana basement, a remnant Archean marginal basin? (abs.): Geological Society of America Abstracts with Programs, v. 13, number 4, p. 230.

Witkind, I. J., 1972, Geologic map of the Henrys Lake Quadrangle, Idaho and Montana, *in* Folio of the Henrys Lake Quadrangle, Idaho-Montana: U.S. Geological Survey Map I-781-A.

Zietz, I., Hearn, Jr., B. C., Higgins, M. W., Robinson, G. D., and Swanson, D. S., 1971, Interpretation of an aeromagnetic strip across the northwestern United States, Geological Society of America Bulletin, v. 82, pp. 3347-3372.

Thirty-Third Annual Field Conference — 1982
Wyoming Geological Association Guidebook

BASIN CREEK UPLIFT AND HEART LAKE CONGLOMERATE, SOUTHERN YELLOWSTONE NATIONAL PARK, WYOMING*

BY J.D. LOVE[1] AND W.R. KEEFER[2]

ABSTRACT

A major, but still largely buried structural feature, the Basin Creek uplift, and a previously undescribed sedimentary sequence, the Heart Lake Conglomerate (a new name in this report), provide new data for interpreting the geologic history of southern Yellowstone National Park. The uplift was initially folded at or near the close of Cretaceous time; in late Cenozoic time a major part of it, at least, was rapidly elevated. This later movement resulted in deposition of the Heart Lake Conglomerate, 70-330 feet thick, composed chiefly of Madison Limestone (Mississippian) fragments in a talc-and chlorite-rich matrix. The conglomerate may be of Pliocene or Pleistocene age.

A structural feature, the Basin Creek uplift, and a sedimentary sequence, the Heart Lake Conglomerate, both in southern Yellowstone National Park, have not previously been described in detail. They provide data necessary to reconstruct the structural and depositional history and one episode of the volcanic history of southern Yellowstone National Park.

BASIN CREEK UPLIFT

An arcuate exposure of Mesozoic and Paleozoic rocks, partly buried by rhyolitic flows and tuffs, has long been known in southern Yellowstone National Park (Hague, 1896;1904). This feature was remapped between 1945 and 1949, and rocks that Hague had named the Sheridan Quartzite and had assigned to the Precambrian were found to be the Muddy Sandstone Member of the Thermopolis Shale and the sandstones of the Cloverly Formation that had been locally silicified (Love and others, 1955).

As a result of these studies and others made by Keefer in 1966-67 during the current U.S. Geological Survey investigation of Yellowstone National Park (Campbell, 1969), the arcuate exposure of pre-Tertiary rocks was recognized as the southeast flank of a major Laramide uplift that had been extensively modified by additional movements in late Cenozoic time. The name "Basin Creek uplift" is here proposed for this structural feature. Its exposed parts extend from Yellowstone Lake southward across the southern boundary of Yellowstone National Park

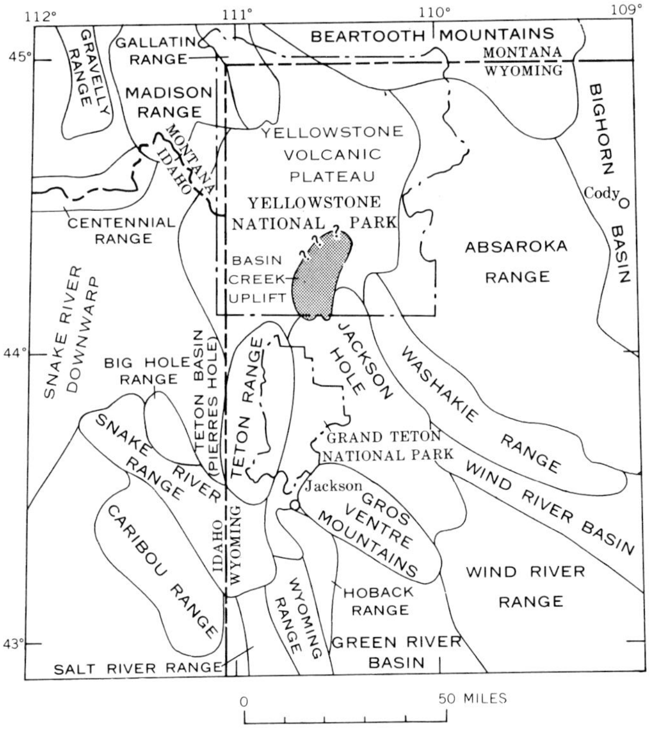

Figure 1: Map showing relation of Basin Creek uplift (patterned) to surrounding uplifts and downwarps.

[1]U.S. Geological Survey, Laramie, Wyoming 82701
[2]U.S. Geological Survey, Denver, Colorado 80225
*Reprint of USGS Professional Paper 650-D, Pages D122-D130.
Reprinted with permission of the authors and the director of the U.S.G.S.

(figs. 1, 2). Mount Sheridan, the Red Mountains, Flat Mountain, and areas to the north and west are capped by younger volcanic rocks that conceal the northern and western sides of the uplift. Basin Creek has cut a canyon across the southeast margin of the uplift south of Mount Sheridan.

The major vertical movement of the Basin Creek uplift occurred during Laramide deformation, probably at or near the end of Cretaceous time, and its amount can be estimated. The Harebell Formation (latest Cretaceous) in northern Jackson Hole (fig. 1) is about 10,000 feet thick; its upper part contains several zones of marine or brackish-water fossils (mollusks identified by E. G. Kauffman, written commun., 1967; hystrichomorphs and associated Cretaceous pollen identified by R. H. Tschudy and E. B. Leopold, written commun., 1966-68). Deposition therefore, took place at or near sea level in a subsiding basin. Inasmuch as there is no evidence that the uplift contributed debris to the Harebell Formation, the initial upwarping apparently followed deposition of the Harebell.

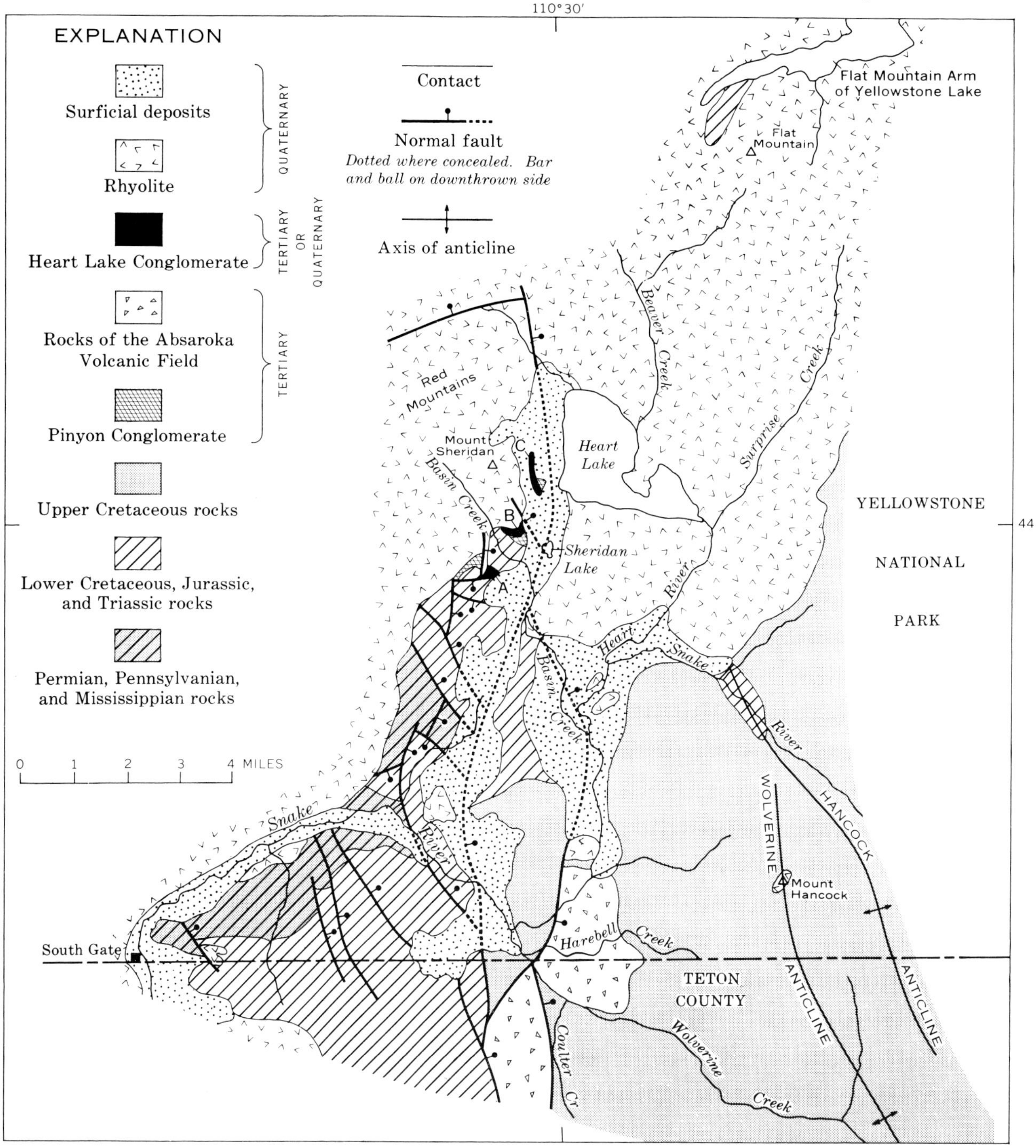

Figure 2: Generalized geologic map of part of southern Yellowstone National Park and the area to the south. Localities A, B, and C, are outcrops of Heart Lake Conglomerate referred to in text.

When uplift of the Basin Creek area began, the base of the Harebell was more than 10,000 feet below sea level in the downwarp directly to the southeast. By the beginning of deposition of the Pinyon Conglomerate (Paleocene), however, all the Harebell and 6,000 feet of the older Mesozoic rocks had been eroded from the crest of the uplift. The Pinyon Conglomerate was then deposited across rocks at least as old as Jurassic along the east flank of the uplift (figs. 2, 3, 4), as well as across all of the Harebell Formation in the adjacent downwarp to the southeast. These relations suggest that the vertical movement in the southeastern part of the Basin Creek uplift during the Laramide orogeny was 12,000-15,000 feet.

At the beginning of the Pinyon Conglomerate deposition the uplift probably was a simple north- or north-northwest-trending elliptical fold, which had a moderately flat top and unfaulted south and east flanks. The nature of the west flank is not known; however, nearby major upwarps, such as the Washakie Range (fig. 1), and lesser folds, such as the Bailey (6 miles south) and Spread Creek (20 miles southwest) anticlines, were thrust westward or southwestward (Love, 1956a). Similar thrusting may have taken place along the west margin of the Basin Creek uplift. A syncline with an amplitude of 6,000 feet developed at the same time as the Basin Creek uplift between the uplift and the ancestral Teton Range to the southwest (fig. 1).

The original thickness of the Pinyon Conglomerate is not known, but one remnant on the southeast face of

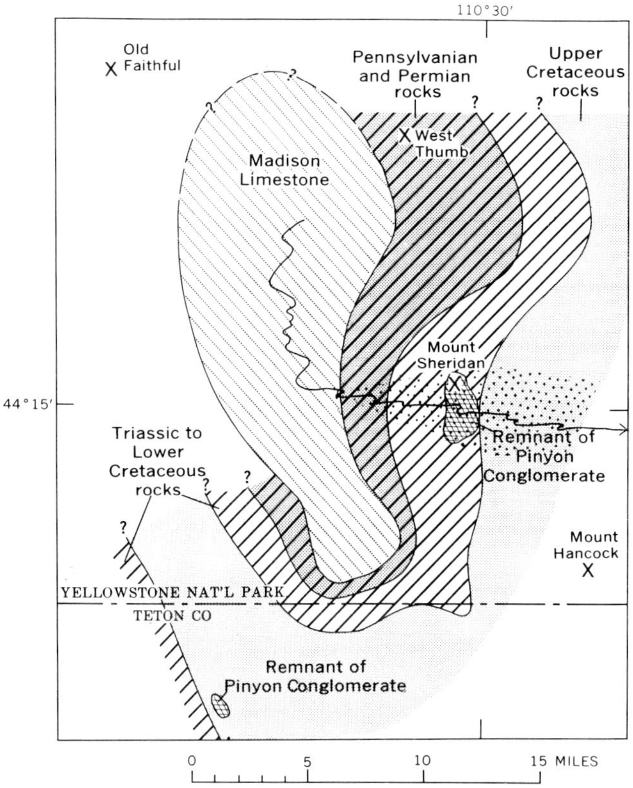

Figure 3: Reconstructed geologic map of Basin Creek uplift at time of deposition of Heart Lake Conglomerate, before caldera formation to the north. Stippled area indicates hypothetical location of Heart Lake Conglomerate.

Figure 4: Contact between highly silicified Pinyon Conglomerate (above), indicated by line; man in center shows scale; locality B, Mount Sheridan measured section.

Mount Sheridan is 450 feet thick. The Basin Creek uplift appears to have been relatively quiescent during deposition of the conglomerate, as well as throughout the remainder of Tertiary time, for the preserved remnants of the Pinyon have not been extensively deformed.

During some part of late Cenozoic time, an area including at least the southern part of the Basin Creek uplift was rejuvenated. The movement triggered deposition of the Heart Lake Conglomerate and may also have been related to the initial stages of volcanism that resulted in vast outpourings of rhyolite which blanketed this and other areas to the north, west, and south. The extensive silicification of sandstones, siltstones, and shales in Lower Cretaceous and Jurassic rocks, as well as the alteration of the Pinyon Conglomerate (fig. 5) and clasts within the Heart Lake Conglomerate, may have been in part related to this volcanism. In many places, silification of the Pinyon progressed to the point where its conglomerate nature is now barely recognizable.

HEART LAKE CONGLOMERATE

A distinctive conglomerate which overlaps Mesozoic rocks and the Pinyon Conglomerate and consists largely of Paleozoic rock fragments and rhyolitic debris was found in the Mount Sheridan area by Love in 1949. Because the overlying rhyolites were then considered to be of Pliocene age, this conglomerate was called "Pliocene and Miocene" on the geologic map of Wyoming (Love and others, 1955).

Figure 5: Pinyon Conglomerate at locality B, Mount Sheridan measured section. Entire rock has been so completely impregnated with secondary silica that it is hardly recognizable as a conglomerate. One quartzite boulder is outlined below knife at upper right.

The name Heart Lake Conglomerate is here proposed for this unique sequence; the Heart Lake overlies the Pinyon Conglomerate and underlies the main mass of rhyolite welded tuff on the steep east and south slopes of Mount Sheridan, west and southwest of Heart Lake. The conglomerate is known from three exposures (locs. *A*, *B*, and *C*, fig. 2), at each of which a section has been measured and sampled. At locality A, the conglomerate is either 70 or 125 feet thick, depending on correlation of beds; at B, about 330 feet thick; and at C, about 110 feet thick. Locality B is designated the type section (fig. 6 and stratigraphic section given below). At that locality the strata are exposed on a steep south-facing headwall at the base of the great bare precipitous slope that rises 1,700 feet to the top of Mount Sheridan. No opportunity was available to trench the type section or to describe it in detail; it is therefore subdivided into gross units with only approximate thicknesses. The rocks strike about north and dip about 15°W.

Type section of Heart Lake Conglomerate on east and south slopes of Mount Sheridan (loc. B, *fig. 2)*

Unit and description	*Thickness (feet)*
Rhyolitic sequence (Pleistocene):	
10. Rhyolite bedded tuff, altered, chalky-white; surface weathers to bright red, orange, and yellow; stands in conspicuous ragged sheer cliff; rock composed of clear crystals of glassy quartz and feldspar in spongy soft matrix containing chalky altered pseudomorphs of plagioclase crystals; many vugs, some with quartz crystals; unit dips west at about same angle as underlying rocks; rests on an irregular surface at top of underlying unit	200±
Sharp change in lithology; exposures too limited to show whether an unconformity is present.	
Heart Lake Conglomerate (Pleistocene or Pliocene):	
9. Conglomerate, yellowish-gray; composed of moderately small fragments of volcanic rock, chiefly rhyolite, and rounded pebbles, cobbles, and boulders of limestone and quartzite; unit is poorly exposed on slope at angle of repose and was not examined in detail	40
8. Conglomerate, yellowish-gray; lithified and forms hard cliff; chiefly rounded fragments of limestone, quartzite, and volcanic rocks, largely rhyolite; at top is 3 feet of bedded tuffaceous sandstone	20
7. Conglomerate, yellowish-gray, soft; forms slope; may contain some claystone beds; coarse debris composed of rounded fragments of limestone, quartzite, and rhyolite(?) in a tuffaceous sandstone and probably claystone matrix	20
6. Conglomerate, yellowish-gray; forms hard ledge; roundstones are chiefly quartzite and are rarely more than 3 inches in diameter	20

Figure 6: View north from locality A across Basin Creek toward the type section (locality B) of the Heart Lake Conglomerate. Indicated are the main mass of rhyolite welded tuff (Qr) on Mount Sheridan, the Heart Lake Conglomerate (QTh), marker bed (4) which is unit 4 in measured section, Pinyon Conglomerate (Tp), Cloverly and Morrison (?) Formations (KJm), dike of rhyolite mixed with Pinyon Conglomerate (Qrd), and landslide debris (Qls).

Type section of Heart Lake Conglomerate on east and south slopes of Mount Sheridan (loc. B, fig. 2)—Continued

Unit and description	Thickness (feet)
5. Claystone, gray, bentonitic; soft, in lower part; grades up to yellowish-gray conglomerate which contains highly fractured roundstones that are smaller than those in overlying unit	20
4. Conglomerate and tuffaceous sandstone, brown, hard, bedded, noncalcareous; forms conspicuous ledge that extends across outcrop area; roundstones are of quartzite and Paleozoic rocks, smaller than in overlying units; many black rounded chert pebbles, probably derived from Madison Limestone (Mississippian); many gray waxy soft talc pebbles and gray clayey matrix	8
3. Conglomerate, mouse-gray, soft, poorly sorted; forms slope at angle of repose; fresh exposures in torrent gullies show that unit is a jumble of large and small fragments dumped together and tightly cemented with limy cement; about 75 percent is Paleozoic limestone (chiefly Madison) and remainder is Paleozoic quartzite and sandstone, Pinyon quartzite, and rhyolite; quartzite is conspicuous as lag gravel on slopes but is a minor lithology in fresh exposures; highly rounded boulders of fossiliferous Madison Limestone 1–2 feet in diameter are common but average size is 4–6 inches; numerous boulders of highly altered rhyolite(?) that disintegrate on outcrop; some highly rounded black chert pebbles, probably derived from the Madison Limestone; matrix is white, limy, with abundant white to pale-green waxy claystone; increasingly tuffaceous in upper part, with quartz crystals 30 feet below top; many pale-green to white rounded pebbles of waxy noncalcareous nondecrepitating talc; talc pebbles are so abundant 50 feet below the top that they make up nearly half the rock and give it a slippery feel	200±
Approximate total thickness of Heart Lake Conglomerate	330

Angular unconformity. Beds above dip 15° W., and those below 20° W. Actual trace of unconformity is covered or distorted by intrusion of rhyolite dike.

Pinyon Conglomerate (Paleocene):

Type section of Heart Lake Conglomerate on east and south slopes of Mount Sheridan (loc. B, *fig. 2)*—Continued

Unit and description	*Thickness (feet)*
2. Conglomerate; green and brown in lower part, rusty brown in upper part; hard, tightly cemented, highly silicified with secondary silica. Roundstones are chiefly quartzite boulders 8–15 inches in diameter; commonly gray, but some are black; others are tan and gray hard rounded boulders of sandstone; some are red and blue-green quartzite; many of the hardest boulders still have polish preserved; many are fractured and recemented; pressure scars are conspicuous; matrix is iron-rich coarse-grained sandstone; bedding planes poorly developed; entire sequence apparently is conglomerate	450±
Approximate total thickness of Pinyon Conglomerate	450
Angular unconformity. Beds above strike about north, dip 20° W.; those below strike N. 45° W., dip 15° SW. The actual unconformity is visible on the east side of the spur that bounds the east margin of the headwall where the overlying section was measured (fig. 4).	
Cloverly and Morrison(?) Formations, undivided (Lower Cretaceous and Upper Jurassic):	
1. Quartzite, white, very fine grained, hard, well bedded	10
Base of measured section.	

The section at locality A is exposed on a steep to sheer south-facing bare outcrop on the high hill nearly half a mile southwest of the sharp bend in Basin Creek canyon and 6,500 feet west-southwest of the outlet of Sheridan Lake. Dips are probably less than 5°W., and the entire section is on a west-tilted down-thrown fault block.

Measured section, locality A *(fig. 2)*

Unit and description	*Thickness (feet)*
Rhyolitic sequence (Pleistocene):	
5. Tuff, altered white, platy, extremely hard; conchoidally fractured in part, granular; some layers have graded bedding; others are finely laminated with as many as 100 alternating light gray and black laminae per inch; some layers are black because of abundant finely comminuted carbonaceous material; samples collected for pollen and diatom study (no diatoms; pollen listed, p. D129); unit forms steep slope littered with white rock chips at top of sheer ragged rusty brown and white cliff	20
4. Rhyolite tuff, white to rusty brown; forms ragged cliff; is soft, spongy, layered, with bedding planes conspicuous in lower part; basal beds appear to merge with much softer underlying unit	135±
Approximate total thickness of rhyolitic sequence	155
Heart Lake(?) Conglomerate (Pleistocene or Pliocene):	
3. Pumice breccia and claystone; upper 40 feet is white to gray pumice breccia composed of angular soft white and gray chunks of highly altered pumice and rhyolite in a soft chalky clayey matrix; many quartz bipyramids; much of rock looks as if it had been intensely altered by hydrothermal action; lower 15 feet is colloidally fine-grained soft plastic claystone with many pumice chunks and quartz bipyramids; white and black laminations, some very closely spaced. Pyrite concentrated in black laminae and in irregular finely crystalline masses; has sulfur odor. Sporadic pebbles and cobbles of quartzite, chert, limestone; black obsidian welded tuff, and other dense volcanic rocks have surficial sheen; some are as much as 6 inches in diameter, but none this large was seen above middle of unit	55
Heart Lake Conglomerate (Pleistocene or Pliocene):	
2. Conglomerate; rusty-brown and iron stained in upper 10 feet, chalky bluish white in remainder; coarser grained near middle. Roundstones are nearly all of Madison Limestone; some are 2 feet in diameter, 1-foot size is common, and average is about 4 inches; most are rounded to subrounded. Various Madison lithologies represented, from chalky white dolomite to dark-gray crinoidal limestone; corals are common; no Darby Formation (Devonian) roundstones noted, but one boulder is of mottled dolomite that might be Bighorn Dolomite (Ordovician); no Cambrian, Precambrian, or red Chugwater Formation (Triassic) observed; Pinyon type quartzite roundstones are present but sparse; some black obsidian welded tuff, hard dense volcanic rocks (chiefly rhyolite), Paleozoic chert, and Pinyon quartzite roundstones commonly have a conspicuous sheen; abundant rounded pebbles of talc and some chloritic claystone; slippery clayey matrix is probably predominantly talc and chlorite	70±
Angular unconformity. No evidence of Pinyon Conglomerate here but a few feet of it is present at this horizon several hundred yards to the northwest.	
Sundance Formation (Upper Jurassic):	
1. Shale, blue-gray, hard, chippy, interbedded with soft plastic gray claystone and mudstone	20+
Base of section.	

Corals and other fossils are common in many of the Madison Limestone boulders. One from unit 2 in the measured section at locality A yielded *Vesiculophyllum* sp. (identified by W. J. Sando, written commun., 1968). This form is of Mississippian age and is typical of the Madison Limestone.

The section at locality C is similar to the sections at localities A and B. Especially notable at locality C are the abundance of waxy gray talc and chloritic claystone pebbles and the presence of a clayey matrix in the limestone conglomerate. The conglomerate is about 75 feet thick. The overlying siltstones and claystones here are like those in the upper part of the type section at locality B but are much thinner. The underlying Pinyon Conglomerate is at least

300 feet thick, with the base not exposed, and is highly silicified and brecciated.

The most significant features of the Heart Lake Conglomerate are: (1) the dominance of rock fragments from sand size to boulders 3 feet in diameter (fig. 7), most of which were derived from the Mississippian Madison Limestone; (2) the abundance of waxy talc and chloritic claystone fragments in the conglomerate; and (3) the presence of rhyolitic debris, which is especially abundant in the middle and upper parts.

Figure 7: Detail of Heart Lake Conglomerate at locality A. south Basin Creek measured section. Nearly all rock fragments are Madison Limestone. Matrix contains abundant waxy very fine grained gray claystone.

Spectrographic analyses of two samples of claystone in unit 3 at locality A show no unusually high amounts of trace elements, and are relatively low in magnesium, calcium, manganese, and sodium.

Samples of the pebbles suspected of being talc gave an X-ray diffractometer pattern of talc and calcite (Theodore Botinelly, written commun., 1968). Botinelly observed pale-green inclusions, possibly pseudomorphs after feldspar, in some talc pebbles. An X-ray diffraction pattern of the green-colored material indicates that it is chlorite.

It has been noted that the Jurassic and Cretaceous shales and sandstones, as well as the Pinyon Conglomerate, are highly silicified at locality B, whereas the limestones appear superficially to be little altered. Marine fossils in the Jurassic limestones are well preserved. In the Heart Lake Conglomerate, limestone roundstones containing Mississippian corals likewise show little conspicuous alteration. Nevertheless, representative samples of limestone from the lower part of the Sundance Formation of Jurassic age in unit 1 and a boulder of Madison Limestone from unit 6 of the Heart Lake Conglomerate in the measured section at locality B were submitted for carbon and oxygen isotopic analysis. The results are as follows (Irving Friedman, written commun., 1967):

Sample	O^{18} *per mil SMOW*	C^{13} *per mil PDB*
Jurassic limestone	−8.4	+1.3
Madison Limestone boulder	−1.4	−0.3

The oxygen analyses are reported relative to Standard Mean Ocean Water (SMOW), a standard water whose isotopic composition is close to that of mean ocean water. The carbon analyses are given in terms of the PDB standard. PDB is a Cretaceous belemnite, *Belemnitella americana*, from the Peedee Formation of South Carolina (Craig, 1957, p. 135). Friedman's communication stated:

> The O^{18} analysis shows that both samples are altered isotopically, the first [Jurassic] more than the second [Madison]. A normal value for Madison limestone would be +20 to +25 per mil SMOW. The carbon values of both samples are in the marine limestone range.

Most occurrences of talc in the Rocky Mountain region are associated with Precambrian rock terranes. A Precambrian metamorphic source for the talc in the Heart Lake Conglomerate, however, is virtually precluded by the fact that, despite careful search, neither Precambrian crystalline rocks nor Cambrian rocks were found in the formation. The close association of the talc and chlorite with abundant fragments of post-Cambrian Paleozoic carbonate rocks, particularly Madison Limestone, suggests strongly that the talc and chlorite were derived from metamorphosed impure dolomites within these strata. (See discussion by Cooper, 1957, p. 582-588). Thick beds of dolomite occur in the Bighorn Dolomite (Ordovician) and Darby Formation (Devonian), as well as in the Madison Limestone. Some of these dolomites could have been partially transformed to talc and chlorite by the low-grade metamorphism that affected the Paleocene, Lower Cretaceous, and Jurassic rocks underlying the Heart Lake Conglomerate in the Mount Sheridan area. If the late Cenozoic rise of the Basin Creek area was related to rhyolitic intrusion, this magma could have been responsible for the metamorphism, but it would have had to occur before the debris of the Heart Lake Conglomerate was eroded from the core of the uplift.

The Heart Lake Conglomerate rests with angular unconformity on the Pinyon Conglomerate of Paleocene age at localities B (fig. 2) and C, and on the Sundance Formation of Late Jurassic age at locality A. No conclusive evidence of an unconformity was found at the top of the conglomerate, and the progressive increase in rhyolitic(?) debris in the upper part suggests the possibility that the

Paleozoic source area was being covered with rhyolite tuff during the later stages of deposition of the Heart Lake strata.

No fossils were found in the Heart Lake Conglomerate. Farther south in Jackson Hole the first abundant rhyolitic debris occurs in Pliocene rocks (Love, 1956a). In the type locality of the Teewinot Formation, 6,000 feet thick and dated as middle Pliocene by vertebrate, invertebrate, and plant fossils, are several zones containing roundstones of black and gray obsidian as well as crystallized rhyolitic lavas and welded tuff and thick beds of vitric and crystal tuff (Love, 1956b). Obsidian clasts from a bed in the upper part of this section yielded a potassium-argon age of 9.2 million years (Evernden and others, 1964, p. 185-186).

No vent areas of a size adequate to supply the middle Pliocene rhyolitic debris are known in Jackson Hole, so it is inferred that the debris came from vents in Yellowstone National Park. The Heart Lake Conglomerate might, therefore, be equivalent to some part of the Teewinot Formation. On the other hand, if additional work should demonstrate that the conglomerate grades into the overlying rhyolitic sequence of possible Pleistocene age (R. L. Christiansen, written commun., 1968) on Mount Sheridan, then it too, might be Pleistocene.

The white platy slightly carbonaceous tuff comprising unit 5 of the measured section overlying the Heart Lake Conglomerate at locality A yielded the following flora of post-Oligocene age (E. B. Leopold, written commun., 1968):

Compositae, cf. *Ambrosia*, 3 grains
Chenopodiaceae, 13 grains
Pinus, 15 grains
Compositae, *Artemisia*, 1 grain

This assemblage is of little help in precise age determination, but it does indicate that pollen is present and that more extensive collecting might be rewarding.

SUMMARY OF INTERPRETATIONS

Because a large part of the Basin Creek uplift is covered by Pleistocene rhyolitic rocks, its reconstruction and its geologic history are necessarily based on indirect evidence and inference. Two episodes of uplift occurred, the first in Laramide time and the second in late Cenozoic time. The Laramide uplift was probably elliptical — elongated in a north-northwest direction — and relatively flat topped, and was eroded to the Jurassic before the deposition of the Paleocene Pinyon Conglomerate. The late Cenozoic area of uplift is believed to be at least partly coincident with the Laramide area of uplift; such a pattern would be consistent with the pattern for the adjacent Teton Range. The later movement of the Basin Creek uplift could have been either a fault block uplift or a simple rejuvenation by folding of a large segment of the original structural feature.

Observations made in the limited outcrop areas at localities A, B, and C (fig. 2) provide data for at least a partial reconstruction of the Paleozoic core of the Basin Creek uplift at the time of deposition of the Heart Lake Conglomerate. The large size and lack of sorting of the rock fragments indicate rapid erosion and transport by powerful streams from a nearby source of high relief; imbrication of slabs in the conglomerate shows that the streams flowed from the west. If it is assumed that the 3,000 feet of strata between the Cloverly Formation and the Madison Limestone all dipped eastward off the uplift at about 10° (because of later movements, the strata now dip 15°-25°W) at this time, large areas of the Madison could have been exposed to erosion as near as 4-5 miles west of the present exposures of the conglomerate (locs. *A*, *B*, and *C*, fig. 2). The postulated shape of the uplift at the time of late Cenozoic tectonism and deposition of the Heart Lake Conglomerate, as shown in figure 3, is based on the above interpretations, on estimates of the volume of conglomerate deposited, and on regional structural trends in the Paleozoic and Mesozoic rocks to the west, south, and east. Data are entirely lacking on rocks to the north.

Following the deposition of rhyolitic in Pleistocene time, the exposed parts of the Basin Creek uplift, as well as the adjacent areas, were offset by many large and small normal faults. These faults tend to disguise the configuration of the uplift before rhyolite deposition and the relation of the Heart Lake Conglomerate to the uplift.

REFERENCES

Campbell, A. B., 1969, Current U.S. Geological Survey research in Yellowstone National Park, *in* Abstracts for 1968: Geol. Soc. America Spec. Paper 121, p. 590.

Cooper, J. R., 1957, Metamorphism and volume losses in carbonate rocks near Johnson Camp, Cochise County, Arizona: Geol. Soc. America Bull., v. 68, no. 5, p. 577-610.

Craig, Harmon, 1957, Isotopic syandards for carbon and oxygen and correction factors for mass- spectrometric analysis of carbon dioxide: Geochim. et Cosmochim. Acta, v. 12, nos. 1-2, p. 133-149.

Evernden, J. F., Savage, D. E., Curtis, G. H., and James, G. T., 1964, Potassium-argon dates and the Cenozoic mammalian chronology of North America: Am. Jour. Sci., v. 262, no. 2, p. 145-198.

Hague, Arnold, 1904, Atlas to accompany U.S. Geological Survey Monograph 32 on the Geology of Yellowstone National Park: U.S. Geol. Survey.

suggestive of a different mechanism and volume for upper-crustal magma emplacement.

A relative perspective of Quaternary volcanism at Yellowstone is shown in Figure 1, where volumes of several well-known eruptions (Williams and McBirney, 1979; Decker and Decker, 1981) are shown for comparison. The 1980 Mt. St. Helens eruption, which dominates our current concept of explosive eruptions, is on the order of a ten thousandth of the volume of the large explosive Yellowstone eruptions. And while the distribution of airborne ash from Mt. St. Helens produced millions of dollars of damage, airborne ash flows from the Quaternary Yellowstone eruptions were measured as far away as Saskatchewan and Mississippi. A comparison with other large explosive eruptions, such as Katmai in 1920, Krakatoa in 1882, Tambora in 1815, and Mazama (about 7000 years b.p.) dramatizes the fact that these catastrophic eruptions were hundred's of times smaller than the Quaternary Yellowstone explosive eruptions and emphasizes the importance to understand this major silicic volcanic system.

The Quaternary tectonic framework of the Yellowstone Plateau is portrayed in Figure 2, which shows the distribution of volcanic flows, collapse calderas (including the most recent 600,000-year-old caldera), the Sour Creek (northeast) and the Mallard Lake (southwest) resurgent domes, and Cenozoic faulting related both to tumescence during thermal expansion and collapse and contraction of the crust following the explosive eruptions. The youthfulness and large volumes of Quaternary volcanics at Yellowstone infer that intermediate- to upper-crustal heat sources may still be present.

Geophysical experiments designed to characterize the crustal properties of the Yellowstone system have been conducted through the last decade. Earthquake monitoring on both a regional and local scale have been conducted by the University of Utah and the U.S. Geological Survey (Eaton *et al.*, 1975; Trimble and Smith, 1975; Smith *et al.*, 1977; Pitt, 1980). Aeromagnetic surveys and gravity compilations have been used to infer the distribution of shallow silicic melts and caldera infill (Blank and Gettings, 1974; Eaton *et al.*, 1975; Smith *et al.*, 1974, 1977). *P*-wave delays of up to 1.8 sec from teleseismic arrivals were interpreted by Iyer *et al.* (1981) to represent up to a 20 percent velocity reduction in the crust and a 5 percent reduction in the upper mantle to depths of 250 km. These delays were interpreted to reflect zones of possible melt, deep in the lithosphere beneath the Yellowstone caldera. While conductive heat-flow measurements are difficult to obtain in Yellowstone, because of the widespread shallow-water circulation, inferences of conductive heat flow from marine-type probes in Yellowstone Lake (Morgan *et al.*, 1977), and the Cl geochemical budget (Fournier *et al.*, 1976) show that the average heat flux through the Yellowstone caldera is approximately 1800 mW/m^2—about 60 times the world average. These data and inferences from the geologic record suggest that crustal sources of heat are driving the Yellowstone hydrothermal systems and provide much of the energy for the contemporary tectonic deformation.

In 1978 a major seismic refraction/reflection experiment was conducted in the eastern Snake River Plain-Yellowstone region as a means of evaluating the lithospheric structure of this continental tectono-volcanic province (Braile *et al.*, 1982; Smith *et al.*, 1982). In this project up to 225 vertical-component seismographs were located on profiles and in two-dimensional arrays throughout the region. Within Yellowstone, five explosions external to the caldera were recorded across the Yellowstone Plateau. Interpretations of *P*-wave refraction and wide-angle reflections from these profiles yielded average crustal velocity structures (Schilly *et al.*, 1982). Three-dimensional velocity distributions of the upper crust were determined from delay-time analyses (Lehman *et al.*, 1982), from simultaneous inversions of travel times from local earthquake and refraction data measurements for the upper-crustal structure (Benz and Smith, 1981), and from University of Utah unpublished refraction data of the southwest Yellowstone Plateau recorded in 1980. Information on the attenuation of *P*-waves was determined from inversion of spectral amplitude ratios obtained from the refraction data (Clawson and Smith, 1981). Together these studies provided a seismic image of the crustal velocity structure, with a highest resolution of a structure of a few kilometers — near the scale of surface tectonic features. Longer-range refraction/wide-angle reflection profiles (Smith *et al.*, 1982) and *S*-wave delay and surface-wave dispersion investigations (Daniel and Boore, 1982) were used to evaluate the averaged *P*- and *S*-wave structure of the crust and upper mantle.

Interpretations of the new seismic refraction-reflection data served as the basis for this discussion and together with other geophysical and geological data provide an evaluation of the locations of possible bodies of hot rock, zones of partial melt, and magma in the crust beneath the Yellowstone Plateau. Seismic models are synthesized with new hypocenter data, with evaluations of contemporary crustal deformation from level-line re-observations (Pelton and Smith, 1979, 1982) and the Quaternary geologic record (Christiansen, in press). Inferences of physical state including composition and temperature are then used to evaluate the sources of the Quaternary volcanism and to speculate on future volcanism at Yellowstone.

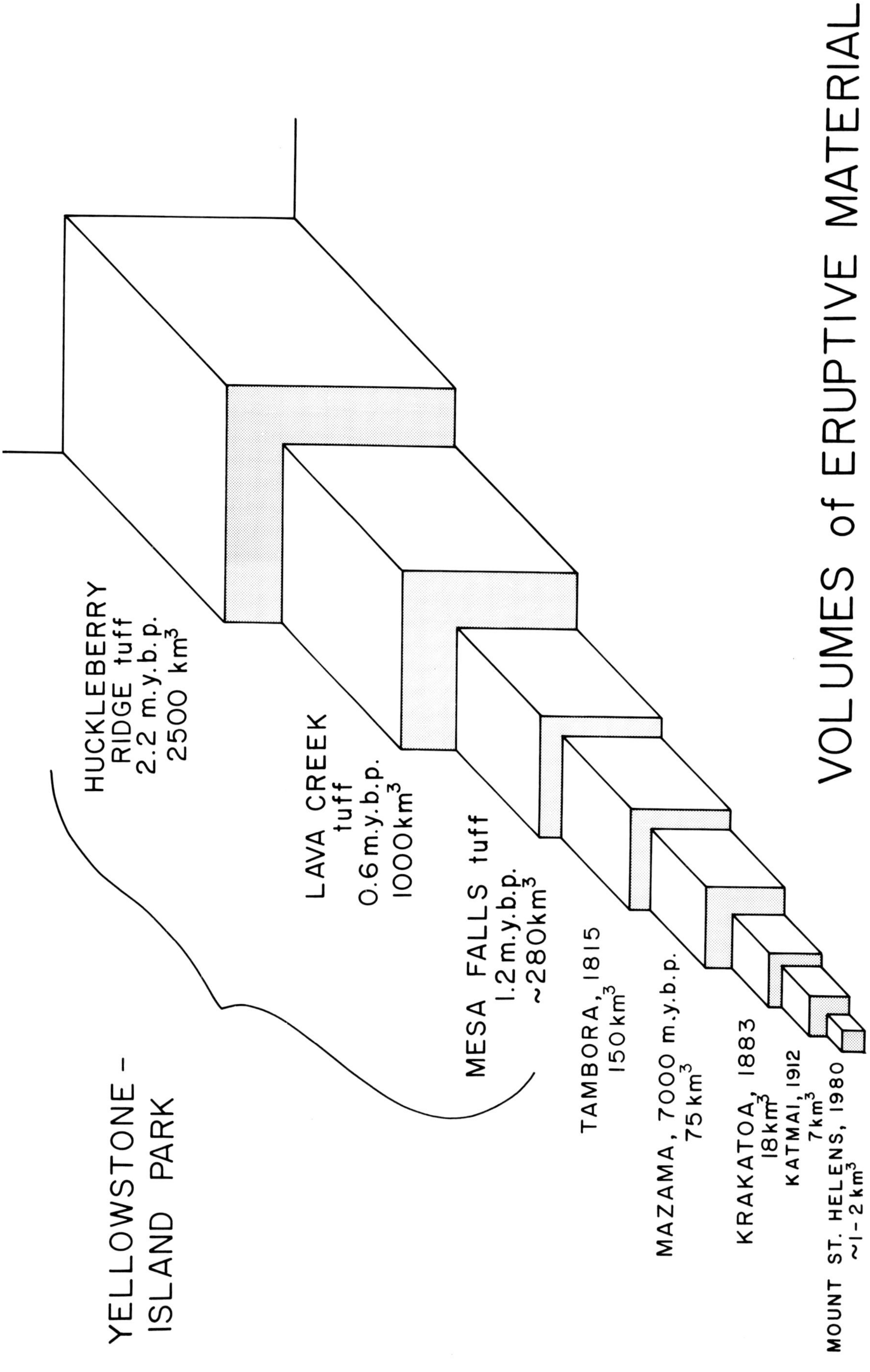

Figure 1: Relative volumes of some well-known volcanic eruptions (from Williams and McBirney, 1979; Decker and Decker, 1981).

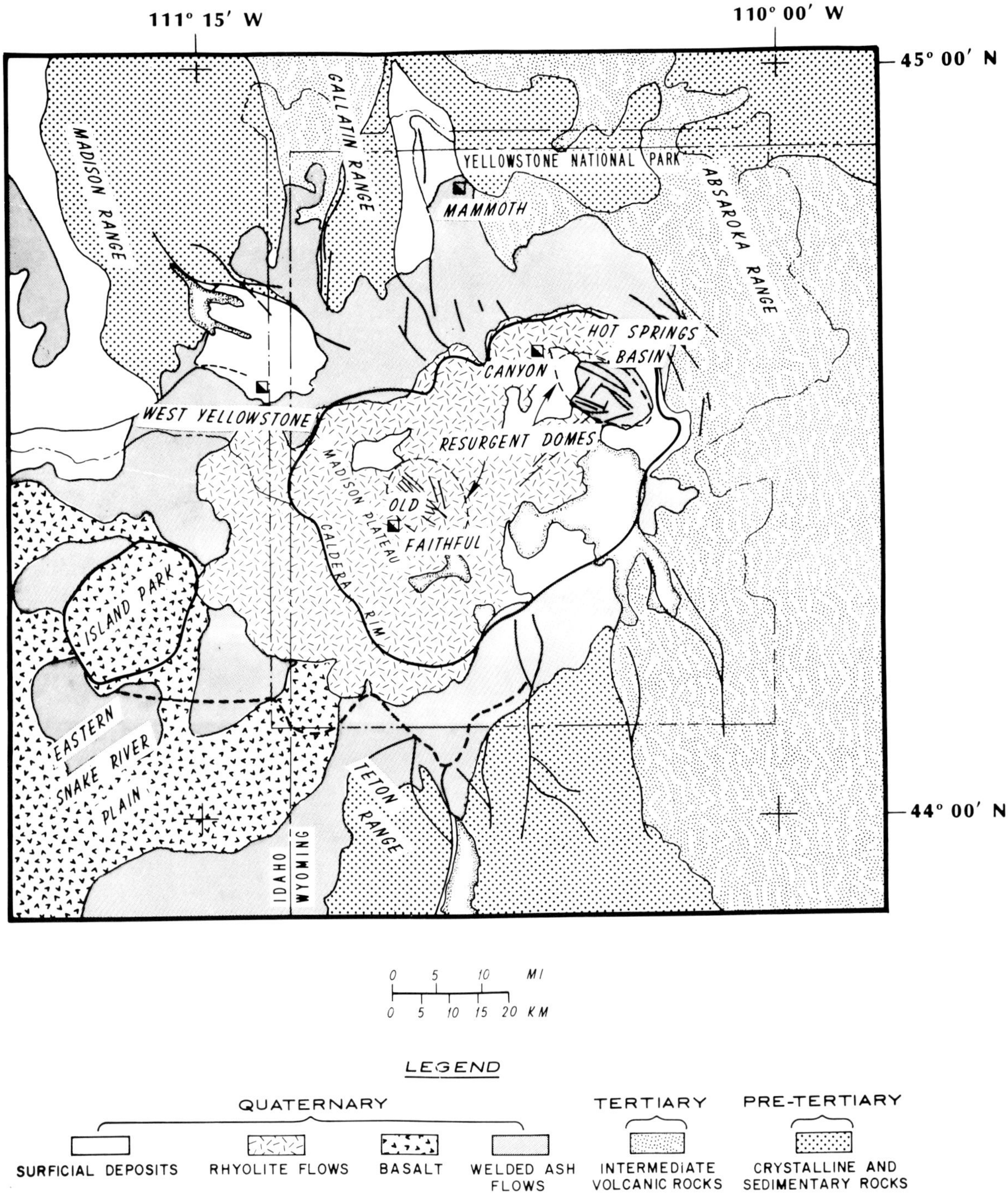

Figure 2: Generalized Late Cenozoic tectonic map of Yellowstone, showing distribution of main volcanic flows, collapse calderas, resurgent domes, and Cenozoic faulting (modified from R. L. Christiansen, U.S. Geological Survey, personal communication, 1978). Heavy line corresponds to boundary of the 0.6-m.y.-old caldera.

GEOPHYSICAL AND GEODETIC MEASUREMENTS

Earthquake Data

Yellowstone National Park and the adjacent Hebgen Lake region to the west have been the most seismically active areas in the Rocky Mountains in historic time with one earthquake of magnitude 7 (the 1959 Hebgen Lake earthquake) and seven magnitude-6 events (Figure 3). Earthquakes have occurred primarily in the western Yellowstone Park area and along the adjacent Hebgen Lake fault zone. The seismicity in the Yellowstone region is part of the Intermountain Seismic Belt, which marks the location where the north-south trending earthquake zone to the south turns toward the northwest. A secondary east-west zone of seismicity including that of the Hebgen Lake fault intersects this regional trend at Yellowstone. This intersection of seismicity zones suggest a major point of intraplate stress reorientation.

Crustal earthquakes are considered to result from stress drops across shear fractures in a domain of brittle failure. Thus their occurrence and distribution can be used to infer the relative state of stress. Maximum focal depths may mark a distinction between brittle and nonbrittle mechanical behavior. To evaluate the contemporary seismicity of the Yellowstone region, a new compilation of hypocenter data has been developed from data of the University of Utah and the U.S. Geological Survey (Smith *et al.*, 1977; Smith and Christiansen, 1980; Pitt, 1981). The new data were selected on the basis of an average rms location error equivalent to less than ±0.2 sec. Most epicenters were located with six or more stations and assumed flat-layered velocity models. More than 3000 epicenters were compiled from over 50 seismograph stations that were located throughout the Yellowstone region for the period 1971-1979. The earthquake map (Figure 3) demonstrates several important characteristics of the Yellowstone seismicity: (1) The seismicity south of the E-W trending Hebgen Lake fault zone extends eastward toward the Yellowstone caldera, then turns southeastward along zones that continue 2 to 4 km into the caldera. (2) Diffuse seismicity occurs throughout the caldera and is characterized by numerous swarms but decreases in frequency of occurrence. Scattered events were distributed throughout the southwest caldera and along N-S and E-W zones that extend radial to the Sour Creek resurgent dome. (3) Epicenters near Yellowstone Lake occurred on N-S trends, and the regional seismicity extends southward beyond the caldera boundary along a zone probably related to the Teton fault system.

The state of stress cannot be determined directly from the earthquake observations, however, directions of the principal components of the stress field can be inferred from the directions of *P*- and *T*-axes of fault-plane solutions (Smith, 1978; Smith and Lindh *et al.*, 1978; Weaver *et al.*, 1979). These data (not shown) suggest that the regional stress distribution of the Yellowstone system is dominated by N-S extension in the Hebgen Lake-Norris Geyser Basin area, but the direction of maximum extension changes to northeasterly in the Yellowstone caldera and eventually rotates to E-W extension in the southern caldera.

While the earthquake data collected to date are not sufficient to evaluate the stress around possible magma bodies, the directions of the minimum compression (*T*-axes) and maximum compression (*P*-axes) are consistent with a concept in which the Yellowstone Plateau is being uplifted, producing horizontal minimum compressive stress and primarily vertical components of maximum compressive stress associated with the gravitational loading. The stress field is also consistent with the inferred for the circular ring fractures that mark the boundary of the Yellowstone caldera as normal faults, downdropped toward the center of the caldera.

Gravity Field and Teleseismic P-*Wave Delays*

Excess heat and advective magma transport from deep lithospheric sources can generate zones of low density and low seismic velocities, parameters that produce measurable gravity anomalies and delays in the travel times of seismic waves. The Earth's gravitational field in the Yellowstone region is shown in Figure 4, which is a complete Bouguer gravity map contoured from data compiled by Blank and Gettings (1974). A striking feature of the gravity data map is the negative, -60 mgal, regional anomaly that closely coincides with the mapped caldera. This large negative anomaly corresponds to a mass deficiency and may reflect shallow low-density material in the crust or larger low-density bodies that extend into the upper mantle. Modeling of the gravity data in conjunction with a velocity model from the seismic refraction-reflection data will be discussed later.

Of particular importance is the delineation of an additional gravity low of -20 mgal at the northeast side of the Yellowstone caldera, which corresponds in location in part with the Sour Creek resurgent dome but extends ~20 km north beyond the caldera. As will be shown later this anomaly coincides with an upper-crustal low-velocity body.

Superimposed on the gravity map are the locations of the regional seismography stations that were used to determine teleseismic *P*-wave delays by Iyer *et al.* (1981). The station locations are shown by large circles that include

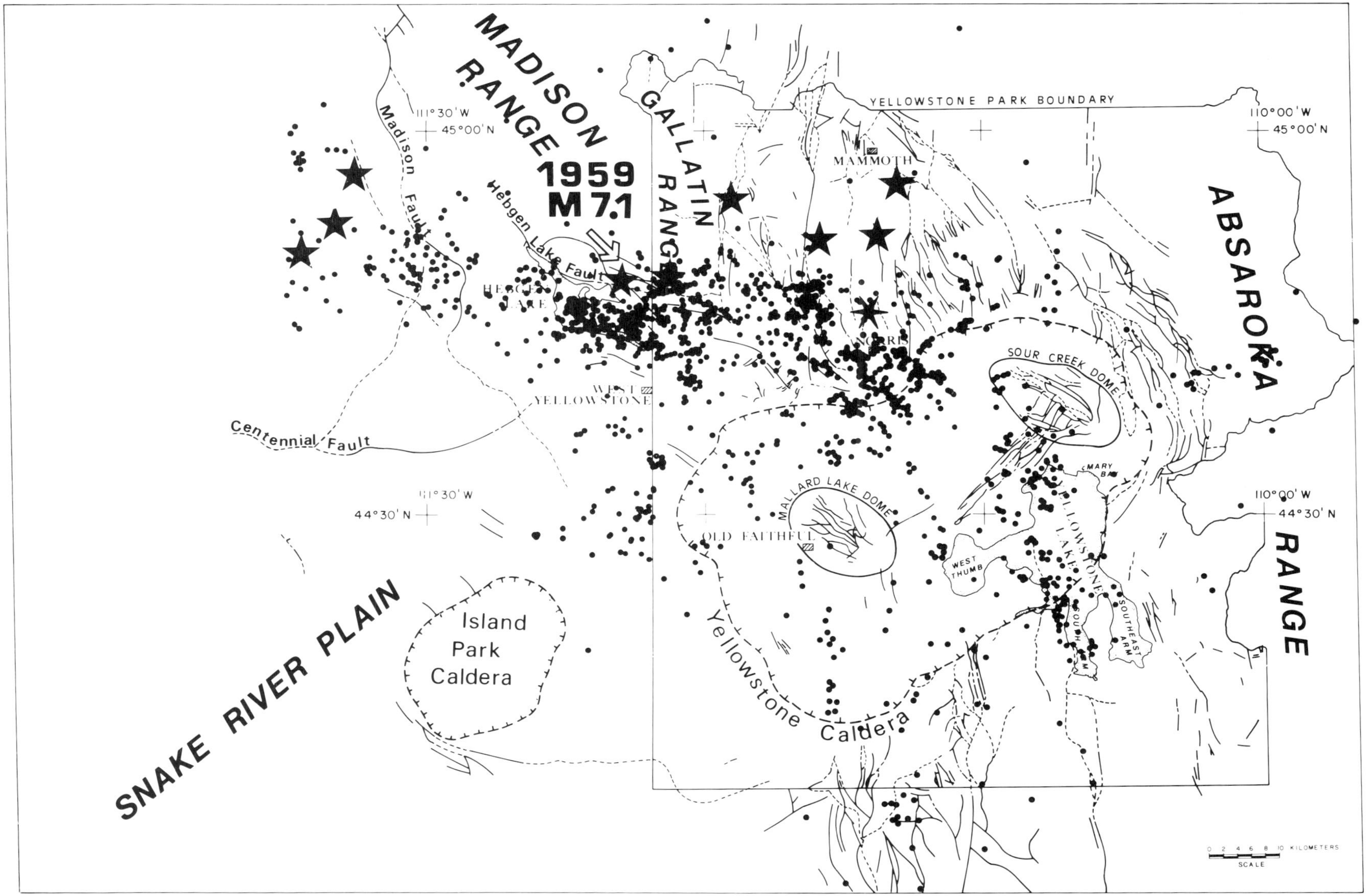

Figure 3: Earthquake distribution of the Yellowstone-Hebgen Lake region (1971-1979). Epicenters are those accurately determined from data of the University of Utah and the U.S. Geological Survey. Faults are taken from a compilation by R. L. Christiansen (U.S. Geological Survey, personal communication, 1978). Calderas are shown corresponding to their respective ages: I, 2.0 m.y.; II, 1.3 m.y.; and III, 0.6 m.y. Large stars correspond to epicenters of magnitude 6-7 historical earthquakes.

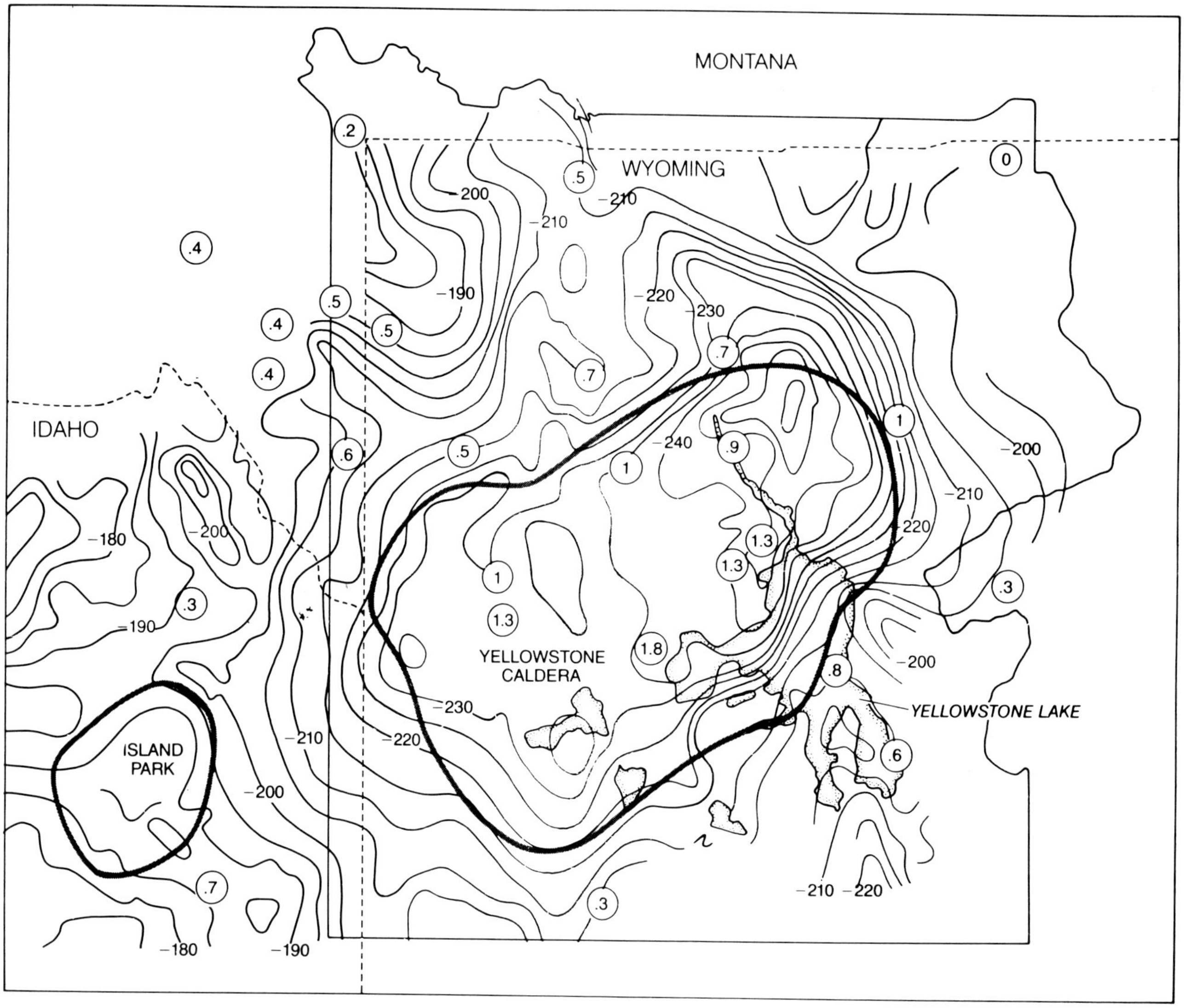

Figure 4: Bouguer gravity anomaly map at a contour interval of 5 mgal (map from Smith and Christiansen, 1980); data recontoured from Blank and Gettings, 1974). Open circles correspond to locations of seismograph stations that recorded teleseismic P-wave delays. Average delays are shown within the circles in seconds (from Iyer et al.,1981). P-wave delays were normalized to the seismograph station at the northeast side of Yellowstone National Park.

the *P*-wave delays (in seconds) relative to a uniform-velocity Earth model. These data clearly show that the Yellowstone Plateau is underlain by material of relatively low seismic velocities necessary to produce *P*-wave seismic delays of up to 1.8 sec. Iyer *et al.* (1981) interpret these delays to represent velocity decreases averaging 10-20 percent throughout the upper crust and 5-10 percent in the lower crust and upper mantle to depths as great as 250 km.

In a joint inversion of the Yellowstone gravity data and the teleseismic *P*-wave delays (Iyer *et al.*, 1981), density and velocity decreases were modeled by Evoy (1977) and demonstrated that the regional gravity low and teleseismic *P*-wave delays could be produced by a concomitant low-density, low-velocity body that corresponds in lateral shape to the outline of the caldera and extends to depths of 100 km. These combined interpretations suggest that the Yellowstone Plateau is underlain by anomalies low-density, low-velocity material that may extend into the asthenosphere and reflect deep magmatic sources. Because of the nonuniqueness of the potential field interpretations the gravity anomalies could also be produced by shallower density anomalies, a topic to be discussed further.

Contemporary Crustal Deformation

If the Yellowstone volcano-tectonic system is tectonically active, then measurements of surface deformation could reveal the causative body. To examine this property, first- and second-order level-line bench marks were re-observed after a 55-year span to calculate relative deformation rates (Pelton and Smith, 1979, 1982). Figure 5 shows a map of orthometric height variations determined from the releveling data portrayed in units of vertical velocity (millimeters/year). These data were determined from level-line reobservations (Pelton and Smith, 1979, 1982) that were converted to displacement rates by fixing a bench mark at the east side of Yellowstone Park as a relative zero base outside the effects of the thermal and seismic deformation. The uplift data were corrected for systematic and random errors and were evaluated for errors resulting from correlation with topography and poor rod calibrations (Pelton and Smith, 1982). These data (Figure 5) demonstrate anomalously high uplift rates over the Yellowstone Plateau with uplift contours that generally parallel the outline of the Yellowstone caldera. Uplift rates averaged greater than about 5 mm/yr within the outer edge of the ring fracture zone and are greatest, up to 15 mm/yr, in the northeast caldera area adjacent to the Sour Creek resurgent dome. An increase from~7 mm/yr to~15 mm/yr northeast across the caldera toward the Sour Creek dome suggests an asymmetry or tilt and may reflect mass transport in the northeast caldera area. The anomalous crustal uplift rates measured in Yellowstone are similar to rates measured in other active volcanic areas of the world such as Hawaii and Iceland.

Over the 55-year monitoring period, a symmetrical uplift of up to 700 mm has occurred coincident with the Yellowstone caldera and raises the question of the longevity of this rapid elevation change. If these high uplift rates continue, within a few million years the inner caldera would be evaluated on the order of kilometers, obviously an unrealistic value. We treat the current uplift to be ephermal as demonstrated by the occurrence of lake terraces along the northwest shores of Yellowstone Lake, which show little evidence of large-scale warping (Pelton and Smith, 1982) and suggest that the deformation on that region is fairly recent. The size and relatively large vertical strain rates that characterize the Yellowstone uplift likely correspond to a causative body in the upper crust that would have a lower viscosity than the cooler, brittle crustal material. Thus the upper crust beneath the caldera appears to behave plastically or visoelastically to long-term strain.

UPPER-CRUSTAL SEISMIC VELOCITY STRUCTURE

Perhaps the most definitive characteristic of the upper crust beneath the Yellowstone Plateau is the low *P*-wave seismic velocity corresponding to the crystalline layer. Information on this layer has been interpreted from delay-time analyses (Lehman *et al.*, 1982) of the critically refracted branch, Pg, that were recorded in the 1978 Y-SRP seismic refraction project and that utilized travel times from a two-dimensional distribution of ray paths that crossed the Yellowstone caldera at various azimuths between the seismic stations and the sources.

The upper-crustal crystalline layer shows an elongate northeast-trending velocity distribution (Figure 6). The 6.0 km/sec velocity layer external to the caldera is interpreted to represent the thermally undisturbed crystalline basement that is interpreted to correspond in composition to rocks like the Precambrian granitic gneiss of the nearby Beartooth Mountains. The boundary of the large 5.7 km/sec body (marked by a 6 percent decrease compared to the thermally undisturbed basement) closely coincides with the outline of the southern Yellowstone caldera but extends to the north approximately 15 km beyond the boundary. Two additional low-velocity zones of 30 percent decrease to 4 km/sec were identified in the southwestern and the northeastern caldera areas.

The crystalline velocity layering is a diagnostic parameter of the upper crust beneath the Yellowstone caldera. The outline of the 5.7 km/sec layer closely corresponds to the large, regional gravity anomaly of -60 mgal (Figure 4) and thus suggests a concomitant density decrease corresponding to the low-velocity body.

The 4.0 km/sec low-velocity layer at the northeast side of the caldera is by itself unusual, but an examination of the Bouguer gravity map (Figure 4) shows that this anomalous body coincides with an additional (-20 mgal) local gravity low, suggesting an additional concomitant low-density, low-velocity volume, whereas the low-velocity body to the southwest shows no corresponding gravity anomaly.

To compare the causative bodies that produced the

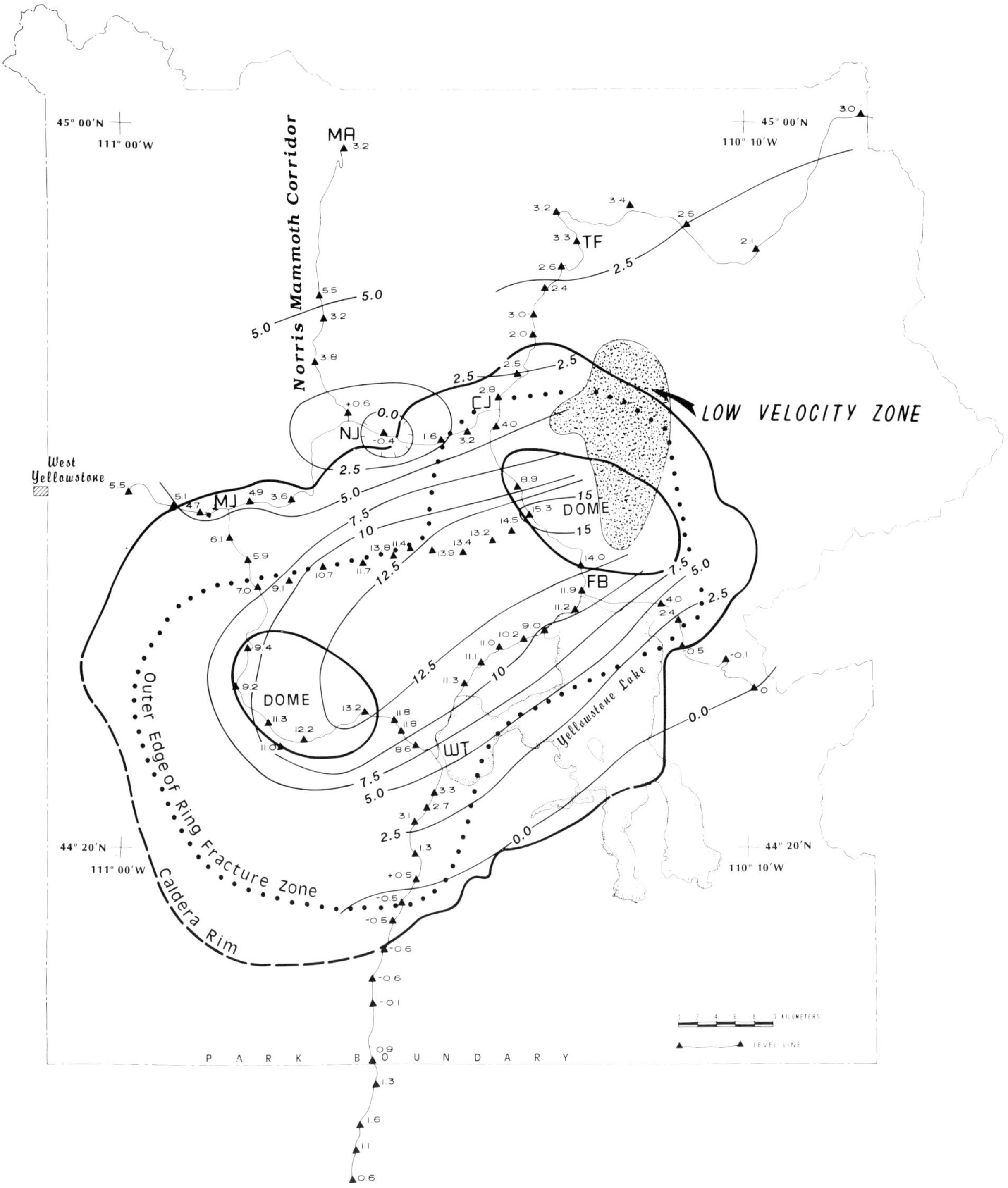

Figure 5: Contour map of relative uplift rates over the Yellowstone Plateau. Contours are in millimeters/year and were computed from the data of Smith and Pelton (1979, 1982). Bench-mark locations shown by triangles; MA, Mammoth; NJ, Norris Junction; CJ, Canyon Junction; FB, Fishing Bridge; WT, West Thumb; TF, Tower Falls.

density and velocity anomalies, three-dimensional gravity modeling was done (Lehman *et al.*, 1982) with the constraints that the density layers corresponded to the general thickness and outline of the velocity layers determined from the seismic refraction data. The seismic data showed that the 5.7 km/sec body did not appear to extend to depths beyond about 10 km.

For the northeast 4.0 km/sec low-velocity body, the assumption of a normal velocity-density decrease was not sufficient to model both parameters, and an additional constraint was required, namely, that the causative body, in order to have a concurrent 30 percent velocity reduction and a 10 percent density reduction, required a fluid constituent or an unlikely compositional variation. Theoretical models of seismic-wave propagation in fluid-filled material (O'Connell and Budiansky, 1977; Mavko, 1980) and experimental measurements of *P*-wave velocities in melts (Murase and Suzuki, 1966; Murase and McBirney, 1973) suggest a decrease in the velocity due to partial melting of the order reported here. The gravity and seismic modeling of the anomalous northeast low-velocity body provide an additional constraint — the density-velocity ratio requires a major percentage of fluid that may range from 10 percent to 50 percent of water or silicic partial melt. While this interpretation is not unique it would be unlikely that low velocities from a highly porous gas- or steam-filled system would produce the relatively small density change. A large compositional variation to low-density, low-velocity material such as high-porosity sediments could partly account for the velocity and density anomalies, but the size and depth extent of the anomalous body make that interpretation also unlikely. As a working hypotheses we consider that the northeast caldera low-velocity, low-density body may correspond to an upper-crustal fluid-filled (staem or partial melt) silicic body.

The southwestern low-velocity body, on the other hand, does not have an accompanying gravity anomaly and was interpreted by Lehman *et al.* (1982) to be more representative of a fractured, fluid-saturated zone producing attenuation, velocity reduction, or both. It may represent a northward extension of a Cenozoic fault system beneath the Yellowstone Plateau.

To examine the anomalous physical parameters of the Yellowstone system, Figure 7 shows a NW-SE upper-crustal cross section through the Yellowstone-Hebgen Lake region that portrays (1) average *P*-wave delays from teleseismic arrivals (Iyer *et al.*, 1981), (2) geodetically measured crustal deformation (Pelton and Smith, 1979), (3) depths of the most-accurately determined hypocenters, and (4) averaged *P*-wave velocities from the combined results of the Y-SRP seismic experiment following the interpretations of Lehman *et al.* (1982), Schilly *et al.* (1982), and Smith *et al.* (1982). These data show a correspondence for the area of the marked increase in *P*-wave teleseismic delays to greater than 1.0 sec with the area of greatest uplift — maxima of both parameters appear centered along the axis of the Yellowstone caldera.

The maximum depths of earthquakes (Figure 7) shallow from ~ 20 km in the Madison Valley west of Yellowstone to 5 km or less beneath the Yellowstone caldera. The lateral extent of deepest foci shallow markedly toward the caldera and appear to decrease in depth to 1-2 km near the top of the 5.7 km/sec low-velocity layer. Note that relatively few earthquakes occurred within the 5.7 km/sec layer with the exception of a cluster at the northwest edge of the caldera, which corresponds to aftershocks of the magnitude-6.0 June 30, 1975, earthquake (Figure 3).

Modeling of plastic-elastic deformation based on strain-rate calculations of quartz-olivine systems has led Smith (1981) to the conclusion that maximum depths of earthquakes in continental crustal environments such as Yellowstone may be controlled by temperature and composition, where cessation of brittle failure is modeled at a critical temperature of 400 ± 50°C at which point the strain release is in the form of plastic deformation. If this model is applicable to the upper crust of Yellowstone, the maximum depth of earthquakes may be used to infer the depth of this boundary isotherm. The shallow focal depths thus appear to be constrained to a shallow layer characterized by brittle failure to depths marked by an isotherm of approximately 400°C (Smith, 1981). This boundary also appears to mark the lateral transition from the higher-velocity, thermally undisturbed 6.0 km/sec crystalline layer to the lower-velocity, 5.7 km/sec layer. The data on this profile do not directly map magma bodies, but they show that the upper 10 km of the crust beneath the 'yellowstone caldera is marked ny a low*P*-wave velocity and apparently a hot crust that restricts brittle fracture.

In Figure 8 a SW-NE upper-crustal cross section parallels the axis of the Yellowstone caldera. In this profile rhyolite flows and caldera fill are shown from the determinations of Lehman *et al.* (1982) and Schilly *et al.* (1982). The 4.0 km/sec body is interpreted to be of granitic composition on the basis of low velocity and low density and correlation with the inferred source for the rhyolite flows. To the northeast, the Precambrian granitic basement and possible buried Paleozoics sediments are overlain by Tertiary volcanics. Also plotted is the relative *Pg* attenuation that has been determined by the spectral ration technique (Clawson and Smith, 1981) for the laterally varying velocity structure. This averaged apparent attenuation profile is shown at the top of Figure 8 in a block

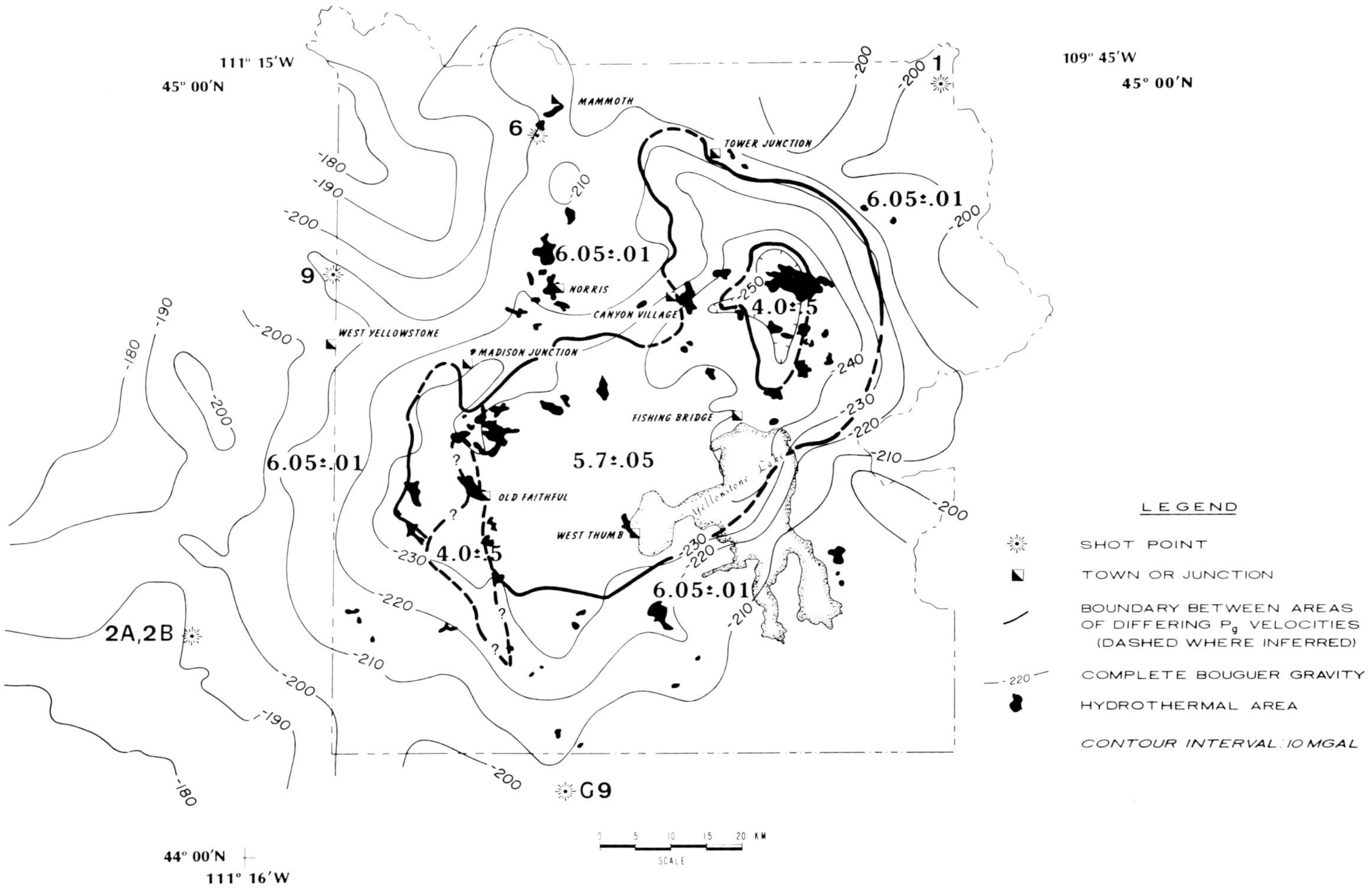

Figure 6: Areal extent of crystalline upper-crustal layer (2-10 km depth). P-wave velocities from Lehman et al. (1982) were determined from delay-time analyses. Velocities are in kilometers/second.

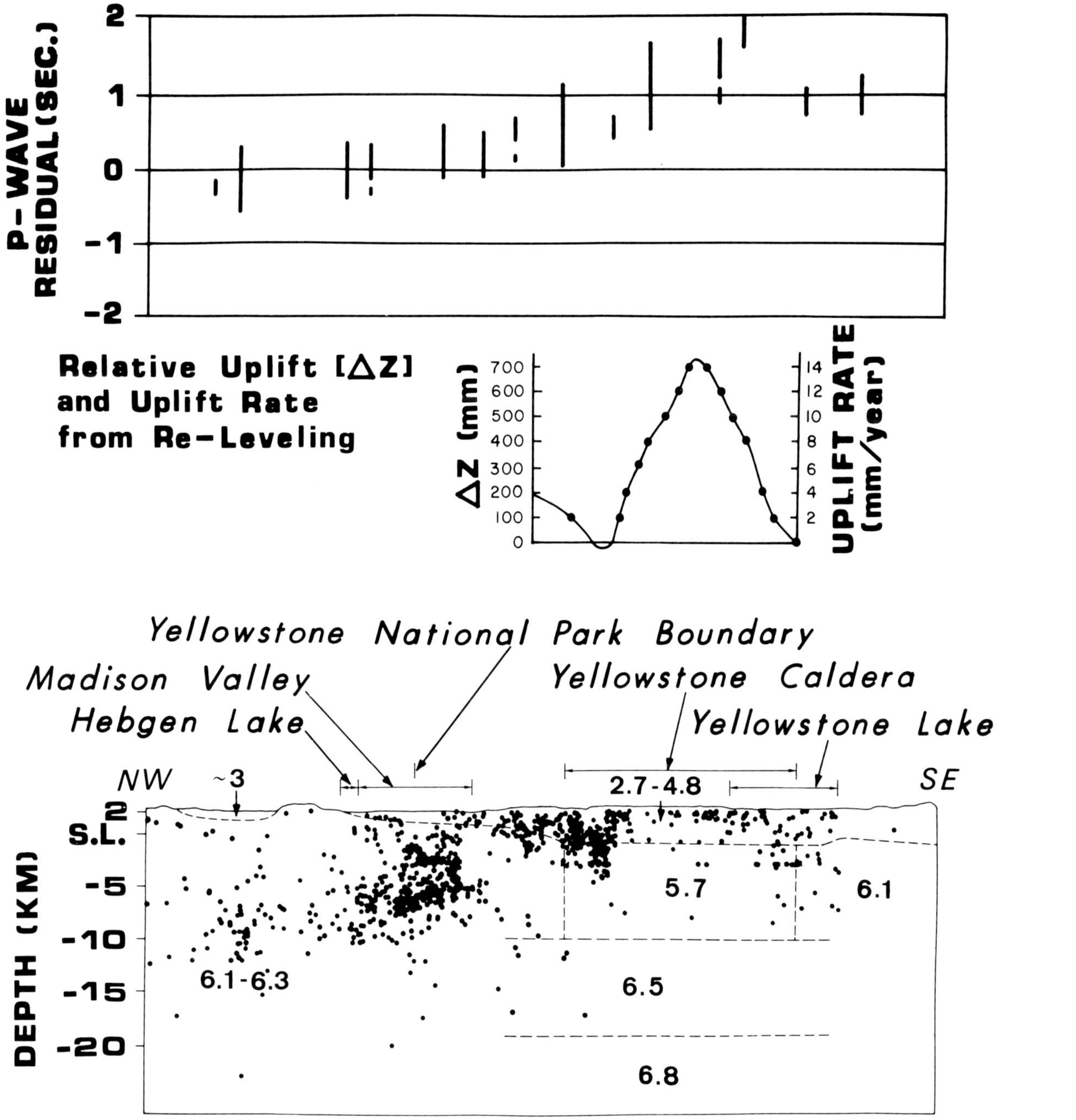

Figure 7: Generalized NW-SE cross section of crustal parameters across the Yellowstone-Hebgen Lake region. Teleseismic P-wave delays from Iyer et al(1981); relative uplift and uplift rates from Pelton and Smith (1979); focal depths from new earthquake compilation (Benz and Smith, 1981; Pitt, 1981); seismic velocities from interpretations of Schilly et al. (1982), Smith et al. (1982), and Lehman et al. (in press). Focal depths are shown by black dots.

diagram. Note that the apparent-Q is high over the northeast low-velocity body and the southwest caldera.

LOCATION AND STATE OF POSSIBLE MAGMA SOURCES

The upper- and intermediate-depth crustal structure of the Yellowstone Plateau is shown in Figure 9. Locations of anomalous bodies are based on the inference of shallow granitic volumes from the relationship to surface geology, from the seismic velocity structure, from the high heat flow, and by distribution of hydrothermal features of the Yellowstone Plateau. Also included is an interpretation of the state and composition based on gravity-velocity modeling. The pervasive 5.7 km/sec layer appears to represent a hot but plastically deforming body of granitic composition, most likely a residium of Quaternary Yellowstone rhyolite flows and ash-flow tuffs. The 4.0 km/sec body near the southwest caldera boundary appears to be related to a water-saturated fracture system, perhaps a buried extension of the Teton fault zone, whereas the 4.0 km/sec low-velocity body of the northeast caldera boundary is interpreted to correspond to a volume of material that is consistent with a silicic body that could range in composition from a highly porous steam-water system to a 10 to 50 percent partial melt — an interpretational model that is not resolvable with the existing data but is considered as a working hypothetical model. It is important to note that the 5.7 km/sec and the 4.0 km/sec layers appear to be bounded at a depth of about 10 km by a 6.5 km/sec layer that has been shown to extend along the length of the Snake River Plain as a pervasive intermediate crustal layer, perhaps the remnants of the undersaturated mafic upper crustal material.

The sources of magma and crustal evolution accompanying the development of the Yellowstone Snake River Plain volcanic-tectonic system have been interpreted by Smith (1981), Braile *et al.* (1982), and Smith *et al.* (1982) based on the seismic velocity structure, the Quaternary volcanic history, petrologic models (Christiansen, in press; Lehman, in press). These data suggest that the P-wave velocity structure of the crust along the Y-SRP volcanic system is characterized by systematic variations that are consistent with a model in which mantle-derived basalts ascend buoyantly to the base of a cold Archean crust heating, partially melting the upper mantle, and producing a P-wave, low-velocity body. In our seismic model for the crust, the basaltic magma ascends into the lower crust but apparently without significant perturbation of the seismic velocity structure. This conclusion is based on the similarity of the lower crustal (6.8 km/sec) layer beneath the Yellowstone-Snake River Plain system and that of the surrounding undisturbed Rocky Mountain crust. This suggests that the mantle derived basalt ascended into the lower crust without mechanically deforming or modifying the seismic velocity structure.

On ascent into the upper crust, partial melts transport heat advectively upward and in turn could produce additional silicic components that further ascend toward the surface. The final stage produces tumescence and thermal expansion prior to explosive eruptions. A residual mafic body may be marked by the ~6.5 km/sec layer that appears well developed beneath the Snake River Plain between 10- and 20-km depth, whereas beneath Yellowstone it is less clearly defined. The shallow 5.7 km/sec layer is interpreted as a body of granitic composition accumulated from numerous crustal intrusions. As the parental magma source is depleted and in its low-temperature components by eruptions, the crust is cooled and becomes refractory, thereby reducing the surface elevation and completing the main silicic volcanic cycle. Finally, basaltic parent magmas follow the silicic components and are extruded giving the biomodal rhyolite/basalt complement; however, we have no evidence for shallow basaltic bodies from the geophysical data gathered at Yellowstone.

Petrologic models by Christiansen (in press) and Leeman (in press) suggest that the Yellowstone rhyolitic magmas were derived from one or more of the following sources: (1) fractional melting, (2) separation of inermediate composition liquids into silicic fractions, and (3) partial fusion of the lower crust producing a felsic melt. Lead and strontium isotope ratios (Doe, *et al.*, 1982) infer that the rhyolitic magmas at Yellowstone did not interact extensively with upper-crustal materials, i.e., they did not form by partial melting of once-sedimentary, upper-crustal rocks but rather by partial melting of high-grade metamorphic rocks of the lower crust. An important implication of the nonunique relationship between composition and velocity is in the lower crust of Yellowstone, where the process as modeled petrologically did not significantly alter the velocity structure as measured by the critically refracted (horizontally propagating) head waves. A quandry thus exists; the lower crust shows normal 6.8-km/sec P-wave velocity (no percentage reduction) from the seismic refraction data but a 5-10 percent decrease in the lower crust from generalized inversion of telesesimic P-wave delays of vertically incident rays (Iyer *et al.*, 1981). Possible explanations of the differences in the seismic and petrologic implications are (1) the residual lower-crustal layer may have a composition different than the original thermally undisturbed crust but with the same P-wave velocity; (2) mafic magmas from the mantle may have

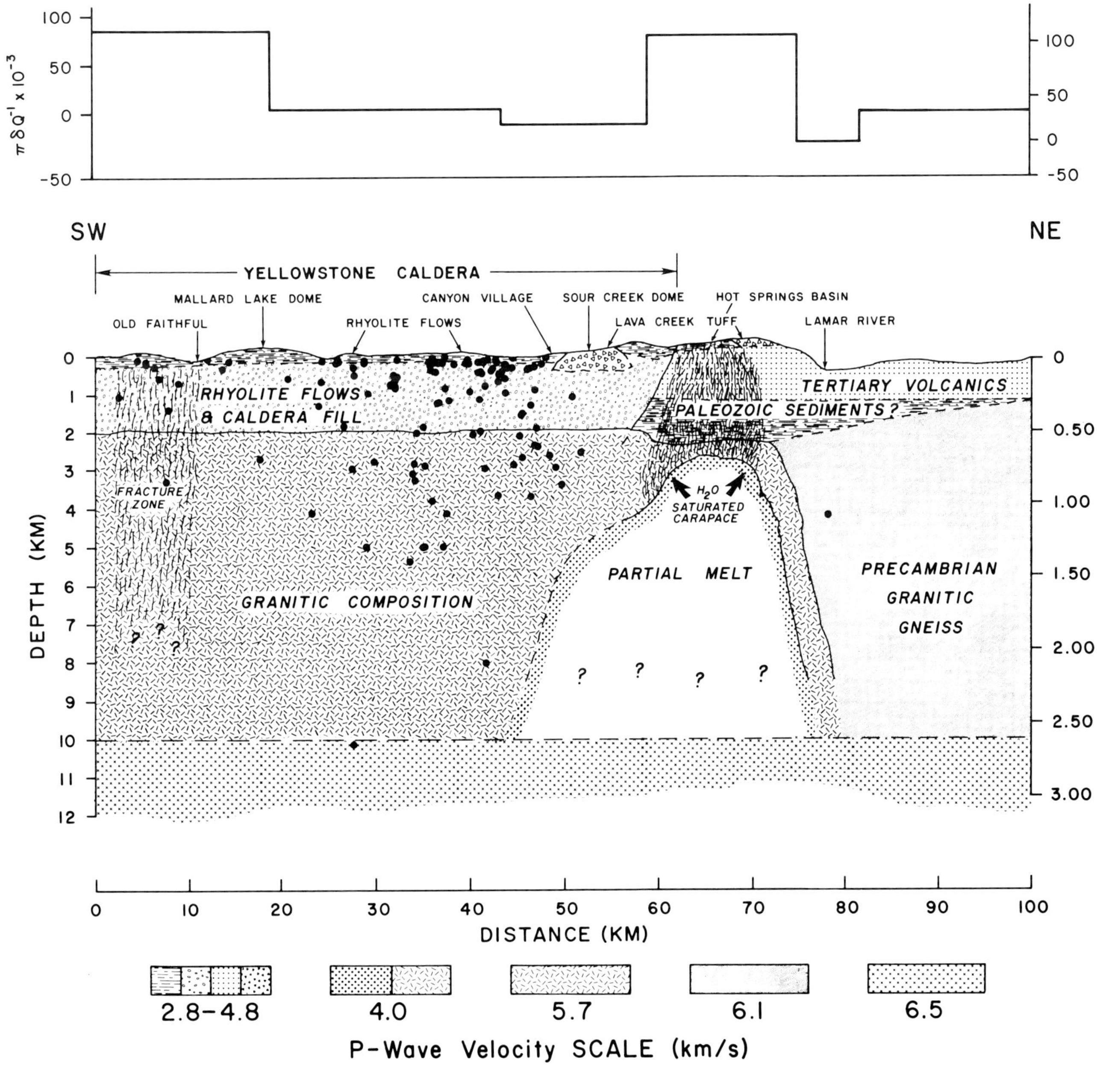

Figure 8: Generalized SW-NE upper-crustal cross section parallel to the axis of the Yellowstone caldera. Focal depths (shown by black dots) were projected from outside the low-velocity body. Velocities correspond to sources in Figure 7. Relative P-wave seismic attenuations are averaged from data of Clawson and Smith (1981).

induced the lower crust along vertically oriented plexes, which are not large enough to be resolved by the refracted waves; (3) the lower crust may be segregated into horizontally layer, sill-like low-velocity bodies in a high velocity matrix that produces an apparent seismic anisotropy, i.e., refracted waves propogate at 6.8 km/sec, whereas the vertically incident teleseismic rays pass through a residium of variable-composition low- and high-velocity material with an average vertical velocity 10-20 percent slower than normal; (4) true vertical-horizontal oriented seismic anisotropy produced by oriented silicate layering; and (5) frequency-dependent *P*-wave velocity — frequencies of telesesimic waves had energy in the 0.1-1 Hz band compared with the frequency band of 3-12 Hz for the refraction/wide-angle reflections. The last explanatiom has important implications to velocity determination of hot material as discussed by Daniel and Boore (1982), who determined averaged crustal shear velocities at Yellowstone from Raleigh-wave dispersion and teleseismic *P*-wave delay by fixing the *P*-wave velocity at 6.8 km/sec and utilizing *S*-wave velocity delays of 2-9 sec, determined Poisson's ratios of 0.32 for the lower crust and 0.42 for the upper crust, values that can normally be attributed to the presence of fluids (water or magma) or to subsolidus relaxation (Bonner *et al.*, 1981). Daniel and Boore (1982) point out that frequency-dependent velocity and attenuation can be produced by (1) melt squirts and phase changes, (2) frequency-dependent softening, (3) subsolidus grain boundary relaxation, and (4) thermally activated lattice defects. The recent laboratory work by Bonner *et al.*,(1981) suggests frequency-dependent moduli of partly melted quartz manganite in the sesimic band at simulated crustal pressure and temperature and may account for mechanisms (1) and (2). However, mechanisms (3) and (4) may permit the existence of a shear-wave velocity reduction in hot but solid rock, a hypothesis not thoroughly investigated but important to seismic velocity determination in thermal environments.

SPECULATION ON POSSIBLE FUTURE EXPLOSIVE ERUPTIVE ACTIVITY AT YELLOWSTONE

The primary question with respect to explosive volcanism is the potential for future eruptions at Yellowstone. The seismic velocity structure (Figures 7-9) taken together with the Quaternary geologic record suggests that melting and crystallization of the upper crust has generated a large (5.7 km/sec) body that was the source for the Quaternary rhyolites and ash flows. On the basis of the crustal deformation, this body appears to be relatively solid but deforming plastically or viscoelastically.

The pocket of low-velocity (4.0 km/sec) material beneath the northeast caldera boundary suggests that a fluid-filled body of steam-water, partial melt, or both, as existed recently. Whether it is in a stage prior to an eruption cannot be determined from the available data. However, the trend of increasing crustal uplift (at rates greater than 15 mm/yr and in excess of normal nonthermal processes) toward the northeast points toward this low-velocity body. The coincidence of the low-density, low-velocity body near an area of increasing uplift suggests that the northeast caldera could be a candidate for eruption. However, on the basis of the relatively small volume of the low-velocity layer, an eruption from a source of the size of the 4.o km/sec body would probably not be explosive like the earlier catastrophic eruptions. On the other hand, the state of stress in the crust may reflect "breathtaking," in which the influx and solidification of upper-crustal melts vary with time, producing inflation and deflation without eruption. We do not have data that can differentiate these models, but we should be aware of the potential hazard suggested by their inferences.

The detailed crustal structure from seismic measurements and the inference of the volcanic evolution of the Yellowstone silicic system in relation to its well-known Quaternary history suggest that the volcanic evolution has not ceased. There is not yet enough information to make specific predictions, but the past volcanism, contemporary uplift, length of time between volcanic eruptions, swarms of earthquakes, recent volcanism in the caldera, high rate of uplift, and presence of the shallow heat source in the upper crust all point to a future of continued volcanism in the Yellowstone Plateau. The very high heat flow, the teleseismic *P*-wave delays, the anomalous upper-crustal low-velocity layers, and the apparent shallowing of an isotherm separating brittle fracture from stable sliding are properties that are consistent with the probable existence of shallow hot bodies at least one of which may contain a large volume of fluid (steam-water, partial melt, or both).

The evidence does not indicate what form the future volcanism might take; it could consist of small eruptions of rhyolite and basalts at the margins of the Plateau, medium-size eruptions of rhyolite magma within the caldera, or a major ash flow as suggested by Smith and Christiansen (1980). We believe that any renewal of volcanism within the Yellowstone Plateau would probably be preceded by phenomena such as localized earthquake activity, variation in seismic velocity, increased gas emission, and increased crustal deformation — properties that can be assessed in a continuous manner by modern instrumentation.

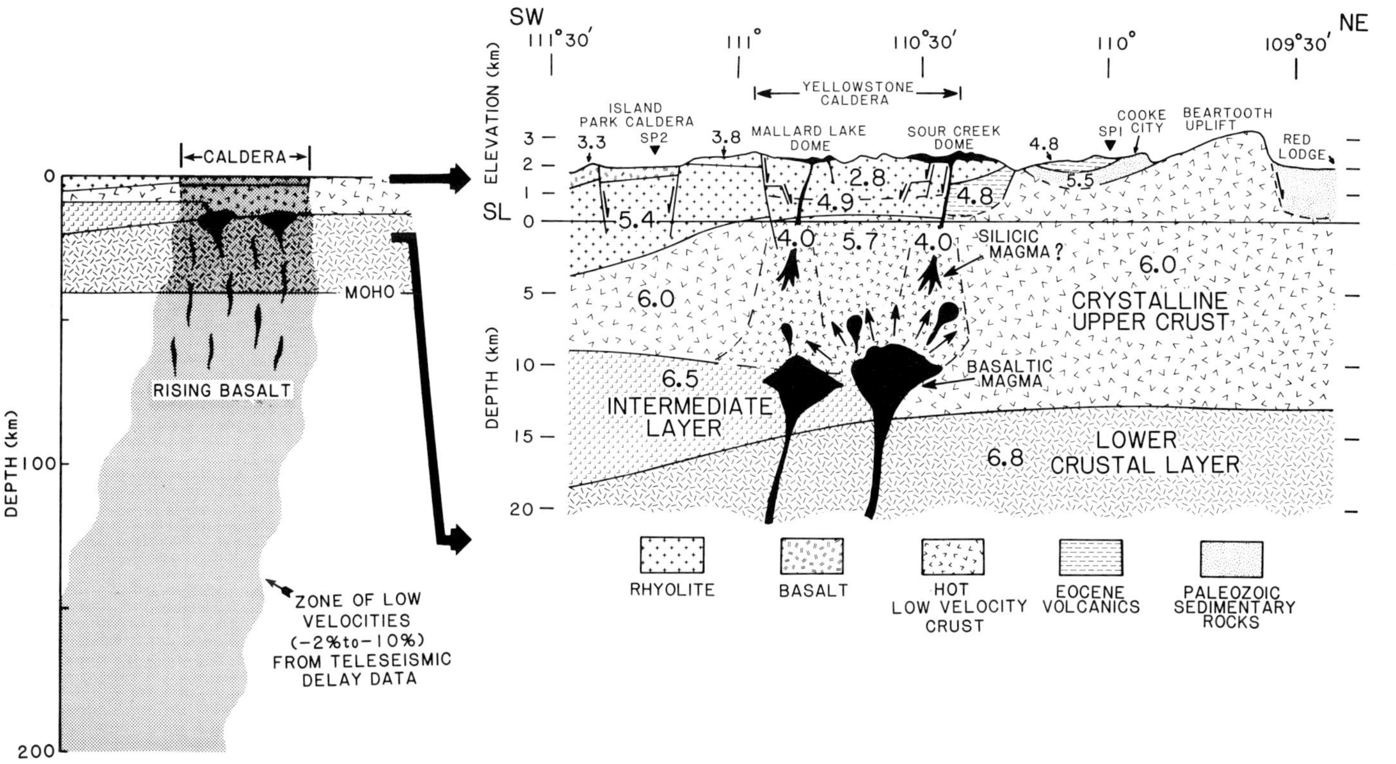

Figure 9: Idealized NE-SW geologic-seismic velocity model for the crustal structure of the Yellowstone-Island Park - Snake River Plain region. P-wave velocities are in kilometers/second (from Braile et al., 1982; Smith et al., 1982). Geologic interpretations are based on constraints of the Quaternary history and the petrologic models of Leeman (in press) and Christiansen (U.S. Geological Survey, personal communication, 1978; in press).

ACKNOWLEDGEMENTS

This paper synthesized geophysical and geologic information obtained by us and our colleagues. We appreciate their cooperation and discussions. In particular, R. L. Christiansen has kindly shared his geologic perspectives, provided preprints, and broadened our geophysical interpretations. H. Benz, M. Brokaw, and S. Clawson assisted in data compilations and figure preparation. A. M. Pitt shared some U.S. Geological Survey earthquake data with us. W. P. Leeman provided a preprint that was helpful for our discussions. Personnel of the National Park Service, Yellowstone National Park (J. Townsley, W. Hamilton, A. Mebane, J. Phillips, and M. Meagher) were very helpful in coordinating and planning our field work. We thank R. L.Christiansen, W. P. Nash, G. A. Thompson, and H. M. Iyer for their critical review and thoughtful comments on this manuscript. Support of the research discussed here was from the National Science Foundation (Grant EAR 77-23707 to the University of Utah and Grant EAR 77-23357 to Purdue University) and from the U.S. Geological Survey, Geothermal Research Program (Grant 14-08-0001-G-771) to the University of Utah and Grant 14-08-0001-G-532 to Purdue University).

REFERENCES CITED

Amstrong, R. L., W. P. Leeman, and H. E. Malde (1975). K-Ar dating, Quaternary and Neogene volcanic rocks of the Snake River Plain, Idaho, *Am. J. Sci. 275*, 225-251.

Benz, H. M., and R. B. Smith (1981). Simultaneous inversion for lateral velocity variations and hypocenters in the Yellowstone region using earthquake and controlled source data, *EOS 62*, 470.

Blank, H. R., Jr., and M. E. Gettings (1974). Complete Bouguer gravity map: Yellowstone-Island Park Region, *U.S. Geological Survey Open-File Map 74-22.*

Bonner, B. P., L. Thigpen, and G. W. Hedstrom (1981). An elastic creep of partly melted quartz monzonite with implications for the frequency dependence of *Q* (abstract), *Earthquake Notes 52*, 33.

Braile, L. W., R. B. Smith, J. Ansorge, M. R. Baker, M. A. Sparlin, C. Prodehl, M. M. Schilly, J. H. Healy, S. T. Muller, and K. H. Olsen (1982). The Yellowstone-Snake River Plain seismic profiling experiment: Crustal structure of the eastern Snake River Plain, *J. Geophys. Res. 81,* 2597-2609.

Christiansen, R. L. (in press). Yellowstone magmatic evolution: Its bearing on understanding large-volume explosive volcanism, in *Explosive Volcanism: Inception, Evolution, and Hazards,* Geophysics Study Committee, National Academy Press.

Christiansen, R. L., and E. H. McKee (1978). Late Cenozoic volcanic and tectonic evolution of the Great Basin and Columbia Intermountain regions, in *Cenozoic Tectonics and Regional Geophysics of the Western Cordillera,* R. B. Smith and G. P. Eaton, eds., Geol. Soc. Am. Mem. 152, 283-311.

Clawson, S. R., and R. B. Smith (1981). The lateral *Q*-structure of the upper crust in Yellowstone National Park from seismic refraction data, *EOS 62,* 970.

Daniel, R. G., and D. M. Boore (1982). Anomalous shear wave delays and surface wave velocities at Yellowstone caldera, Wyoming, *J. Geophys. Res. 87,* 2731-2744.

Decker, R., and B. Decker (1981). *Volcanoes,* W. H. Freeman and Co., San Francisco, Calif., 255 pp.

Doe, B., W. P. Leeman, (1982).

Eaton, G. P., R. L. Christiansen, H. M. Iyer, A. N. Pitt, D. R. Mabey, H. R. Blank, Jr., I. Zeitz, and M. E. Gettings (1975). Magma beneath Yellowstone National Park, *Science 188,* 787-796.

Evoy, (1977).

Fournier, R. O., D. E. White, and A. H. Truesdell (1976). Convective heat flow in Yellowstone National Park, in *Proceedings of the Second U.N. Symposium on Development and Use of Geothermal Resources 1,* U.S. Government Printing Office, Washington, D. C., pp. 731-739.

Hill, D. P. (1976). Structure of Long Valley caldera, California, from a seismic refraction experiment, *J. Geophys. Res. 81,* 745-753.

Iyer, H. M., J. R. Evans, G. Zandt, R. M. Stewart, J. M. Cookley, and J. N. Rolof (1981). A deep low velocity body under the Yellowstone caldera, Wyoming: Delineation using teleseismic *P*-wave residuals and tectonic interpretations: Summary, *Geol. Soc. Am. Bull. 92,* 792-798, 1471-1646.

Leeman, W. P. (in press). Development of the Snake River Plain-Yellowstone Plateau Province: An overview and petrologic model, *Bull. Idaho Bur. Mines and Geology.*

Lehman, J. A., R. B. Smith, M. M. Schilly, and L. W. Braile (1982). Upper crustal structure of the Yellowstone caldera from delay time analyses and gravity correlations, *J. Geophy. Res. 87, 2713-2730.*

Mavko, G. M. (1980). Velocity and attenuation in partially molten rocks, *J. Geophys. Res. 85,* 5173-5189.

Morgan, P., D. D. Blackwell, R. E. Spafford, and R. B. Smith (1977). Heat flow measurements in Yellowstone Lake and the thermal structure of the Yellowstone caldera, *J. Geophys. Res. 82,* 3719-3732.

Murase, T., and A. R. McBirney (1973). Properties of some common igneous rocks and their melts at high temperatures, *Geol. Soc. Am. Bull. 84,* 3563-3592.

Murase, T., and T. Suzuki (1966). Ultrasonic velocities of longitudinal waves in molten rocks, *J. Fac. Sci. Hokkaido U. Serv. 11 VII 2,* 272-285.

O'Connell, R. J., and B. Budiansky (1977). Viscoelastic properties of fluid-saturated cracked solids, *J. Geophys. Res. 82,* 5719-5735.

Pelton, J. R., and R. B. Smith (1979). Recent crustal uplift in Yellowstone National Park, *Science 206,* 1179-1182.

Pelton, J. R., and R. B. Smith (1982). Contemporary vertical surface displacements in Yellowstone National Park, *J. Geophys. Res. 87,* 2745-2751.

Pitt, A. M. (1981). Catalogs of earthquakes in Yellowstone National Park — Helsgen Lake region from November to December 1975, *U.S. Geol. Survey. Open-File Report 80-2006.*

Reasenberg, P., W. Ellsworth, and A. Walter (1980). Teleseismic evidence for a low-velocity body under the Coso geothermal area, *J. Geophys. Res. 85,* 2471-2483.

Sanford, A. L., R. P. Mott, Jr., P. J. Shuleski, E. J. Rinehart, F. J. Caravella, R. M. Ward, and T. C. Wallace (1977). Geophysical evidence for a magma body in the crust in the vicinity of Socorro, New Mexico, in *The Earth's Crust,* J. G. Heacock, ed., Geophysical Monograph 20, American Geophysician Union, Washington, D.C., pp. 385-403.

Schilly, M. M., R. B. Smith, L. W. Braile, and J. Ansorge (1982). The 1978 Yellowstone-eastern Snake River Plain seismic profiling experiment: Data and detailed crustal structure of the Yellowstone region, *J. Geophys. Res. 87,* 2692-2704.

Smith, R. B. (1978). Seismicity crustal structure, and intraplate tectonics of the interior of the Western Cordillera, in *Cenozoic Tectonics and Regional Geophysics of the Western Cordillera,* R. B. Smith and G. P. Eaton, eds., Geol. Soc. Am. Mem. 152, pp. 111-144.

Smith, R. B. (1981). Listric faults and earthquakes, What evidence? *EOS 62,* 961.

Smith, R. B., and R. L. Christiansen (1980). Yellowstone Park as a window on the Earth's interior, *Sci. Am. 242,* 194-117.

Smith, R. B., and A. Lindh (1978). A compilation of fault plane solutions of the western United States, *in Cenozoic Tectonics and Regional Geophysics of the Western Cordillera,* R. B. Smith and G. P. Eaton, eds., Geol. Soc. Am. Mem. 152, pp. 107-110.

Smith, R. B., R. T. Shuey, R. Freidline, R. Otis, and L. Alley (1974). Yellowstone hot spot: New magnetic and seismic evidence, *Geology 2,* 451-455.

Smith, R. B., R. T. Shuey, J. R. Pelton, and J. P. Bailey (1977). Yellowstone hot spot: Contemporary tectonics and crustal properties from earthquakes and aeromagnetic data, *J. Geophys. Res. 82,* 3665-3676.

Smith, R. B., M. M. Schilly, L. W. Braile, H. A. Lehman, J. Ansorge, M. R. Baker, C. Prodehl, J. H. Healy, S. Muller, and R. W. Greenfelder (1982). The 1978 Yellowstone-eastern Snake River Plain seismic profiling experiment: Crustal structure of the Yellowstone region and experiment design, *J. Geophys. Res. 87,* 2583-2596.

Trimble, A. E., and R. B. Smith (1975). Seismicity and Contemporary tectonics of the Hebgen Lake-Yellowstone Park region, *J. Geophys. Res. 80,* 733-741.

Weaver, C. S., A. M. Pitt, and D. P. Hill (1979). Crustal spreading direction of the Snake River Plain-Yellowstone system (abstract), *EOS 60,* 946.

Williams, H. P., and A. R. McBirney (1979). *Volcanology,* Freeman, Cooper and Co., San Francisco, 397 pp.

Thirty-Third Annual Field Conference — 1982
Wyoming Geological Association Guidebook

ROCKY MOUNTAIN LARAMIDE-TERTIARY SUBSURFACE SOLUTION VS. PALEOZOIC KARST IN MISSISSIPPIAN CARBONATES

BY L.W. DAN BRIDGES[1]

INTRODUCTION

The extent and power of Laramide-Tertiary subsurface solution has heretofore not been recognized. High-pressure, heated subsurface waters related to thermal springs, tectonism, mineralization, vulcanism, and dolomitization is invoked to account for all Rocky Mountain Mississippian carbonate solution.

This concept, which is in its formative stage, is thought to be important to develop a proper understanding of geothermal plumbing systems. It appears to have an adverse effect on the preservation of hydrocarbons. Unexpectedly, it strongly suggests major late chert formation and late dolomitization.

Whether this concept may imply a much longer history of thermal solution in Yellowstone National Park is open to discussion.

PREVIOUS INTERPRETATIONS

Paleozoic karst solution has been the most common interpretation for solution deposits and phenomena within or at the top of the Mississippian carbonates in the Rockies (Sando, 1974 (Wyoming); De Voto, 1980 (Colorado)). Because of the rounding of limestone clasts and boulders in some Mississippian solution breccias, Berry, 1843 (Montana), and Sutherland and Land, 1959 (New Mexico) concluded that they were limestone conglomerates of Mississippian age.

Severson (1952) was the first to recognize that several fairly continuous, widespread breccia zones in the upper part of the Madison Group in central Montana outcrops are correlative with anhydrite zones in the subsurface. He concluded that these solution-breccia beds formed in Laramide-Tertiary time. However, neither Severson (1952) nor Roberts (1966) invoked heated subsurface waters.

Lovering, Tweto, and Lovering (1978) reported that the Gilman breccia at Gilman, Colorado suffered more intense dissolution near Laramide-Tertiary ore bodies indicating that much of this soultion was related to heated subsurface waters. However, they (p. 32) leaned strongly toward some prior karst modification in Paleozoic time.

OUTCROP DESCRIPTIONS AND DISCUSSION

Preliminary Statement

Figure 1 shows the selected locations which are discussed in this article. There are many more Mississippian outcrops with solution deposits. Sando (1974) summarized 23 solution outcrops from the Wind River, Bighorn, and Absaroka Mountains of Wyoming.

The outcrops south of Livingston and west of Cody will be discussed first because they are closest to Yellowstone National Park and most pertinent to this field conference. However, the best evidence for my Laramide-Tertiary heated-water subsurface solution concept is at Guernsey, Wyoming and Treasure Mountain south of Marble, Colorado.

Livingston, Montana

North-dipping Mississippian Mission Canyon limestone and dolomite disconformably overlain by Pennsylvanian Amsden calcareous siltstone and Quadrant sandstone crops out prominantly in the canyon walls of the Yellowstone River to the east and west of highway 89 two miles south of Livingston.

Approximately 300 feet below the top of the Mission Canyon is a major solution breccia interval 50 to 75 feet thick (Fig. 2) which has been correlated into anhydrite in the subsurface to the north (Roberts, 1966, fig. 8). This same thick breccia interval is reported by Roberts (p. 20) to crop out in the northwestern part of Yellowstone Park near Bannock Peak. On the basis of Severson's 1952 dissertation work, this solution breccia is Laramide-Tertiary in age.

On the east side of the canyon across the Yellowstone River, which is not readily accessible, is a small, discontinuous rubble-filled deposit 45 feet below the top of the Mission Canyon. Roberts (figs. 6 and 10) concluded that this is a Paleozoic karst deposit. He reported (p. 20-21):

[1]Independent Geologist, Aurora, Colorado

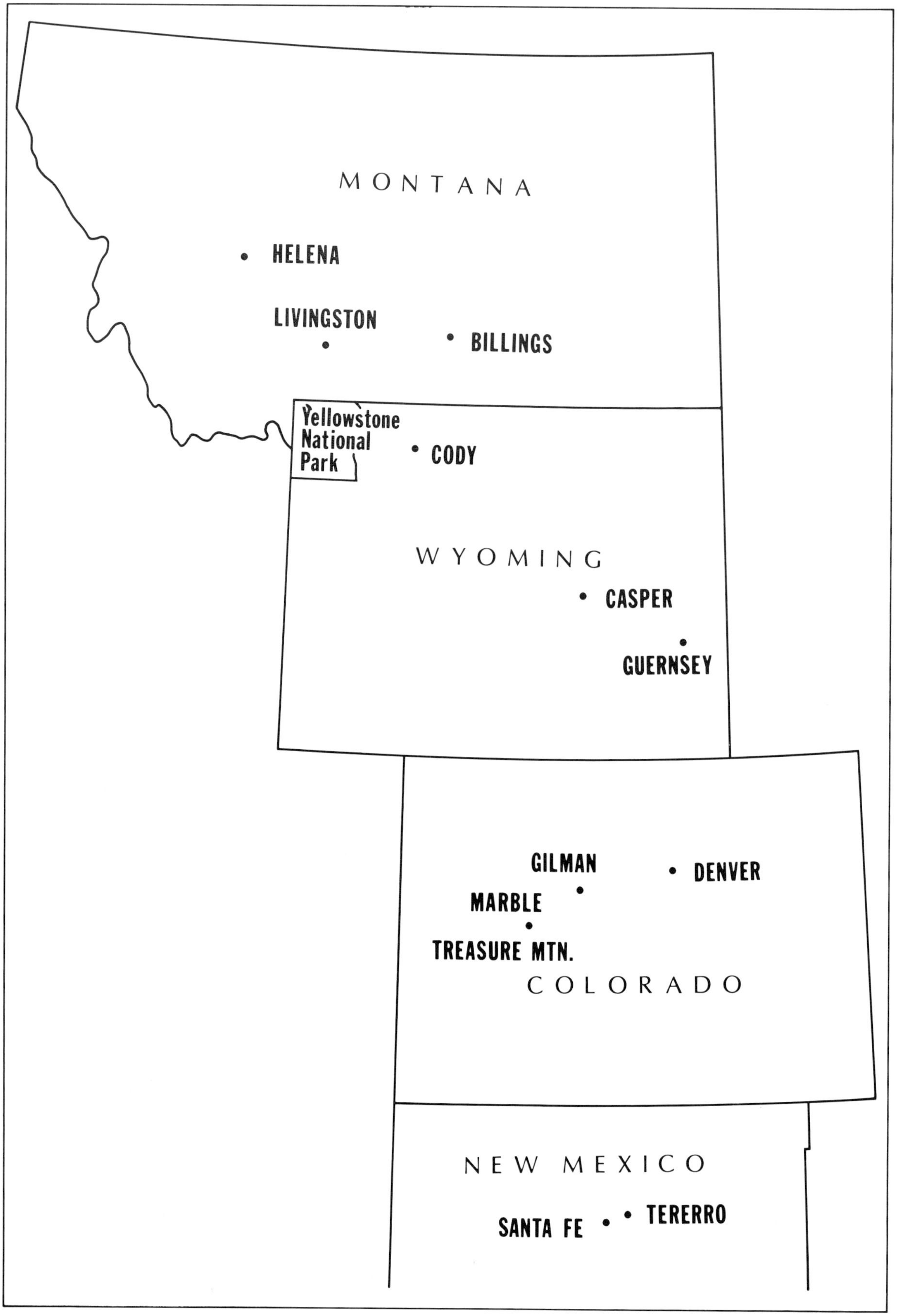

Figure 1: Location map.

"Near Livingston the karst deposits are litholographically very similar to the solution-breccia beds. Both are unstratified fragmental deposits that consist primarily of carbonate rocks and lesser amounts of chert in a matrix that weathers reddish or yellowish gray. The sorting is poor, and the range in size of the fragments is wide. The matrix is composed of indurated clayey siltstone containing limestone fragments, calcite, and smaller amounts of quartz. The fragments in both the karst deposits and the solution-breccia beds are mostly angular to subangular, although some carbonate fragments are rounded; these rounded fragments may have been shaped by the action of solutions. The chert in both types of deposits is sharply angular."

If these so-called karst solution cavities truly were formed during late Mississippian weathering, the infill material should have been brought in by transgressing Pennsylvanian seas. It seems to me that such Pennsylvanian deposits should contain rounded chert fragments; the coarsest material should be at the bottom; and the deposits should be stratified. This is not the case, so the writer is forced conclude that the so-called Paleozoic karst deposits are also Laramide-Tertiary dissolution deposits.

The only difference Roberts (p. 21) found between his so-called karst deposits and the solution-breccias is the clay mineralogy. Kaolinite is the chief clay mineral in the karst deposits; and illite is the chief clay mineral in the solution-breccias. However, Lovering, Tweto, and Lovering, 1978 (p. 30) pointed out that kaolinite is a hydrothermal alteration product of illite. Therefore it seems logical that the smaller cavities and joints containing kaolinite near Livingston simply were solution-conduits which carried hotter water for a longer time than did the main solution beds where anhydrite was removed. These smaller cavities and joints were dissolved out of carbonate, not just anhydrite.

Cody, Wyoming

Three miles west of Cody on the north side of the Shoshone River, at the entrance to Shoshone Canyon, the Madison carbonate has a major solution feature. This V-shaped solution-breccia lies in a south-facing erosional re-entrant high on the canyon wall in an area which can readily be viewed from highway 14-16-20 (Fig. 3), but is difficult to reach by foot on the far side of the river canyon. A smaller, related, solution-breccia sag is present farther west and farther north of the highway.

Sando (1974) catalogued the outcrop as a Paleozoic karst deposit. This solution-breccia is composed largely of carbonate boulders with less than 1% chert. Both the Livingston and Cody solution-breccias contain about 1% dog-toothed calcite spar which is not present in the other solution deposits discussed.

West of Cody, the solution-breccia extends downward from the top of Madison to the base of the exposed carbonate which is about 600 feet. How could ground waters, supposedly related to karst weathering, have developed the power to erode so deeply? The lowest carbonate beds on either side of the V-shaped solution-breccia have slumped downward into the center of the V indicating a solution cavity below the exposure. This same sag persists to the top of the Madison; and upon careful inspection, it is obvious that Amsden beds 100 feet above the Madison have also slumped into this sag which proves that the solution-brecciation must post-date the lower Amsden. Even a few gray Amsden sandstone boulders can be found in the Madison solution-breccia.

The basal Pennsylvanian Amsden beds in this area are mostly brown limey shales and siltstones with a pale-purple tint. Above the Madison solution-breccia, the shale and siltstone which fill the sag as well as Amsden beds above and on either side of the sag have been altered to a deep red. The iron-staining of Amsden beds is further proof of post-Amsden solution.

Thirty feet above the Madison and mostly farther east and downhill from the solution-breccia in the Madison is a 5- to 20-foot limestone lense which is partly solution-brecciated and cut with solution channels filled with non-bedded, pale-purple siltstone with some angular chert fragments. Disconformably overlying the limestone lense, which is now partly solution-brecciated, are thin limestone conglomerates that grade upward and laterally into cross-bedded sandstones.

It seems likely that the solution-breccia bed in the lower Amsden formed at the same time as the solution-brecciation in the Madison. Obviously this soultion-brecciation cannot be karst because it involves the overlying Amsden. In my opinion, the only other reasonable answer is Laramide-Tertiary subsurface solution by heated waters.

Guernsey, Wyoming

The Guernsey solution surface is figured prominently on page 110 of the Rocky Mountain Association of Geologists *Geologic Atlas* (see Craig, 1972). In addition to this lower solution surface on the Mississippian Guernsey Limestone, there is an upper solution surface in the Pennsylvanian Hartville sandstone (Figs. 4 & 5) which is so

Figure 2: Mission Canyon limestone and dolomite south of Livingston, Montana. The thick, irregular bench in the middle of the photo is solution-breccia. The Quadrant sandstone is on the skyline.

Figure 3: V-shaped solution-breccia west of Cody, Wyoming.

irregular that it cannot be a depositional unconformity.

The solution outcrops lie north and northeast of Guernsey in quarries and farther north in Guernsey State Park on both sides of the North Platte River downstream from the dam. These outcrops cover about 2 square miles. The best outcrop is on the northeast side of the river several hundred yards downstream from Guernsey Reservoir dam.

There are 3 types of solution deposits between the upper and lower solution surfaces (Fig.6): 1) pale yellowish- and pinkish-gray carbonate, which is mostly non-bedded, that has been dissolved from the Guernsey Limestone and redeposited, 2) red, mostly non-bedded sandstone which has been dissolved from the Hartville sandstone, redeposited, and strongly cemented, 3) chert breccia composed of slump deposits of grayish-red sandstone and 5-10% angular chert fragments. There are also a few cave deposits at the top (Fig. 7). Generally speaking, the greatest volume of and oldest solution deposits are carbonate; and the smallest volume of and youngest deposits are chert breccia in Guernsey State Park. Farther south, redeposited sandstone is most abundant.

Unlike the solution-breccias at Livingston and Tererro (see Fig. 1), there is very little brecciated carbonate material in the redeposited carbonate at Guernsey. As amplified under Late Chert, there are no angular chert fragments in the redeposited carbonate. Apparently solution began at Guernsey as a sheet flow in the carbonate, because the lower solution surface generally follows the same stratigraphic horizon as it does at Tererro. There is some marble just below this lower solution surface in one area, and some marble in vugs in the Guernsey Limestone suggesting greater heating in some areas.

Liesegang banding is prominent in some of the redeposited sandstone and in some altered Hartville sandstone. Steep dips and a general slumping toward the river on both sides of the river suggest the possibility that the present course of the river in Guernsey State Park may have been the site of thermal hot springs or a geyser in Tertiary time.

The most complex area involves a major sinkhole (Fig. 8). The sandstone at the top of the sinkhole is Hartville sandstone which has hinged and faulted against Guernsey Limestone. The fact that Pennsylvanian Hartville sandstone has been faulted against Mississippian Guernsey Limestone is proof that this soultion cannot be karst (i.e. late Mississippian), but must have occurred after lithification of the Hartville sandstone.

Beneath the Hartville sandstone, in the lower lefthand corner of the sinkhole (Fig. 9) are some nearly vertical, highly contorted sandstone beds with some chert. It is concluded that these beds were deposited as solution was progressing. They became partly lithified, and more solution of the Guernsey Limestone beneath them caused them to slump into their present attitude. Bretz, 1940, (figs. 28 & 30) illustrated some similar penecontemporaneous slump and collapse features which must have resulted from progressive solution beneath them in his "Southern Cavities in the Silurian Joliet Limestone in northeastern Illinois" paper.

Higher up in the sinkhole fill is a sandstone bed with a prominent black-stained bedding-plane surface which dips 20° toward the river and generally has a trough-shaped attitude (Fig. 9) because it has also partially slumped into the cavity caused by more solution below it. The bedding-plane surface has grooves resembling slickensides. Bretz (figs. 15 & 16) termed these features slickolites and

Figure 4: Chert breccia, Guernsey, Wyoming. The jacob staff lies on the upper solution surface.

Figure 5: Chert breccia, Guernsey, Wyoming. Shirts and knapsack lie on redeposited carbonate. The jacob staff lies on the upper solution surface. Note slumped chert breccia in the left center.

noted (p. 357) that these features required solution under pressure conditions for their formation.

Above the prominent bedding-plane surface are thin-bedded sandstones which progressively fill the trough. These sandstones are micro-laminated and highly micro-fractured which indicates penecontemporaneous deformation that is interpreted to result from further settling into the solution cavity below.

The upper solution surface in the Hartville sandstone is very obscure but generally marked by bleaching of the red sandstone. This bleaching, which indicates heated solutions, is present to a lesser degree just above much of the upper solution surface in the Hartville sandstone. The obscurity of this upper solution surface undoubtedly has been a major factor which prevented recognition in the past and will continue to be the focus of major debate in the future. However, the upper solution surface above much of the chert breccia (Fig. 4) and above the cave deposits (Fig. 7) is undeniable. As viewed from across the river from the west side, several 3- to 5-foot solution arches can be seen along this upper solution surface.

South of the sinkhole is a major slot (Fig. 6) eroded into the Guernsey Limestone. As Bretz (p. 381) pointed out, these slots are difficult to explain as a surface-weathering phenomenon, but are more readily explained as solution along subsurface joints.

Obviously the solution at Guernsey, Wyoming must be post-Hartville sandstone which is Morrowan (lower Pennsylvanian). The slickolites imply a considerable amount of overburden. In my opinion, the only logical conclusion for the solution is to invoke heated subsurface waters of Laramide-Tertiary age. Using the Tererro, New Mexico model (Fig. 16) (discussed below) as a guide, there is a strong possibility that these heated waters were directed by and persisted for a long period of time because of a major fault system to the northeast of Guernsey and that the waters moved updip toward the southwest (Fig. 10). The fault system must have acted as a pressure sink toward which subsurface waters moved in a tangential direction. McCarthy and Bridges (in preparation) will attempt to demonstrate similarities between the Laramide-Tertiary subsurface solution concept and present-day geothermal systems in Colorado. Fault control seems to be a key element.

Bretz (1940) provided only a lengthy, descriptive, partial negation of karst in his northeastern Illinois paper. It never occirred to him that heated Laramide-Tertiary subsurface waters related to Mississippi Valley Pb-Zn mineralization might be the cause of the solution he investigated in Joliet Limestone quarries. The similarity of the features in the Guernsey sinkhole to Bretz's solution features is striking. It seems likely to me that the cause and timing of solution at Guernsey, Wyoming and northeastern Illinois was similar. If so, this puts another nail in the coffin of those who invoke sedimentary Pb-Zn deposits in the Mississippi Valley province.

Treasure Mountain south of Marble, Colorado

Treasure Mountain was domed in late Miocene time by granite intruded about 12.5 million years ago (Obradovich et al, 1969). The intrusion and doming caused extensive contact metamorphism of all Paleozoic and Cretaceous rocks. Mississippian Leadville Limestone was altered to exceptionally pure white marble which has been quarried at the Yule Mine.

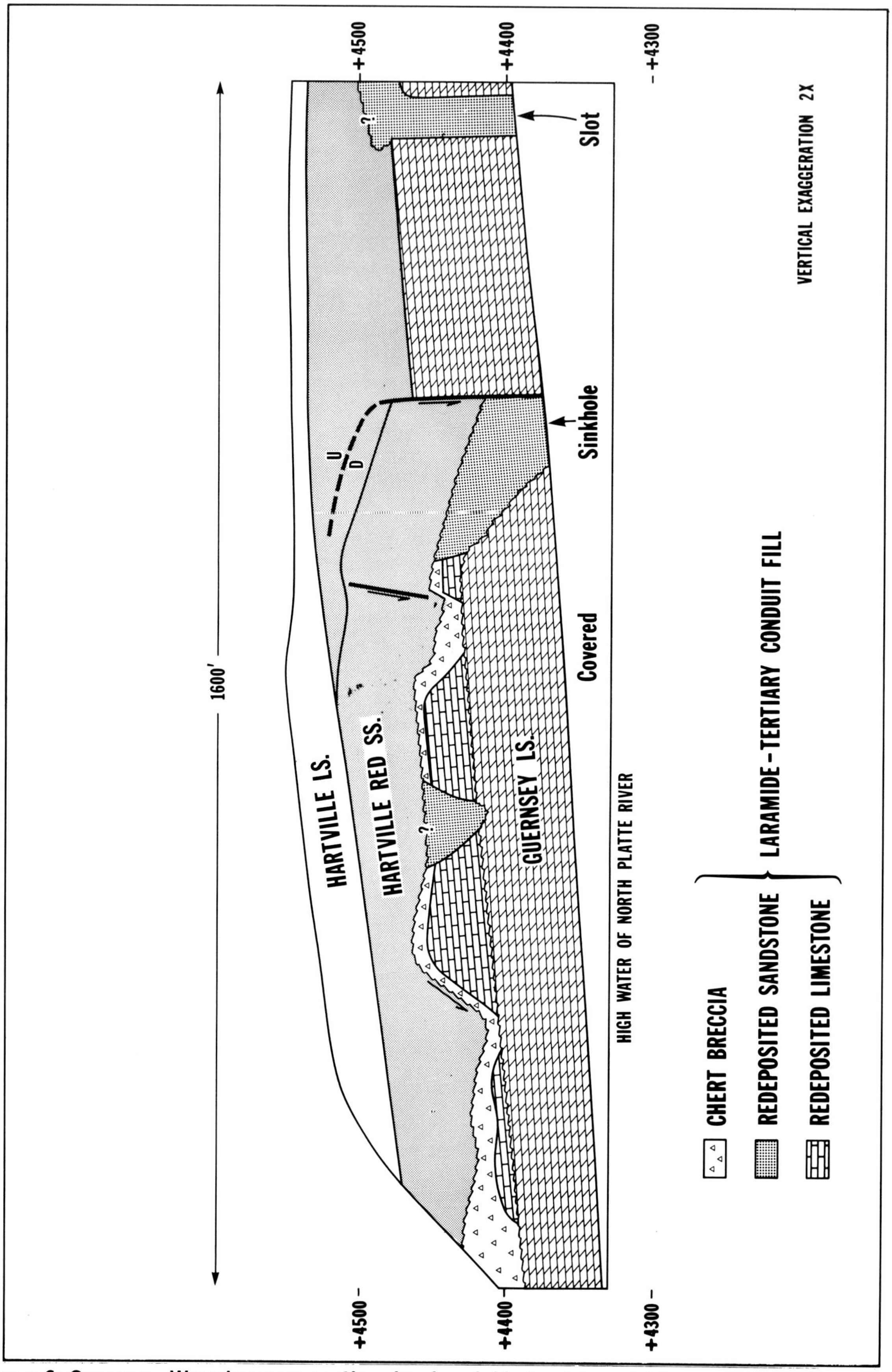

Figure 6: Guernsey, Wyoming cross section showing sinkhole, slot, and 3 types of solution deposits.

Figure 7: Cave deposit at the top of solution deposits, Guernsey, Wyoming. Draped material under hammer is sandstone, not carbonate. Angular chert fragments to the right.

Figure 8: Sinkhole, Guernsey, Wyoming. This photo covers the central half of Figure 6.

Above the marble is a discontinuous thickness of 0 to 50 feet of reddish- to greenish-gray argillite (Fig. 11) with angular chert fragments (Figs. 12 & 13). The argillite has been called the Molas Formation based on the premise that the angular chert is part of an errosional regolith on karst topography of Mississippian age (Gaskill and Godwin, 1966). Although this chert is concentrated in the lower part of the argillite, it is not bedded. In fact, many of the chert fragments are completely isolated like plums in a pudding; and some chert fragments have patinas suggesting growth in place (Fig. 12). Since there are no limestone of dolomite fragments from the Leadville in the so-called regolith and the angular chert fragments show no sign of abrasion, it is concluded that the chert is of Late Tertiary origin related to contact metamorphism, soultion, alteration, and replacement. As the chert was being formed above the marble, Treasure Mountain was still tectonically active and the chert was fractured in place. Thus there is no evidence for a karst surface on the Leadville at Treausre Mountain.

Heretofore it has been assumed that all of the deposits above the so-called Molas Formation at Treasure Mountain are lower Pennsylvanian (Morrowan) and part of the Belden Formation. My regional stratigraphic field work indicates that the 33 meters of carbonate (including the replacement material called the Molas) is Osagean (Fig. 14). Therefore the disconformity time gap above the Leadville is not 25 million years, but probably on the order of 2 to 3 million years. Thus there was not sufficient time to develop a significant karst erosion surface at Treasure Mountain above the Leadville.

Some 21 miles southeast of Treasure Mountain at Cement Creek (Fig. 14), there is a limestone conglomerate on top of the Leadville with 5% Precambrain clasts of quartzite and gneiss. Thus the entire Paleozoic section and Precambrian terrane was being eroded. Undoubtedly this conglomerate, which is atypical of the basal Belden sandstones, shales, and limestones, is detritus from the emerging Uncompahgre Uplift farther to the southwest. This conglomerate contains no chert which makes it even more impossible for the so-called regolith at Treasure Mountain to have developed on the Leadville and contain nothing but chert. Instead of correlating the limestone conglomerate at Cement Creek with the angular-chert zone at Treasure Mountain, the limestone conglomerate is correlative with 6 meters of sandstone and quartzite some 33 meters above the Leadville. Clearly, the angular chert above the Leadville at Treasure Mountain is not eroded Paleozoic chert.

Gilman, Colorado

As discussed above, the Gilman breccia at the base of the Leadville Limestone is a Laramide-Tertiary solution breccia. Several small faults in the basal Gilman sandstone do not fault the Gilman breccia. Thus the Gilman breccia is post-faulting. Where bedding is present in the Gilman breccia it tends to be horizontal. Thus the Gilman breccia post dates the regional eastward tilting of the east flank of the Sawatch Uplift in the Gilman area. To the contrary, the breccias south of Livingston and west of Cody both appear to be pretilting.

Lovering, Tweto, and Lovering, 1978 (p. 10) concluded that the top of the Leadville Limestone underwent major Paleozoic karst erosion and solution. Unfortunately, the only surface evidence is a poorly exposed slot cut in the top of the Leadville where the Belden

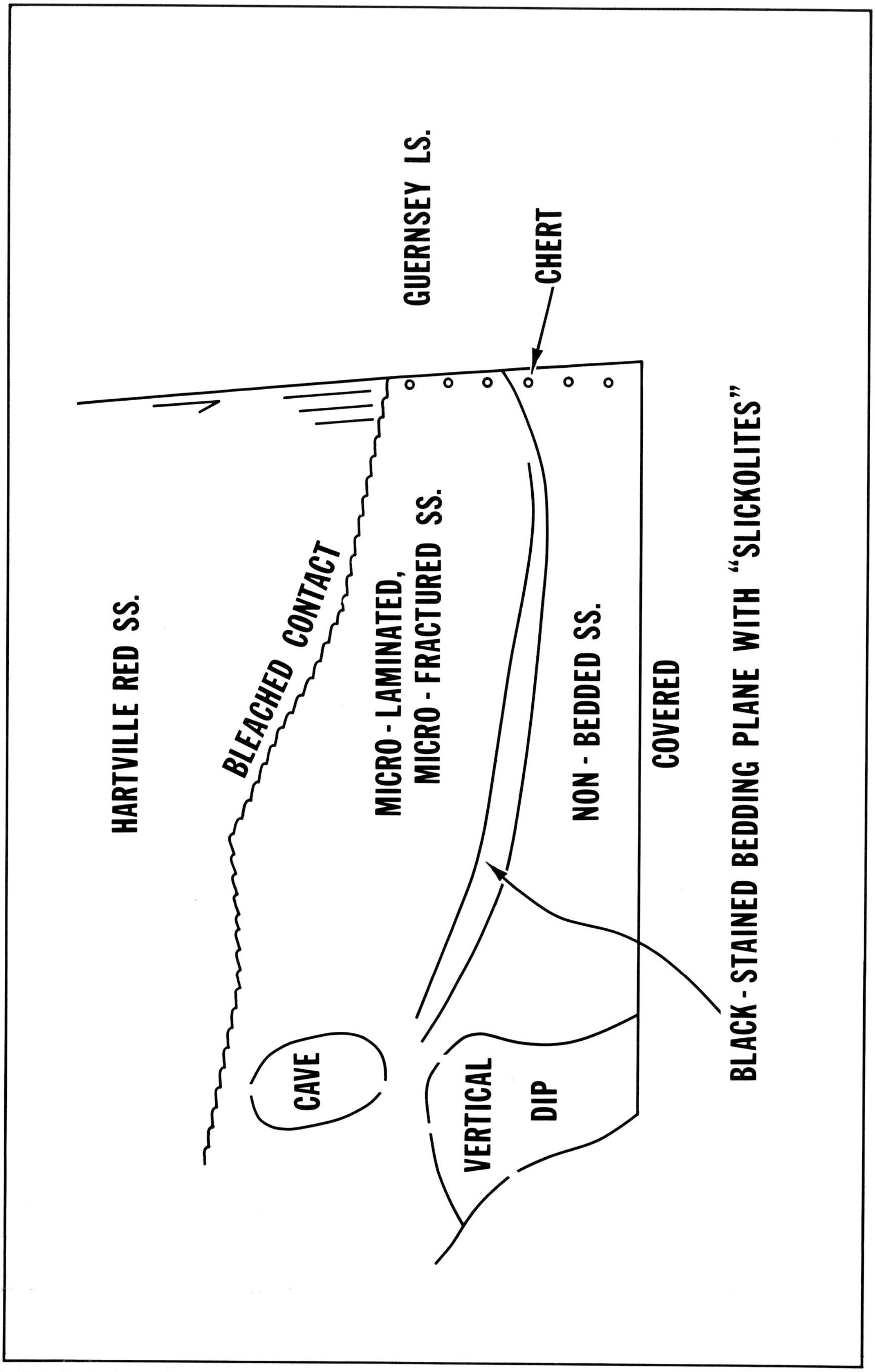

Figure 9: Detail of sinkhole deposits, Guernsey, Wyoming. This is the same sinkhole shown in Figures 6 & 8.

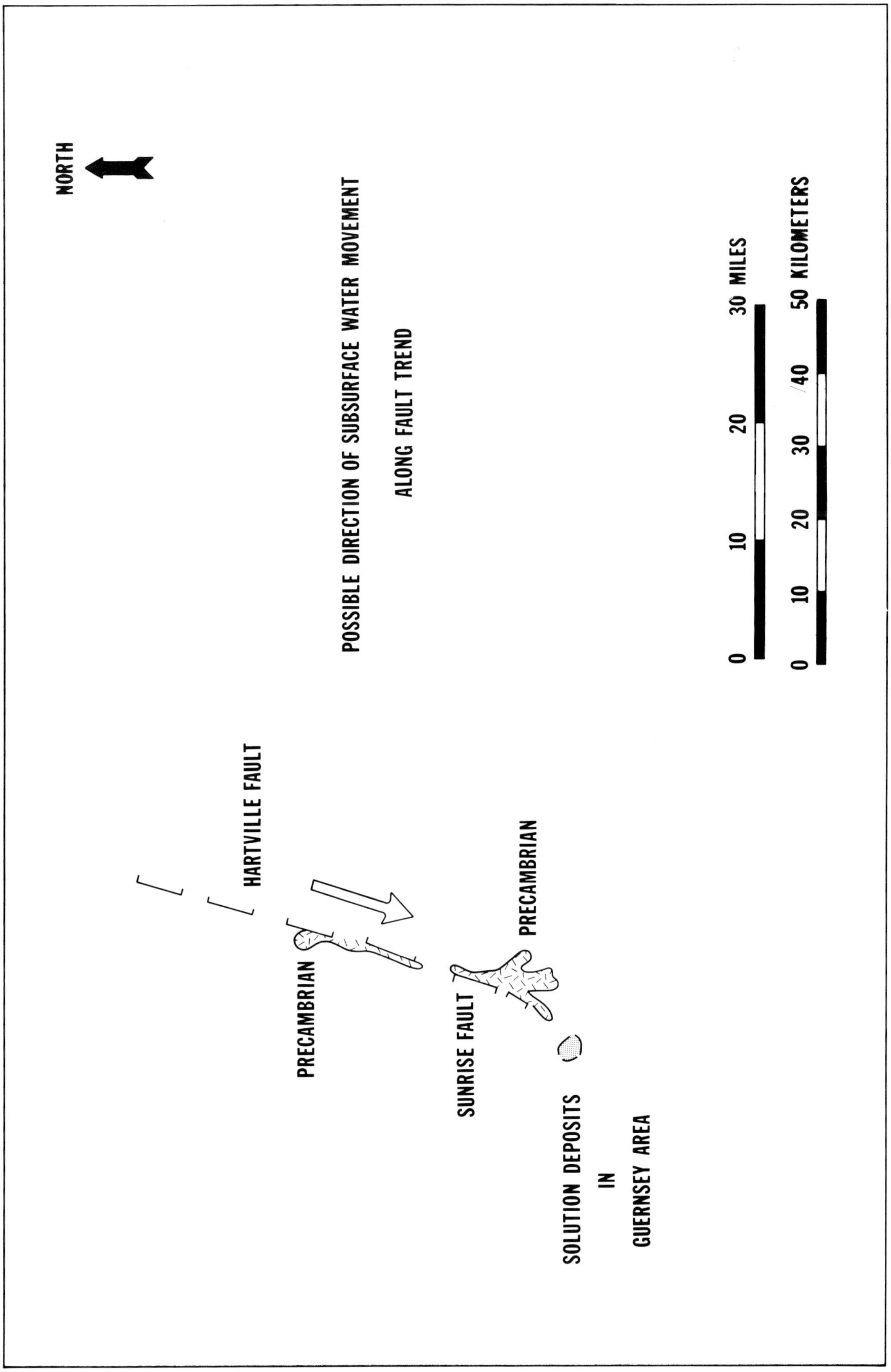

Figure 10: Guernsey, Wyoming map showing possible direction of subsurface water movement along fault trend.

Figure 11: Treasure Mountain, Colorado. Leadville Limestone marble below. Osage carbonate above. Irregular, so-called "Molas Formation regolith" in center.

Figure 12: Treasure Mountain, Colorado. Chert in "Molas" above marble. Key case for scale. Note patina around chert fragments.

has been completely removed by erosion. Thus there is no way to determine whether the Pennsylvanian Belden Formation was involved in the solution or not. This is a persistent problem in Colorado because the basal Belden is most frequently composed of soft shale which is easily eroded.

Most of the evidence at Gilman is underground in the Eagle Mine. Rotting timbering, cave-ins, and a general cessation of mining has resulted in permanent closure of the Eagle Mine so this evidence is no longer available for re-examination. Because the lower Belden is soft, sloughing shale and is not mineralized it is rarely available for inspection in Colorado mines. Thus the extent of Belden involvement in the solution has not been properly evaluated, even in the mines, in my opinion.

Tererro, New Mexico

In the Pecos River valley, 15 miles east of Santa Fe, the Mississippian carbonate called the Arroyo Penasco Formation is 60 to 80 feet thick. The most extensive outcrop is on the edge of a small park at Tererro (Fig. 15). In this area the middle 40 feet of the Arroyo Penasco Formation has been described as a limestone conglomerate (Sutherland and Land, 1959) and as a karst solution breccia (Armstrong, 1967). The redeposited carbonate at Tererro is remarkably similar in its general non-bedded, yellowish-gray weathered appearance to the solution-breccia south of Livingston, Montana and the redeposited carbonate at Guernsey, Wyoming.

I would agree with Armstrong (p. 21) that the breccia zone became a channel for increased movement of water. The amazing thing is the persistence of this same thick

Figure 13: Treasure Mountain, Colorado. Angular chert fragments in "Molas" argillite above marble.

breccia zone over 50 to 100 square miles. If this were the result of karst erosion, it would seem likely to me that there should be much more thickening and thinning of the solution breccia, areas where the solution involved the upper beds of the Arroyo Penasco Formation, and areas where no solution took place at all. None of these conditions are met.

I am convinced that we have not been thinking big enough, and conclude that this breccia lense is part of a Tertiary heated-water subsurface-plumbing system that allowed hot waters from the Rio Grande graben to funnel updip northeastward along a major fault system and finally come to the surface as hot springs and perhaps geysers which are now extinct in this part of New Mexico (see Fig. 16).

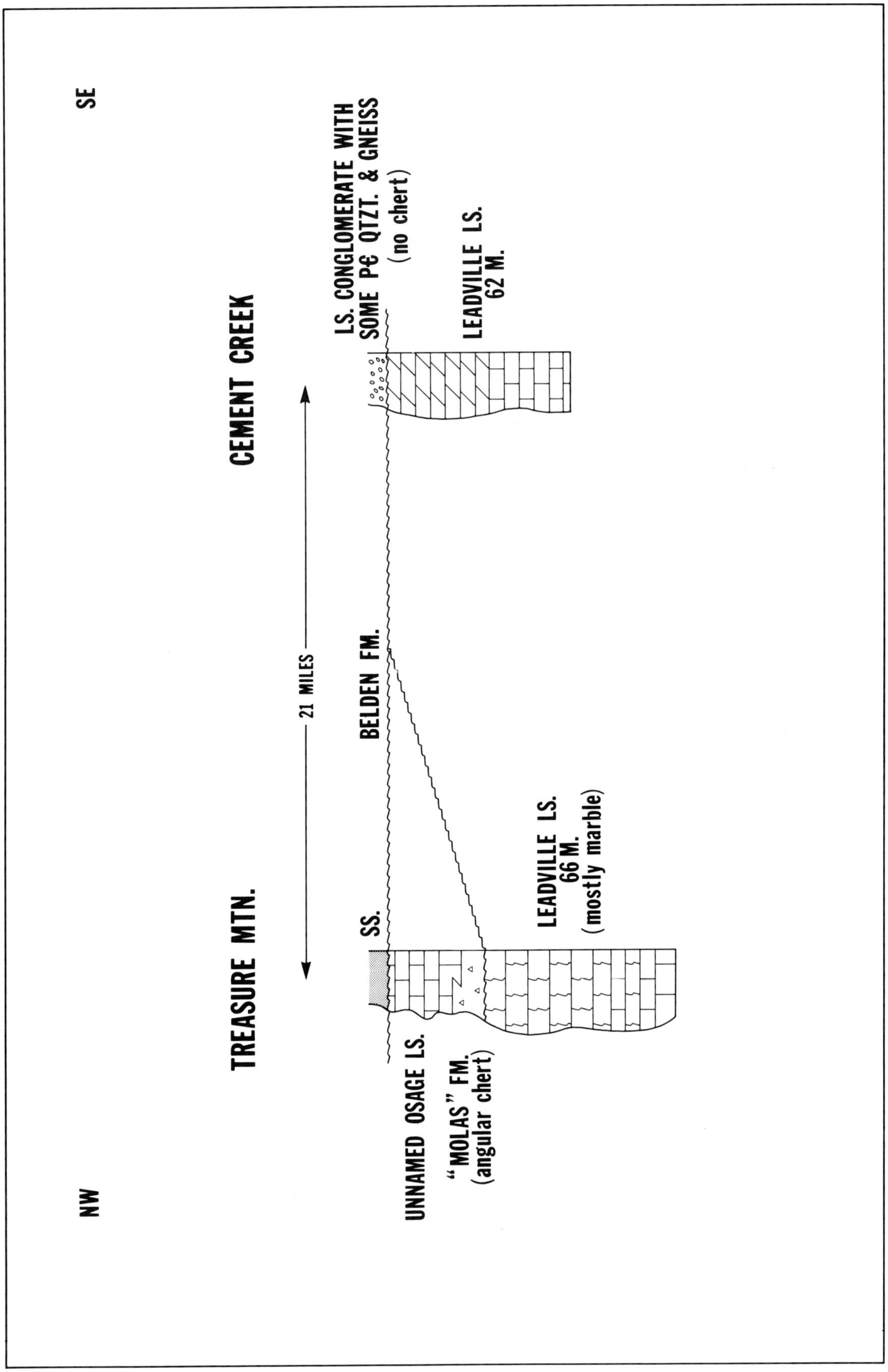

Figure 14: Treasure Mountain to Cement Creek, Colorado cross section showing stratigraphic relations.

Figure 15: Tererro, New Mexico solution-breccia in Arroyo Penasco Formation. Note bedded Mississippian carbonate above and below the solution-breccia. The long, narrow cave trends N60°E which probably indicates the general direction of water movement.

LATE CHERT

Obviously the chert at Treasure Mountain south of Marble, Colorado is Tertiary chert as discussed above. The chert at Guernsey, Wyoming is also angular, shows no sign of abrasion, and has a decidedly preferential distribution. There is one large 3-foot long, 8-inch thick flat-lying chert nodule in the chert breccia which must have developed in place; and one thin bed of chert which extends, without interruption, from Mississippian Guernsey Limestone into redeposited carbonate. Thus it must be younger than the redeposited carbonate. Nearly all the chert in the redeposited limestone is bedded chert at the contact above the Mississippian carbonate. If the chert in the Guernsey solution deposits is eroded Paleozoic chert, why are there no chert fragments in the redeposited limestone? Similarly, the only chert fragments in the redeposited sandstone are found along the edges and bottom of the redeposited sandstone where circulation was restricted. It is therefore concluded that at Guernsey, the heated Laramide-Tertiary subsurface waters dissolved a considerable amount of silica that originally cemented the Hartville sandstone. Then this silica was reprecipitated as chert where circulation was restricted. The chert breccias are slump deposits that resulted from earthquakes triggered by spasmodic subsidence of Hartville sandstone-blocks into solution cavities as in the sinkhole. Thus the Guernsey chert is also late chert of Laramide-Tertiary age; not Paleozoic chert.

In the Gilman area, Banks (1970) concluded that there was not enough silica within the Leadville to account for all of the chert. Therefore the silica must be from pre-Mississippian beds. Banks (p. 3045) invoked mass transport of silica via major water movement through the Leadville. He noted angular chert fragments in the overlying "Molas regolith" and concluded that the silica was precipitated from ground waters prior to, or duing karst erosion in Late Mississippian or Early Pennsylvanian time.

Lovering, Tweto, and Lovering (p. 33) reported that the Cambrian Sawatch Quartzite was subjected to hydrothermal dissolution on an extensive scale before and, to a lesser extent, during mineralization" at Gilman, Colorado. They also noted (p. 34) that none of the channels in the quartzite contained any rubble so the debris "was evidently dissolved and removed in solution, suggesting as the agent, a very hot and very active fluid. On the other hand, Engel et al (1958) reported that the chert in the Leadville was deposited at low temperatures.

No one has explained what happened to the silica that was removed from the Sawatch Quartzite by dissolution. It seems most likely to me that the silica which was dissolved by heated waters was later precipitated as chert higher in the section by cooler waters in Laramide-Tertiary time in the Gilman area.

A huge mass of chert some 3-feet wide and irregularly 10-feet long is present in the solution breccia at Terrero. Large flat-lying nodules of black chert are present in the solution-breccia south of Livingston. The Tererro and Livingston chert in the solution-breccias also must be Laramide-Tertiary in age.

LATE DOLOMITIZATION

Associated with and an integral part of the subsurface-redeposition story is a major amount of late dolomitization. Lovering, Tweto, and Lovering concluded that Laramide hydrothermal activity was responsible for dolomitization of the Leadville Limestone at Gilman, Colorado.

HYDROCARBONS

In my opinion, Laramide-Tertiary heated subsurface waters destroyed many oil and gas accumulations by continual flushing over a long period of time. This is most easily visualized in areas of major tectonism and vulcanism like the Basin-and-Range Province of the western U.S.

However, the area around Guernsey, Wyoming on the Hartville Uplift is not an area that one would expect has suffered major flushing. Yet there are several good structures east of the Guernsey area on the southeast edge of the Hartville Uplift where there are good Pennsylvanian

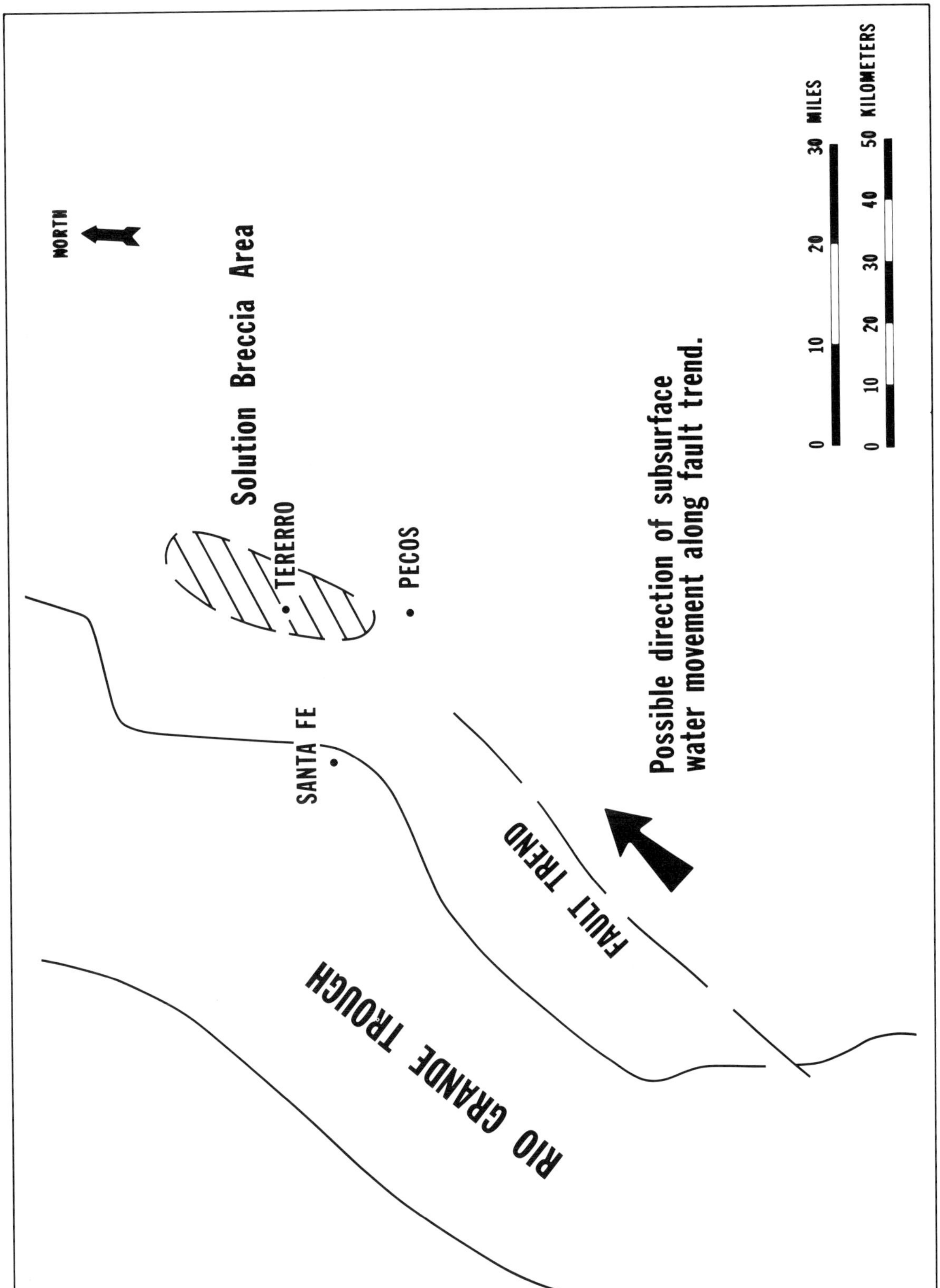

Figure 16: Tererro, New Mexico map showing possible direction of subsurface water movement along fault trend.

source beds in the adjacent basin but no hydrocarbons in the structures. I think these structures have been flushed by the heated subsurface waters which caused the Laramide-Tertiary solution in the Guernsey area.

EXTINCTION OF DINOSAURS AND AMMONITIES

The sudden and simultaneous disappearance or mass extinction of dinosaurs on land and ammonities in the oceans toward the end of Cretaceous (Laramide) time has been very perplexing. As Mintz, 1977 (p. 476) stated: "Only something that would cut across many environments and food chains could cause the Late Mesozoic extinction". One hypothesis advanced in Mintzs basic geology text is a change in the atmospheric O_2/CO_2 ratio.

Could it be that this Laramide-Tertiary subsurface solution story, which is unfolding, was a world-wide phenomenon that was so huge that all land and ocean waters were polluted and/or heated sufficiently to cause the mass extinction of dinosaurs and ammonities?

CONCLUSIONS

The primary conclusion is that Laramide-Tertiary subsurface solution from heated subsurface waters is an important geologic phenomenon whose scope vastly exceeds present geologic thinking. Many low-grade metamorphic phenomena in sedimentary rocks have gone unnoticed or been misinterpreted. It is suggested that much of the chert in Paleozoic rocks, which we have taken for granted to be Paleozoic chert, may be Tertiary chert.

In my opinion, the late Mississippian karst story in the Rocky Mountains is completely fallacious. Granted, my judgement is based primarily on evidence which I think is conclusive at Guernsey, Wyoming and at Treasure Mountain south of Marble, Colorado. This is a difficult problem because much, if not most, of the evidence could be interpreted either way. Certainly much work must be done, but at least let's start giving Laramide-Tertiary subsurface solution a fair hearing rather than covering it up and pretending that it is not a viable theory.

SELECTED REFERENCES

Armstrong, A. K., 1967, Biostratigraphy and carbonate facies of the Mississippian Arroyo Penasco Formation, northcentral New Mexico: New Mexico Bur. Mines and Mineral Resources Memoir 20, 89 p.

Banks, N. G., 1970, Nature and origin of early and late cherts in the Leadville Limestone, Colorado: Geological Society of America Bulletin, v. 81, p. 3033-3048.

Berry, G. W., 1943, Stratigraphy and structure at Three Forks, Montana: Geological Society of America Bulletin, v. 54, p. 1-29.

Bretz, J. H., 1940, Solution cavities in the Joliet Limestone of northeastern Illinois: Journal of Geology: v. 48, p. 337-384.

Craig, L. C., 1972, Mississippian system, in Geologic atlas of the Rocky Mountain Region, Rocky Mountain Association of Geologists, p. 100-110.

De Voto, R. H., 1980, Mississippian stratigraphy and history of Colorado in Rocky Mountain Association of Geologists, Colorado Geology Symposium: p. 57-70.

Engel, A. E. J., Clayton, R. N., and Epstein, S., 1958, Variations in isotopic composition of oxygen and carbon in Leadville Limestone (Mississippian, Colorado) and its hydrothermal and metamorphic phases: Journal of Geology, v. 66, p. 374-393.

Foster, N. H., 1966, Stratigraphic leak: American Association of Petroleum Geologists Bulletin, v. 50, p. 2604-2606.

Gaskill, D. L., and Godwin, L. H., 1966, Geologic map of the Marble quadrangle, Gunnison and Pitkin Counties, Colorado: U.S. Geological Survey Quadrangle Map GQ-512.

Lovering, T. S., Tweto, O., and Lovering, T. G., 1978, Ore deposits of the Gilman district, Eagle County, Colorado: U.S. Geological Survey Professional Paper 1017, 90 p.

Maslyn, R. M., 1976, Late Mississippian paleokarst in the Aspen Colorado area: unpublished M.S. thesis, Colorado School of Mines, Golden, 96 p. (T 1811).

______, 1977, Fossil tower karst near Molas Lake, Colorado: The Mountain Geologist, v. 14, p. 17-25.

McCarthy, K. P., and Bridges, L. W. D., in preparation, Similarities between the Laramide-Tertiary subsurface solution concept and present-day geothermal systems in Colorado.

Mintz, L. W., 1977, Historical Geology, The science of a dynamic earth; 2nd edition: Columbus, Ohio, Charles E. Merrill Publishing Company, 588 p.

Obradovich, J. D., Mutschler, F. E., and Bryant, B., 1969, Potassium-argon ages bearing on the igneous and tectonic history of the Elk Mountains and vicinity, Colorado: A preliminary report: Geological Society of America Bulletin, v. 80, p. 1749-1756.

Quinlan, J. F., Jr., 1978, Types of Karst, with emphasis on cover beds in their classification and development: unpublished Ph.D. thesis, University of Texas, Austin.

Roberts, A. E., 1966, Stratigraphy of Madison Group near Livingston, Montana, and discussion of karst and solution-breccia features: U.S. Geological Survey Professional Paper 526-B, 23 p.

Sando, W. J., 1974, Ancient solution phenomena in the Madison Limestone (Mississippian) of north central Wyoming: U.S. Geological Survey Journal of Research, v. 2, no. 2, p. 133-141.

Severson, J. L., 1952, A comparison of the Madison Group (Mississippian) with its subsurface equivalents in central Montana: unpublished Ph.D. thesis, University of Wisconsin, Madison.

Sutherland, P. K., and Land, C. B., Jr., 1959, Mississippian limestone boulder conglomerates in the southernmost Sangre de Cristo Mountains, New Mexico (abstract): Geological Society of America Bulletin, v. 70, p. 1683.

ThirtyThird Annual Field Conference — 1982
Wyoming Geological Association Guidebook

HYDROCARBONS IN THERMAL AREAS, NORTHWESTERN WYOMING[1]

BY J.D. LOVE,[2] AND JOHN M. GOOD[3]

ABSTRACT

Five natural occurrences of hydrocarbons in thermal areas lie in an arcuate southeastward- to eastward-trending area 70 miles long in northwestern Wyoming. They are, from northwest to east: Tower Bridge, Calcite Springs, Rainbow Springs, Sweetwater Mineral Springs, and Cedar Mountain. The western most three are in Yellowstone National Park. All are associated with abundant sulfur and thermal springs and vents (active in four localities, recently extinct in one). Four are surrounded and underlain by volcanic rocks, which range in age from Eocene to Pleistocene. The fifth emerges from Paleozoic strata.

The source of the hydrocarbons could be Paleozoic or Mesozoic sedimentary rocks that underlie the surficial volcanics, but at Rainbow Springs the oil might have been distilled from Pleistocene nonmarine algae at shallow depth. Sulfur isotope analyses do not indicate conclusively whether the sulfur is of igneous or bacteriogenic origin.

In at least two of the localities, hot water and steam that derive their heat from an igneous source below the sedimentary sequence are believed to relate directly to the extraction of the hydrocarbons from the strata, to their subsequent transport upward along conduits, and their extrusion on the surface.

INTRODUCTION

The purpose of this report is to call attention to five natural occurrences of liquid hydrocarbons in thermal areas. Two (loc. 1 and 3, fig. 1) offer promise of yielding significant data on the geochemical processes that are responsible for the origin of oil in an unusual environment. This knowledge should, in turn, lead to a better understanding of the characteristics of the oil that results from these processes and of the mechanics of its emplacement. Two of the localities of hydrocarbons in thermal areas have already been obliterated by man (loc. 2 and 5, fig. 1), and a third (loc. 4) has been considerably altered. It is hoped that this preliminary discussion will stimulate the search for new localities and encourage more sophisticated studies of those that remain before they, too, are damaged or destroyed and the opportunity for recording and interpreting this type of data is lost.

Hydrocarbons in thermal areas of northwestern Wyoming have been known or suspected ever since the verbal reports of John Colter, who observed "the great tar spring" at Cedar Mountain (loc. 5, fig. 1) in 1807 (Chittenden, 1895, p. 23; 1901, p. 3781). Casual references to springs that contained oil, tar, bitumen, or asphalt in other parts of northwestern Wyoming were made by several explorers and writers. For example, Denig (1961, p. 141), in a manuscript written about 1854 and published over a ahundred years later, stated, "Some of the springs near the head of the Yellowstone are bituminous, sending forth a substance like tar, which is inflammable." Father DeSmet (letter of Jan. 20, 1852, quoted by Chittenden, 1895, p. 388) likewise commented on the abundant "bituminous, sulphurous and boiling springs" between the sources of the Madison and Yellowstone Rivers in what is now northwestern Wyoming.

Despite the general knowledge among early trappers

[1]Reprint of USGS Professional Paper 644-B originally published in 1970
[2]U.S. Geological Survey
[3]National Park Service, present address: P.O. Box 354, Grand Canyon, Arizona, 86023

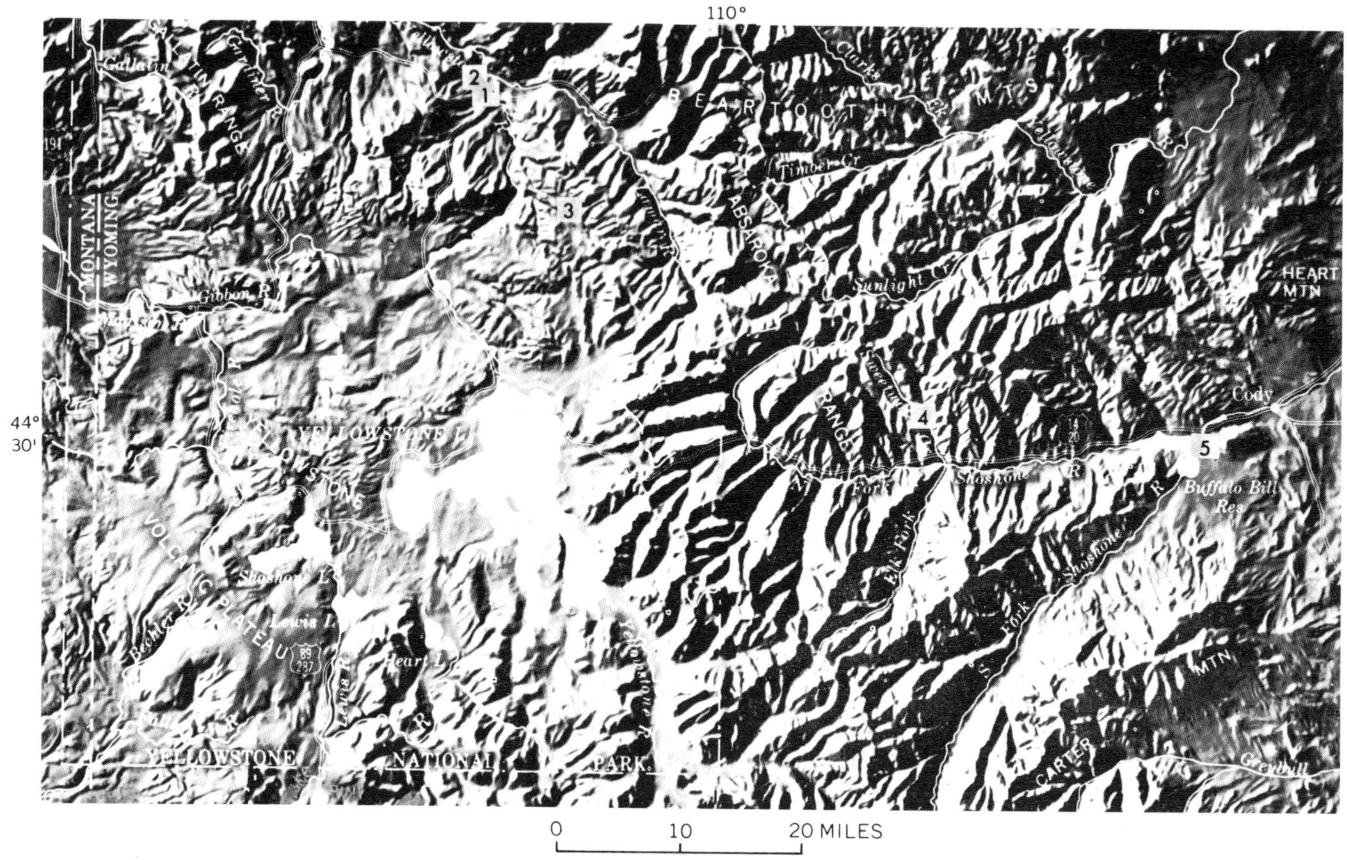

Figure 1: Relief map of northwestern Wyoming showing described localities of hydrocarbon occurrences in thermal areas. Indicated are (1) Calcite Springs, (2) Tower Bridge, (3) Rainbow Springs, (4) Sweetwater Mineral Springs, and (5) Cedar Mountain.

and explorers that oil seeps were present in thermal areas of this region, we have found no reference to these phenomena in any of the scientific studies by the Hayden (1833) or the Hague (1904) groups or by any of the other pioneer geologists such as Comstock (1875) and Eldridge (1894) who worked in the region. As far as we have been able to determine, Allen and Day (1935) were the first scientists to make mention, and that only brief, of bituminous material at localities 1 and 3. Hewett (1913, 1914) gave a short account of the oil at locality 4. Our data were obtained on brief reconnaissance trips between 1963 and 1968.

We are grateful to many people for help on various aspects of this study. The following U.S. Geological Survey paleontologists studied the fossils: E. B. Leopold identified pollen, K. E. Lohman and G. W. Andrews identified diatoms from the Rainbow Springs locality, and J. A. Wolfe identified leaves from the early basic breccia. We are especially grateful to M. F. Skinner, of the American Museum of Natural History, for comparing a bison skull from one of our localities with his reference collection. J. W. Smith, U.S. Bureau of Mines, and G. J. Schrayer, Gulf Research and Development Co., provided not only interest and encouragement but many analyses during a period of several years. W. E. Haines and A. B. Hubbard, U.S. Bureau of Mines, were most helpful in supplying both counsel and analyses. J. G. Palacas, U.S. Geological Survey, studied a sample of oil-impregnated silt from the Sweetwater Mineral Springs locality. K. L. Pering, Ames Research Center, National Aeronautics and Space Administration, collected oil samples and made laboratory comparisons of the samples with biogenic and possible abiogenic oils from other parts of the world. Irving Friedman, U.S. Geological Survey, provided two carbon isotope analyses. M. L. Jensen, University of Utah, sampled sulfur in the Calcite Springs area and made a series of sulfur isotope analyses from both his samples and ours.

This study was conducted with the cooperation of R. S. Houston, Chairman, Department of Geology, University of Wyoming, and D. L. Blackstone, Jr., State Geologist of Wyoming.

CHARACTERISTICS AND GEOLOGIC SETTING OF THE LOCALITIES

The five described localities lie in an arcuate southeastward- to eastward-trending area 70 miles long (fig. 1). They are, from northwest to east, Tower Bridge (loc. 2), Calcite Springs (loc. 1), Rainbow Springs (loc. 3), Sweetwater Mineral Springs (loc. 4), and Cedar Mountain (loc. 5). Characteristics common to most of these localities, in addition to the occurrence of hydrocarbons, are the abundance of sulfur and its close association with the hydrocarbons, the presence of thermal springs and vents (active in four localities, recently extinct at loc. 4), and the presence of volcanic rocks that surround and underlie localities 1-4. The water in many springs at localities 1 and 3 is boiling, but it is cooler at 2, cold at 4, and reportedly 90°F at 5.

Localities 1. 2, and 4 are in the early basic breccia of Hague (1904). At locality 2 is a dark-colored facies of volcanic-rich strata equivalent to the lower part of the Cathedral Cliffs Formation (H. J. Prostka, oral commun., 1969). Locality 3 is in Pleistocene lacustrine tuff adjacent to and probably underlain by rhyolite. Below the rhyolite, presumably, is early basic breccia. Locality 5 is probably in Pennsylvanian or Permian sandstone and carbonate rocks. Geologic details are given in the description of each locality. The regional structural pattern and distribution of prevolcanic rocks in areas adjacent to localities 1-4 suggested that Paleozoic and Mesozoic sedimentary rocks underlie the volcanics; the type of thermal features at localities 1 and 3 suggests that at still greater depth there are bodies of hot magma or only partly cooled igneous rocks. The actual depths of these sedimentary rocks and igneous material have not been determined; they could be hundreds to thousands of feet below the surface. The attitude of the sedimentary rocks is not known, and geophysical data are not definitive as yet. No drilling has been done in the vicinity of the localities.

SOURCES OF OIL AND SULFUR

The characteristics of the oil at each locality are given in the detailed discussions. Only at localities 1 and 3 was sufficient oil obtained for even a preliminary interpretation as to source. The gas chromatograms of the alkane portion of these two oils (fig. 11) are so different that they may well have been derived from unrelated sources. The source of oil at locality 1 is unknown. The oil at locality 3 appears to have no highly unusual characteristics and may have been derived from underlying organic-rich Cretaceous rocks and carried upward along conduits by thermal water and steam. On the other hand, aided by the available heat, it might have been distilled from Pleistocene nonmarine algae at shallow depth. This possibility is discussed later.

The sulfur from locality 1 was analyzed in some detail, but the resulting data were not sufficiently definitive to assign it an igneous or bacteriogenic origin. At all localities, the sulfur was hydrothermally concentrated.

RELATION OF THERMAL PHENOMENA TO EMPLACEMENT OF OIL

Hot water and steam are closely associated with the conduits along which oil traveled at localities 1 and 3 and almost certainly are responsible for the surface and near-surface emplacement of the oil. At the other localities, because of floods or tampering by man, conclusive evidence of the process of transport and emplacement has been obliterated, but it is believed that the process was somewhat similar to that at localities 1 and 3. When more surface and subsurface information on thermal features elsewhere in and near Yellowstone National Park (Campbell, 1969) is obtained, an interpretation can probably be made concerning the depth at which the thermal activity originated, how far the oil was transported, and what chemical changes occurred in it en route from the source area to the surface.

INDIVIDUAL LOCALITIES

CALCITE SPRINGS (LOC. 1)

The Calcite Springs emerge near the bottom of Yellowstone Canyon, about 1 mile down the Yellowstone River from the Tower Fall area (fig. 2). The geology of the springs area has been described by Howard (1937). Allen and Day (1935, p. 495-500) discussed the springs and gave the first description of tarry bitumen at that locality.

Oil emerges from the early basic breccia, as well as from Quaternary deposits. The lower 200 feet of the Yellowstone Canyon at this locality is cut in the early basic breccia, a sequence of drab dark-green to brown mafic tuff, conglomerate, and possibly basalt flows (fig. 3). These rocks have been extensively altered by thermal action to produce the yellow to white canyon walls. The age of the early basic breccia is considered to be Eocene on the basis of a small flora collected from the breccia at the base of Tower Fall and identified by J. A. Wolfe (written commun., 1962) as *Celastrus inaequalis* Knowlton, *Ficus densifolia* Knowlton, and *Magnolia spectabilis* Knowlton.

The upper 130 feet of rocks on the east wall was

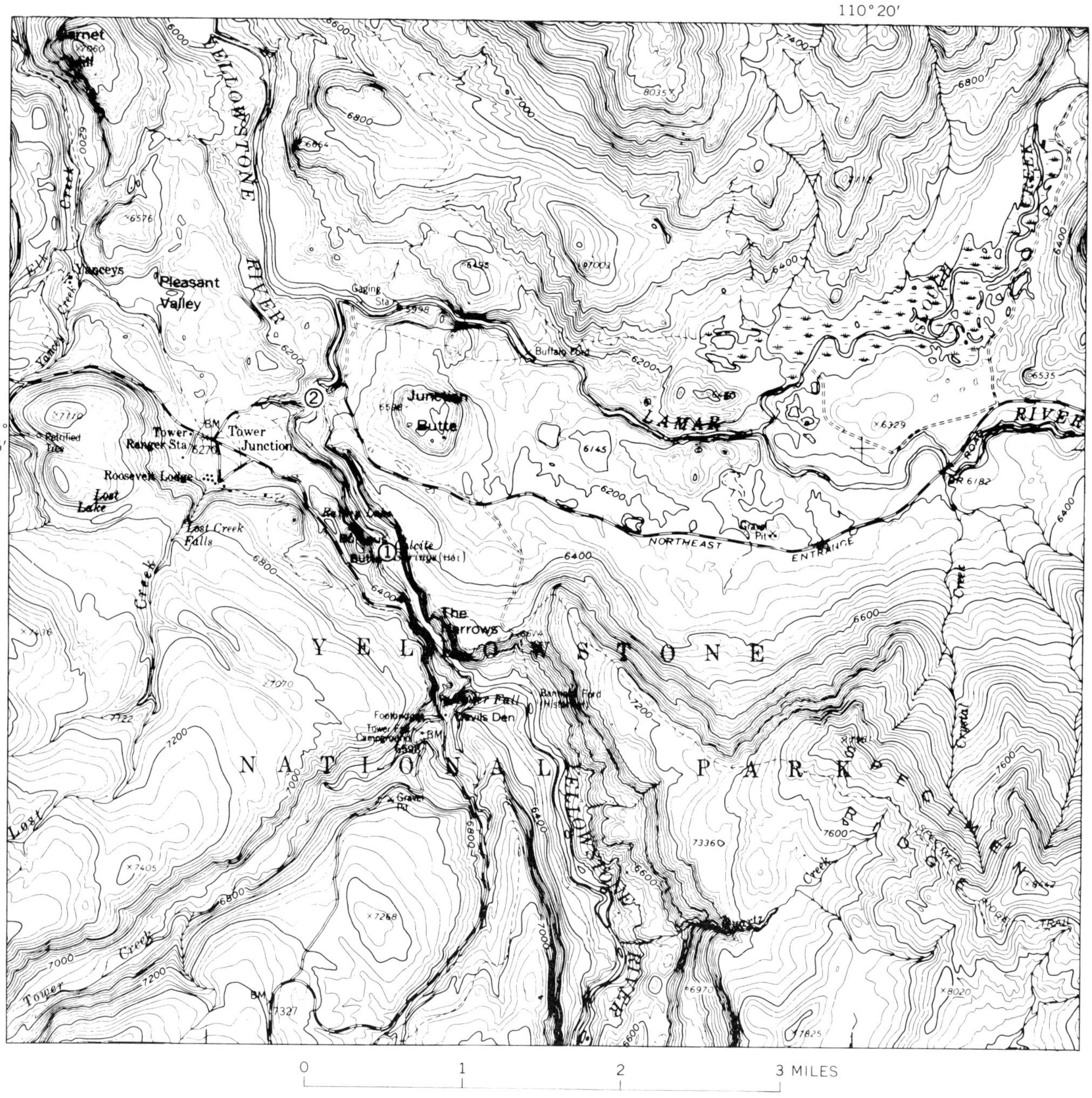

Figure 2: Location of Calcite Springs (loc. 1) and Tower Bridge (loc. 2). Base from U.S. Geological Survey Tower Junction quadrangle, 1959.

called "Canyon basalts and associated gravels" by Howard (1937). This sequence consists of lenticular basalt flows separated by dark-brown conglomerate made up chiefly of basalt and rhyolite pebbles, cobbles, and boulders. There are, however, many light-colored highly rounded quartzite pebbles and cobbles, similar to those in the Pinyon Conglomerate farther south, that were extensively fractured after deposition. The conglomeratic units thin rapidly from east to west near Calcite Springs; Bumpus Butte (figs. 2 & 3) is composed of basalt with a thin conglomerate at the base. The relation of this conglomerate to the Tower Creek Conglomerate (Hague, 1904) near Tower Fall, 1 mile southeast, has not been established. The Tower Creek was considered to be of Pliocene age on the basis of a horse tooth identified by O. C. Marsh (in Hague, 1904). No other fossils have been found in it. An unsuccessful attempt was made by Love in 1937 to locate the horse tooth at the U.S. National Museum and in the Marsh collection at Yale University in order to check its identification.

A flat terrace on the west side of the Yellowstone River at Calcite Springs (fig. 3, lower right corner) is composed of alluvial boulders cemented with white siliceous hot-spring deposits and sulfur and impregnated with oil.

The hot springs have been adequately described by Allen and Day (1935); their data are not repeated here. Chemical analyses of the water and identification of some of the associated minerals are likewise given by Allen and Day. Regarding the origin of the water, they stated (p. 498):

> The thermal waters are best explained as of mixed origin; the high sulphate and free sulphuric acid probably washed down the slope in a superficial circulation, the chloride, bicarbonate and carbonate brought up from greater depths. In those spring waters of this group that show an acid reaction, the infiltration of superficial drainage is in excess, while in the alkaline springs the carbonate and bicarbonate of the deeper water are more than equivalent to the acid of the shallow supply.

The north-south areal extent of oil seeps at this locality has not been determined. The seeps occur for at least one-fourth mile along the west side of the river and emerge at many elevations between water level and a horizon 250 feet up the west side of the canyon (fig. 3). No streams of oil flow at present, but they have at several times in the moderately recent past. This conclusion is based on the occurrence of at least 10 lithified flows, one of which is shown in figure 4; the oil emerged as a fluid, cascaded down the steep slopes, incorporated rock debris, then hardened, and finally broke up into blocks and cakes of bituminous residue. These flows are of different ages. Some are old and gray, whereas others are black and fresh appearing and must have been extruded in the last few years. The rapidity of breakup and obliteration of these flows was demonstrated by comparing photographs of the flow shown in figure 4 (taken July 16, 1964) with a similar view taken 3 years later. During this interval of time, much of the hard crust had broken up and had been scattered by water and snowslides.

The orifices from which the oil emerges are along fractures in soil and rock. Temperatures of vapors 1 foot inside the orifices range from warm to hot. Several of the hottest (fig. 5) that were measured were 93°C, but the average was 80°-85°C. The oil is highly fluid, has a strong but pleasant

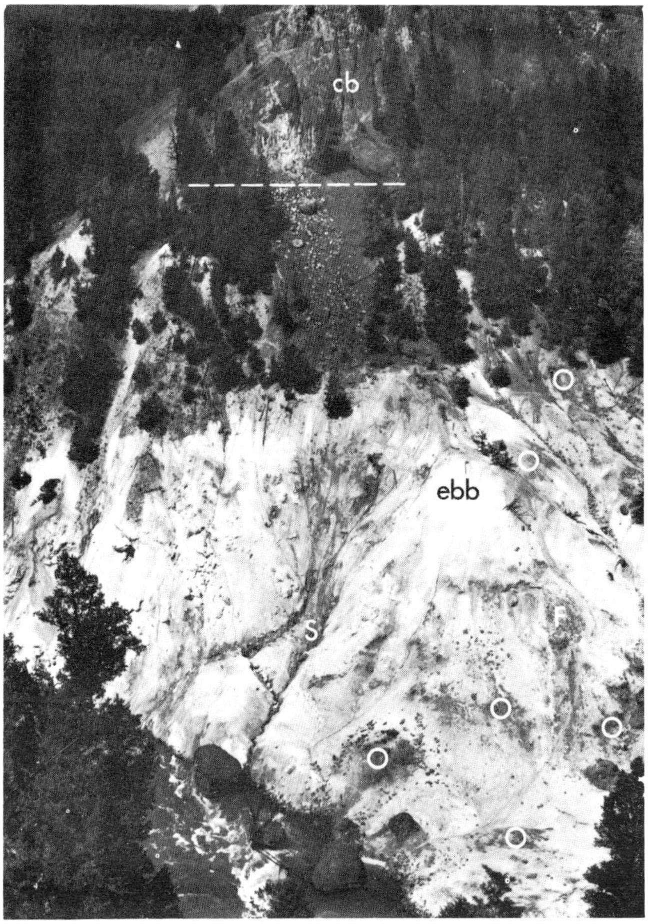

Figure 3: View west across the Yellowstone River toward Calcite Springs locality. Indicated are highly altered early basic breccia (ebb); Canyon basalt (cb) of Howard (1937), capping Bumpus Butte at top center; boiling springs (S); oil seeps (O); and lithified tarry residue from an old oil flow (F) shown in figure 4. Height of scarp is about 500 feet.

petroleum odor, and is light brown. It is accompanied by sulfurous gas and steam in the hottest vents and by small amounts of water in the cooler ones. The conduits and fractures are invariably lined with layers of beautiful euhedral yellow sulfur crystals, some as much as 2 inches long. Many crystals contain globules of oil and are generally oil stained. Some fractures, such as the one shown in figure 5, widened gradually during a considerable period of time; and as they widened successive layers of sulfur crystals with interstitial hydrocarbons were deposited.

After the oil reaches the surface in the active vents, it mixes with soil, rock, and dead vegetation to form spongy, bulbous, flexible masses. These deposits, resulting from the slow extrusion of oil, are different from the lithified flows, which were apparently emplaced much more rapidly.

The Calcite Springs, so named because of the nearby masses of milky-white calcite crystals, are directly adjacent to the oil seeps near river level. Barite crystals are associated with the calcite. In addition, there are deposits as much as 50 feet thick (fig. 6) of finely crystalline calcite interlayered with spongy, fibrous mats of gypsum. A geochemical and mineralogical study of these deposits would contribute much to an understanding of the nature of the underlying rocks, including those from and through which the oil, sulfur, and thermal water came. Elemental

Figure 4: View west at part of Calcite Springs locality where oil that emerged from the early basic breccia at point A flowed downslope and then was lithified into a tarry crust. Dashed line indicates margin of flow. For general location see figure 3. Ledge of crystalline sulfur is at S. Temperature of active oil and sulfur vents is 80°-85°C. Oil sample PC65-3 is from seep at point A, and PC65-4 is from lithified flow.

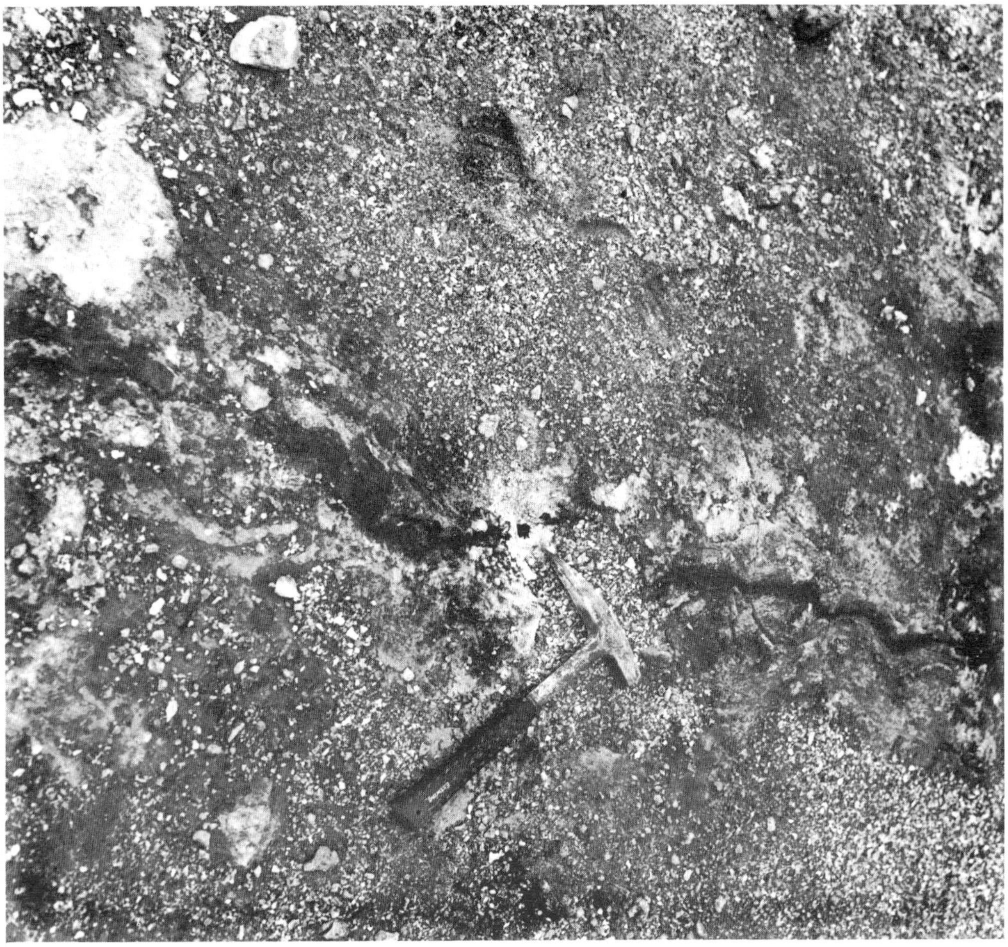

Figure 5: One of the hottest (93°C at point of hammer) sulfur and oil vents at Calcite Springs. Layered deposits on both sides of dark fissure are oil-saturated crystalline sulfur, and fissure is filled with brown fluid oil.

sulfur is invariably associated with the oil but it is not known if this sulfur and that in the gypsum and barite had a common origin.

Allen and Day's comments (1935, p. 496) regarding the oil seeps are very perceptive:

> Breaking into fumaroles with the pick, one finds well-crystallized masses of it [sulfur], mixed with bitumen of tarry appearance and odor. On the bench above the river's edge, sinuous trails, resembling the marks of burrowing moles, follow the lines of cracks where the same products have been deposited. No evidence regarding the origin of this tarry substance in the fumaroles is at hand. A unique occurrence in the Yellowstone Park, it seems to be best explained as a distillate from organic matter in buried sediments.

No analyses of the oil were made by Allen and Day. Therefore, as a part of the present study, samples of oil-saturated silt and sand were collected from vents between the river level and the highest occurrences of such material, 250 feet above the river (fig. 3). A representative sample (sample PC65-3) of oil from one of the hottest vents about 100 feet above the river contained (A. B. Hubbard, U.S. Bur. Mines, written commun., 1965) 15.5 percent benzene-soluble oil and 64.1 percent free sulfur (because the oil and free sulfur are both benzene soluble and difficult to separate, the figures may not be exactly correct). The sample also contained 20.1 percent benzene-insoluble matter. Total carbon analysis of this benzene-insoluble residue showed that it contained 13.62 percent organic carbon (J. W. Smith, written commun., 1965). Concerning this insoluble residue Smith commented:

> Presence of this amount of organic carbon not soluable in benzene is a very surprising feature of your samples. While many petroleums carry heavy asphalts, the asphalts are almost entirely benzene-soluble. The insoluble organic carbon in the

Figure 6: View northeast across the Yellowstone River at Calcite Springs. Indicated are tilted light-colored interbedded calcite and gypsum (C), early basic breccia (ebb), and partly solidified dark-colored oil flow (O) in center and left foreground that comes up through the breccia. Shovel (S) at bottom indicates scale.

Yellowstone samples may well be a parent material for the oil, carried along in the oil by the hot water.

It is interesting to compare the composition of the oil with that of the lithified flow shown in figure 4. The lithified material (sample PC65-4) contained only a trace of benzene-soluble oil (detected by ultraviolet fluorescence and infrared spectroscopy), 59.6 percent free sulfur, and 40.4 percent benzene-insoluble residue (A. B. Hubbard, written commun., 1965). This residue contained 1.42 percent organic carbon (weight percent; J. W. Smith, written commun., 1965).

M. L. Jensen, University of Utah, made sulfur isotope studies of water and sulfur from the Calcite Springs area, with results as follows (written commun., 1967):

Sample	δS^{34} per mil		
Water	−1.9	-----	-----
Sulfur	−1.9	+4.7	+7.8
	+ .3	+ .7	+4.7
	+1.1	+3.0	+5.4

With reference to these data, Jensen commented:

This spread in values is a little greater than I would like to see for undoubted magmatic hydrothermal sulfur. On the other hand, it is not quite as broad as I would like to see to be diagnostic of a bacteriogenic origin. Even considering that the δS^{34} values are near zero per mil does not enhance the magmatic hydrothermal origin of the sulfur, as a bacteriogenic origin would involve the reduction of $SO_4^{=}$ to H_2S, and exothermic oxidation of H_2S to S^0 with the $SO_4^{=}$ being typically enriched in S^{34} with a δS^{34} value of +20 with ±10 per mil. Such reduction could readily provide the δS^{34} values obtained, suggesting biogenic sulfur enriched to a variable content in S^{32} in comparison to the S^{32} content of the $SO_4^{=}$ that was reduced. A magmatic hydrothermal origin would more closely average zero per mil and should exhibit a more narrow spread in δS^{34} values than those obtained. In conclusion, therefore, with a strong assist from the geologic occurrence, I favor the bacteriogenic origin of the sulfur.

K. L. Pering, research chemist for the National Aeronautics and Space Administration, sampled oil seeps throughout the 250 feet of vertical section in which they occur. She commented on the nature of the oil as follows (written commun., 1968):

> A gas chromatogram of the alkanes in the bitumen from Calcite Springs [fig. 11*B*, this report] is characterized by a largely unresolved hump, which suggests the presence of an unresolvable mixture of alkane isomers. Individual peaks on the hump may be normal alkanes. Because of the abundance of free sulfur in the oil sample, the amount of alkanes was difficult to estimate but a rough guess is that less than 1 percent is present.

A comparison of the gas chromatogram of this oil with that of oil from Rainbow Springs (loc. 3, fig. 11*A*) is given in the discussion of the latter locality. It is worth noting that of the two oil samples on which gas chromatograms were run, the Calcite Springs oil probably contains less than 1 percent alkanes, whereas the Rainbow Springs oil contains about 99 percent.

No data are available as to the composition, age, depth, and attitude of the underlying rocks that might be supplying the oil and sulfur but, as mentioned in the first part of this report, some guesses probably can be made when regional tectonic, stratigraphic, and structural studies currently in progress in Yellowstone National Park are completed (Campbell, 1969). The reason for the sporadic flows of oil down the canyon slope is not known. The flows may be related to earthquake activity that caused rapid extrusion of oil that had accumulated underground. The unique occurrence of crystalline calcite and barite and the volume and source of the associated sulfur and gypsum warrant further investigation.

TOWER BRIDGE (LOC. 2)

The Tower Bridge oil seeps (fig. 1, loc. 2; fig. 2) are adjacent to the westernmost pylon of the new Tower Bridge across the Yellowstone River, about half a mile northeast of Tower Junction, Yellowstone National Park. No oil was known in this area prior to 1962. During January 1962, a construction company dug a pit about 30 feet deep and 20 feet square in which to put the base of the westernmost pylon of the bridge. The pit was entirely in a dark-colored facies of volcanic-rich strata equivalent to the lower part of the Cathedral Cliffs Formation (H. J. Prostka, oral commun., 1969). This sequence had previously been mapped as the early basic breccia by Hague (1904). It consists of dark-green to dark-brown coarse-grained mafic tuff, conglomerate, and basaltic breccia with a few partings of plastic gray claystone. The claystone was checked for pollen but proved to be barren. The rock has been deeply altered by hydrothermal action so that most outcrops appear white, yellow, and light green. Near the bottom of this pit, light-brown translucent oil seeped out and dribbled down the walls. At first it was thought that this was oil from construction machinery used in digging the pit, but excavation back along the conduits showed that the oil actually came from bedrock. The matrix of the rock is soft and porous because of the extensive alteration.

A few small samples of the oil-stained claystone and tuff were collected before the pylon was built and the excavation filled. Unfortunately, however, no oil sample was obtained, and no photographs were taken of the actual seeps. The rock samples were too small for adequate oil analysis, so during the following summer a bulk sample of petroliferous rock fragments that had been thrown out of the excavation was collected and analyzed for hydrocarbons. J.W. Smith (written commun., 1963) extracted it for approximately 32 hours using a Soxhlet extractor that had a 60- by 180-mm thimble, with benzene solvent, and recovered 0.24 percent benzene-soluble hydrocarbon and 0.10 percent sulfur.

The gray claystone, taken from the original pit while it was open and checked for pollen (barren), seemed unusually heavy, so a spectrographic analysis was run. The only anomalously high amount of any heavy metal was 0.2 percent molybdenum. Because the rock was slightly petroliferous, carbon and sulfur analyses were made (I. C. Frost, written commmun., 1963), with the following results: 0.29 percent total carbon and 0.02 percent mineral carbon (total carbon determined by induction furnace and mineral carbon determined gasometrically); total sulfur is 1.1 percent (determined by pyrometric-titration method).

There is no present thermal activity in or adjacent to the pylon pit, although the intense alteration of the volcanic-rich strata indicates past activity. On the east bank of the Yellowstone River, on the upstream side of the bridge, however, active thermal vents are present. These emit lethal amounts of gas, probably some type of sulfur gas. The most specific record of this gas occurence is given in a National Park Service accident report. According to W.A. Whitmore (written commun., 1939), a test pit 26 feet deep was dug in 1939 in connection with construction of a highway bridge over the Yellowstone River at the site of the present bridge. Gas with a slightly sulfurous odor was encountered. One morning, shortly after the pit was dug, several men in it were overcome by gas in a matter of minutes, and one of them died shortly thereafter. A second pit, 10 feet from this one and dug to a depth of 47 feet, encountered only small amounts of sulfurous gas. No gas was noted in a third pit, which was on the west side of the river. No hydrocarbons were noted in any of these pits, but they could easily have been overlooked, as were those in the 1962 excavations.

The original source of the oil is not known. Precambrian granite along the southwest margin of the Beartooth Mountains is exposed 2,800 feet northeast of Tower Bridge and may have been thrust southwestward over Paleozoic and Mesozoic rocks prior to deposition of the volcanic-rich strata. Farther northwest, Brown (1961) mapped the Gardiner and Everts thrust faults along the southwest margin of the Beartooth Mountains. Both faults disappear

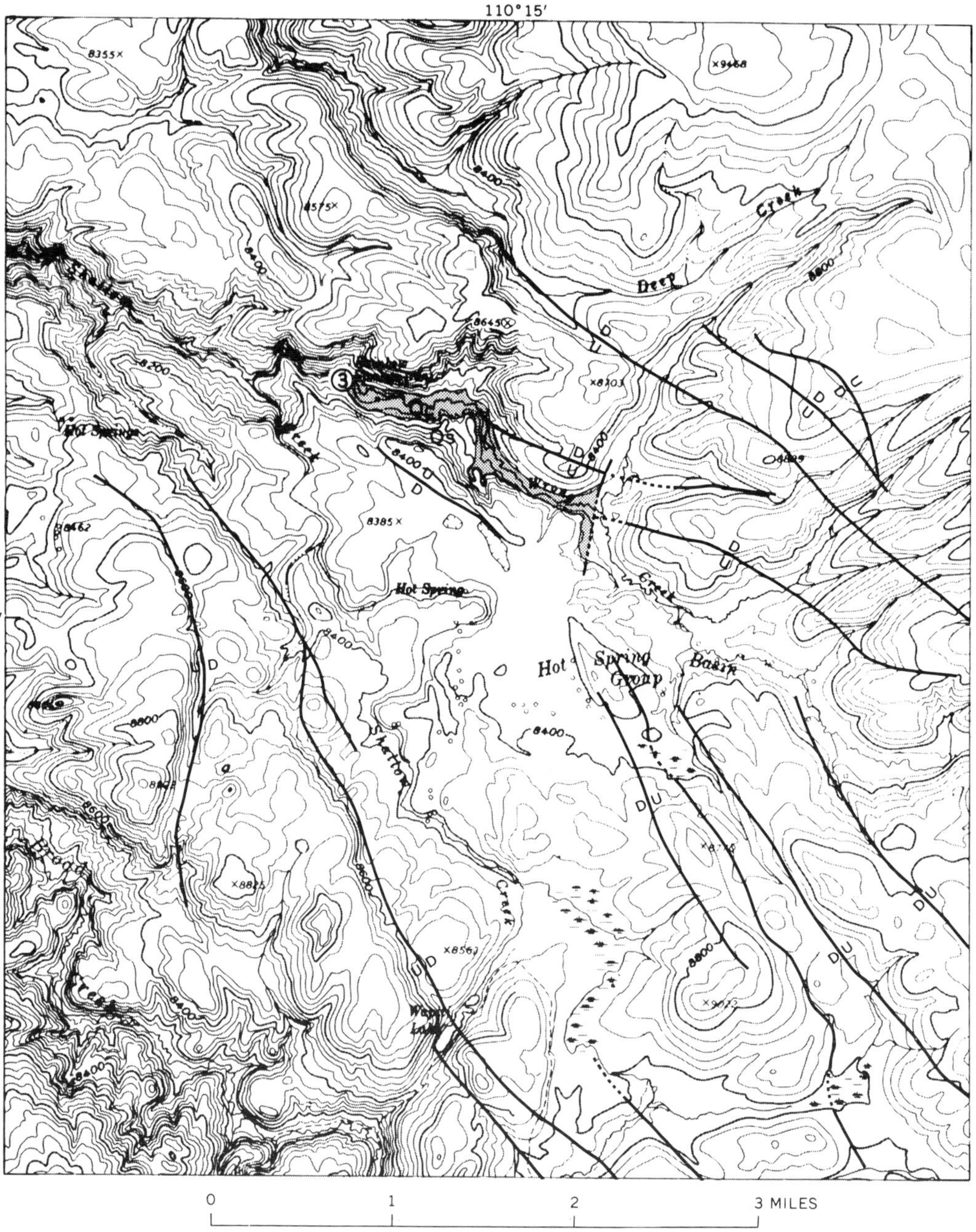

Figure 7: Map of Rainbow Springs area (loc. 3). Indicated are some of the Pleistocene faults and two Pleistocene sedimentary sequences. Oil seeps emerge from the claystone sequence (Qc, stippled area), which is overlain by a sand (Qs). Position of faults is taken from mapping by Love (1961); U, upthrown block; D, downthrown block; dotted where concealed. Base from U.S. Geological Survey Canyon Village, Tower Junction, Abiathar Peak, and Pelican Cone quadrangles, 1959.

southeastward beneath igneous and pyroclastic rocks, so their extent in that direction is not known. The oil might be migrating along these or other buried faults and reaching the surface at this locality because of the porosity and permeability of the highly altered volcanic-rich strata.

RAINBOW SPRINGS (LOC. 3)

Rainbow Springs were named by Everett (Ben) Arnold, an oldtime ranger in Yellowstone National Park (Lee Coleman, written commun., 1964). The name appeared for the first time on the 1959 Tower Junction 15-minute topographic quadrangle map (fig. 7). This descriptive name could have been derived from the rainbows of oil on the hot-water pools and streams or from the unusually brightly colored algae in the thermal area. Of all the localities, this one contains the most oil, is the most fossiliferous, and is the least accessible.

The only previous mention of oil in this area was by Allen and Day (1935, p. 356-357), and it was based on notes by Allyn Hanks, a park ranger. Allen and Day stated: "A short distance down stream from the area described is another 30 by 50 yards in size, with a brown oily deposit smelling of petroleum." The locality is designated "Deep Creek Springs (Eastern Group)." In order to determine if there was actually a set of petroliferous springs on Deep Creek or if this described locality was Rainbow Springs on Wrong Creek, which flows into Shallow Creek rather than into Deep Creek, we sent the 1959 topographic maps and aerial photographs of the area to Allyn Hanks. After examining them, he stated (written commun., 1964) that his observations that were published by Allen and Day were probably of Rainbow Springs. The confusion of drainages is understandable, because the stream pattern is unusually complex as a result of Pleistocene faulting (fig. 7), and the 1884-85 topographic map of this part of the Canyon quadrangle, which was the only map available at the time of Hanks' examination, contained extensive errors, not only in the positions and relations of the streams, but also in the direction of their flow.

Two widespread sequences of Pleistocene strata are present in the Rainbow Springs area, a blue-gray claystone and mudstone at least 75 feet thick and an overlying pebbly sandstone and sand at least 125 feet thick. These are discussed in some detail because of the bearing of stratigraphy and environment of desposition on the interpretation of the possible Pleistocene origin of the oil. The lower sequence, from which the oil seeps emerge, is exposed only in the bottom part of the valley of Wrong Creek near Rainbow Springs but is more widespread in areas to the south and southeast. The claystone and mudstone is bright to dull blue gray, soft to hard, blocky, and noncalcareous; it is locally interbedded with gray fine- to medium-grained poorly cemented volcanic sandstone containing abundant clear glass shards, black magnetite grains, and larger clear angular quartz and feldspar grains. Much of the rock is peppered with tiny black specks that have not been identified. Irregularly distributed angular grains of clear quartz and feldspar are characteristic of the claystone as well as the sandstone. The mudstone has a persistent acrid odor. Analysis of a representative sample (lab. No. D110908) showed 76 ppm (parts per million) arsenic, 0.5 ppm selenium, 0.10 percent fluorine, and 1.1 percent sulfur (determined by A. J. Bartel, G. T. Burrow, W. D. Goss, and I. C. Frost, respectively, written commun., 1963).

Spectrographic analyses show that this claystone and mudstone sequence contains less iron, calcium, manganese, strontium, and nickel than the other lacustrine Pleistocene sequences that we have studied in the area north of Yellowstone Lake.

The claystone and mudstone sequence in adjacent localities laps across Tertiary and possibly Quaternary igneous and pyroclastic rocks and is unconformably overlain by the sandstone and sand sequence and in places by lithified conglomeratic hot-spring deposits (fig. 8) or by glacial debris.

The Pleistocene age of the claystone and mudstone sequence is based on fossil diatoms and pollen. The following diatoms were identified from a sample in the lower part of a section 6,500 feet upstream from the main part of Rainbow Springs (U.S.G.S. diatom loc. 5589; identified by K. E. Lohman and G. W. Andrews, written commun., 1966). Occurrence is indicated by R, rare, and F, frequent:

Cyclotella comta (Ehrenberg) Kützing	F
cf. *C. operculata* (Agardh) Kützing	F
sp	F
Melosira cf. *M. italica* (Ehrenberg) Kützing	F
sp	R
Rhopalodia gibba (Ehrenberg) O. Müller	R
Stauroneis anceps Ehrenberg	F
Stephanodiscus astraea (Ehrenberg) Grunow	F
astraea var. *minutula* (Kützing) Grunow	F

Concerning this assemblage, as well as others from Pleistocene lacustrine strata in the general area, they commented, "The environment in which these diatoms lived was probably a cool to cold, fairly shallow, clear lake, very near neutral, as the rare species usually preferring slightly acid water are balanced by those usually preferring slightly alkaline water, and many are indicative of water of low mineral content."

Pollen assemblages from oil-saturated claystone from the Rainbow Springs locality consist of the following forms

Figure 8: Ledge of Pleistocene brown oil-stained conglomerate; darker appearing unvegetated spots are active oil seeps. View is of north bank of Wrong Creek, east of bison skull site at Rainbow Springs locality. Base of ledge marks contact with blue-gray claystone sequence which yielded pollen collection D3366.

(identified by E. B. Leopold, written commun., 1964, 1966). Collection D3310 is from the Rainbow Springs, and collection D3366 is from an area one-fourth mile upstream from Rainbow Springs and stratigraphically a few feet higher.

	Pollen collection D3310	*Pollen collection D3366*
Pinus	42	57
Artemisia	35	12
Picea	2	12
Ephedra and *Ephedra*?	1	1
Botryococcus	1	7
Quercus?	-----	1
Juniperus	-----	1
cf. *Acer*	-----	1
Eriogonum	1	-----
Polygonum cf. *P. bistortoides*	1	-----
Sarcobatus vermiculatus	1	-----
Gramineae	1	1
Compositae	2	2
Cyperaceae	1	2
Dicots, undet	10	1
Chenopodiaceae, undet	3	-----
Total number of grains	101	98

These collections represent pollen rains that are nearly identical with modern pollen accumulations present in lakes of this general area today.

The sandstone and sand sequence is about 125 feet thick. The basal bed is a hard rounded pebble conglomerate. It is overlain by sandstone that is sufficiently lithified to form a weak ledge about 15 feet thick. The remainder of the sequence is crudely stratified uncemented coarse- to fine-grained sand; much of it consists of clear shards and white pumice fragments. Black obsidian shards, clear angular quartz and feldspar grains, and some rounded granules of pink rhyolite are present. These soft strata support a dense growth of trees, and exposures are rare. The sequence extends westward along both sides of Wrong Creek at elevations beginning just above the thermal vents. No fossils were found in it.

Along the bottom of Wrong Creek within the thermal area, there are conspicuous bare to grassy outcrops of hard rusty-brown conglomerate cemented with siliceous hot-spring deposits (fig. 8). Some white to rusty-brown hot-spring deposits contain little or no conglomerate. These outcrops are exposed in weak ledges along the banks of Wrong Creek, which in Holocene time cut through them into the underlying claystone. Partly rotted and partly silicified logs are embedded at various horizons in the conglomerate and in the hot-spring deposits. If the logs are within the range of carbon-14 dating methods, analyses could provide a precise age for the conglomerate besides data on the rate of accumulation of these deposits.

In connection with an earlier study (Love, 1961), Quaternary faults were mapped in this area, and these are shown in figure 7. Because of dense timber, landslides, and poor exposures, the traces of less conspicuous faults were determined in part by low-level helicopter traverses at times when light was very oblique. Under these conditions, lines that are related to drainage and topographic anomalies and thermal features became visible in unconsolidated debris. Several faults trend into the Rainbow Springs area, but exposures are such that we were not able to determine if the thermal conduits are along faults.

The ground throughout the thermal area is warm. Many springs boil violently, although the actual volume of water that enters Wrong Creek from the north is not large. On the south side, however, one thermal complex emits about 1 cubic foot (about 7½ gallons) of hot water per second. Some springs have built siliceous cones as much as 5 feet high. Others, not quite as hot, are rimmed with white siliceous deposits and harbor algae which are red, pink, brilliant green, and inky black. Trees that have fallen into some pools are partly silicified, indicating the abundance of silica in the water and the rapidity of petrifaction. A few springs contain blue-black iron sulfide. Sulfur is not as abundant as at Calcite Springs, but some pools and rivulets have long white streamers of sulfur, and the sulfurous odor is moderately strong. More sulfur is present on the south side of the creek than on the north. Hissing steam vents are present, and gas bubbles up violently in some pools. Most of the gas that was tested was nonflammable, but a few seeps ignited explosively.

Oil seeps are numerous on the north side of Wrong Creek from Rainbow Springs upstream for at least three-

Figure 9: View northeast showing western part of Rainbow Springs locality. In foreground, unvegetated spot is dark-brown oil-saturated siliceous hot-spring deposit (site of oil sample SBR63-359X). Hole is mold of partly silicified tree that was conduit for the oil. In grass at upper left is swamp that yielded Bison skull (site of oil-saturated sand and silt sample SBR63-149X, highly fluid oil sample for Rainbow Springs gas chromatogram in fig. 11, and dark oil sample analyzed for carbon by Irving Friedman). Steam vents are at upper right.

eights of a mile. No attempt was made to determine the maximum areal distribution of the seeps, and they may be considerably more widespread. None was observed on the south side of Wrong Creek; there, the color of the brown deposits that superficially resemble oil seeps was caused by iron staining.

Some seeps are entirely oil (figs. 8, 9), whereas others are combination of oil and warm-to-boiling water. Rainbows of oil are conspicuous, streaming along the surface of water that moves out of the more aqueous seeps.

A few feet west of the oil seep shown in figure 8 is an oil-impregnated warm-water swamp in which both ancient and modern bison have mired and died. While exploring this swamp, we excavated an unusual oil-soaked bison skull that had long, slender, back-swept horns with a greater spread than that of modern bison in the region. This skull was sent to Morris F. Skinner, of the American Museum of Natural History, who has studied in detail the fossil bison of North America (Skinner and Kaisen, 1947). He made comparative measurements of various parts of the skull and concluded that it was most like *Bison (B.) bison athabascae* (the Mountain bison). This form is still living in northern Canada but had a prehistoric range as far south as Colorado. This skull is of special interest to our study because its presence near the surface indicates that the water and oil level in this swamp has changed little since prehistoric time.

As a result of this discovery, we used an iron bar, 4 feet long and one-half inch in diameter, to probe the swamp for bones of other animals. During the process of probing, we found several places in the swamp where explosively flammable gas bubbled up after the iron bar was withdrawn. Following the emergence of this gas, a creme-colored waxy substance with the consistency of a soft butter emerged on the surface of the warm water. This was collected immediately, and after 2 days, even in 50°F temperature, it

Figure 10: Untreated transparent amber -colored oil and associated water (hot at time of collection) from Rainbow Springs swamp that yields Bison skull.

changed to a dark-reddish-brown transparent highly fluid oil. In other parts of the swamp, a light-amber-colored transparent oil (fig. 10) with a strong but pleasant petroliferous odor accumulated on the surface of pools of water; at 70°F it also was highly fluid, but at lower temperatures it congealed.

Two types of samples were studied: oil-saturated sand and silt, and fluid oil. Table 1 (G. J. Schrayer, Gulf Research and Development Co., written commun., 1965) presents data on representative samples of oil saturated sand and silt from several parts of the Rainbow Springs locality. These samples, collected from within 6 inches of the weathered outcrop surface, were brown and had a strong petroleum odor. J. W. Smith (written commun., 1964) analyzed benzene extracts of two of these samples with the following results:

	Sample	
	SBR63-364X (oil-impregnated sand)	*SBR63-366X (oil-impregnated sand)*
Extract, weight percent rock	4.8	8.6
Nitrogen, weight percent extract	.03	.03
Sulfur, weight percent extract	.16	.16
Carbon residue, weight percent extract	.2	1.2
Nitrogen to carbon residue ratio	.15	.025

Table 1: Composition of oil, in weight percent, from seeps at Rainbow Springs locality, Yellowstone National Park, Wyo.

[Material analyzed is benzene-soluble extract from oil-saturated sands and silt. Sample SBR63-149X, midway between sites of figs. 8 and 9; sample SBR63—364X, locality shown in fig. 9; sample SBR63—366X, locality shown in fig. 8. Analysis by G. J. Schrayer]

	Sample		
	SBR63-149X	SBR63-364X	SBR63-366X
Gross composition of crude oil:			
Volatile hydrocarbons	8.4	16.5	22.1
Asphaltenes	0	0	0
Maltenes (Asphaltene-free, hexane-soluble non-volatile residue)	91.6	83.5	77.9
Gross composition of maltenes:			
Saturated hydrocarbons	84.3	79.1	73.1
Aromatic hydrocarbons	10.8	15.3	20.3
Nonhydrocarbons (resins)	3.9	5.6	5.8
Group composition of saturated hydrocarbons:			
Alkanes	75.2	58.7	56.9
Noncondensed cycloalkanes	17.0	27.7	28.5
Condensed cycloalkanes	7.7	13.5	14.5
Group composition of aromatic hydrocarbons:			
Substituted benzenes	39.0	38.0	32.0
Substituted naphthalenes	13.0	13.0	13.0
Acenaphthylenes	11.4	11.6	14.5
Indanes	6.9	7.2	7.3
Acenaphthenes	10.3	10.3	11.7
Pyrenes	4.0	3.9	4.5
Dinaphthenobenzenes	9.2	9.6	10.3
Phenanthrenes and (or) anthracenes	3.8	3.6	4.0
Chrysenes	2.0	2.1	2.1
Distribution of cycloalkanes:			
Monocyclic	68.7	67.2	66.2
Bicyclic	16.0	17.0	16.0
Tricyclic	6.6	7.5	7.7
Tetracyclic	4.9	5.0	5.5
Pentacyclic	2.0	2.0	2.9
Hexacyclic	2.0	0.7	1.0

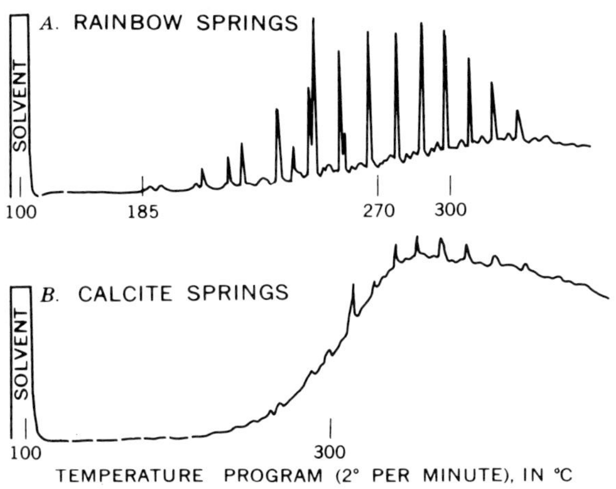

Figure 11: Gas chromatograms of the alkanes in unprocessed oils. A, Rainbow Springs; B, Calcite Springs. 50-ft Micropak SE-30 column; 2°C per min; Aerograph 660; flow rate 6 ml per min. Data by courtesy of K. L. Pering, National Aeronautics and Space Administration.

Two samples of fluid oil were collected from the locality of figure 9; one was a split of that shown in figure 10, from the surface of pools of water, and the other was a darker, more highly fluid oil, described earlier, from a probe hole. Preliminary examination of these fluid oil samples showed that they were not appreciably different from the oil in the nearby sand and silt (table 1; G.J. Schrayer, oral Commun., 1968).

A gas chromatogram of the alkanes extracted from the light-amber-colored oil shown in figure 10 was made by K. L. Pering and is reproduced as part of figure 11. She commented on the nature of this oil sample as follows (written commun., 1968):

> Preliminary examination of gas chromatogram of the alkanes from the Rainbow Springs oil [fig. 11A] shows that it consists almost entirely of saturated, acyclic alkanes. Normal alkanes predominate, and peaks with the retention time of the isoprenoids, pristane and phytane, are evident. These alkanes exhibit a distribution seen in many sediment alkanes. Lack of major alkane components below nC_{16} suggest a possible loss of volatile constituents.
>
> The disparity between the gas chromatographic distribution of alkane mixtures from Calcite Springs and Rainbow Springs [figs. 11*A*, *B*] suggests either that they came from two different sources, or that one or both represent altered fractions of the same oil.

Figure 12, reproduced from figure 2 of Ponnamperuma and Pering (1967), is included for the purpose of comparing the gas chromatograms of oil from Calcite Springs and Rainbow Springs (fig. 11) with those of a normal marine shale, the Posidonian (Jurassic) of Europe, and a nonhydrothermal deposit, the Trinidad Lake asphalt. On the basis of similarities between these two sets of gas chromatograms, it would be easy to reach conclusions that might be invalidated when suites of gas chromatograms are made of many samples from various parts of the Calcite Springs and Rainbow Springs areas. Therefore, the ones presented here are meaningful only in the sense that they call attention to a line of investigation that may yield significant data.

As mentioned earlier, the contrast in alkane content of the two oil samples on which gas chromatograms were run is very striking (about 99 percent alkane in the Rainbow Springs oil and less than 1 percent alkane in the Calcite Springs oil). More analyses are needed to confirm this difference and, once confirmed, to determine its significance.

Carbon samples from a light-colored highly fluid oil (fig. 10) skimmed off the surface of a spring and from a dark-colored even more fluid oil from a probe hole were analyzed by Irving Friedman (written commun., 1968), with results as follows:

Oil sample	C^{13} *per mil* PDB[1]	*Deuterium/ Hydrogen percent-* SMOW[2]
Light	−28.9	−18.6
Dark	−26.4	−19.5

[1]The carbon analyses are given in terms of the PDB standard. PDB is a Cretaceous belemnite, *Belemnitella americana*, from the Peedee Formation of South Carolina (Craig, 1957, p. 135). A value of −28.9 per mil indicates that the sample contains 2.89 percent less C^{13} than does the standard belemnite.

[2]The deuterium results are given relative to Standard Mean Ocean Water (SMOW), a standard water whose isotopic composition is close to that of mean ocean water. A value of −18.6 percent SMOW indicates that the sample has 18.6 percent less deuterium than does SMOW (Craig, 1961).

The carbon data are suggestive but not definitive; they suggest that the oil at Rainbow Springs probably is of marine origin, but there is, nevertheless, a small possibility that it is of fresh-water origin.

Examination of the dark-colored highly fluid oil sample by infrared spectroscopy and nuclear magnetic resonance shows it to have a very high paraffin content with only 1.5±0.5 percent aromatics. The ratio of CH_3 to CH_2 is 0.7. Content of even the simplest cycloalkanes is low. The oil was already extensively refined before it reached the near-surface probe hole, for it contains a minimum of unsaturates, nitrogen, sulfur, and oxygen (J. W. Smith, written commun., 1965).

Oil in Pleistocene and Holocene sediments is known in a few places in the Rocky Mountain region, but is common-

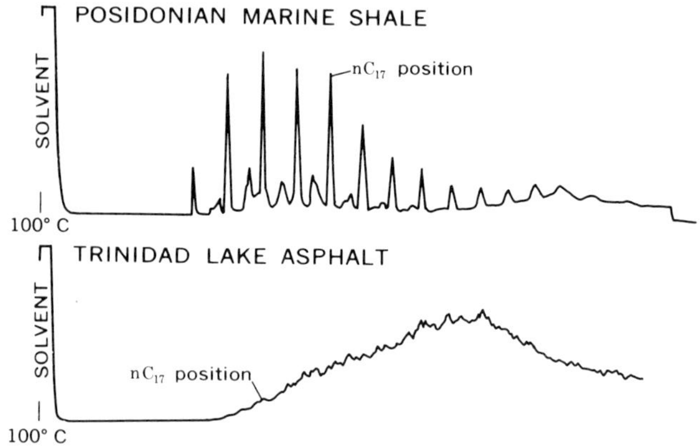

Figure 12: Comparison of the gas chromatogram of the aliphatic hydrocarbons extracted from Posidonian marine shale with that of the aliphatic hydrocarbons extracted from Trinidad Lake asphalt. Temperature progression 2°C per min; 12 ft by 1/8 in., 3 percent SE-30 column; Aerograph 660. (From Ponnamperuma and Pering, 1967, fig. 2.)

ly believed to have migrated into these young strata from older, buried petroliferous rocks. New studies of modern sediments (Swanson and others, 1968) and algal communities (Parker and Leo, 1965; Leo and Parker, 1966), however, indicate that under certain conditions, such as shallow-water and nearshore environments favorable to abundant growth of plants of various types, significant local accumulations of bitumen can and do occur.

The oil-secreting alga, *Botryococcus*, was identified, though it is not abundant, in the claystone and mudstone sequence at several localities in the Rainbow Springs area. (See pollen list, p. B12.) Greater concentrations of plant material may be present at depth. The diatoms do not indicate an unusually saline environment of deposition for these strata, but it is not known if such an environment is necessary for accumulation of organic material that could yield oil in the amount and of the type observed at Rainbow Springs. As far as is known now, it is possible for oil to have been distilled, with the help of thermal conditions, from dead algae and other plant debris at shallow depths, perhaps less than 100-200 feet, in the Rainbow Springs area. A more detailed investigation of this possibility should be made.

An alternate interpretation of origin is just as likely. The PDB and SMOW data suggested a marine origin, and the sulfur and nitrogen contents and ratio of nitrogen to Ramsbottom carbon residue indicate post-Jurassic age. If these data are indicative, the hydrocarbons are probably Cretaceous and therefore must have come up from buried Cretaceous rocks at some unknown depth.

SWEETWATER MINERAL SPRINGS (LOC. 4)

The Sweetwater Mineral Springs (fig. 1, loc. 4; fig. 13) are located on Sweetwater Creek, about 3.5 miles north of the North Fork of the Shoshone River and 18 miles east of Yellowstone National Park.

The general geology of the region was described by Hague (1899). Later, Hewett (1914) studied sulfur deposits adjacent to the springs and mentioned the occurrence of oil as an item of special interest (1913). There are no subsequent publications describing the area.

The rock exposed at the surface is the early basic breccia of Eocene age. It is a brown mafic volcanic conglomerate and agglomerate several thousand feet thick and nearly horizontal in position. One of the conspicuous features of this sequence at this locality is the presence of angular masses, as much as 20 feet in diameter, of light-colored dolomite and limestone of Paleozoic age.

The mineral springs emerge an estimated 500 feet above the base of the volcanic rocks, a few hundred feet upstream (north) from the hydrocarbon seeps, between the upper and lower sulfur deposits described by Hewett (1914). The water in these springs differs from that at the localities previously discussed, because it is cold. The presence at and near the present stream level, however, of bedded sulfur, siliceous sinter, travertine, and gas vents suggests that this is a thermal area that has recently cooled. Some latent heat apparently remains in the rocks, for in winter the snow melts quickly from some spots near the sulfur deposits but accumulates elsewhere (Mel Stonehouse, owner, in 1968, of the Sweetwater Minerals Springs Resort, oral commun., 1968).

The water is so highly charged with iron that the channelways are lined with a bright limonite yellow iron precipitate. The surface of water in quite pools and seeps has a thick iridescent coating of iron which, unlike oil, breaks up into chunks when a twig is dragged through it. At one time, because of its reported curative attributes, the water was bottled and sold. In recent years, the springs have become so choked with alluvial debris that they flow only a few tens of gallons per day. Analysis of the water by the U.S. Geological Survey (O. J. Feist, Jr., analyst, 1968) is as follows:

Chemical components

Cations:	*Milligrams per liter*	*Milli-equivalents per liter*
SiO_2	27	-----
Fe	.31	-----
Ca	45	2.25
Mg	12	.99
Sr	.48	.01
Na	16	.70
K	1.1	.03
Li	< .01	.00
Total		3.98

Anions:		
HCO_3	97	1.59
CO_3	0	.00
SO_4	131	2.73
Cl	1.3	.04
F	.3	.02
NO_3	.5	.01
PO_4	.00	-----
Total		4.39

Physical characteristics and computed values

Dissolved solids:		
Residues on evaporation at 180° C	milligrams per liter	340
Calculated	milligrams per liter	283
Hardness as $CaCO_3$:		
Total	milligrams per liter	163
Noncarbonate	milligrams per liter	83
Specific conductance	micromhos at 25°C	449
pH		6.2

Hewett did not mention the mineral springs. Regarding the springs deposits, he stated (1914): "Siliceous sinter and travertine, such as in many places accompany sulphur deposited by hot springs, are conspicuously absent."

The sulfur deposits described by Hewett (1914) are on the east side of Sweetwater Creek and extend from water level to more than 400 feet up the slope (fig. 14). He apparently did not see the bedded sulfur and travertine, referred to later, on the west side of the creek. His description of the oil and sulfur is as follows (1913, p. 51):

> The [oil] spring is unique in that it lies within 100 yards of several small sulfur deposits, situated on the east side of Sweetwater Creek. The sulfur deposits * * * embrace two classes of material: (1) sulfur which lies along the walls of open fractures in the lavas, and from which gases containing CO_2, CH_4, and H_2S are issuing, and (2) sulfur filling the interstices of gravels and surface debris along stream channels. The second class of material probably covers fracture zones. The oil spring is opposite an area 100 feet square into which prospect pits have been sunk showing sulfur cementing angular rock debris to a depth of 9 feet.

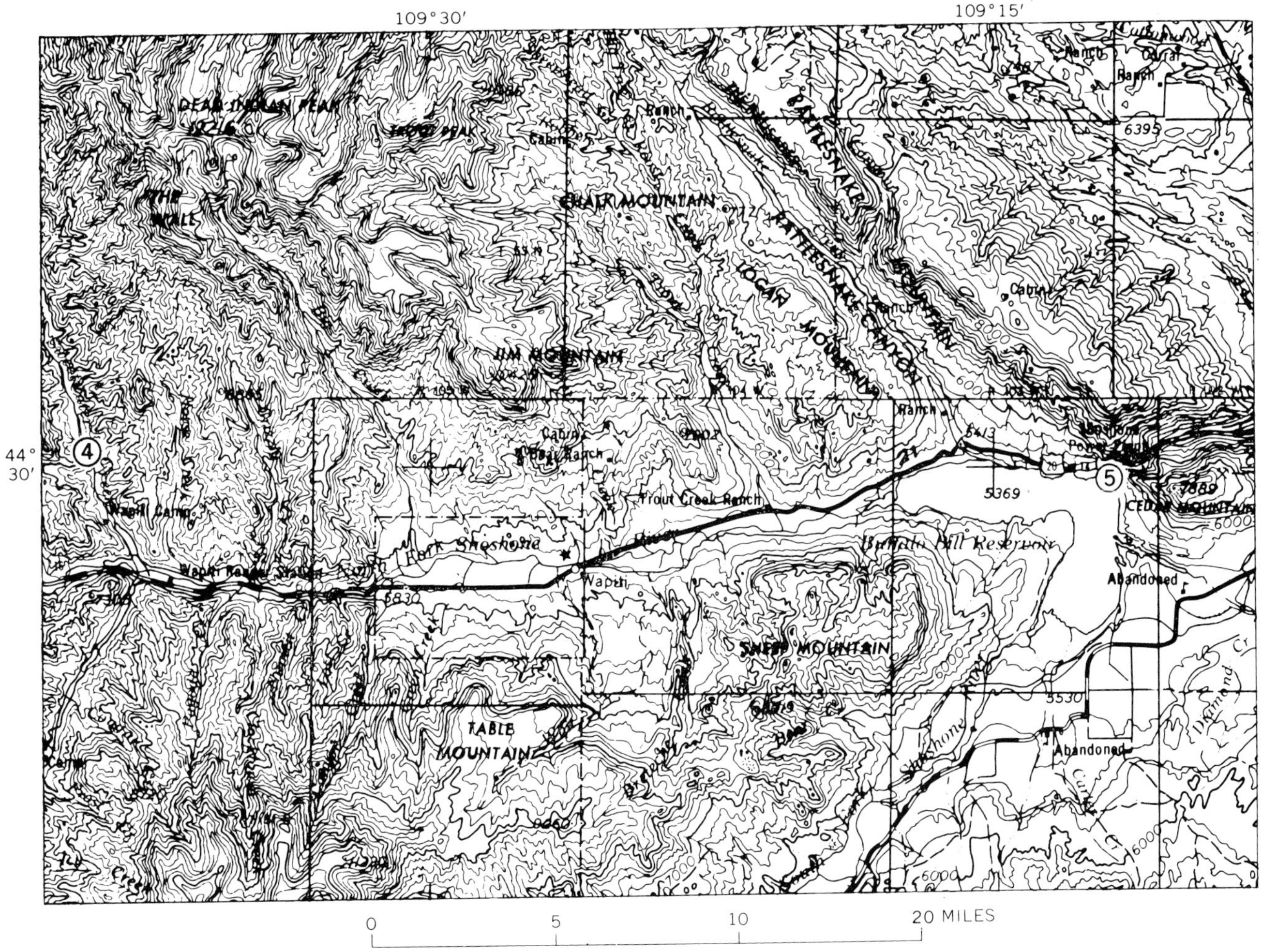

Figure 13: Locations of Sweetwater Mineral Springs (loc. 4) and Cedar Mountain (loc. 5). Base from U.S. Geological Survey Cody, 1: 250,000, 1955-62.

Other smaller deposits of sulfur lie 700 feet farther up the creek.

Oil was first recognized issuing from the sands adjoining the creek.

In the hole which was dug along the bank, water and oil slowly accumulated several quarts of clear light oil were thus collected during the summer of 1911. During the writer's visit a pit was dug near the location of the old one which had been destroyed by freshets. The sand at this point is dark brown and has an asphaltic odor, but otherwise is such as would form bars along rapid mountain streams. In the short time at the writer's disposal only enough oil was collected to give assurance of its identification, but not enough for analysis.

Tho the superficial rocks of this region are igneous flows and breccias, these rocks overlie a great thickness of sedimentary rocks ranging in age from Cambrian to Eocene, the Mesozoic section along being approximately 14,000 feet thick 25 miles east.

Another occurence of sulfur not mentioned by Hewett is on the west side of Sweetwater Creek beneath a flat alluvial valley fill covered with trees 100 feet west of the camera point for figure 14. The sulfur is yellow, finely crystalline, vuggy, and bedded; associated with it is a breccia of white laminated fine-grained siliceous sinter similar to that deposited by thermal water at localities 1 and 3. Overlying this is about 3 feet of peculiar alluvium consisting of soft brown sand and silt that is not bouldery as is alluvial debris in the adjacent stream channel. Carbonized twigs are abundant, the darkest brown parts have a strong petroliferous and sulfur odor, and some layers look like a lapillia tuff. It may be that the deposit is residual debris left after a fire that burned the forest cover and ignited the oil and sulfur. Subsequently, enough oil reimpregnated the silt to give it the petroliferous odor.

Laboratory studies by J. G. Palacas (written commun., 1969) of a single sample of this brown silt indicate that the retained saturated hydrocarbons, restricted to those molecules containing 19 carbon atoms and over, which make up only about 1 percent of the total benzene-extractable organic matter, appear to be derived from indigenous organic substances, such as algal organisms, rather than from petroleum sources. Petroleum hydrocarbons, especially saturated hydrocarbons with carbon numbers less than 18, may still be present, as is indicated by the petroleumlike odor of the alluvium, but were not detected by the analytical procedures used. These procedures included pulverization, extraction with benzene at room temperature, removal of elemental sulfur by use of a copper-amalgam chromatographic column, and fractionation of sulfur-free bitumen by silica-gel-column chromatography into saturated hydrocarbons, aromatic hydrocarbons, and asphaltic compounds. The saturated hydrocarbon fraction was studied further by gas-liquid chromatography. In addition, ultimate analysis was made of an aliquot of the bitumen extract. The negative results of these procedures on the alluvial sample, despite its petroliferous odor, suggest that future studies should concentrate on fluid oil from pits such as those described by Hewett. A puzzling feature of the alluvial sample is its surprisingly low sulfur content, which was 0.79 percent though the sample was taken less than 3 feet above bedded sulfur deposits.

When locality 4 was visited, both in 1963 and in 1968, a series of torrential floods had recently plastered much of the banks and stream bottom with mud so that the oil occurrences at water level could not be studied or sampled adequately. Bulk samples of fetid gravel, sand, and fine-grained alluvium were taken, but they did not yield sufficient oil for analysis. The oil is extremely volatile (C. A. Burk, written commun., 1963), and even the residue is not tarry or gummy as is that at Calcite Springs and Rainbow Springs. Sporadic traces of oil were observed within an area more than 100 yards in diameter on the west side of the creek. In 1967, petroliferous alluvium was found on the east side in a ditch for a waterline between the cabin and the sulfur bank shown in figure 14 (Mel Stonehouse, oral commun., 1968). A strong petroleum odor was detected in ferruginous cemented alluvial gravel at several places along both sides of the flood channel (fig. 14).

Two samples, one from a sulfur pit west of the creek (SBR63-362X) and the other from a ferruginous cemented gravel at the west edge of the creek (SBR63-363X), 100 feet north of the pit, were submitted to the U.S. Bureau of Mines for analysis. They were extracted for approximately 32 hours by means of a Soxhlet extractor with 60- by 180-mm thimble, using chloroform solvent. Sample SBR63-362X yielded 3.21 percent extract, and SBR63-363X yielded 0.01 percent (J.W. Smith, written commun., 1963). The extract of petroliferous material from the sulfur pit was brown and had a rubbery consistency and an unusually pungent odor.

The position of the sulfur deposits in the bottom of the valley on the west side of Sweetwater Creek (water has a gradient of 150 feet per mile), only 5-10 feet above the present flood channel, indicates that this episode of sulfur deposition and thermal activity was in Holocene time. When larger samples of oil are analyzed, the source rock from which they originated may be determinable. Isotope studies of sulfur from the active and extinct vents should

Figure 14: View east across Sweetwater Creek at Sweetwater Mineral Springs locality. Indicated are early basic breccia (ebb); exposed yellow sulfurous deposits (S); black sulfurous deposit (B); brown petroliferous iron-cemented gravel and sand (P); and stream gravel in which pits were dug that yielded oil (G).

provide a clue concerning the source of the sulfur, the depth and nature of the heat source, and the relation of the heat source to the oil.

CEDAR MOUNTAIN (LOC. 5)

The description of the oil spring at the Cedar Mountain locality is based entirely on observations made by others more than 60 years ago; the spring was obliterated in 1907 by water of the Buffalo Bill Reservoir. Nevertheless, the scattered data can be used to reconstruct a general description that is pertinent to this study when considered within the context of the other localities. This locality is included primarily because the oil spring emerged directly from Paleozoic strata, yet is in a thermal area that previously had been buried by several thousand feet of igneous, pyroclastic, and nontuffaceous sedimentary rocks. It may, therefore, provide some data as to what might be expected at depth at the other localities.

This is the locality (loc. C, fig. 15) of John Colter's "great tar spring" or "boiling tar spring" (Harris, 1952, p. 95). Several thermal areas are present along the Shoshone River near here. The one that was shown on many maps (for example, Wheat, 1957-1963) is now called De Maris Springs; it is at the east end of the Shoshone Canyon (loc. E, fig. 15), where colorful sulfur deposits have long been recognized and intermittently exploited. The thermal area at the west end of the canyon received no scientific attention, however, and as far as we have been able to determine was never studied by a geologist prior to its destruction.

Comstock (1875) came close to the oil spring but apparently did not know of its existence. One of his men (Assistant Surgeon C. L. Heizmann), however, described two small adjacent thermal areas (loc. A and B, fig. 15 of this report) as follows (jones, 1875, p. 295-296):

> Camp 32, July 26, was made on the Stinkingwater, near which we found a large spring raised two or three feet above the ground [loc. *A*, fig. 15], with an abundant flow of a blackish water of great specific gravity, of strong sulfurous smell (temperature 56°; of air 75°), and depositing black. The water contained soda, sulphates, and sulphides. The deposit was made up of sulphur and hyposulphite of soda. As we approached the river, by Cretaceous bluffs, an odor of sulphuretted hydrogen became distinctly perceptible in the air, observable several miles away. A number of extinct springs occur here [loc. *B*, according to Jones' route map, sheet 24], but the existence of numerous active ones near, besides the last mentioned, was indubitable.

Jones (1875, p. 17) added a few more details:

> We passed the remains of a large depositing sulfur spring [loc. *A*, fig. 15]. The water oozed from a cylindrical mound of soft mud, a little hardened on its rim and held together by the roots of a rank growth of water-grasses. It lies close to the Ish-a-woo-a River, on the south side near where our trail crosses, and probably at one time contributed largely to the odorific title of the main river. A few miles lower down, below the

> canon, a mass of sulphur springs occur [loc. *E*, fig. 15] which still give cause for the river's name. On the North Fork we found the extinct remains of a mass of depositing springs [loc. *B*, fig. 15]; the whole hill-side was covered with the large sedimentary mounds of soft black and brown earths, having a good deal of transparent gypsum in crystals and small pieces scattered through them.

Eldridge (1894) made a remarkably detailed geologic map of the Bighorn Basin region and described the sulfur spring at locality E (fig. 15) but did not mention those up the river to the west. Chittenden (1895, p. 23), however, was aware of the oil spring, for he stated:

> * * * after [Colter] had passed through Pryor's Gap, he took a south-westerly direction as far as Clark's Fork, which stream he ascended for some distance, and then crossed over to the Stinkingwater. Here he discovered a large boiling spring, strongly impregnated with tar and sulphur, the odor of which, perceptible for a great distance around, has given the stream its "unhappy name."

Later, in an official report to the War Department, Chittenden (1901, p. 3781) amplified his earlier remarks:

> This stream has heretofore been known as the Stinkingwater River. It was so named in 1807 by its discoverer, John Colter, who came upon it where there is a large tar spring near the junction of the two forks and gave it the name on account of the impression received at this particular point.

Fisher (1906) was the last geologist to map the area covered by the Buffalo Bill Reservoir before it was filled in 1907. He showed a small area of hot-spring deposits overlapping Paleozoic rocks at the west end of the Shoshone Canyon downstream from the forks of the Shoshone River (loc. D, fig. 15). He did not describe them, although he did discuss those at the east end of the canyon (loc. E. fig. 15) in moderate detail. He made no mention of the oil spring. The geology of the entire area shown in figure 15 was mapped in detail in recent years by Pierce and his associates (Pierce, 1966, 1970; Pierce and Nelson, 1968, 1969). On the basis of their mapping, it seems probable to us that the oil spring emerged from the Tensleep Sandstone (Pennsylvanian) or from Permian rocks. Pierce investigated the reservoir area during an especially low water stage in 1941 and noted travertine and sulfur on both sides of the valley west of the canyon and sulfurous gas bubbling up through the reservoir water on the south side (W. G. Pierce, written commun., 1968).

By far the most detailed description of the oil spring at locality C was by J. K. Rollinson, a rancher and writer who published several books (for example, 1948) on the human history of the region. As a preface to Rollinson's observations, Harris (1952, p. 95) stated:

> The legend of Colter's Hell calls for boiling springs, sulphur lying on the ground, flames shooting from the earth, and the great tar spring. Much fun has been poked at the story of the tar springs, since none are known to the present-day residents of the Shoshone valley. Nevertheless, the maps of Drouillard and Clark, by showing the trail winding around Cedar Mountain prove that Colter, if not Drouillard, actually saw at least one of the tar springs. The "boiling tar springs" was not the product of Colter's imagination but was located on the South Fork of the Stinking Water just above the junction with the North Fork and at the bottom of Cedar Mountain. The entire area has been at the bottom of Shoshone, or Buffalo Bill, Lake for more than forty years. The late J. K. Rollison, who punched cattle on the upper Stinking Water before the irrigation dam was constructed in 1906, is authority for this statement. Mr. Rollison described the boiling tar spring, which he saw several times, as being situated with several other hot-water springs that were surrounded by a number of small semi-geyser spouters.

Rollinson's statement, quoted by Harris (1952, p. 170), is as follows:

> In the years prior to the government Shoshone Reclamation Project * * * there was a spring of water highly impregnated with sulphur and magnesium located on the extreme north end of the ranch then known as the Buffalo Meadows, which lay at the junction of the Forks of the Stinking Water River. There were also a half-a-dozen little "puffer" or semi geyser spouts * * * these small sulpho-sodium geysers which steamed up all the time and bubbled up mud and minerals. Then exactly at the very foot of Cedar Mountain close to these incipient geysers there was a forty foot cut bank or cave-in, and from the bottom of this came a considerable flow of asphaltum, which travelled by gravity about eighty feet before it disappeared for a few feet (about 30 feet) where it again appeared in one of the hot Sulpho-Magnesia Springs. While the temperature was (in my time) never hot

enough to boil, still the gasses came up and gave the spring the bubbling appearance * * * I suppose the temperature was about 90 degrees and with the foam and gas from the subterranial passage, it gave the complete effect of a boiling tar spring. This spring and all others on the side of Cedar Mountain were inundated by the impounding of the flood waters.

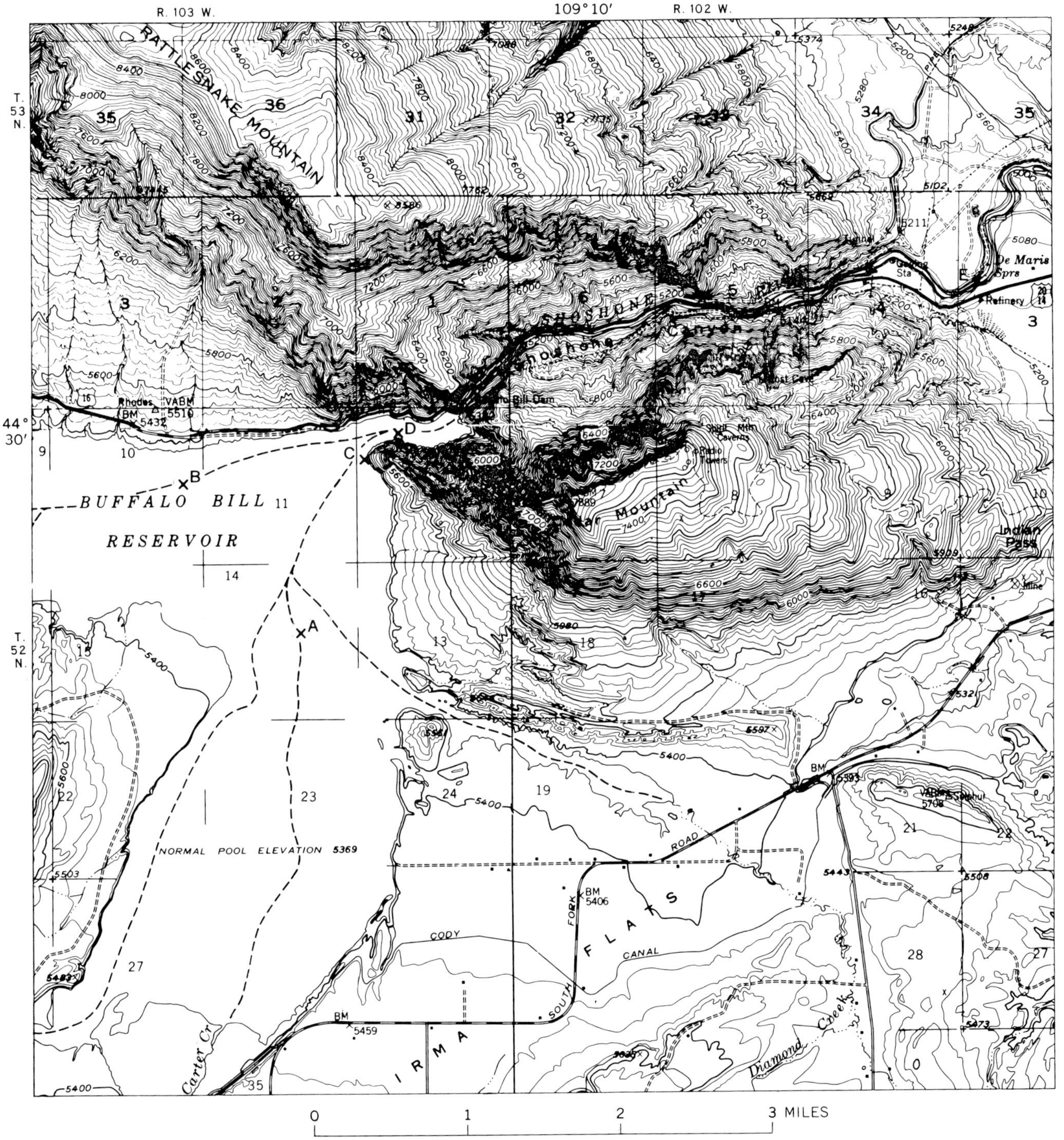

Figure 15: Cedar Mountain oil seep (loc. C) and adjacent thermal localities near forks of the Shoshone River, reconstructed from data by Comstock (1875), Fisher (1906), and J. K. Rollison (in Harris, 1952, p. 170). Indicated are: A, sulfur spring described by Comstock; B, extinct sulfur and iron spring described by Comstock; C. Colter's "great tar spring" described by Rollison; D, hot-spring deposit mapped by Fisher; and E. De Maris sulfur springs at mouth of Shoshone Canyon. Heavy dashed lines indicate original stream courses prior to flooding by reservoir water. Base from U.S. Geological Survey Cody quadrangle, 1949, and Devils Tooth quadrangle, 1957.

The various descriptions of the oil as "tar" or "asphaltum" by nontechnical observers suggest that it is much more like the viscous high-sulfur oils characteristic of Paleozoic rocks in the Bighorn Basin than like the typically lighter colored more fluid Cretaceous oils (Stone, 1967; Wenger and Lanum, 1954; open-file crude-oil analyses, Rocky Mountain regional office, Laramie Petroleum Research Center, U.S. Bureau of Mines). This is compatible with the interpretation that the "tar spring" emerged from Pennsylvanian or Permian rocks.

Fisher (1906) postulated that the origin of the hot water (he says about 98°F) in the sulfur springs which emerge chiefly from Permian rocks at locality E could be accounted for on the basis of thermal gradient. Woodruff (1908), however, disagreed. On the basis of his study of the sulfur deposits, he concluded (p. 454):

> As the source of the water is not known, the cause of the heat remains undiscovered. There is no evidence of solfataric origin and the indications are that only a small amount of the heat is derived from chemical reactions, because but little of the material produced by chemical changes below is brought to the surface. It is known that intensive activity must necessarily take place to maintain the water at a temperature above 98°F., the temperature found in these springs. It is supposed, therefore, that the heat is derived from slowly cooling magmatic bodies, probably intrusions, and increased slightly by heat from chemical reactions.

Pierce (written commun., 1968), as well as Fisher, considered a thermal gradient as a sufficient explanation for the hot water. His interpretation used evidence of high temperatures in wells 5 miles to the southeast. Water heated by thermal gradient in the deep part of the Dry Creek syncline (pierce and others, 1947; Zapp, 1956), 15 miles southeast of Cedar Mountain, could have flowed northwestward up the trough of the syncline, then divided, continued around both sides of Cedar Mountain, and emerged in the hot springs at localities A, B, C, and E (fig. 15). The temperature at De Maris Springs (loc. E) ranges from 75° to 100°F. No detailed figures are available on magnitude of flow, but one spring that formerly provided water for a swimming pool emitted 50-100 gallons per minute.

It is apparent that an interpretation of the geologic history of the Cedar Mountain area must include an explanation of the oil spring closely associated with the hot-water springs, sulfur, open conduits, and caverns. The hot-spring conduits would have to be so open that the observed volume of water could reach the surface rapidly enough to retain the heat acquired at depth, regardless of the nature of the heat source. Active solution of the Madison Limestone (Mississippian) has gone on for a long time, as is indicated by the location and elevation of an extensive network of caves, such as Spirit Mountain caverns and Frost Cave (Jones, 1875, p. 17; fig. 15, this rept.) on Cedar Mountain, more than 1,000 feet above the oil spring to the west and De Maris springs to the east.

It is probable that cavern solution began before the pyroclastic rocks were stripped from the area. At least locally, the caves developed first, and sulfur was then precipitated in them. For example, great bulbous masses of crystalline sulfur, forming stalactites and stalagmites, were encountered in some of the caves cut by the Bureau of Reclamation tunnel on the south side of the Shoshone Canyon, 200 feet or more above the present river level. Descriptions of this sulfur have not been published, but samples of the sulfur collected by Love in 1935 at the time the tunnel was being driven are in the University of Wyoming mineral collections.

It seems unlikely that these caves would have been formed by cold-water solution on an anticline that had hot springs emerging from both sides and that the sulfur found in the tunnel that cut these caves would be unrelated to the adjacent thermal phenomena.

In summary, the source of oil at locality C was probably Paleozoic rocks at a relatively shallow depth. The close association of the oil and sulfur with hot water suggests that all three were boosted to the surface together. The absence of oil from locality E might be accounted for by the emergence of these springs from Paleozoic rocks that had previously been flushed of any oil that was originally present. On the other hand, oil may be present but could have been overlooked.

All these various shreds of evidence indicate or suggest a long and complex episode of solution and thermal activity, at least some of which is involved in the emergence of the oil. Sulfur isotope studies of the various occurrences of sulfur in the Cedar Mountain area may determine whether the deposits are of magmatic-hydrothermal or bacteriogenic origin. More detailed studies of the composition, temperature distribution, and flow of the thermal springs and of the adjacent wells should provide additional pertinent information.

CONCLUSIONS

Localities 1, 3, and 4 should be studied in detail before they are destroyed or damaged further and vital evidence is lost. As the foregoing discussion indicates, data are not sufficient as yet at any locality to justify unequivocal conclusions about the origin or the geochemical processes involved in emplacement of the oil. Nevertheless, the data are so encouraging that they provide impetus for detailed studies which may lead to a reasonable explanation for these occurrences. It is pertinent to consider whether the problems of origin and emplacement of oil in thermal areas are worth investigating. We think they are. The literature on origin and emplacement of hydrocarbons in general is voluminous, and research on these subjects is continually expanding. However, oil in thermal areas is rare and in most places has not been studied in detail. These occurrences may offer previously overlooked clues as to the role of heat in natural processes of oil extraction from source rocks, the subsequent migration and emplacement of the oil, and the changes that occurred in it along the route.

The five described occurrences of oil in thermal areas are the only ones of any type that are known to us in this part of Wyoming. This suggests that hot water is an integral part of the process that liberates the oil from the source sediments adjacent to the hot-water conduits. The primary source of heat at localities 1 and 3 is believed to be igneous rocks or possibly magma at depth.

Publication of descriptions of the known localities will, hopefully, result in the discovery of additional ones that will broaden our knowledge of what types of phenomena are related to oil in thermal areas, what characteristics can be predicted for these occurrences, and how the data can be utilized elsewhere.

Locality 1 (Calcite Springs) shows evidence that there were periods of spasmodic outflow of oil, separated by periods of quiescence. Observations of these petroliferous vents over a considerable period of time may show whether these intervals of outflow are related to earthquakes. If such a relation is established, the remnants of semilithified and partly buried old oil flows here and perhaps in other areas may give clues to the age, magnitude, and frequency of prehistoric earthquakes. It is possible that the youngest flow at Calcite Springs, now rapidly breaking up, may have been formed at the time of the severe earthquake of 1959 in and near Yellowstone National Park.

The source of the abundant sulfur at localities 1 and 4 has not been explained. When studies of other sulfur occurrences in Yellowstone National Park and areas to the east are assembled, the significance of those associated with oil may become more apparent.

REFERENCES CITED

Allen, E. T., and Day, A. L., 1935, Hot Springs of the Yellowstone National Park: Carnegie Inst. Washington Pub. 466, 525 p.

Brown, C. W., 1961, Cenozoic stratigraphy and structural geology, northeast Yellowstone National Park, Wyoming and Montana: Geol. Soc. America Bull., v. 72, no. 8, p. 1173-1193.

Campbell, A. B., 1969, Current U.S. Geological Survey research in Yellowstone National Park, *in* Abstracts for 1968: Geol. Soc. America Sprc. Paper 121, p. 590.

Chittenden, H. M., 1895, The Yellowstone National Park: Cincinnati, Ohio, Robert Clarke Co., 397 p.

____1901, Improvement of the Yellowstone National Park, including the construction, repair, and maintenance of roads and bridges, *in* Report of the chief of engineers: U.S. War Dept. Ann. Rept., year ended June 30, 1901, v. 2, pt. 5, App. FFF, p. 3777-3798.

Comstock, T. B., 1875, Geological report, *in* Jones, W. A., Report upon the reconnaissance of northwestern Wyoming, including Yellowstone National Park, made in the summer of 1873: Washington, U.S. Govt. Printing Office, p. 85-292.

Craig, Harmon, 1957, Isotopic standards for carbon and oxygen and correction factors for mass-spectrometric analysis of carbon dioxide: Geochim. et Cosmoshim. Acta. v. 12, nos 1-2, p. 133-149.

____ 1961, Standard for reporting concentrations of deuterium and oxygen-18 in natural waters: Science, v. 133, no. 3467, p. 1833-1834.

Denig, E. T., 1961, Five Indian tribes of the Upper Missouri: Norman, Okla., Oklahoma Univ. Press, 218 p.

Eldridge, G. H., 1894, A geological reconnaissance in northwest Wyoming, with special reference to its economic resources: U.S. Geol. Survey Bull. 119, 72 p.

Fisher, C. A., 1906, Geology and water resources of the Bighorn Basin, Wyoming: U.S. Geol. Survey Prof. Paper 53, 72 p.

Hague, Arnold, 1899, Description of the Absaroka quadrangle [Crandall and Ishawooa quadrangles, Wyoming]: U.S. Geol. Survey Geol. Atlas, Folio 52, 6 p.

____ 1904, Atlas to accompany Monograph 32 on the geology of Yellowstone National Park: U.S. Geol. Survey, 15 sheets.

Harris, Burton, 1952, John Colter, his years in the Rockies: New York, Charles Scribner's Sons, 180 p.

Hayden, F. V., 1883, Twelfth annual report of the United States Geological and Geographical Survey of the Territories; a report of progress of the exploration in Wyoming and Idaho for the year 1878, Part 1: Washington, U.S. Govt. Printing Office, 809 p.

Hewett, D. F., 1913, An occurrence of petroleum near Cody, Wyoming [abs.]: Washington Acad. Sci. Jour., v. 3, no. 2, p. 51-52.

____1914, Sulphur deposits in Park County, Wyoming: U.S. Geol. Survey Bull. 540-R, p. 477-480.

Howard, A. D., 1937, History of the Grand Canyon of the Yellowstone: Geol. Soc. America Spec. Paper 6, 159 p.

Jones, W. A., 1875, Report upon the reconnaissance of northwestern Wyoming, including Yellowstone National Park, made in the summer of 1873: Washington, U.S. Govt. Printing Office, 331 p.

Leo, R. F., and Parker, P.L., 1966, Branched-chain fatty acids in sediments: Science, v. 152, no. 3722, p. 649-650.

Love, J. D., 1961, Reconnaissance study of Quaternary faults in and south of Yellowstone National Park, Wyoming: Geolo. Soc. America Bull., v. 72, no. 12, p. 1749-1764.

Parker, P. L., and Leo. R. F., 1965, Fatty acids in blue-green algal mat communities: Science, v. 148, no. 3668, p. 373-374.

Pierce, W. G., 1966, Geologic map of the Cody quadrangle, Park County, Wyoming: U.S. Geol. Survey Geol. Quad. Map GQ-542.

____ 1970, Geologic map of the Devils Tooth quadrangle, Park County, Wyoming: U.S. Geol. Survey Geol. Quad. Map GQ-817. (In press.)

Pierce, W. G., Andrews, D. A., and Keroher, J. K., 1947, Structure contour map of the Big Horn Basin, Wyoming and Montana: U.S. Geol. Survey Oil and Gas Inv. (Prelim.) Map 74.

Pierce, W. G., and Nelson, W. H., 1968, Geologic map of the Pat O'Hara quadrangle, Park County, Wyoming: U.S. Geol. Survey. Geol. Quad. Map GQ-755.

____ 1969, Geologic map of the Wapiti quadrangle, Park County, Wyoming: U.S. Geol. Survey Geol. Quad. Map GP-778.

Ponnamperuma, Cyril, and Pering, K. L., 1967, Aliphatic and alicyclic hydrocarbons isolated from Trinidad Lake asphalt: Geochim. et Cosmochim. Acta, v. 31, no. 8, p. 1350-1354.

Rollinson, J. K., 1948, Wyoming cattle trails: Caldwell, Idaho, Caxton Printers, 366 p.

Skinner, M. F., and Kaisen, O. C., 1947, The fossil *Bison* of Alaska and preliminary revision of the genus: Am. Mus. Nt. History Bull., v. 89, art. 3, p. 123-256.

Stone, D. S., 1967, Theory of Paleozoic oil and gas accumulation in Big Horn Basin, Wyoming: Am. Assoc. Petroleum Geologists Bull., v. 51, no. 10, p. 2056-2114.

Swanson, V. E., Palacas, J. G., Love, A. H., Ging, T. G., Jr., and Gerrild, P.M., 1968, Hydrocarbons and other organic fractions in recent tidal-flat and estuarine sediments, northeastern Gulf of Mexico, *in* Am. Chem. Soc., Div. of Petroleum Chem., Inc., Symposium on hydrocarbons from living organisms and recent sediments, Atlantic City, N.J., September 1968: Preprints, v. 13, no. 4, B35-B37.

Wenger, W. J., and Lanum, W. J., 1954, Characteristics of crude oils from Big Horn Basin fields: Petroleum Engineer, v. 26, no. 2, p. A52-A60.

Wheat, C. I., 1957-1963, Mapping the Trans-Mississippi West, 1540-1861: San Francisco, Inst. Hist. Cartography, 5 v.

Woodruff, E. G., 1908, Sulphur deposits at Cody, Wyoming: U.S. Geol. Survey Bull. 340-L, p. 451-456.

Zapp, A. D., 1956, Structure contour map of the Tensleep sandstone in the Big Horn Basin, Wyoming and Montana: U.S. Geol. Survey Oil and Gas Inv. Map OM-182.

Thirty-Third Annual Field Conference — 1982
Wyoming Geological Association Guidebook

PETROLEUM GEOLOGY OF THE NORTHERN PART OF YELLOWSTONE NATIONAL PARK

BY: JOHN J. TONNSEN[1]

ABSTRACT

Over 10,000 feet of Paleozoic and Mesozoic sedimentary rocks are present in the park region. The strata contain several possible oil and gas source rocks and some potential reservoir rocks. These rocks were folded and thrust faulted during the Late Cretaceous and Early Tertiary, and the region was moderately eroded prior to a major episode of volcanism during the Eocene. The Middle and Late Tertiary was a period of great normal faulting. The region has been subjected to intrusion and repeated volcanic episodes but remains of great interest to petroleum geologists.

INTRODUCTION

The Yellowstone National Park region is situated just east of the Southwestern Montana segment of the Overthrust Belt and the Snake River Plains volcanic field. The region is south and west of the Crazy Mountains and Big Horn oil and gas basin complex of Montana and Wyoming. The most famous geological features of the park were formed by Tertiary volcanic and Quaternary glaciological episodes. The initial explosive events began about 2,000,000 years ago when the caldera of the park was formed. The caldera became more violently explosive about 600,000 years ago and a great outpouring of rhyolitic lavas, ash-flow tuffs, breccias, hot spring deposits, and basalt lava flows occurred.

The geologist must look through these Quaternary events and work with the Late Cretaceous, Paleocene, and Eocene orogenic and volcanic events in order to understand the geological setting of the park from the petroleum geology point of view.

Stratigraphy, geological structure, and oil and gas activity is reviewed in this paper. Plate 1 (in the pocket part) is a map showing the major structures, and Plate 2 consists of a regional correlation cross-section.

STRATIGRAPHY

The stratigraphic section contains approximately 3000 feet of Paleozoic and 4000 feet of Mesozoic rocks. This section has been correlated north of the Townsend Valley near Helena, Montana, west to recent drilling in the Beaverhead Basin near Dillon, Montana, and east to the section near Cody, Wyoming (Plate 2).

The Paleozoic Section

The Paleozoic rocks are predominantly dolomites and limestones with some marine shales, coastal or transitional sandstones, and continental sediments. The principal source beds are the black, organic-rich, marine shales at the Devonian-Mississippian contact, the Chester-Morrow interval at the end of the Mississippian, and in the Permian Phosphoria Formation (Plate 2).

The late Devonian Sappington Member of the three Forks Formation consists of some 10 to 28 feet or more of light brown to orange to yellow calcareous siltstone. Just above this unit in some places is a very dark gray to black fissile shale up to 2.5 or more feet thick. The shale is present in a few localities in the northwestern part of the park (Witkind, 1969), but throughout most of the park it is probably absent (Gutschick, McLane, and Rodriquez, 1976). The shale is present more generously to the north

[1]Exploration Manager, General Hydrocarbons, Inc., 200 Transwestern II Building, Billings, Montana 59107

Plates 1 and 2 in pocket.

and west, however, as can be observed on the correlations. This unit corresponds with the Bakken Shale Formation of the Williston Basin, and probably would be an excellent source rock

The Big Snowy Formation of Late Mississippian age is present in the north and west sections and this formation contains one or two light gray to grayish black fissile shale zones up to 10 or more feet thick. These shales appear to correlate with the Heath Shale of Central Montana (Hadley, 1980). Again, as with the Sappington shale, these shales are not present in the Gardiner area and to the east and south of there but they develop only to the north and west of the park (Maughan and Roberts, 1967). These shales are thought to be the source rocks for the Tyler Sandstone (Chester-Morrow) oil production in Central Montana north of Billings.

The Permian Phosphoria Formation contains two thick (to 80 feet) oil-bearing marine shale zones to the west in the Dillon area (Claypool, Love, and Maughan, 1978). It is very significant that there were some very thin dark shales at this interval in the Phillips well in Carrot Basin west of the park (Plate 1). E. K. Maughan of the U.S.G.S. (Claypool et al., 1978) has mapped these shales on the basis of the organic carbon content. This unit also thins in a southeasterly direction but it probably is present throughout much of the northern part of the park.

The shales described above constitute the principal Paleozoic source rock candidates for the region. The thinning is due to the region being situated at the northwestern part of the Paleozoic "Wyoming Shelf" (Figure 1). The shelf is so named because most of the Paleozoic sediments reflect a depositional pattern suggesting that the Wyoming area was a bit higher than the adjacent parts of Montana north and west of Wyoming. The transgressing Paleozoic seas found entry into the miogeosyncline of southwestern Montana and a trough in west central Montana. The shale deposits, therefore, are better developed in the Dillon section than they are in the Gardiner and Cody areas.

Other potential source rocks are dark, fetid dolomites of the Upper Devonian Jefferson Formation and dark dolomite zones of the Mississippian Mission Canyon Formation. In addition, there are occasional thin dark calcareous shale zones at the base of the Ordovician Big Horn Dolomite Formation that might have generative value.

A few reservoir rocks are present. The Jefferson Formation contains two to five benches of brecciated and internally fractured sucrosic dolomite. In a few localities there are small (to 2 or 3 inches) "vugs" or solution-collapse features that might constitute a reservoir. The Madison Formation contains several brecciated zones but these are filled with recrystallized calcite and probably are not good reservoirs. There are a few thin sandstone stringers in the Amsden Formation and the Quadrant Formation contains nearly 35 to 50 net feet of porous sandstone. These beds together with the overlying Permian Shedhorn Sandstone Formation hold potential as reservoir rocks.

Throughout the region the Paleozoic section is observed to have bedding plane faults. The age of the faults is probably associated with the compressional tectonism of the Laramide events, but there is the possibility that the faults predate the Laramide disturbance. These complexities benefit the opportunity for extensive fracturing and reservoir development.

The Mesozoic Section

The Triassic and Jurassic rocks total only about 400 to 500 feet out of a total Mesozoic section of over 4000 feet. These rocks do not contain a clearcut source rock. They are marine and near shore deposits and might be good reservoir rocks.

The Triassic Dinwoody Formation is about 35 to 48 feet thick and consists of a thin-bedded, flaggy, fine grained, gray limestone. Overlying the Dinwoody are 100 feet of red beds termed the Woodside Formation. The top of the Triassic is formed by the Thaynes Formation, which is a 12 to 40 foot thick, clean, calcareous, quartz sandstone. The grains are well sorted and very fine grained and the beds are a few inches to a few feet in thickness. The Triassic beds are near shore transitional sediments between the clastic red beds of the Chugwater in the Big Horn Basin and the more lagoonal shaly facies of Southwestern Montana.

The Jurassic rocks consist of a marine sedimentary sequence overlain by the continental Morrison beds. The marine units are called the Ellis Group, which is comprised of the Sawtooth Formation (red beds and limestone), the Rierdon Formation (oolitic limestone), and the Swift Formation (quartz and chert sandstone and coquina). These units total about 215 feet in thickness and constitute a western marginal facies of the Sundance Shale Formation found in the Cody area and throughout much of Wyoming.

The Morrison Formation is over 200 feet thick and these continental variegated shales and silty mudstones are gradational into the overlying basal Cretaceous Kootenai conglomerate, "salt and pepper" sandstones, and fresh water limestones. The Kootenai, in turn, is somewhat gradational upwards into the lower marine sandstone unit of the Thermopolis Formation. All of these transitional rocks taken together comprise the Dakota Formation, which is so familiar in electric log character throughout the Crazy Mountains-Big Horn Basin region (Plate 2). The

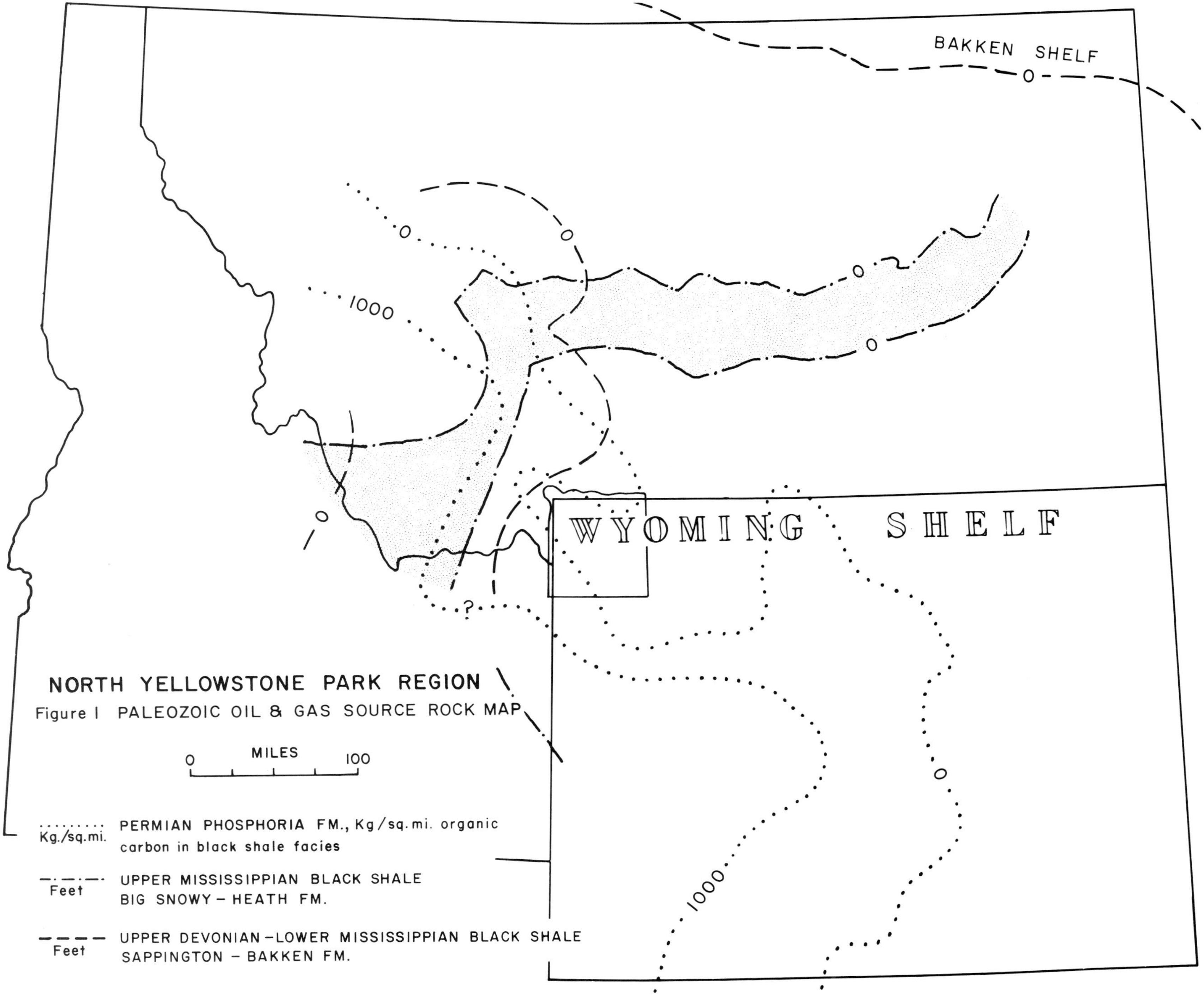

NORTH YELLOWSTONE PARK REGION

Figure I PALEOZOIC OIL & GAS SOURCE ROCK MAP

.... Kg./sq.mi.	PERMIAN PHOSPHORIA FM., Kg/sq.mi. organic carbon in black shale facies
—·—·— Feet	UPPER MISSISSIPPIAN BLACK SHALE BIG SNOWY – HEATH FM.
— — — Feet	UPPER DEVONIAN – LOWER MISSISSIPPIAN BLACK SHALE SAPPINGTON – BAKKEN FM.

total thickness of the Morrison through Lower Thermopolis interval is over 600 feet.

The Lower Cretaceous Thermopolis Formation contains a 300-foot thick, gray to black, fissile marine shale. The upper part of the Thermopolis shale is gradational into a glauconitic, dirty, cross-bedded sandstone. The lower sandstone, middle shale, and upper sandstone of the Thermopolis total over 500 feet in thickness. The formation is overlain by about 320 feet of gray to black, bentonitic shale with fish scales called the Mowry Shale Formation.

The Upper Cretaceous Frontier Formation ranges in thickness from about 50 to over 300 feet in the region. The Frontier is composed of transitional sandstones, including coal beds and bentonitic shales, but north and east of the region there are thick marine sandstone deposits that are very productive of oil and gas.

Overlying the Frontier are over 1220 feet of the Cody Shale Formation, which is a gray, marine shale. The Cody Formation contains a lower shale unit (350 feet thick), a middle glauconitic sandstone unit (45 feet thick), and an upper shale unit (over 600 feet thick).

Overlying the Cody Shale are over 1200 feet of calcareous, slightly arkosic, thin- to massive-bedded, sandstones interstratified with coaly units and silty shale. These comprise the Telegraph Creek, Virgelle, and Eagle formations. The sediments are indicative of coastal, swampy environments. Coal, for example, has been commercially mined from these sediments in the Cokedale area near the northern end of the Paradise Valley. In the southern part of Yellowstone Park these sediments are called the Bacon Ridge Sandstone Formation (Love and Keefer, 1975) and consist of sandstones, carbonaceous shale zones, and oyster beds.

The top of the Cretaceous consists of the Everts Formation (over 1000 feet of medium and coarse grained sandstone) and the Landslide Creek or Harebell Formation (coarse, poorly sorted sandstone, conglomerate, and volcanic detritus).

The Cretaceous section contains a number of source beds: the Thermopolis shale, the Mowry shale, the Lower and Upper Cody shale, and the carbonaceous shaly interbeds in several of the sandstone units. Reservoir rocks are less well developed than to the east in the Big Horn Basin where the Cretaceous sandstones were dominated by marine processes and good barrier and near shore marine sandstone deposits exist. In the park area and west of the park the Cretaceous section is more deltaic and coastal and, though the sandstone lithology is present in great volume, the lateral proximity to marine shales is poor. The sandstones contain very fine grained volcanic detritus and many are very bentonitic. The basal beds of the individual sandstone units are thin bedded zones interstratified with bentonitic shales and siltstones. The more massive sandstone units with good permeability and porosity do not occur until several feet upwards within the sandstone body. These sandstones exhibit some permeability in the horizontal direction but poor vertical permeability. A number of attempts have been made to produce natural gas from Cretaceous sandstones of similar depositional environment west of Driggs, Idaho (T5N, R44E, Teton County). The section is greatly disturbed by thrust faulting (Pampeyan, Schroeder, Schell, and Cressman, 1967) and each drilling effort seems to discover gas but production is not achieved. Extensive tectonic fracturing is anticipated to increase the reservoir quality of the Paleozoic limestones but perhaps these disturbances diminish the reservoir potential of the Cretaceous sandstones.

The influence of intrusives on the Cretaceous rocks is also important. All along the Gardiner thrust trend and along the South Madison thrust trend significant intrusions occurred. The Tobacco Root Stock, the Lone Mountain intrusive, the Electric Peak sills, and possibly the Sunlight Basin Mineralization Complex evidence that the reverse and thrust faulting tapped deep seated magma sources. The magma penetrated the Paleozoic and Triassic-Jurassic section but expanded into the Cretaceous rocks along many bedding planes as sills and lacoliths. The heat might have actually matured the sediments but the extensive intrusions into the bedding planes has probably distorted the reservoirs.

The Cenozoic Section

It has been estimated that the floor of Paradise Valley contains nearly 4000 feet of Eocene andesite breccias, tuffs, basalt flows and related volcanic sediments, and that these volcanic rocks cover the Paleozoic and Mesozoic section (Bonini, Kelley, and Hughes, 1972). This section is overlain by nearly 2500 feet of Oligocene, Miocene and Pliocene fluvial and lake bed sediments (see Thompson, Fields, and Alt, 1981). Extensive Quaternary volcanic and glacial deposits form distinctive surface features.

STRUCTURE

Gardiner Thrust Fault Trend

The Gardiner Thrust Fault has been mapped along a northwest to southeast trend across the north and northwestern part of the park (Fraser, Waldrop, and Hyden, 1969, and Ruppel, 1972). The more regional extent of the thrust trend is shown on Plate 1. The thrust correlates northwest to the Spanish Peaks and Bismarck faults, a total distance of nearly 100 miles. The southeastern extension of the fault is more difficult to correlate. It has been correlated to the Cody area (Wilson, 1934), and, alternatively, to the Buffalo Fork Thrust in the southern part of the park (Ruppel, 1972).

The Gardiner thrust has at least brought the Precambrian rocks into contact with the Upper Cretaceous formations, and the actual displacement might be greater. The fault plane appears to dip steeply to the northeast about 75° (Fraser, Waldrop, and Hyden, 1969). The Buffalo Fork Fault also dips to the east-northeast and the dip of the plane is only about 30° (Love and Keefer, 1975). These faults are parallel with the Wind River and Owl Creek foreland thrusts. The direction of transport is to the southwest and is opposite to the eastward relative movement in the Overthrust Belt.

Grayling Creek Reverse Fault Trend

The Grayling Creek trend is composed of a series of reverse faults with fault planes dipping to the south-southwest. This dip is coincident with the Overthrust Belt. East to northeast tectonic transport is more clearly delineated along the South Madison Thrust Fault trend (Plate 1). The South Madison fault brings the Precambrian rocks up through most of the stratigraphic section. The magnitude of this thrust system is best expressed west of Lone Mountain (the Big Sky Ski Area) in the Madison Range southeast of Ennis where the Precambrian has been faulted up and eastward into contact with the Upper Cretaceous Frontier Formation (Hadley, 1980).

The Grayling fault is mapped as a high angle reverse fault but the Gardiner and South Madison faults are mapped as thrust faults (Plate 1). The Gardiner and South Madison faults appear to be reverse faults on the surface and to not be greatly thrusted. However, oil and gas drilling in the Rocky Mountain region has disclosed that there can be greater overthrusting on these foreland features than is possible to be observed on the surface (Gries, 1981).

Age of Tectonism and the Sphinx Conglomerate Problem

At the top of the Cretaceous is the Landslide Creek Formation and it is composed of coarse sandstone and pebble conglomerate, andesite volcanic detritus, and greenish gray to carbonaceous shales and silty claystones (Plate 2). These lithologies suggest that active tectonism in this area post-dates the sediments.

The Paleocene rocks are thought to be absent in the region extending from the Beartooth Mountains westward across the park to the Ruby-Tobacco Root area. They are present in great thicknesses around the periphery of the park. South of this region is the Pinyon Conglomerate, to the west is the Beaverhead Conglomerate, along the north are the Elkhorn Mountains Volcanics and the Livingston Formation, and on the east there is the Fort Union Formation (Lindsey, 1972). These conglomerates and volcanic rocks constitute the Late Cretaceous and Paleocene sediments of a wide radius around the region. The writer has sought conglomerates in the area between the South Madison thrust and the Gardiner thrust that might be tectonic debris from the leading edges of the two opposing thrust faults. A tectonic rubble should have been formed in this area as the two opposing thrust systems "closed" together.

A patch of conglomerate is situated within this aperture and it is composed of sandstone, pebbles, and boulders that were derived from Paleozoic and Mesozoic strata and from the Precambrian basement rocks. This is the Sphinx Conglomerate Formation and it is shown on Plate 1. Fossil evidence to date the rocks has not yet been found. The prevailing view is that the conglomerates are of Eocene age (Hadley, 1980 and Beck, 1960), although Beck commented that they might be of Late Paleocene age. The sediments exhibit a crude, high angle stratification in thick (several hundred feet) cross-bedded zones on the east side of Sphinx Mountain. It is suggested that this unusual pattern is due to an interfingering of alluvial material from the two opposing directions. The Sphinx is folded into a syncline with dips as great as 32° or more. The age of the conglomerate might be more squarely within the Paleocene than has been previously considered. The rocks are nearly 2000 feet thick and unconformably overlie folded Paleozoic, Mesozoic, and Livingston Group sediments. The Livingston Group rocks were derived principally from the Elkhorn Mountains volcanic complex near Butte, Montana. The volcanism was concluded before the end of Campanian time (74 m.y.) and the Livingston Group sedimentary cycle was over by about 65 m.y. (Roberts, 1972). The Sphinx Formation is composed of limestone, quartzite, and crystalline clasts in a red sandstone matrix and these sediments bear striking similarity to the adjacent tectonic conglomerates observed all around the periphery of the park region (Lindsey, 1972). The structural relationships within the framework described by the Gardiner and South Madison thrust features suggest that the Sphinx Conglomerate might be a "missing link" Paleocene deposit in the region.

After Late Cretaceous to Paleocene folding and thrust faulting, the region was eroded. Subsequent to the orogenic events there was extrusion of the great Absaroka Volcanics of Eocene age. These volcanics were deposited over eroded hogback-shaped unconformities of Paleozoic and Mesozoic strata and Precambrian rocks in many places. Petroleum prospecting in the region must take into consideration the erosional era that occurred prior to the Eocene volcanism.

Extensive strike valleys, flaring gaps, and other cross-cutting geomorphic features greatly scored the folded sediments and the Paleozoic and Mesozoic rocks were removed from many of the structural crests. The Eocene volcanic rocks are as thick as 4000 feet and have been age dated at about 50 m.y. (Chadwick, 1981).

The folding and thrusting, therefore, was initiated in the region after the deposition of the Late Cretaceous Landslide Creek Formation, continued during deposition of the Sphinx Conglomerate, and was concluded prior to the Eocene Absaroka volcanic episode. During Middle and Late Tertiary time a great system of normal faults developed including a distinctive north-south trend extending from Paradise Valley south to Jackson Hole. These structures have not been reviewed in this paper.

Oil Seeps and Drilling

A few significant oil and gas wells have been drilled in the region adjacent to the park (Figure 2). The nearest production is from the Dakota and Frontier Formations along the series of faulted anticlines west of Red Lodge and from several fields in the western Big Horn Basin to the southeast (Plate 1).

There are at least three known occurrences of oil seeps in the park. One seep is at Calcite Springs about one mile north of Tower Falls and on the Yellowstone River bank, a second seep is on the west side of the Tower Bridge over the Yellowstone, and the third seep is remote at Rainbow Springs about eight miles east of Mount Washburn (east of the Yellowstone Canyon). Speculations regarding their origin are varied. There is evidence that they might be Tertiary organic matter that created local, small scale oily occurrences. However, analytical studies (Love and Good, 1970) permit speculation that the oils originated from Cretaceous or other pre-Tertiary shales and the petroleum implications might be great. The seeps are shown on Plate 1, and their alignment along the leading edge of the Gardiner Thrust trend is a very interesting coincidence.

The Phillips test in the Carrot Basin and the oil wells southwest of Cody are important positive indications that Paleozoic and Mesozoic formations in the park area might contain commercial amounts of petroleum. The Pennsylvanian Tensleep, Permian-Triassic units, and Cretaceous rocks are all productive near Cody. Live oil shows were observed in the Cretaceous Kootenai, Triassic Dinwoody, and Permian Phosphoria formations in the Carrot Basin test. ARCO Oil & Gas Company recently completed a stratigraphic test well in Paradise Valley. The well was spudded in August 1980 and was drilled to 1800 feet when a lost circulation problem caused the operator to abandon the well. The completion report has not yet been

FIGURE 2 - OIL AND GAS WELL DATA - NORTH YELLOWSTONE PARK REGION

CONTINENTAL OIL COMPANY, UNIT #1, SECTION 25 - T50N - R105W,
PARK COUNTY, WYOMING
6705' K.B. 4842' T.D. (1947)

PHILLIPS PETROLEUM CO., #1 CARROT BASIN, SECTION 11 - T10S - R3E,
GALLATIN COUNTY, MONTANA
9200' G.R. 2140' T.D. (1949)

MAY PETROLEUM, #1 HAGENBARTH, SECTION 12 - T5S - R8W,
BEAVERHEAD COUNTY, MONTANA
4957' K.B. 4040' T.D. (1977)

AMERICAN QUASAR, #29-1 REBISH, SECTION 29 - T5S - R7W,
BEAVERHEAD COUNTY, MONTANA
4864' K.B. 6320' T.D. (1977)

AMERICAN QUASAR, #27-22 HAGENBARTH, SECTION 27 - T5S - R9W,
BEAVERHEAD COUNTY, MONTANA
5431' K.B. 12,048' T.D. (1980)

ARCO OIL & GAS COMPANY, #1 PARADISE VALLEY, SECTION 16 - T4S - R8E,
PARK COUNTY, MONTANA
5951' G.L. 1800' T.D. (1980)

released but it has been learned that the well penetrated 300 feet of volcanic rocks, 350 feet of Jurassic-age sediments, and 950 feet of the Pennsylvanian and Mississippian section (R. E. Barker, ARCO Oil & Gas, personal communication).

The drilling results together with reasonable speculations regarding the significance of the seeps certainly suggest that the region is an interesting petroleum prospecting area. The intrusive igneous events and the extreme difficulty encountered in trying to obtain reliable seismic records through the volcanic and glacial deposits are serious negative factors.

CONCLUSION

The purpose of this paper has been to demonstrate that the northern part of Yellowstone National Park is the keystone of several regional structural and stratigraphic transitions that are of interest to the petroleum geologist. The writer has done extensive fieldwork in Southwestern Montana, Central Idaho, and Northwestern Wyoming and has had the pleasure to observe the geology of the park while ferrying from Billings to the field for several years.

Appreciation is extended to the several USGS geologists who have published many fine papers on the geology of the region. Also, numerous excellent discussions with geophysicist Al Artzis of Billings were of great benefit and are acknowledged. Thanks, too, to my secretary, Robin Scilley, for her effort and patience in typing the manuscript.

BIBLIOGRAPHY

Beck, F. M. (1960), Geology of the Sphinx Mountain Area, Madison and Gallatin Counties, Montana: Mont. Geol. Society Guidebook (11th), pp. 129-134.

Bonini, W. E., Kelley, W. N., Jr., and Hughes, D. W. (1972), Gravity Studies of the Crazy Mountains and the West Flank of the Beartooth Mountains, Montana: 1972 Mont. Geol. Society Guidebook (21st), pp. 119-127.

Chadwick, R. A. (1981), Chronology and Structural Setting of Volcanism in Southwestern and Central Montana: 1981 Mont. Geol. Society Guidebook (Southwestern Mont. Field Conf.), v. 1, pp. 301-310.

Claypool, G. E., Love, A. H., and Maughan, E. K. (1978), Organic Geochemistry, Incipient Metamorphism, and Oil Generation in Black Shale Members of Phosphoria Formation, Western Interior United States: Bull. Am. Assoc. Petroleum Geol., v. 62, no. 1, pp. 98-120.

Foose, R. M., Wise, D. V., and Garbarini, G. S. (1961), Structural geology of the Beartooth Mountains, Montana and Wyoming: Geol. Soc. America Bull., v. 72, no. 8, pp. 1143-1172.

Fraser, G. D., Waldrop, H. A., and Hyden, H. J. (1969), Geology of the Gardiner area, Park County, Montana: U.S.G.S. Bull. 1277, p. 118.

Freeman, V. L., Ruppel, E. T., Klepper, M. R. (1958), Geology of Part of the Townsend Valley, Broadwater and Jefferson Counties, Montana: U.S.G.S. Bull. 1042-N, pp. 481-556.

Gries, R. (1981), Oil and Gas Prospecting Beneath the Precambrian of Foreland Thrust Plates in the Rocky Mountains: The Mountain Geologist, v. 18, no. 1, p. 1-18.

Gutschick, R. C., McLane, M., and Rodriquez, J. (1976), Summary of Late Devonian-Early Mississippian Biostratigraphic Framework in Western Montana: Mont. Bur. Mines and Geology, Spec. Publication 73, pp. 91-124.

Hadley, J. B. (1980), Geology of the Varney and Cameron Quadrangles, Madison County, Montana: U.S.G.S. Bull. 1459, p. 108

Keefer, W. R. (1972), The Geologic Story of Yellowstone National Park: U.S.G.S. Bull. 1347, p. 92.

Lindsey, D. A. (1972), Sedimentary Petrology and Paleocurrents of the Harebell Formation, Pinyon Conglomerate, and Associate Coarse Clastic Deposits, Northwestern Wyoming: U.S.G.S. Prof. Paper 734-B, p. 68.

Love, J. D., and Keefer, W. R. (1975), Geology of Sedimentary Rocks in Southern Yellowstone National Park, Wyoming: U.S.G.S. Prof. Paper 729-D, p. 60.

Maughan, E. K., and Roberts, A. E. (1967), Big Snowy and Amsden Groups and the Mississippian-Pennsylvanian Boundary in Montana: U.S.G.S. Prof. Paper.

Montagne, John et al. (1960), Preliminary Geomorphic Map of Madison Valley, Montana: 1960 Mont. Geol. Society Guidebook (#11).

Nelson, W. H., Prostka, H. J., Williams, F. E., Elliott, J. E., and Peterson, D. L. (1980), Geology and Mineral Resources of the North Absaroka Wilderness and Vicinity, Park County, Wyoming: U.S.G.S. Bull. 1447, p. 101.

Pampeyan, E. H., Schroeder, M. L., Schell, E. M., and Cressman, E. R. (1967), Geologic Map of the Driggs quadrangle, Bonneville and Teton Counties, Idaho, and Teton County, Wyoming: U.S.G.S. Mineral Inves. Field Studies Map MF-300.

Pierce, W. G., (1960), The "break-away" point of the Heart Mountain detachment fault in northwestern Wyoming, in Geological Survey research 1960: U.S.G.S. Prof. Paper 400-B, pp. B236-B237.

Roberts, A. E. (1972), Cretaceous and Early Tertiary Depositional and Tectonic History of the Livingston Area, Southwestern Montana: U.S.G.S. Prof. Paper 526-C, p. 120.

Ruppel, E. T. (1972), geology of Pre-Tertiary Rocks in the Northern Park of Yellowstone National Park, Wyoming: U.S.G.S. Prof. Paper 729-A, p. 66.

Samuelson, K. H., and Schmidt, C. J. (1981), Structural Geology of the Western Tobacco Root Mountains, Southwestern Montana: 1981 Mont. Geol. Society Guidebook (Southwestern Montana Field Conf. Volume), pp. 191-199.

Thompson, G. R., Fields, R. W., and Alt, D. (1981), Tertiary Paleoclimates, Sedimentation Patterns and Uranium Distribution in Southwestern Montana: Mont. Geol. Society Guidebook (Southwestern Montana Field Conference), pp. 105-110.

Wilson, C. E., Jr. (1934), Geology of the thrust fault near Gardiner, Montana: Jour., Geology, v. 42, no. 6, pp. 649-663.

Witkind, I. J. (1969), Geology of the Tepee Creek Quadrangle, Montana-Wyoming: U.S.G.S. Prof. Paper 609, p. 101.

ThirtyThird Annual Field Conference — 1982
Wyoming Geological Association Guidebook

GENERAL GEOLOGIC SETTING AND MINERALIZATION OF THE PORPHYRY COPPER DEPOSITS, ABSAROKA VOLCANIC PLATEAU, WYOMING

BY W. DAN HAUSEL[1]

INTRODUCTION

Several volcanic centers in the Absaroka Mountains of northwestern Wyoming exhibit disseminated, stockwork, and vein mineralization, and hydrothermal alteration, characteristic of many porphyry copper deposits of the Basin and Range physiographic province, southwestern U.S. For the most part, the Absaroka porphyry systems approximate the Lowell and Guilbert (1970) porphyry copper model, although similarities to the Diorite Model (Hollister, 1974; 1978) may also be present, but less common.

The mineralized centers are characterized by a central intrusive complex. Adjacent to the intrusive complex are vent-facies autoclastic flow breccias, lava flows, mudflows, avalanche debris, and tuffs. The vent-facies rocks commonly are domed, altered, and radially fractured and, in the southern Absarokas, grade laterally into volcaniclastics. The appearance of these volcanic complexes is that of the dissected stratovolcano, or shield volcano.

Mineralization associated with these volcanic centers generally occurs as disseminations and stockworks in intensely altered rock, and in fractures and veins that transect both the mineralized intrusive and the vent-series wall rocks. The effect of hydrothermal alteration processes has been to produce zoned mineralization and alteration. Generally, copper, molybdenum, and trace gold occupy the central portion of the complex. These metals give way to zinc, lead, and silver ores laterally away from the center of the intrusive complex.

Both mineralized and barren intrusive centers are the source of the great pile of volcanic rock that forms the Absaroka Mountains. The Absaroka volcanics extend over a surface area of more than 8,000 square miles and compositionally consist of calc-alkaline flows, flow breccias, breccias, and reworked sedimentary facies. A small volume of flows in the northern Absarokas is shoshonitic.

[1]Deputy Director, Geological Survey of Wyoming, Laramie, 82071.

The Absaroka volcanics form a northwesterly trend of igneous rock along the eastern border of Yellowstone Park that extends north into Montana (Figure 1). The trend of the volcanics is assumed to be structurally controlled by some deep-seated fracture system.

Overviews of the Absaroka volcanic field have been published by Chadwick (1970), Hague (1899), Parsons (1939), Rouse (1937), and Wilson (1971). Several excellent papers and theses are available which describe the known mineralized centers. Drier (1967), Nelson and others (1980), Osterwald and others (1966), Parsons (1937), Pedersen (1968), and Rich (1974) discuss the geology and mineralization of the Sunlight area. The Eagle Creek region was the subject of a thesis by Galey (1971). Mineral deposits and alteration were examined by Fisher (1972) in the Stinkingwater region. The mineralized centers in the

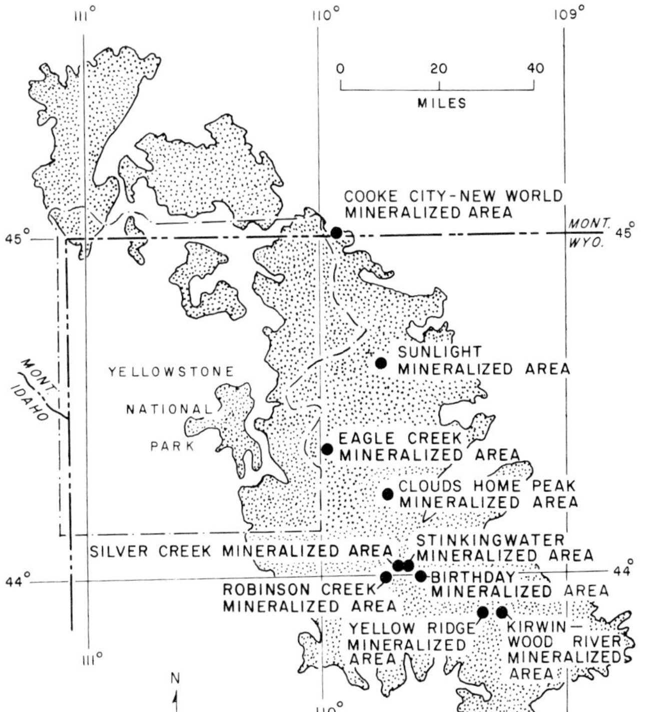

Figure 1: Location map of the known porphyry mineralized areas of the Absarokas.

Kirwin-Wood River area were mapped and summarized by Wilson (1954, 1960, 1963, 1964a, 1964b, 1975, 1982), discussed by Nowell (1971), and Osterwald and others (1966). The Cooke City deposits are the subject of Butler (1965), Lovering (1929), Nelson and others (1980), and Reed (1950). The Silver Creek and Yellow Ridge porphyry centers were briefly examined by Fisher and others (1977); and Fisher and Antweiler (1980) conducted reconnaissance surveys of mineralization and alteration associated with the Robinson Creek, Birthday, and Clouds Home Peak porphyry systems. More recently, an abstracted review of the Absaroka mineralized centers was published by Hausel (1982).

This paper is written as a brief overview of the porphyry copper deposits of the Absaroka volcanic plateau, and is written to hopefully alleviate a few editorial errors printed in Hausel (1981). The paper is primarily a summarzation of the previous listed published literature, with supportive cursory field reconnaissance. Essentially, all of the following discussions are abstracted from the published literature, and the interested reader will want to refer to the bibliography.

THE ABSAROKA VOLCANIC FIELD

The Absaroka volcanic field covers more than 8,000 square miles of surface area in northwestern Wyoming and southern Montana. The bulk of the field consists of calc-alkaline andesites with lesser basaltic and dacitic rock. These occur as flows, flow breccias, and breccias that grade into volcaniclastic sediment in the southern half of the field; the northern half of the field is formed predominantly of volcanic flows and related igneous rocks. A small volume of the Absaroka field (about 10%) is formed of potassic alkalic mafic flows of the shoshonite suite. The genesis of these ultra-potassic flows is discussed by Prostka (1973), Carmicheal and others (1974, p. 511-517 and 657-659), and Turner and Verhoogen (1960, p. 235-256), and their genesis will not be pursued here.

The tremendous volume of volcanic material (greater than 7,000 cubic miles) erupted from several volcanic centers that lie along a north-northwesterly trend. The locations of these centers are marked by intrusive stocks and plugs, with dikes and veins radiating outward from the central intrusive complex. These intrusive complexes contain stocks and plugs commonly of granodiorite, quartz monzonite, monzonite, diorite, syenogabbro, or syenite composition that exhibit equigranular to porphyritic texture. Peripheral to these intrusive complexes are volcanic flows, flow breccias and breccias that in the southern Absarokas grade outward into well-bedded deposits of epiclastically reworked volcanic debris comprising well-sorted volcanic breccia, conglomerate, sandstone and tuff (Lipman and others, 1972). These volcaniclastics predominate in the southern Absaroka field (Wilson, 1971). The entire Absaroka Range has the appearance of a deeply dissected volcanic plateau, with the volcanics (Eocene-Obligocene?) unconformably overlying Paleozoic and, in places, Mesozoic sediments.

The Absaroka volcanics and volcaniclastics are termed the Absaroka Supergroup. Generally, the Absaroka Supergroup averages greater than 5,000 feet thick (Smedes and Prostka, 1972), and in places is as much as 6,500 feet thick (Wilson, 1964b). Facies relationships and thicknesses of volcanic-rich strata along the east edge of the field indicate that the Absaroka volcanics, at one time, may have extended across much of the Bighorn Basin (Love, 1939).

The Absaroka Supergroup is divided into three groups: from oldest to youngest — the Washburn, Sunlight and Thorofare Creek groups (Figure 2). The Washburn

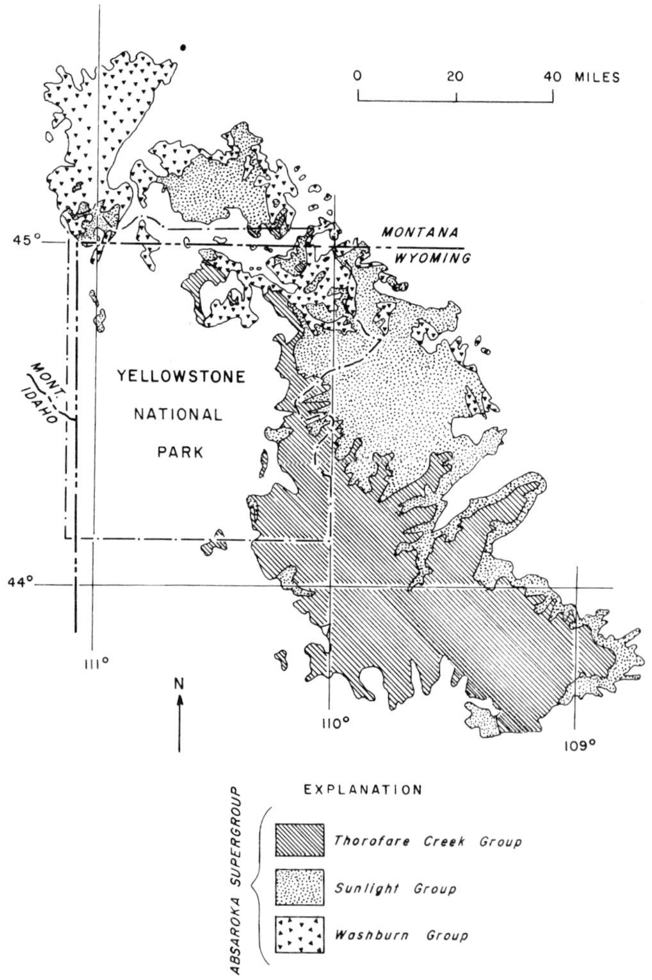

Figure 2: Generalized geologic map of the Absaroka Supergroup (after Smedes and Prostka, 1972).

Group is restricted in the north of the field and represents the oldest volcanic rocks. It is overlapped in most places unconformably by the Sunlight and Thorofare Creek groups. Nearly everywhere, the Thorofare Creek Group overlies the Sunlight Group.

Petrologically, the Sunlight Group is the most mafic unit and contains the highest proportion of basaltic potassic rock. Most of the rocks of the other two groups are generally lighter colored and of andesitic and dacitic composition.

Each of the three groups is subdivided into formations. Stratigraphic descriptions of the formations are presented by Smedes and Prostka (1972) and will not be discussed here.

PORPHYRY COPPER MODELS

Two general porphyry copper models are recognized by most exploration geologists, and these are the Lowell and Guilbert (1970) Model, and the Diorite Model (Hollister, 1974). Nearly all of the Absaroka porphyry systems are of the Lowell and Guilbert (1970) type, although Dioritic-type porphyry systems are also apparently present.

Very generally, the Lowell and Guilbert Model requires the mineralizing intrusive, or intrusives, to be oversaturated with silica. Intrusive rocks are calc-alkaline plutons in the granodiorite-quartz monzonite range. Alteration zones produced in this model include all or some of the following alteration halos progressing outward from the mineralized intrusive: (1) a potassic core (secondary orthoclase-biotite); (2) a phyllic zone (quartz-sericite-pyrite); (3) an argillic zone (kaolinite-illite-montmorillonite-pyrite) and an outer propylitic zone (epidote-chlorite-calcite). Hypogene disseminated copper sulfides are commonly developed at or near the interface of the potassic and phyllic zones and the greater amounts of pyrite are more commonly enclosed by the phyllic zone (Lowell and Guilbert, 1970; Titley, 1982; Beane and Titley, 1981; Hollister, 1978).

The Diorite Model encompasses quartz-free plutons. Included are alkalic suites as well as calc-alkaline diorite to syenite stocks. Alteration zones developed within the Diorite Model generally contain a potassic core assemblage with either biotite or chlorite dominant. The phyllic and argillic hydrothermal altered zones are generally absent or poorly developed. The potassic zone generally abuts against a propylitic zone. Pervasive pyrite generally is confined to the potassic zone (Hollister, 1974, 1978). Porphyry deposits occurring entirely in andesite and diabase are likely to have alteration expressions similar to the Diorite Model (Hollister, 1978).

The Lowell and Guilbert (1970) Model is charateristic of essentially all of the Absaroka porphyry deposits such as the Kirwin mineralized area, although none of the porphyry districts other than possibly Kirwin, Stinkingwater and Sunlight Basin have been studied in sufficient detail to assign a characteristic model with confidence. Both the Kirwin and Stinkingwater appear to more closely approximate the Lowell and Guilbert Model, and the Sunlight Basin complex has similarities to the Diorite porphyry model.

THE MINERALIZED CENTERS

W. H. Wilson was the first to recognize the potential of the vent areas in the Absarokas for porphyry copper deposits. These mineral deposits in the Absarokas (Figure 1) occur as: (1) disseminated and stockwork mineralization in intensely altered stocks or composite stocks, and, at a few localities, in country rock adjacent to mineralized stocks; (2) fracture-filled veins and veinlets extending outward from the mineralized centers; (3) fracture-and fissure-filled veins and replacement deposits hosted by Paleozoic carbonates (recognized only in the New World district); (4) supergene enriched deposits; and (5) placer gold deposits below porphyry districts.

Mineralization of the stocks is associated with hydrothermal alteration of varying intensity (Table 1). All of the districts exhibit widespread deuteric propylitic alteration. Near the intrusive centers, the deuteric alteration takes on the effects of hydrothermal propylitic alteration. Generally, the presence of veins or veinlets of calcite, epidote, chlorite and pyrite in the propylitically altered rock signifies that the deuterically altered rock has been overprinted by hydrothermal alteration.

Phyllic alteration (quartz-sericite-pyrite) is recognized near disseminated copper mineralization at all of the known mineralized areas except the Sunlight Basin, where this alteration halo is absent. Argillic alteration has been identified at several of the intrusive centers, although in most cases, the described alteration appears to be supergene rather than hypogene alteration. Potassic alteration has been identified in several of the districts, although the potassic halo is generally poorly defined.

Kirwin - Wood River Mineralized Area

The Kirwin - Wood River district contains three mineralized centers and north-northwest trending veins. Each of the three centers exhibits varying degrees of copper-molybdenum mineralization and hydrothermal alteration.

Table 1: Characteristics of the Absaroka porphyry copper deposits.

Deposit	Intrusion close to ore	Intruded rock or host* rock	Structural trends
Kirwin	rhyolitic (felsic) tuff and breccias (Nowell, 1971)	Wiggins Fm. andesites (Thorofare Creek Group) (Wilson, 1964b)	northwesterly and northeasterly striking joints and fissures (Wilson, 1964b)
Meadow Creek	Meadow Creek granodiorite (Wilson, 1975)	Wiggins Fm. andesites (Thorofare Creek Group) (Wilson, 1975)	intersected by N-S and E-W trending faults. NW trending veins. Radial dike pattern. (Wilson, 1960, 1975)
Brown Mtn.	Brown Mtn. granodiorite (Wilson, 1964b)	Wiggins Fm. andesite (Thorofare Creek Group) (Wilson, 1964b)	N-NW trending veins along Galena Ridge. Joint sets in intrusion: EW; NE-SW; and NW-SE. (Wilson, 1960)
Silver Creek	dacite - and rhyodacite-porphyry (Fisher and others, 1977)	Wiggins Fm. (Thorofare Creek Group), Trout Peak Trachy-andesite and Wapiti Fm. (Sunlight Group) (Fisher and others, 1977)	well developed fractures trend N35°E to N80°W (Fisher and others, 1977)
Yellow Ridge	granodiorite, diorite, andesite porphyry, hornblende-andesite porphyry (Fisher and others, 1977)	Wiggins Fm. (Thorofare Creek Group) (Fisher and others, 1977)	no data
Birthday	granodiorite-dacite (Fisher and Antweiler, 1980)	Wiggins Fm. (Thorofare Creek Group) (Fisher and Antweiler, 1980)	NW-trending stockworks and dikes, and NE-trending dikes. (Fisher and Antweiler, 1980)
Robinson Creek	rhyodacite-dacite (Fisher and Antweiler, 1980)	Wiggins Fm. (Thorofare Creek Group) (Fisher and Antweiler, 1980)	Major fracture zones trend N20°W to N60°W. Less developed fractures trend NE. (Fisher and Antweiler, 1980)
Clouds Home Peak	dacite, rhyodacite, granodiorite, and quartz monzonite (Fisher and Antweiler, 1980)	Wiggins Fm. (Thorofare Creek Group) (Fisher and Antweiler, 1980)	NE and NW trending dikes and faults. (Fisher and Antweiler, 1980)
Eagle Creek	latite porphyry (Galey, 1971)	Langford Fm. (Thorofare Creek Group) and Trout Peak Trachy-andesite (Sunlight Group) (Galey, 1971)	N-NW and NE striking joint and N32°E and N64°W quartz vein trends. (Galey, 1971)
Stinking-water	Needle Mtn. granodiorite and Crater Mtn. dacite (Fisher, 1972)	Wiggins Fm. (Thorofare Creek Group), Trout Peak Trachy-andesite, and Wapiti Formation (Sunlight Group) (Fisher, 1972)	NW, and E to NE trending fracture sets. (Fisher, 1972)
Sunlight	syenite (Drier, 1967; Pedersen, 1968; Rich, 1974)	Wapiti Fm. (Sunlight Group) (Nelson and others, 1980)	N39°W fracture set and dominant radial dike pattern. (Nelson and others, 1980)
Cooke City	Henderson Mtn. monzonite porphyry, and Goose Lake syenite (Lovering, 1929)	*Paleozoic sediments (Pilgrim Limestone and Gros Ventre Fm. host) (Elliott, 1980b)	————

Alteration types	Critical alteration assemblages	Major sulfides	Supergene enriched zones
propylitic argillic-phyllic potassic (Nowell, 1971)	quartz, calcite, epidote, montmorillonite. sericite, mixed layer illite-montmorillonite, quartz, biotite, kaolinite, chlorite, pyrite. orthoclase, quartz, sulfide veinlets. (Nowell, 1971)	chalcopyrite molybdenite pyrite (Wilson, 1964b)	chalcocite, digenite, covellite blanket (Wilson, 1964b)
propylitic phyllic (Wilson, 1975)	epidote, chlorite, calcite. quartz, sericite, pyrite. (Wilson, 1975)	chalcopyrite pyrite molybdenite (?) (Wilson, 1975)	no data
propylitic (Wilson, 1960)	pyrite, calcite, chlorite, epidote, sericite. (Wilson, 1960)	chalcopyrite pyrite	no data
propylitic phyllic potassic (Fisher and others, 1977)	chlorite, calcite, epidote. quartz, sericite, pyrite. quartz-potassium feldspar veinlets and secondary biotite and magnetite. (Fisher and others, 1977)	chalcopyrite molybdenite bornite pyrite (Fisher and others, 1977)	no data
propylitic phyllic potassic (Fisher and others, 1977)	chlorite, calcite, epidote. quartz, sericite, pyrite. quartz-potassium feldspar veinlets and secondary biotite and magnetite. (Fisher and others, 1977)	pyrite, chalcopyrite, bornite (?), molybdenite (?) (Fisher and others, 1977)	no data
propylitic phyllic potassic (Fisher and Antweiler, 1980)	chlorite, calcite, epidote, saussurite, pyrite. sericite, pyrite, quartz. actinolite/tremolite, K-feldspar, magnetite, quartz, chalcopyrite veinlets. (Fisher and Antweiler, 1980)	pyrite chalcopyrite (Fisher and Antweiler, 1980)	no data
propylitic phyllic argillic (supergene) (Fisher and Antweiler, 1980)	calcite, epidote, chlorite. quartz, sericite, pyrite. clay minerals. (Fisher and Antweiler, 1980)	pyrite chalcopyrite (Fisher and Antweiler, 1980)	no data
propylitic phyllic (Fisher and Antweiler, 1980)	epidote, calcite, chlorite, saussurite. quartz, sericite, pyrite. (Fisher and Antweiler, 1980)	pyrite chalcopyrite (Fisher and Antweiler, 1980)	no data
propylitic phyllic(?) potassic (?) (Galey, 1971)	chlorite, calcite, epidote, sericite. sericite, calcite, clay, chlorite, quartz. pyrite, K-feldspar (?). (Galey, 1971)	pyrite chalcopyrite galena, sphalerite (Galey, 1971)	none
propylitic phyllic potassic argillic (supergene) (Fisher, 1972)	chlorite, sericite, magnetite, calcite, epidote. quartz, sericite, pyrite. clay minerals. biotite, sericite, quartz. (Fisher, 1972)	chalcopyrite pyrite molybdenite chalcocite (Fisher, 1972)	100 to 200 foot thick zone (Fisher, 1972)
propylitic potassic argillic (Elliott, 1980a)	epidote, calcite, chlorite. K-feldspar, sericite, clay, calcite, quartz. clay minerals, chlorite. (Rich, 1974; Pedersen, 1968; Drier, 1967)	chalcopyrite (Elliott, 1980a)	minor covellite, chalcocite, malachite (Elliott, 1980a)
silicification dolomitization (Elliott, 1980b)	dolomite, quartz, calcite, sulfides. (Butler, 1965)	pyrite, chalcopyrite, galena (Elliott, 1980b)	no data

The Kirwin - Wood River area is located in the southern Absaroka Mountains near the headwaters of the Wood River, 33 miles southwest of Meeteetse — the nearest town. Early development of the district took place near the turn of the century. More than 12,000 feet of drifts, shafts and adits were developed in veins, attempting to intersect ore shoots. Production was limited to one carload of ore reported to yield a net value of $65 a ton after smelter and transportation charges were deducted (Hewett, 1912).

The three intrusive complexes (Kirwin, Brown Mountain grandiorite, and Meadow Creek granodiorite) penetrate volcanics of the Wiggins Formation (Eocene) (Figure 3). The Wiggins Formation is one of a group of formations belonging to the Thorofare Creek Group (Smedes and Prostka, 1972). In this area, the Wiggins Formation forms a series of deuterically propylitized hornblende- biotite and pyroxene andesite porphyry flows, tuffs, breccias, and volcaniclastic sediments. The vent facies layered volcanics have been domed, hydrothermally altered, and radially fractured. Andesite porphyry dikes occupy radial fractures, as well as the dominant north-northwesterly structural trend recognized throughout the Kirwin - Wood River area. With a few exceptions, mineralized veins are localized in north-northwesterly fractures and predominantly are lead-silver-zinc veins. In the Kirwin mineralized-altered zone on the northern flank of Bald Mountain, the veins are copper-molybdenum dominant.

Brown Mountain Granodiorite. The Brown Mountain granodiorite is the least mineralized of the three intrusive complexes (Figure 3). The intrusive consists of fine- to medium-grained equigranular granodiorite grading into porphyritic phases near its northern margin. Hydrothermal alteration is expressed as weak propylitization with disseminated pyrite. Although the strongest manifestations of mineralization are north-northwesterly trending veins along Galena Ridge to the north of the intrusion (Wilson, 1960), it is not known if these veins are an expresssion of the Brown Mountain granodiorite or of the Kirwin mineralized complex (Nowell, 1971). Three prominent joint trends: (1) east-west; (2) northeast-southwest; and (3) northwest-southeast have been measured in the central portion of the intrusive (Wilson, 1960).

Meadow Creek Granodiorite. The Meadow Creek granodiorite (Figure 3) is a composite of at least two separate intrusives that vary from granite to diorite and average as granodiorite. The older granodiorite (forming the western third of the intrusive mass) is propylitized, weakly pyritized, and relatively unmineralized. The later granodiorite (on the east) is propylitized with more intense alteration along two elongated zones (Figure 3). The

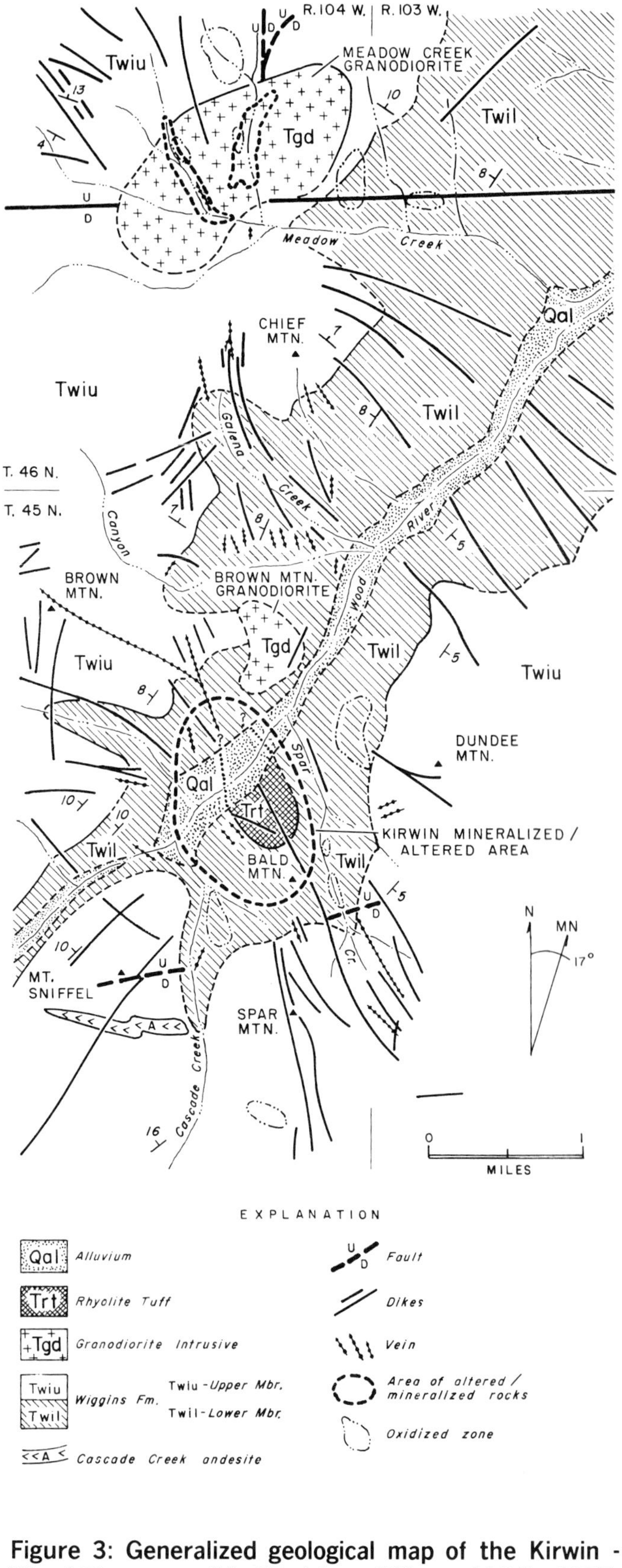

Figure 3: Generalized geological map of the Kirwin - Wood River area (Modified from Wilson, 1964b, 1975; Nowell, 1971).

intensely altered zone in the eastern portion of the complex extends across the intrusive along the projected trend of the north-south fault, and is delineated by phyllic alteration with disseminated copper mineralization. A second intensely altered zone is localized along the contact between the two granodiorite intrusives. This zone is outlined by intense silicification, iron-staining, and disseminated pyrite.

Spotty chalcopyrite and malachite occur as disseminations and as fracture fillings in limited exposures in both the altered granodiorite and adjacent Wiggins Formation. Other expressions of mineralization include copper, lead, and molybdenum geochemical anomalies within the intrusive, and mineralized quartz veins located both north and south of the intrusive. These veins occur as galena (argentiferous), sphalerite, tetrahedrite, minor pyrite and chalcopyrite mineralization in quartz-carbonate gangue (Wilson, 1975).

The Meadow Creek granodiorite is located at the intersection of two faults. The emplacement of the intrusive complex undoubtedly was controlled by this intersection. Dikes form rough radial patterns around the intrusive, and vein mineralization occurs in a north-northwest trending zone between the Brown Mountain and Meadow Creek intrusives (Wilson, 1964b; 1975).

Kirwin. The Kirwin mineralized area, located on the northwest flank of Bald Mountain, is expressed by a rough oval-shaped intense zone of hydrothermal alteration (Figure 3). The mineralized area is associated with a volcanic vent complex with exposures of intrusive rhyolitic tuff breccia (Wilson, 1964b; Nowell, 1971). Brecciation appears more commonly along the edge of the vent.

This mineralized area exhibits hydrothermal alteration with increasing intensity toward the center of alteration. Outside of the alteration zone, Wiggins Formation andesites are deuterically altered and exhibit propylitic alteration products — calcite, chlorite, and clays. Within 1,500 feet of the mineralized center, Wiggins andesites exhibit hydrothermal propylitic alteration characteristics and, in addition to the typical alteration minerals — quartz, calcite, epidote, and montmorillonite, these rocks contain chalcopyrite-bearing quartz-calcite veinlets and disseminated chalcopyrite blebs. Nearer to the volcanic center, alteration assemblages are more characteristic of argillic and phyllic hydrothermal alteration. Major alteration products include sericite, mixed-layer illite-montmorillonite, quartz, and biotite with lesser kaolinite and chlorite. Epidote and calcite are nonexistent in this zone.

The potassic zone is not well-defined, but is suggested by the presence of secondary orthoclase with quartz and veinlet sulfides. This zone lies near the mineralized center (Nowell, 1971).

Drill hole data in the Kirwin area show stockwork mineralization of pyrite, chalcopyrite and molybdenum. A secondary enriched blanket of chalcocite, digenite and covellite overlies a portion of the stockworks, which in turn, is overlain by a barren leached cap (Wilson, 1964b). Veins in the altered area are pyrite-chalcopyrite-molybdenum-quartz veins (Wilson, 1960).

The U.S. Bureau of Mines reports that the Kirwin mineralized-altered area at Bald Mountain, hosts a resource of at least 70 million short tons of 0.75 percent copper (Rosenkranz and others, 1979).

Yellow Ridge

The Yellow Ridge area lies within the boundaries of the Francs Peak 7½-minute quadrangle approximately 22 miles from the end of the Greybull River Road. A pack trail follows the Greybull River to the base of Yellow Ridge. The intrusive is also accessible from the Kirwin mineralized area. From Kirwin, the Greybull River trail crosses Greybull Pass and lies within three miles west-northwest of Kirwin.

Two stocks crop out through Wiggins breccias and flows at Yellow Ridge (Wilson, personal comm., 1982). These are rhyolite and a composite intrusive formed of granodiorite to diorite, andesite porphyry, and hornblende andesite porphyry (Figure 4). Much of the mineralization is associated with the andesite porphyry (dike?) although the other intrusive masses are weakly mineralized.

A narrow (200 to 300 feet) mineralized zone is exposed along a 2,500 feet northeasterly trend in the southeastern part of the stock. Malachite is abundant and occurs as fracture coatings and as disseminated grains. Pyrite, and lesser chalcopyrite, bornite(?) and molybdenum(?) also are found in the mineralized zone.

Hydrothermal alteration consists of an outer propylitic zone that includes the intrusive margin and adjacent extrusive rocks. Near the intrusive center, propylitic minerals grade into phyllic altered rock, and near heavily mineralized zones, potassic altered rock predominates (Fisher and others, 1977).

Stinkingwater Mineralized Area

The Stinkingwater mineralized area lies near the confluence of Needle Creek and the South Fork of the Shoshone River (sec. 18, T.47N., R.106W.) near the center of the southern Absaroka volcanic field. Layered, per-mineral, volcanic rocks in the district include, from oldest to youngest: the Wapiti Formation, the Trout Peak Trachyandesite, and the Wiggins Formation. These are composed principally of basalt and andesite flows, volcanic breccias and flow breccias, and are intruded locally by the

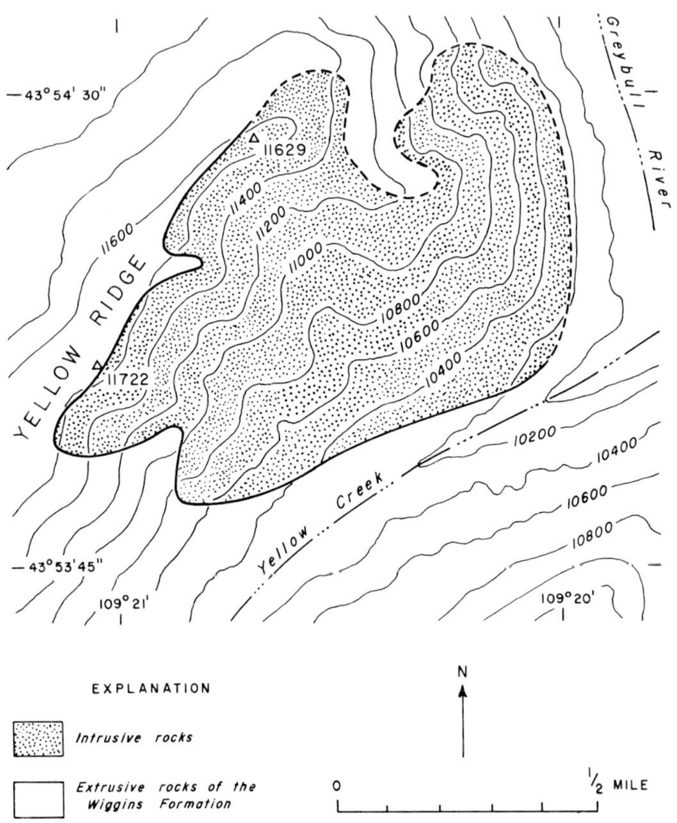

Figure 4: Sketch map of the Yellow Ridge intrusive (after Fisher and others, 1977).

Needle Mountain granodiorite and Crater Mountain dacite and numerous dikes which extend outward from the intrusive complex in a radial pattern (Figure 5). The Crater Mountain dacite intrudes the Needle Mountain granodiorite (Fisher, 1967, 1972).

The layered volcanic rocks adjacent to the intrusive complex are slightly domed and dip gently (maximum of 10°) away from the intrusive center (Fisher, 1972; Osterwald and others, 1966). Two major lineaments are recognized in the district; these trend northwest and east to northeast, and intersect at the intrusive complex. These fracture sets probably controlled the localization of the intrusive and the associated hydrothermal solutions.

Mineralization in the district includes an altered area (approximately three-quarters of a square mile) of disseminated copper-molybdenum near the southwestern edge of the intrusive complex. The greater mineralization is associated with a zone of intense hydrothermal alteration centered in the Crater Mountain dacite; the Needle Mountain granodiorite is poorly mineralized except along the northern edge of Needle Creek. All of the outcrops in the mineralized area are highly fractured and bleached, and are altered to varying degrees. Narrow, steeply dipping veins occur near, and, in places, as much as a mile away from the intrusive complex.

Chalcopyrite is the major disseminated ore mineral in the intensely altered zone followed in abundance by molybdenite and minor bornite. Beyond the central chalcopyrite-molybdenite zone, pyrite is more dominant and is associated with low copper-molybdenum values. Chalcopyrite and molybdenite also occur in narrow quartz veinlets and coat fractures in the area of disseminated sulfide minerals.

Vein mineralization extends outward from the altered mineralized zone for a few yards, and, in places, for nearly a mile. The veins are commonly one to two inches wide and reach a maximum of nearly one foot in width locally. These veins contain galena, chalcopyrite, sphalerite, pyrite, and minor arsenopyrite and tetrahedrite in quartz, calcite,

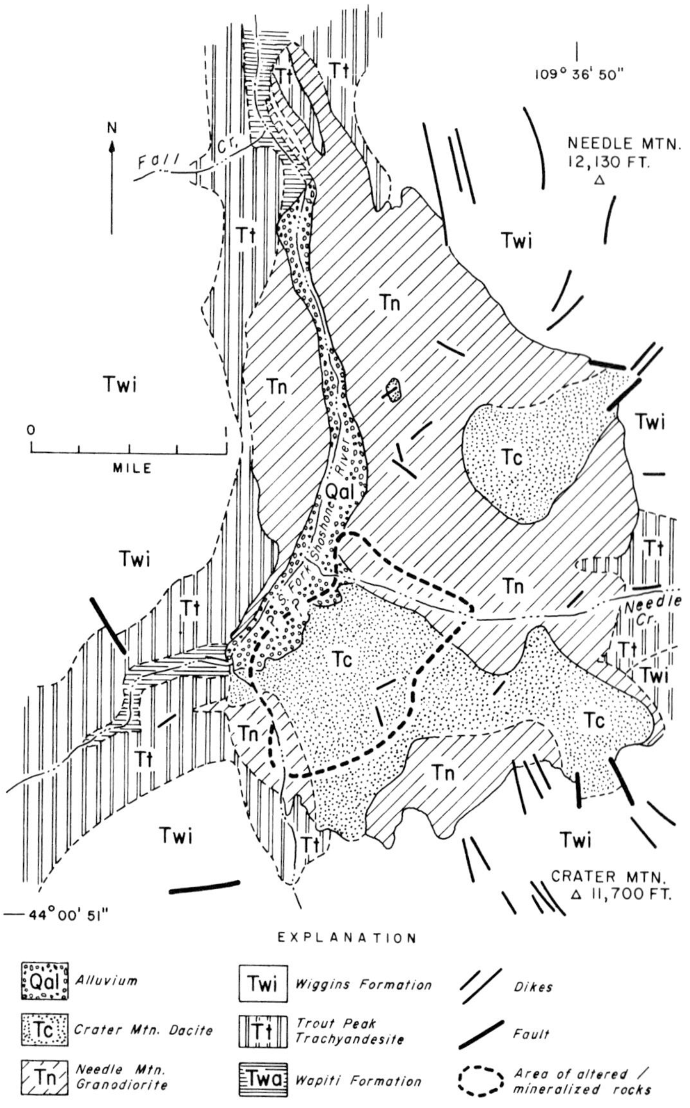

Figure 5: Generalized geological map of the Stinkingwater porphyry copper deposit (after Fisher, 1972).

dolomite, and minor siderite gangue. The veins are commonly crustiform and banded and are considered simple fracture fillings.

A supergene enriched chalcocite blanket, as much as 200 feet thick, was intersected by drill holes near the center of alteration at Crater Mountain. The extent of the supergene zone is not known. This zone is overlain by a leached and oxidized cap (Fisher, 1972).

Although an ore body of 65 million tons of 0.35 percent copper has been delineated by drilling, mineralization in the potassic core has not yet been entirely outlined by drilling (Geological Survey of Wyoming files, 1982).

Hydrothermal alteration increases in intensity towards the center of disseminated mineralization. Propylitization of both the intrusive and layered rocks is common throughout the district. The intensity of the propylitization increases around the fringe area of disseminated sulfide minerals. Propylitization gives way to phyllic alteration towards the center of alteration. The phyllic zone produces highly altered, bleached and silicified rock. In places, the original porphyritic texture is preserved by sericitic pseudomorphs; however, the original texture is completely destroyed in places of pervasive phyllic alteration. Potassic alteration (characterized by development of secondary biotite) is expressed in irregular ill-defined zones in the disseminated mineralized area.

Within the areas of supergene enrichment, there is evidence of argillic alteration which probably is of supergene origin.

Streams in the vicinity of the Stinkingwater mineralized area carry traces of gold (½-1 ppm in pan concentrates) and mercury (Fisher, 1972).

Silver Creek

Copper and minor molybdenite have been identified in altered rocks of an intrusive complex in the Silver Creek area. The complex is formed by dacite and rhyodacite rocks which intrude volcanics of the Wapiti Formation, Trout Peak Trachyandesite, and Wiggins Formation (Figure 6). The intrusive, as well as adjacent volcanics, shows effects of hydrothermal alteration (Fisher and others, 1977).

The Silver Creek area is located within the Washakie Wilderness on the Fall Creek 7½-minute quadrangle, and is accessible from the road's end at Valley along 11 miles of the South Fork pack trail to the mouth of Silver Creek. The intrusive is accessible within one-quarter mile from the mouth of Silver Creek, and lies within the Silver Creek stream valley (approximately 1½ miles southwest of the Stinkingwater mining region).

Hydrothermal alteration has affected the stock and adjacent host volcanics and generally preceded mineralization. Widespread propylitic alteration laterally gives way to localized phyllic and potassic alteration near the center of the intrusive. Phyllic alteration is expressed by sericite, quartz and pyrite, which grade inward into small zones of alteration characterized by small quartz-potassium feldspar-veinlets and secondary biotite and magnetite. Alteration boundaries are indistinct and overlap. The major sulfides are localized within the phyllic and potassic alteration zones.

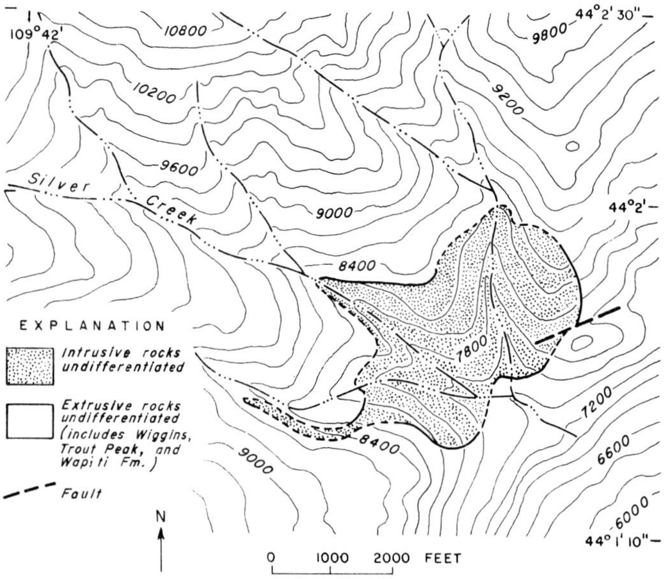

Figure 6: Generalized geological map of the Silver Creek mineralized stock (after Fisher and others, 1977).

A well-developed pyrite halo is localized in the phyllic zone and encloses a highly mineralized 2,000 by 1,000 feet stockwork complex near the center of the intrusive. The highly mineralized zone contains disseminated malachite, chalcopyrite, bornite, and minor molybdenum, along with chalcopyrite-pyrite-magnetite quartz veinlets. Fractures in the stockwork area trend roughly N35° to N80°W, and dip steeply. These commonly are stained by malachite (Fisher and others, 1977).

Molybdenum is poorly expressed in rock outcrops although drilling has intersected strong molybdenum mineralization in 300 feet of core. Estimates of resources in the Silver Creek porphyry suggest that an ore body of at least 27 million tons of 0.5 percent copper is enclosed by the intrusive complex (Geological Survey of Wyoming files, 1982).

Birthday Area

The Birthday mineralized stock was discovered during a U.S. Geological Survey mineral evaluation survey of the Washakie Wilderness. The Birthday area is located on the

Needle Mountain and Emerald Lake 7½-minute quadrangles along the eastern flank of Crater Mountain. Access is by the South Fork Shoshone River pack trail to Saddle Creek. The intrusive lies at the headwaters of Saddle Creek.

Cody, Wyoming, is located about 45 miles northeast, Valley lies about 12½ miles north, and the Stinkingwater mining region is three miles north-northwest of the Birthday area.

Exposures of the intrusive crop out at three localities and cover an area of less than one-half square mile (Figure 7). The largest intrusive is granodioritic to dacitic in composition, and intrudes Wiggins Formation andesites and breccias.

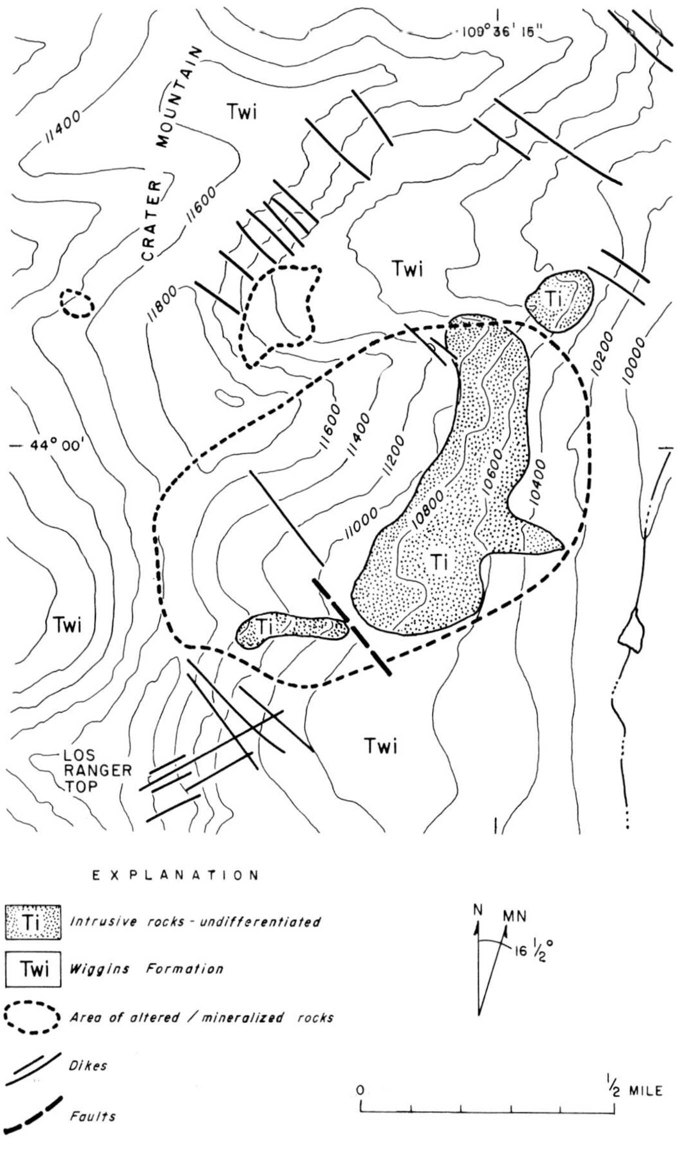

Figure 7: Generalized geological map of the Birthday mineralized intrusive complex (after Fisher and Antweiler, 1980).

Pyrite and chalcopyrite are widespread. Pyrite is disseminated, coats fractures, and occurs in quartz-pyrite veinlets. In some areas, pyrite comprises as much as 10 percent of the mineralized rock. Chalcopyrite occurs as disseminations and in veinlets forming stockworks. These stockworks are northwest-trending zones from 10 to 14 feet wide, that are developed in andesite dikes(?) and, to a lesser extent, in the dacite. Quartz-pyrite-calcite veins and altered rock occur adjacent to many of the dikes in the Birthday region. Oxidation is prominent, and zones of limonite, malachite, and azurite have replaced hypogene minerals. Both anomalous copper and molybdenum have been detected in assays.

Propylitic, phyllic, and potassic hydrothermal altered zones have been recognized in the Birthday area. The propylitic altered rocks are characterized by the development of chlorite, calcite, epidote, saussurite and pyrite. Phyllic altered rocks are present in scattered patches, and are mainly found in fractured zones, adjacent to dikes and veins, and along faults. Variable amounts of sericite, pyrite, and secondary quartz are present in the matrix, or replace phenocrysts, or are found in veinlets in the phyllic altered rocks. Potassic altered rock is present in two localized areas and is expressed by veinlet clusters containing actinolite/tremolite, potassium feldspar, magnetite, quartz, and chalcopyrite. Potassic altered zones are associated with high copper concentrations and mineralized stockworks (Fisher and Antweiler, 1980).

Robinson Creek

The Robinson Creek mineralized area was discovered by the U.S. Geological Survey during mineral evaluation studies of the Washakie Wilderness. The mineralized area lies nearly 50 miles southwest of Cody and 15 miles southwest of Valley. Access to the intrusive complex is by pack trail along the South Fork Shoshone River to Robinson Creek.

The mineralized area lies at the headwaters of Robinson Creek and includes a roughly one-square mile, hydrothermally altered and mineralized rhyodacite to dacite composite stock emplaced in flows and flow breccias of the Wiggins Formation (Figure 8). Texturally, the intrusive has both porphyritic and phaneritic phases.

Wiggins Formation andesites adjacent to the Robinson Creek intrusive are propylitized. Selectively pervasive propylitic alteration resulted in chlorite and lesser epidote replacing hornblende. Where propylitization is more intense, calcite, epidote, and minor chlorite are widespread. Secondary quartz, sericite, and pyrite form phyllic altered zones near the intrusive center. Some clay minerals replace phyllic zone minerals, and are apparently of supergene genesis. Other manifestations of supergene alteration are bleached and limonite-stained rocks.

Mineralization is coextensive with the phyllic altered zone, and is exposed as discontinuous outcrops of chalcopyrite, pyrite, and malachite stockworks. Chalcopyrite also occurs as disseminated grains and is found as fracture coatings with malachite. The mineralized rocks are extensively shattered with fracture zones trending from N20°W to N60°W, with less developed trends to the northeast. Anomalous lead, zinc, silver, and molybdenum have been detected by assay in the mineralized area (Fisher and Antweiler, 1980).

Clouds Home Peak

A complex pluton ranging from porphyritic dacite and rhyodacite to equigranular granodiorite and quartz monzonite intrudes Wiggins Formation flows and breccias in the Clouds Home Peak area. The mineralized pluton was discovered during a mineral evaluation of the Washakie Wilderness by the U.S. Geological Survey (Fisher and Antweiler, 1980).

The area of interest lies within the boundaries of the Lake Creek and Clouds Home Peak 7½-minute quadrangles. The area is accessible by pack trail. The trail head intersects a paved and graded road that services Cody and Valley. Cody is located 25 miles to the northeast of the trail head, and Valley is eight miles to the south. From the Ishawooa Creek trail head, the base of Clouds Home Peak lies nearly 10 miles to the west. The intrusive complex is located near the top of the peak.

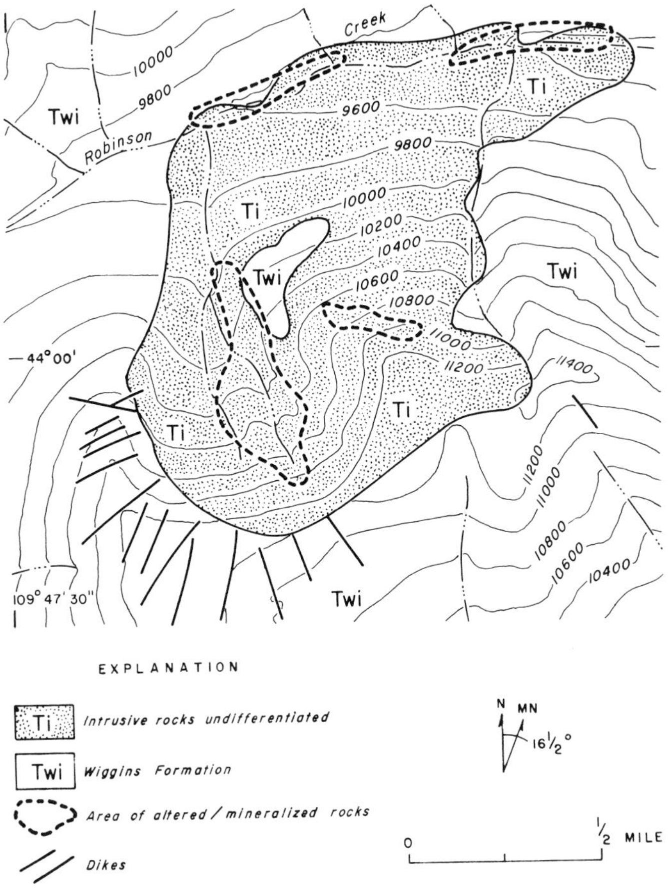

Figure 8: Generalized geologic map of the Robinson Creek mineralized stock (after Fisher and Antweiler, 1980).

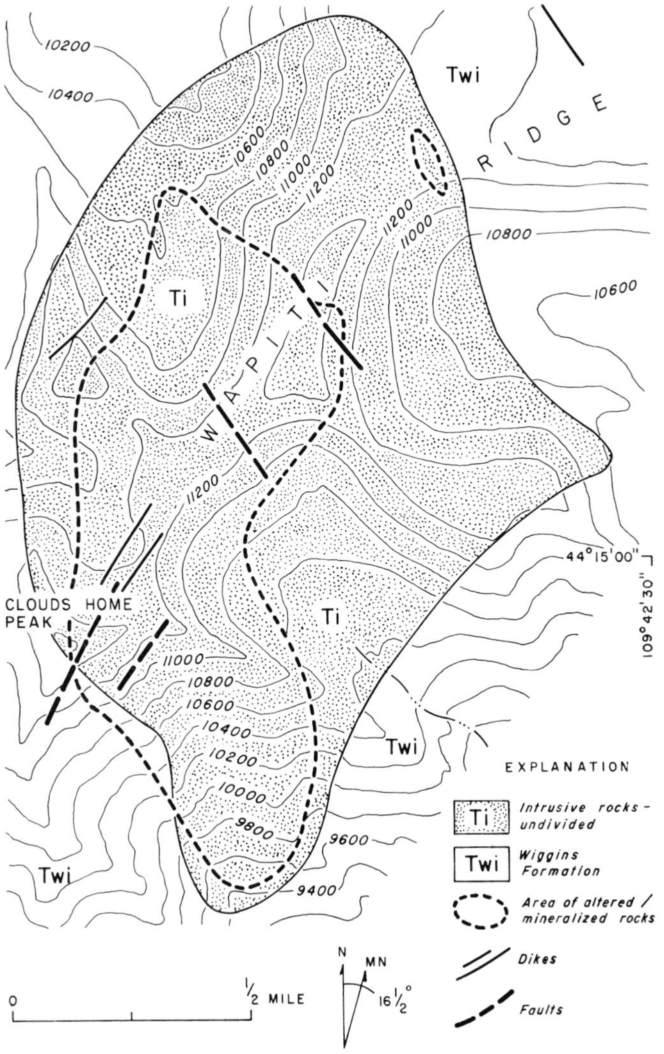

Figure 9: Generalized diagram of alteration and geology of the Clouds Home Peak intrusive complex (after Fisher and Antweiler, 1980).

The intrusive complex crops out over an area of more than one-square mile (Figure 9). Mineralization and alteration occur sporadically along a one-mile north-south trend located within the composite intrusive. According to Fisher and Antweiler (1980), about five percent pyrite and traces of chalcopyrite occur as disseminations and as fracture coatings in the intrusive, and also in some dikes. Other expressions of mineralization include: small pyrite, chalcopyrite, rhodochrosite, quartz veins (two inches and less, in width) that cut granodiorite rocks in a one-yard-wide zone; and a 30-feet-wide stockwork hosting malachite, azurite, chalcopyrite, quartz, pyrite, chlorite, and epidote. Anomalous copper, molybdenum, lead, and zinc are detectable in assays.

Alteration includes both propylitic and phyllic assemblages. The phyllic altered rock is restricted and closely associated with veins, shears, fractures, and dikes. Propylitic alteration is widespread. Oxidation has produced limonite staining, minor malachite staining, and bleaching of the rocks (Fisher and Antweiler, 1980).

Eagle Creek Mineralized Area

The Eagle Creek area lies near the edge of the Washakie Wilderness near the Yellowstone Park boundary, and about 40 air miles west of Cody, Wyoming. The district is accessible by 12 miles of pack trail (Eagle Creek trail) into the wilderness, and lies within T.51N., R.109W. on the Eagle Peak 15-minute quadrangle.

The region was prospected in 1911; and by 1955, 640 feet of drifts and 50 feet of shaft were developed on what is called the Crouch prospect, located near the south edge of the intrusive complex (Figure 10) (Osterwald and others, 1966; Wilson, 1956). The only recorded production was two ore shipments delivered to the Denver Mint in the 1930's that were valued at $1,000 in gold (Wilson, 1956).

The country rock, in the Eagle Creek region, is formed of andesite flows and flow breccias of the Trout Peak Trachyandesite (Sunlight Group). These two formations, in the immediate area of interest, are vent-series facies and are mapped as undifferentiated andesites.

Mineralization is directly related to the emplacement of an irregularly-shaped intrusive (post-Oligocene?) averaging latite (monzonite) composition. This latite porphyritic intrusive has been selectively fractured and mineralized, and exhibits stockwork mineralization near the east-central edge of the intrusive, where the intrusive lies within a topographic saddle (Figure 10). The stockwork is expressed by narrow N32°E and N64°W trending quartz veinlets carrying pyrite, chalcopyrite, and galena. A further expression of mineralization extends out from the stockwork and into silicified and altered latite porphyry and also into relatively fresh latite porphyry (Tlp) on the northwestern edge of the saddle. This mineralization occurs as pyrite disseminations and as narrow pyrite, galena, sphalerite, chalcopyrite quartz veins (less than one inch wide) in the altered-silicified latite, and as fine pyrite and chalcopyrite(?) disseminations in the gray latite porphyry (Tlp). There is no evidence that supergene enrichment is important in the district (Galey, 1971). Some placer gold is reported along Eagle Creek and along Crouch Creek (Wilson, 1956).

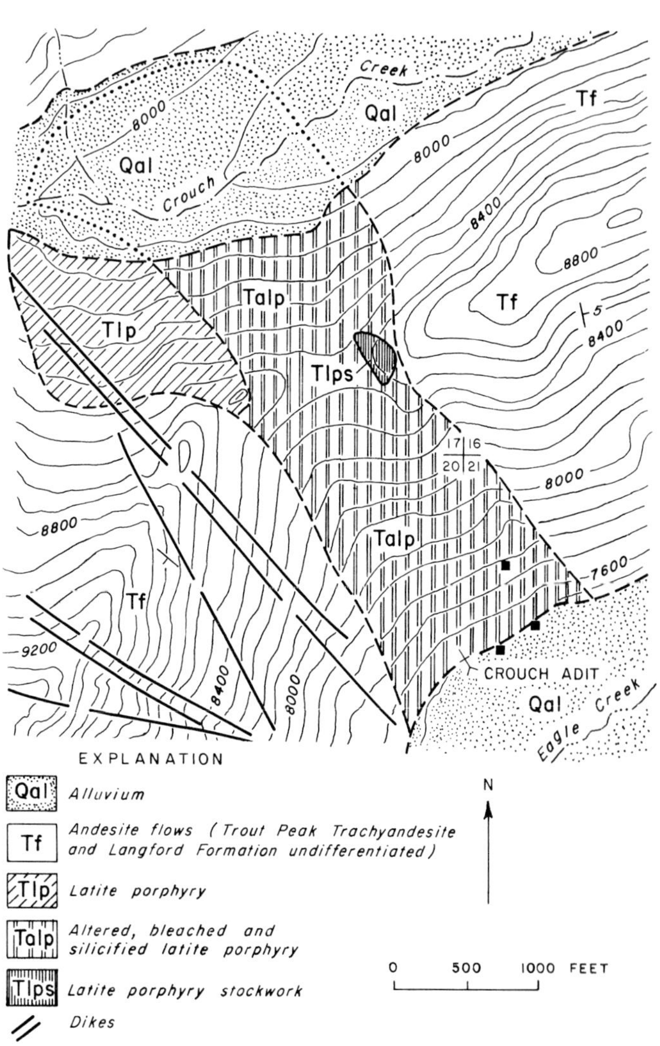

Figure 10: Generalized geological map of the Eagle Creek mineralized area (T.51N., R.109W.). Base metal deposits of the Eagle Creek region are associated with intrusive latite porphyry. There is little evidence for supergene enrichment (after Galey, 1971).

Hydrothermal alteration is intense near the central portion of the stockworks, and is characterized by phyllic and potassic alteration assemblages. The intense alteration, laterally gives way to regional prophylitic alteration (Galey, 1971).

Sunlight Mineralized Area

The Sunlight mineralized area is located on U.S. Geological Survey 15-minute topographic quadrangles Dead Indian Peak (1956) and Sunlight Peak (1956). The nearest population center is Cody, Wyoming, located 40 miles east of the mineralized area, although the Sunlight Basin area houses numerous private cabins and summer homes. Osterwald and others (1966), Parsons (1937), Williams (1980a), Elliott (1980a) and Hausel (1980) provide overviews of the Sunlight deposits. Rich (1974), Pedersen (1968), and Drier (1967) conducted thesis investigations in the area.

The first report of prospecting in the Sunlight area occurred in 1890 and the only report of ore production was 100 tons of gold, silver, and copper ore mined in 1903 (Williams, 1980a, p. 34). Several hundred feet of exploratory workings have been developed by shaft and adits at several mine sites. All of these historic workings were developed along vein mineralization.

Two types of mineralization are recognized in the district; these are disseminated deposits and vein mineralization. The disseminated deposits consist of (1) pyrite in altered zones in all rock types, (2) chalcopyrite in intrusive rocks, and (3) copper-bearing stockworks (Elliott, 1980a). Fairly well-developed stockworks are developed in syenite stock at the headwaters of Sulfur Creek (Drier, 1967), and in andesites in contact with syenite (Wilson, personal comm., 1982). The stockworks consist of less than one inch wide veins and veinlets hosting chalcopyrite, bornite, covellite and chalcocite in quartz, calcite gangue (Figure 11).

Vein mineralization is zoned, and forms copper-rich veins near the mineralized centers, and lead-silver or barren veins at a distance from the stocks.

Three alteration zones are recognized in the district — propylitic, argillic, and potassic. Propylitic alteration is widespread and increases in intensity near the intrusive stocks and veins. While potassic alteration is localized and restricted to the stocks, argillic alteration appears to be supergene in origin and overprints the other two alteration phases.

Primary controls on the mineralization were the intrusive masses and associated fracture systems. Disseminated mineralization is localized in and around intrusives, and veins are controlled by joint sets and shear zones (Elliott, 1980a) (Figure 11).

New World - Cooke City

The New World district (often referred to as the Cooke City district) is located along the Montana-Wyoming border adjacent to Cooke City, Montana. Cooke City is nearly 50 air miles northwest of Cody, and lies on the eastern edge of the Absarokas. Most of the district is within the State of Montana, although 22 mining claims, and a portion of the underground workings of the Irma-Republic mines, penetrate into the North Absaroka Wilderness of Wyoming (Williams, 1980b).

The ore deposits in the New World district are related to a deeply dissected intrusive-volcanic complex. Due to uplift, this complex has been more deeply eroded than any of the other known mineralized centers in the Absarokas. Precambrian gneiss, Paleozoic sediments, and Tertiary intrusives (diorite to syenite) are exposed throughout most of the district. On the western and southern edges, andesitic flows are still preserved and unconformably rest on Paleozoic sediments (Elliott, 1980b).

The Goose Lake and Henderson Mountain stocks, located north of Cooke City, are recognized centers of mineralization. With a few exceptions, the major ore bodies are localized adjacent to the stocks.

The Goose Lake center, located north of the Henderson stock, exhibits mineralization somewhat different from the Henderson lodes. The host rocks for the Goose Lake stock are mafic and felsic igneous rocks. The veins are localized and form principally chalcopyrite, and galena-sphalerite vein deposits adjacent to the intrusive.

The Henderson Mountain mineralization primarily occurs as fissure-replacement deposits. The ore shoots and lodes are best developed at fracture intersections and in sheeted zones. Alteration associated with the mineralization is expressed as silification and dolomitization (Lovering, 1929).

The mineralization extends out from Henderson Mountain in irregular metallogenic zones. Contact-metamorphic gold-copper deposits are developed adjacent to the stock and these grade outward to copper-lead ore. The copper-lead mineralization is further zoned to copper-lead-zinc, and, at greater distances from the stock, to complex lead-silver-zinc, then to an aureole of silver-bearing sideritic calcite veins and finally to barren carbonate veins at greater distances. The better deposits are developed where these zoned veins cut limestone beds of the Pilgrim Limestone (Gallatin Formation) of upper Cambrian age to form replacement lodes, although the underlying Gros Ventre Limestone (middle Cambrian) is an untested potential host (Lovering, 1929; Reed, 1950; Butler, 1965).

Irma-Republic Mines. The Irma-Republic deposits are a vein system of the Henderson Mountain complex. These deposits are formed by fracture fillings and replacements enclosed by oolitic beds of the Pilgrim Limestone. The system is near vertical and strikes N30° to 40°W. Mine workings in the Irma-Republic lode are over 2,700 feet in length and extend to a depth of 250 feet. Silver, lead, and zinc have been extracted from the deposits. Reported production from these mines is more than 18,000 tons of gold, silver, copper, lead, and zinc concentrates (Williams, 1980b).

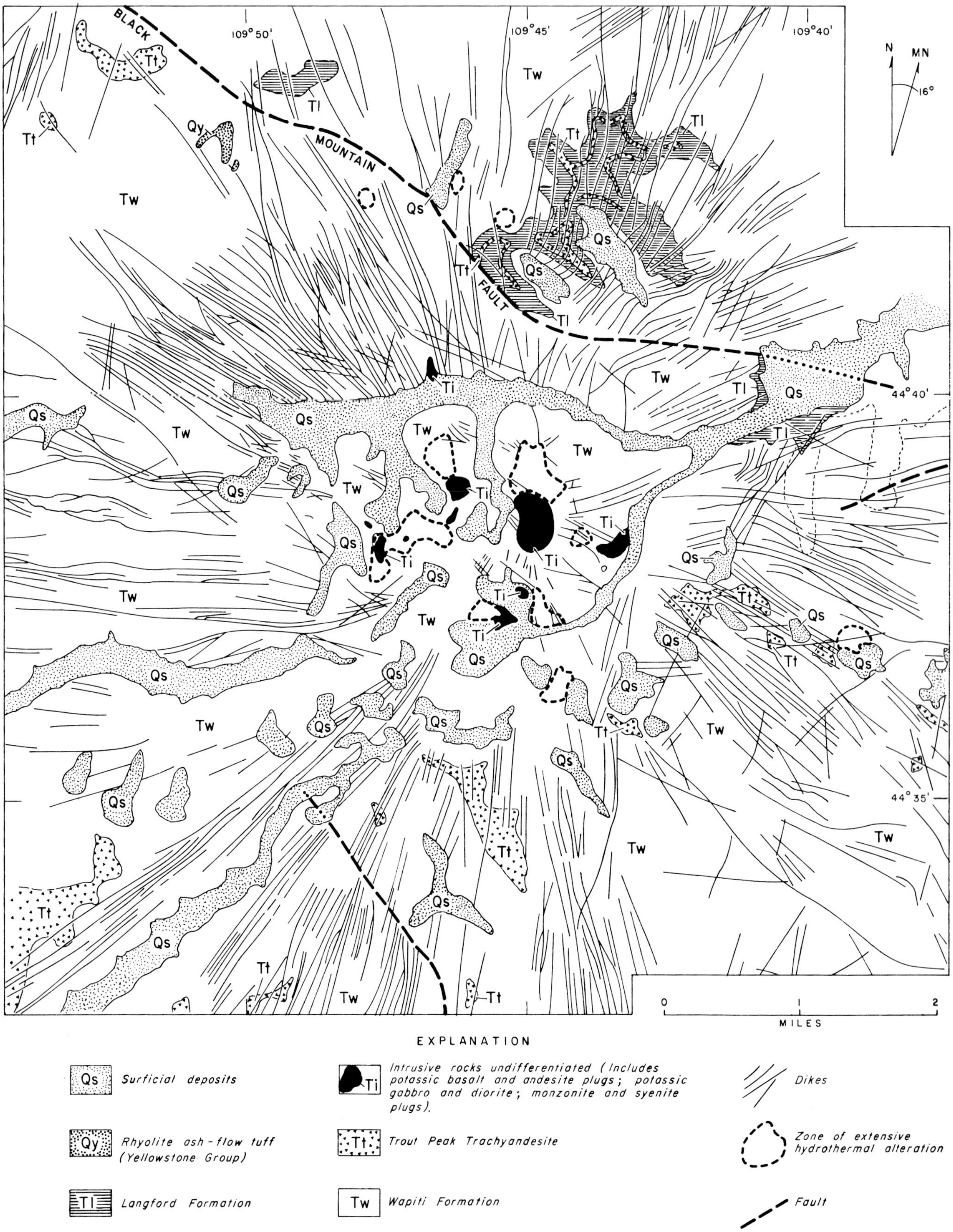

Figure 11: Geological map of the Sunlight intrusive center (after Nelson and others, 1980).

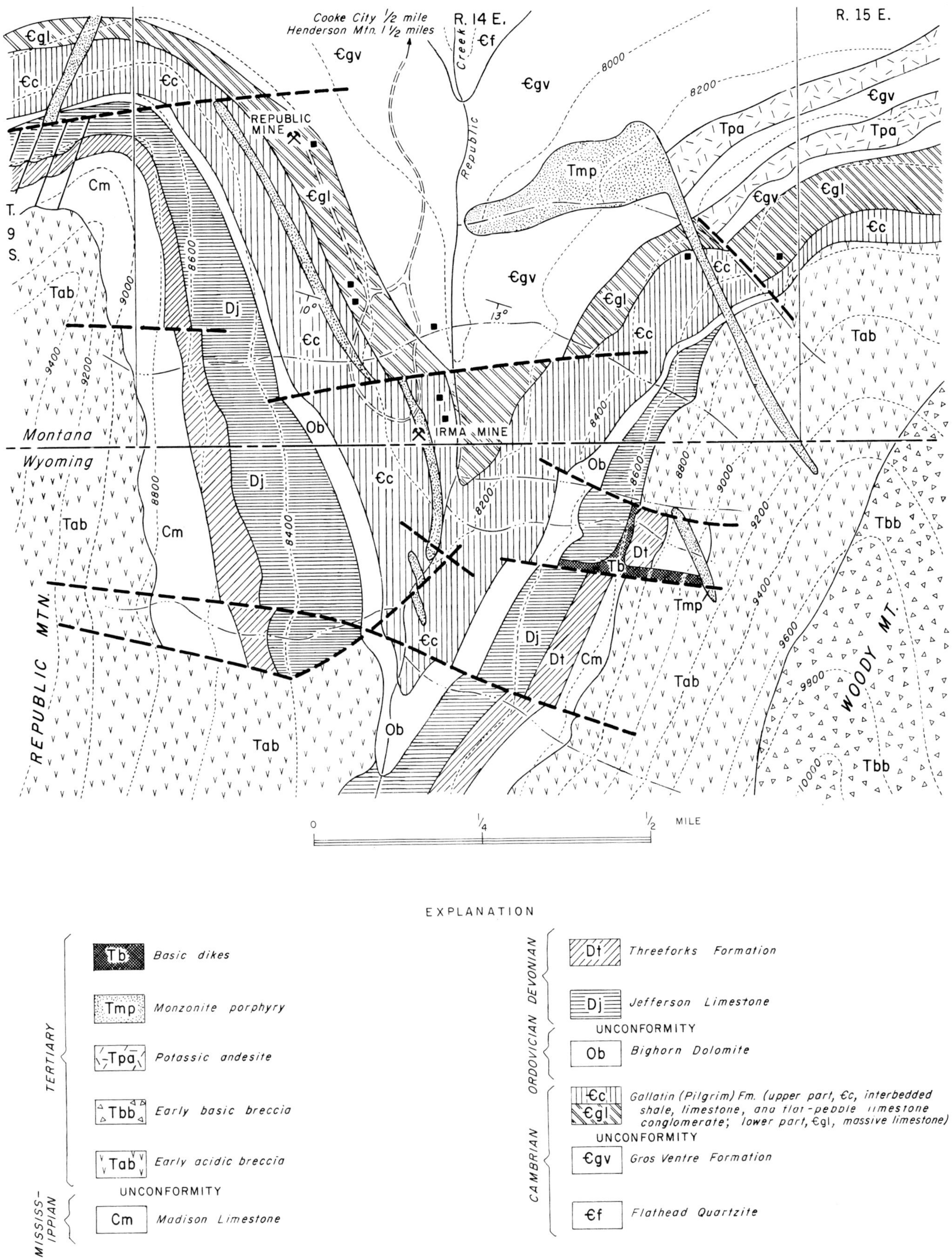

Figure 12: General Geological map of the southern New World district (after Lovering, 1929).

SUMMARY

The Absaroka volcanic plateau contains a well-defined trend of porphyry copper-molybdenum deposits. This trend forms a north-northwesterly lineament marked by mineralized composite intrusives possibly controlled by a deep-seated fracture.

These mineralized centers represent a large potential resource of copper, molybdenum, and other associated base metals and precious metals, but the inclusion of these deposits within and immediately adjacent to wilderness boundaries will greatly restrict mining and production.

There is potential for discovery of additional mineralized porphyry systems, and the search for these should concentrate on felsic composite intrusive complexes in areas of vent facies volcanics (andesite flows and flow breccias).

ACKNOWLEDGMENTS

The author wishes to express sincere appreciation to William H. Wilson for his critical comments and review of this paper, and to Gary B. Glass and David A. Copeland for their editorial comments. Both Rusty Dersch of the U.S. Forest Service and John Wells of Timberline Minerals, Incorporated, provided several property reports that greatly contributed to this manuscript. Finally I would like to thank Phyllis Ranz for her excellent work on the illustrations.

REFERENCES CITED

Beane, R. E., and Titley, S. R., 1981, Porphyry copper deposits — Part II, hydrothermal alteration and mineralization: Econ. Geol. 75th Ann. Vol., p. 235-269.

Butler, T. A., 1965, Geology of the Irma-Republic mines and vicinity, New World Mining District, Montana and Wyoming: unpub. M.S. thesis, Univ. of Idaho, 48 p.

Carmichael, I. S. E., Turner, F. J., and Verhoogen, J., 1974, Igneous Petrology: McGraw-Hill, 739 p.

Chadwick, R. A., 1970, Belts of eruptive centers in the Absaroka-Gallatin volcanic province, Wyoming-Montana: Geol. Soc. Amer. Bull., v. 81, p. 267-274.

Drier, J. E., Jr., 1967, Economic geology of the Sunlight mineralized region, Park County, Wyoming: unpub. M.S. thesis, Univ. of Wyoming, 81 p.

Elliott, J. E., 1980a, Mineralization of the Sunlight mining region, *in* North Absaroka Wilderness and vicinity, Park County, Wyoming: U.S. Geol. Survey Bull. 1447, p. 51-53.

Elliott, J. E., 1980b, Geology and mineralization of the Cooke City mining district, *in* North Absaroka Wilderness and vicinity, Park County, Wyoming: U.S. Geol. Survey Bull. 1447, p. 57-58.

Fisher, F. S., 1972, Tertiary mineralization and hydrothermal alteration in the Stinkingwater mining region, Park County, Wyoming: U.S. Geol. Survey Bull. 1332-C, 33 p.

Fisher, F. S., and Antweiler, J. C., 1980, Preliminary geological and geochemical results from the Robinson Creek, Birthday, and Clouds Home Peak mineralized areas in the Washakie Wilderness, Wyoming: U.S. Geol. Survey Open File Rept. 80-781, 12 p.

Fisher, F. S., Antweiler, J. C., and Welsch, E. P., 1977, Preliminary geological and geochemical results from the Silver Creek and Yellow Ridge mineralized areas in the Washakie Wilderness, Wyoming: U.S. Geol. Survey Open File Rept. 77-225, 11 p.

Galey, J. T., Jr., 1971, Mineralization in the Eagle Creek area, Park County, Wyoming: unpub. M.S. thesis, Univ. of Wyoming, 51 p.

Geological Survey of Wyoming files, 1982, Miscellaneous property reports of Timberline Minerals, Inc.

Hague, A., 1889, Absaroka Folio: U.S. Geol. Survey Geol. Atlas of the United States, Folio No. 52.

Hausel, W. D., 1980, Gold districts of Wyoming: Geol. Survey of Wyoming Rept. Invest. 23, 71 p.

Hausel, W. D., 1981, Economic mineral deposits of Wyoming — a review: Wyoming Geol. Assoc., 32nd Ann. Field Conf. Guidebook, p. 131-159.

Hausel, W. D., 1982, Ore deposits of Wyoming: Geol. Survey of Wyoming Prel. Rept. 19, 39 p.

Hewett, D. F., 1912, The ore deposits of Kirwin, Wyoming: U.S. Geol. Survey Bull. 540, p. 121-132.

Hollister, V. F., 1974, Regional characteristics of porphyry copper deposits of South America: Trans., SME-AIME, v. 256, p. 45-53.

Hollister, V. F., 1978, Geology of the porphyry copper deposits of the Western Hemisphere: AIME, 219 p.

Lipman, P. W., Prostka, H. J., and Christiansen, R. L., 1972, Cenozoic volcanism and plate-tectonic evolution of the western United States. I. Early and Middle Cenozoic: Phil. Trans., R. Soc. Lond. v. 271, p. 217-248.

Love, J. D., 1939, Geology along the southern margin of the Absaroka Range, Wyoming: Geol. Soc. Amer., Special Paper No. 20, 134 p.

Lovering, T. S., 1929, The New World or Cooke City mining district, Park County, Montana: U.S. Geol. Survey Bull. 811-A, 87 p.

Lowell, J. D., and Guilbert, J., 1970, Lateral and vertical alteration mineralization zoning in porphyry ore deposits: Econ. Geol., vol. 65, no. 4, p. 373-408.

Nelson, W. H., and others, 1980, Geology and mineral resources of the North Absaroka Wilderness and vicinity, Park County, Wyoming: U.S. Geol. Survey Bull. 1447, 101 p.

Nowell, W. B., 1971, The petrology and alteration of the Kirwin mineralized area, Park County, Wyoming: unpub. M.S. thesis, Univ. of Montana, 72 p.

Osterwald, F. W., Osterwald, D. B., Long, J. S., Jr., and Wilson, W. H., 1966, Mineral resources of Wyoming: Geol. Survey of Wyoming Bull. 50, 287 p. (revised by W. H. Wilson).

Parsons, W. H., 1937, The ore deposits of the Sunlight mining region, Park County, Wyoming: Econ. Geol., v. 32, no. 6, p. 832-854.

Parsons, W. H., 1939, Volcanic centers of the Sunlight area, Park County, Wyoming: Jour. Geol., v. 47, no. 1, p. 1-26.

Pedersen, K., 1968, Sulfide mineralization of the northwestern part —Sunlight mining region, Park County, Wyoming: unpub. M.S. thesis, Univ. of Wyoming, 50 p.

Prostka, H. J., 1973, Hybrid origin of the absarokite-shoshonite-banakite series, Absaroka volcanic field, Wyoming: Geol. Soc. Amer. Bull., v. 84, no. 2, p. 697-702.

Reed, G. C., 1950, Mines and mineral deposits (except fuels), Park County, Montana: U.S. Bur. Mines Inf. Circ. 7546, 68 p.

Rich, D. H., 1974, Economic geology of Silvertip Basin, Sunlight mining region, Park County, Wyoming: unpub. M.S. thesis, Miami Univ., Ohio, 73 p.

Rosekranz, R. D., Davidoff, R. L., and Lemons, J. F., Jr., 1977, Copper availability — domestic: U.S. Bureau of Mines Inf. Circ. 8809, 31 p.

Rouse, J. T., 1937, Genesis and structural relationship of the Absaroka volcanic rocks, Wyoming: Geol. Soc. Amer. Bull., v. 48, no. 9, p. 1257-1296.

Smedes, H. W., and Prostka, H. J., 1972, Stratigraphic framework of the Absaroka Volcanic Supergroup in the Yellowstone National Park region: U.S. Geol. Survey Prof. Paper 729-C, 33 p.

Titley, S. R., 1982, The style and progress of mineralization and alteration in porphyry copper systems, *in* Titley, S. R., editor, Advances in Geology of the porphyry copper deposits, Southwestern North America, Univ. of Arizona Press, Tucson, p. 93-116.

Turner, F. J., and Verhoogen, J., 1960, Igneous and Metamorphic Petrology: McGraw Hill, 694 p.

Williams, F. E., 1980a, Metallic mineral deposits, *in* North Absaroka Wilderness and vicinity, Park County, Wyoming: U.S. Geol. Survey Bull. 1447, p. 35-51.

Williams, F. E., 1980b, Cooke City mining district, *in* North Absaroka Wilderness and vicinity, Park County, Wyoming: U.S. Geol. Survey Bull. 1447, p. 54-56.

Wilson, W. H., 1954, A preliminary study of the Kirwin mining district, Park County, Wyoming: unpub. Geol. Engineer thesis, Univ. of Utah.

Wilson, W. H., 1956, Crouch gold prospect, Park County, Wyoming: unpub. rept., Geol. Survey of Wyoming files, 6 p.

Wilson, W. H., 1960, Petrology of the Wood River area, southern Absaroka Mountains, Park County, Wyoming: unpub. Ph.D. thesis, Univ. of Utah, 121 p.

Wilson, W. H., 1963, Correlation of volcanic rock units in the southern Absaroka Mountains, northwest Wyoming: Univ. of Wyo. Contr. to Geology, v. 2, no. 1, p. 13-20.

Wilson, W. H., 1964a, Geologic reconnaissance of the southern Absaroka Mountains, northwest Wyoming: Part I — the Wood River-Greybull River area: Univ. of Wyo. Contr. to Geol., v. 3, no. 2, p. 60-77.

Wilson, W. H., 1964b, The Kirwin mineralized area, Park County, Wyoming: Geol. Survey of Wyoming Prel. Rept. no. 2, 20 p. (revised 1980).

Wilson, W. H., 1971, Volcanic geology and mineralization, Absaroka Mountains, northwest Wyoming: Wyo. Geol. Assoc., 23rd Ann. Field Conf. Guidebook, p. 151-155.

Wilson, W. H., 1975, The copper-bearing Meadow Creek granodiorite, upper Wood River area, Park County, Wyoming: Wyo. Geol. Assoc., 27th Ann. Field Conf. Guidebook, p. 235-241.

Wilson, W. H., 1982, Geologic map of the Dick Creek Lakes, Dunrud Peak, Francs Peak, Noon Point and Twin Peaks quadrangles, Fremont, Hot Springs and Park counties, Wyoming: Geol. Survey of Wyoming Map Series MS-10, in press.

Thirty-Third Annual Field Conference — 1982
Wyoming Geological Association Guidebook

THE JACKSON HOLE COAL FIELD

RICHARD W. JONES[1]

INTRODUCTION

The Jackson Hole Coal Field is located in the eastern and northern parts of Teton County, northwestern Wyoming, approximately 23 miles northeast of the town of Jackson. The field is an elongate, north-northwest-trending basinal area approximately 60 miles long and nine to 18 miles wide and is underlain by coal-bearing rocks over an area of about 672 square miles (Figure 1). The field lies almost entirely within the boundaries of the Teton National Forest, with the exception of the extreme northern part in Yellowstone National Park, and the western part in Grand Teton National Park.

FIELD BOUNDARIES

The field is bounded by the Yellowstone Volcanic Plateau to the north, the Washakie Range to the northeast, the Wind River Basin and Wind River Range to the east and southeast, the Gros Ventre Range to the southwest, and the Teton Range to the west. The actual field boundaries (Figure 1) are defined by formation contracts, faults, and the inferred extent of coal-bearing rocks in the subsurface. The boundaries were compiled from published U.S. Geological Survey maps (Love, 1947, 1956, 1973a & b, 1975a & b; Love and Keefer, 1975; Rohrer, 1969) and unpublished mapping by J. D. Love (written communication, 1982).

The eastern boundary of the field is the Buffalo Fork thrust fault on the southwest side of the Washakie Range. This fault places older noncoal-bearing rocks in fault contact with coal-bearing and younger rocks. Several small outcrops of steeply dipping coal-bearing rocks are found to the east of this fault; the field boundary here is based on these outcrops. The southeastern boundary of the coal field east of Purdy Basin is based on the inferred extent of a coal-bearing unit in Eocene rocks (Rohrer, 1969). The southern, southwestern, and northern boundaries of the field are defined by the base of the coal-bearing Bacon Ridge Sandstone. The western boundary is inferred below younger noncoal-bearing rocks and is based on the westward projection of discontinuous, isolated outcrops of known coal-bearing rocks. The northwest part of the coal field is defined by the Heart River and West Sheridan faults (Love and Keefer, 1975) and the Pilgrim Creek and Rodent Creek faults (Love, 1974).

STRUCTURE

The Jackson Hole Coal Field is one of the more structurally complex coal fields in the state. Many structural features are present in the field, including normal and reverse faults, thrust faults, steeply dipping to overturned coal-bearing rocks, and unconformities of various magnitudes of age and angular discordance. These structural features have great effect on the coal field, in that erosion and faulting can drastically alter the thickness of coal-bearing rocks or even remove the coal-bearing units, coal-bearing facies in the rocks may change to noncoal facies through structural influences at the time of deposition, and coal-bearing rocks are often overlain by younger rocks (including volcanic rocks) of great or highly variable thickness.

From a structural standpoint, the least complicated areas in the coal field are located along the southwestern part of the field northeast of the Gros Ventre River, where coal-bearing rocks are well exposed and dip from about 10 to 25 degrees to the northeast, and in the southeast part of the field, where coal-bearing rocks in the Wind River Formation dip from about 10 to 30 degrees to the northeast. Coal-bearing rocks in the southern and northern parts of the field occur on the flanks of the following anticlines: the Bacon Ridge anticline in the south, the Hancock and Wolverine anticlines to the north, and the Wildcat and Kitten anticlines to the northwest. Coal-bearing rocks also occur on the Spread Creek and Whetstone anticlines in the central part of the field. The dip of the coal-bearing rocks on all the above anticlines is quite variable, ranging from less than 10 degrees to vertical or overturned.

COAL-BEARING ROCKS

The coal-bearing rocks in this coal field (Figure 2) are the Bacon Ridge Sandstone, the Sohare sequence (Love and Christiansen, 1980), and the Mesaverde Formation of

[1]Staff coal geologist, Wyoming Geological Survey

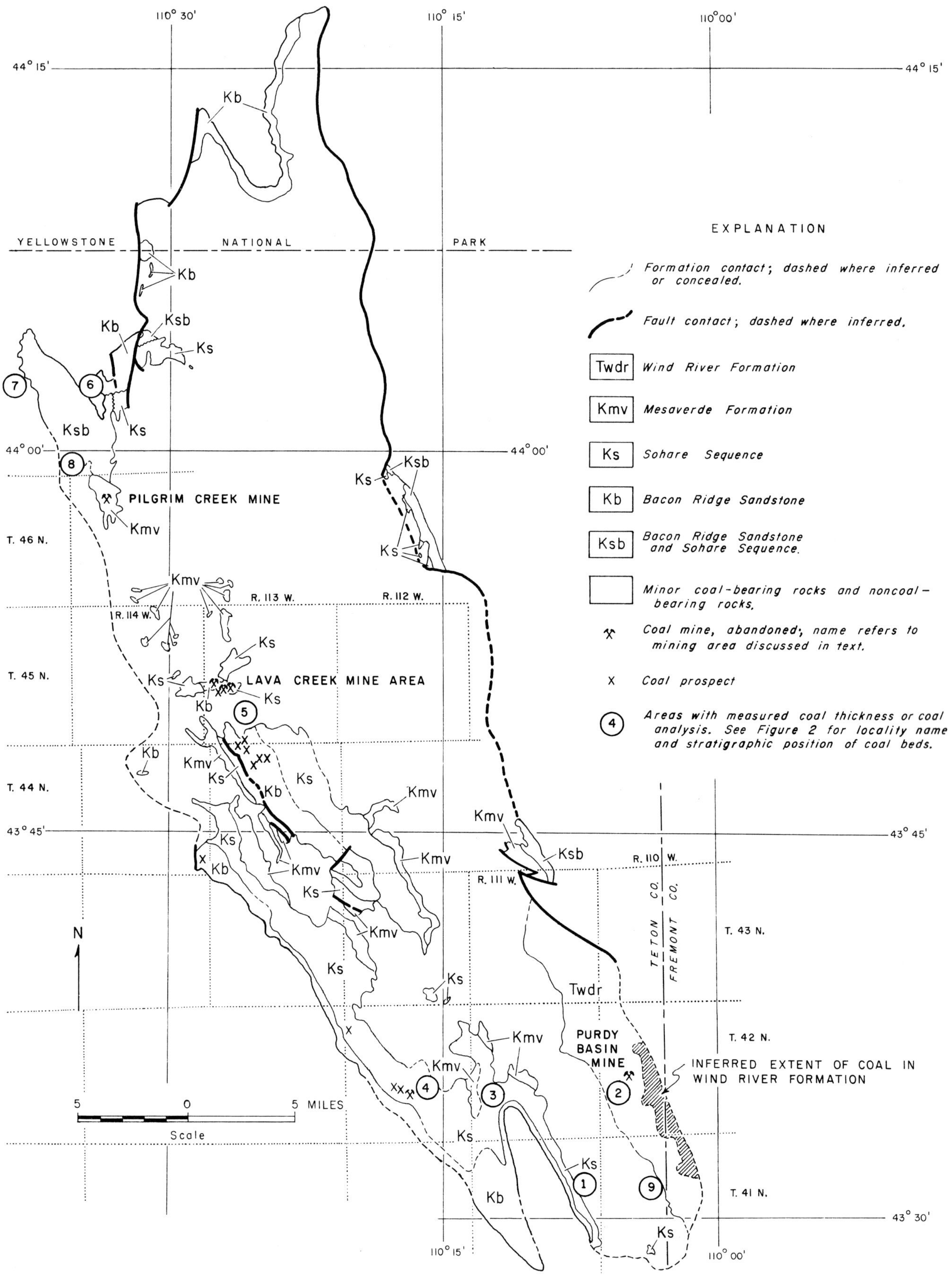

Figure 1: Boundaries and major coal-bearing rocks of the Jackson Hole Coal Field.

Upper Cretaceous age, the coal member of the Pinyon Conglomerate of Upper Cretaceous and Paleocene age (Love, 1973), the Devils Basin Formation" (proposed new name) of Paleocene age (Love, oral communication, 1982) and the Wind River Formation (lower and upper variegated sequence (Love and Christiansen, 1980)) of Eocene age. The Upper Cretaceous Frontier Formation contains an impure coal bed two feet thick near the northern part of the field (Love, Antweiler, and Williams, 1975); other thin, impure coal beds may also be present near the field. However, the coals in this formation are not considered minable and are not included in the coal field.

The Bacon Ridge Sandstone is the oldest, most important, and most extensive coal-bearing formation in the Jackson Hole Coal Field. The Bacon Ridge Sandstone overlies the Cody Shale and includes a lower unit 500 to 1,250 feet thick and an upper unit of about 1,000 feet in thickness. The upper unit, known as the coaly sequence in Love, et al. (1948), is now mapped with the lower unit because both units are coal-bearing, have similar lithologies, and a reliable or meaningful contact between the two units cannot be determined (J. D. Love, oral communication, 1982). The lower unit of this formation contains several zones that have coal beds of minable thickness. A coal zone between the base of the Bacon Ridge and an overlying pearl-gray marker zone contains coal beds up to seven feet thick (Figure 2). This coal zone is best developed in the Lava Creek-Spread Creek area (locality 5, Figue 1), extends northward from this area to Arizona Creek (locality 7), and is absent to the northeast, east, and southeast of locality 5 (Love, et al., 1951). Coal in this zone was mined at Lava Creek and possibly Spread Creek to the south (see Coal Production section below). According to Berryhill, et al. (1950), 11.2 feet of coal in several closely spaced beds occurs near the base of the Bacon Ridge Sandstone at an unspecified location in the field. However, the original references for this coal field cited by Berryhill, et al., as well as other reports published later, make no mention of this coal occurrence. Above the pearl-gray marker zone in the northern part of the field are several coal beds between two and 3.7 feet thick (Love, Antweiler, and Williams, 1975). A coal bed at locality 4 is over four feet thick. In general, coal beds above the pearl-gray marker zone and below the "coaly sequence" are thickest and most numerous in the south-central part of the Jackson Hole Coal Field (Love, et al., 1951).

The upper 1,000 feet of the Bacon Ridge Sandstone contains most of the coal beds of minable thickness in the coal field. Because the information available on these coal beds is so widely spaced and because the coal beds that occur have not been mapped in detail, individual coal beds cannot be correlated. The coal beds shown in Figure 2 for the "coaly sequence" are "equivalent" only in the sense that they occur in approximately the same stratigraphic position; actual physical equivalency determined by tracing and mapping individual coal beds in the field has not been done.

On the basis of four measured sections in the southwestern part of the field (Love, et al., 1948), Berryhill, et al. (1950), reported 13 coal beds 2.5-5 feet thick and five coal beds 5-10 feet thick in this interval, with the thickest coal (8.3 feet) a short distance south of locality 3 (Fish Creek). It seems peculiar that only one mine on one coal bed in the "coaly sequence" is reported in the literature (Love, et al., 1948). The lower half of the "coaly sequence" appears to contain thicker and more abundant coals than the upper half (Love, et al., 1951).

The Sohare sequence, a new name for the lecticular sandstone and shale sequence, lies above the Bacon Ridge Sandstone. The Sohare and the Bacon Ridge are equivalent to the upper part of the Cody Shale in the Wind River and Bighorn Basins to the east and the Hilliard Shale and Adaville Formation in the Thrust Belt to the south (Love and Christiansen, 1980). Coals in the Sohare are relatively thin and widely spaced throughout the section. The thickest coal bed reported is only three feet thick at locality 3; only four other coal beds are over two feet thick (Love, et al., 1948).

The Mesaverde Formation overlies the Sohare sequence. The thickness of the Mesaverde is highly variable; part of all of the formation has been removed by erosion. Where the Mesaverde is present, it is usually composed of a lower white sandstone sequence which is coal-bearing and an upper conglomeratic sandstone. It is believed that these two units are separated by an unconformity (Love, et al., 1951). Coal beds are reported by Love, et al., (1948) at locality 3 (maximum thickness 1.5 feet) and locality 5. In addition, coal was mined from a bed in the Mesaverde at Pilgrim Creek. The stratigraphic position and thickness of this bed are not known, but analyses of two samples of coal from this bed are reported in Love, Antweiler, and Williams (1975).

The Meeteetse and Harebell Formations overlie the Mesaverde Formation in the coal field, but, apparently, do not contain coal beds of minable thickness. Thin, impure coal beds are reported from an outcrop of the Meeteetse Formation near locality 5, but their thicknesses have not been documented and the extent of the Meeteetse is quite limited. The Harebell Formation does not contain minable coal, although Love (1973b) reports several coal beds less than one foot thick near the base of the Harebell and a coal bed 1.5 feet thick 2,240 feet above the base.

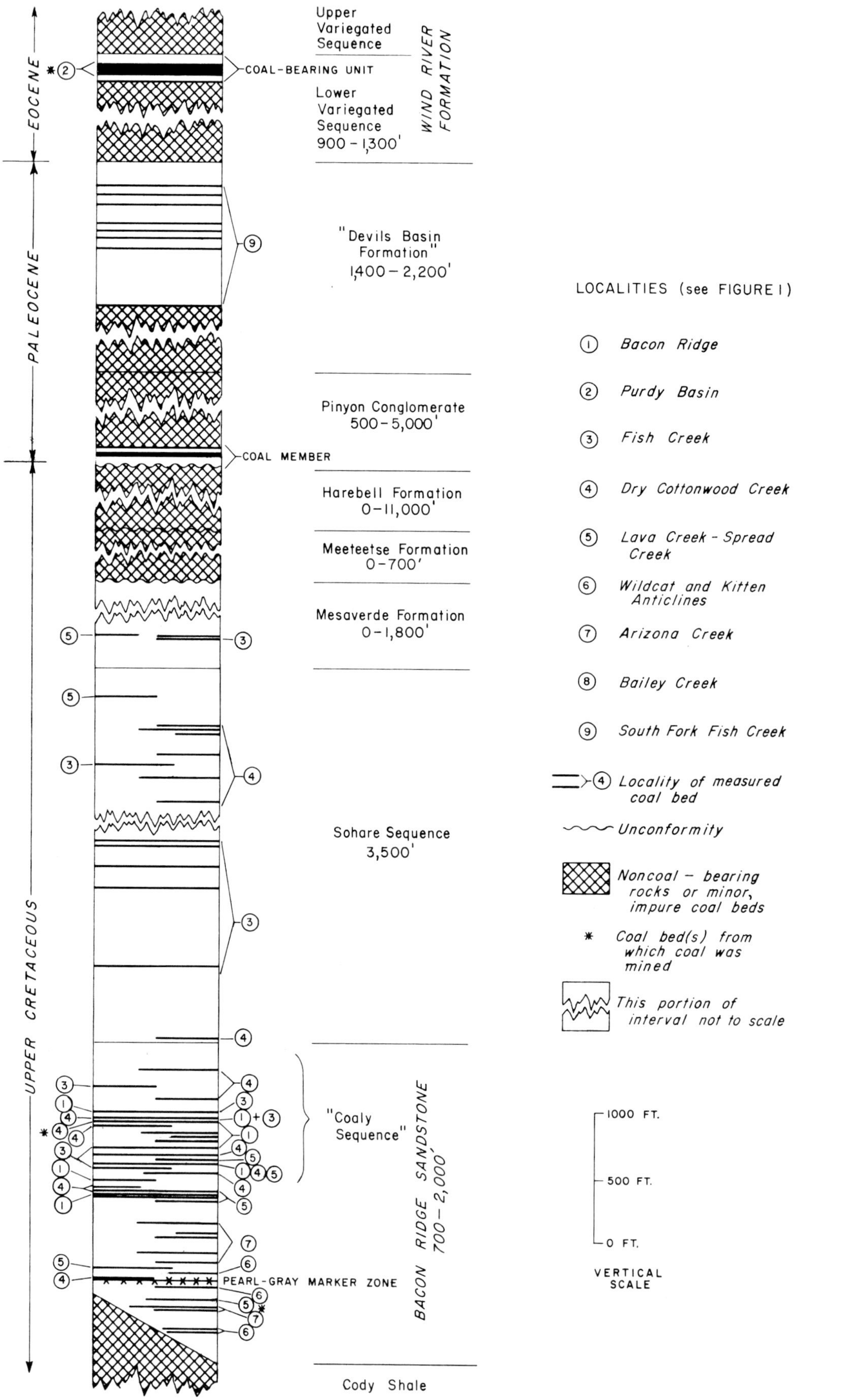

Figure 2: Approximate Stratigraphic Position of Major Coal Beds in the Jackson Hole Coal Field.

The Pinyon Conglomerate overlies the Harebell and/or older formations either disconformably or unconformably in the Jackson Hole Coal Field. A lenticular coal-bearing sequence called the coal member occurs only in the southwestern part of the field at the base of the Pinyon at locality 4 and north of Mount Leidy in T.44N., R.113W. (Love, 1937b). It ranges in thickness from 50 to 140 feet, contains fossils of early Paleocene age, and is equivalent in part to the Fort Union Formation in the Wind River Basin, the Hoback Formation in the northern Green River Basin, and the Evanston Formation in the Thrust Belt (Love, 1973b; Love and Christiansen, 1980). Farther north, at the principal reference section on Pinyon Peak, the lower part of the formation is considered to be Cretaceous in age (Love, 1973b). At locality 4, the coal member contains up to 12.2 feet of coal; the thickest coal beds are 2.5, 2.8 and 4.0 feet thick, but are separated by shale or siltstone partings 1.5 to 2.0 feet thick. Although these coals are major coal beds in the Jackson Hole Coal Field, the coal member is only a localized coal occurrence within the Pinyon; therefore, the Pinyon Conglomerate as a whole is considered a minor coal-bearing unit and is not shown on Figure 1.

Overlying the Pinyon Conglomerate in the southeastern part of the field is the "Devils Basin Formation," a proposed new name (Love, oral communication, 1982) for a coal-bearing unit of Paleocene age in the southeastern part of the coal field. The formation was mapped in detail by Rohrer (1969) as the Fort Union (?) Formation. The upper half of the formation contains eight to ten relatively thin coal beds at locality 9 (Love, 1947). In the same area, Rohrer (1969) has mapped up to eight coal beds ranging in thickness from 0.8 to 2.6 feet. Because this unit is limited in extent and contains few, if any, minable coal beds, it is, overall, a minor coal-bearing unit and is not shown on Figure 1.

Overlying the "Devils Basin Formation" in the southeastern part of the coal field is the Wind River Formation of early Eocene age. At various times, this formation has been called the lower variegated sequence (Indian Meadows), the coal-bearing sequence, and the upper variegated sequence (Wind River) by Love (1947); the Wind River Formation (Love, et al., 1955); the "Member at Purdy Basin" (Rohrer, 1969); the lower and upper variegated sequence (Love and Christiansen, 1980); and, finally, the Wind River Formation (Love, oral communication, 1982). The coal-bearing unit, which is from 40 to 330 feet in thickness (Rohrer, 1969), occurs in about the middle of the Wind River Formation and separates the noncoal-bearing upper and lower variegated sequences. The coal-bearing unit contains the thickest coal bed(s) in the Jackson Hole Coal Field. A 63-foot coal bed with numerous shale partings is reported in the Purdy Basin at locality 2, Figure 1 (Love, 1947; Berryhill, et al., 1950). Detailed mapping of the coal-bearing unit has shown that the thickest coal section occurs at an old coal mine adit along Purdy Creek and contains a total of 46.1 feet of coal in a 51-foot-thick interval (Rohrer, 1969). The thickest individual coal bed within this interval is 27.5 feet thick. The coal beds in this area occur in an elongate, northwest-trending lentil about eight miles long and one to two miles wide. The coal sequence wedges out to the southeast, where it is replaced by arkose; to the northwest, the coals are replaced by carbonaceous shales and the sequence intertongues or merges with variegated rocks; and to the northeast, the coal sequence gradually merges with arkosic rocks (Rohrer, written communication, 1982).

COAL QUALITY

The quality and rank of the coal in the Jackson Hole Coal Field are unknown. Based on the appearance and weathering properties of the coal in the field, Berryhill, et al. (1950) assumed the coal to be subbituminous. The little analytical data that exists on the coals is not sufficient to characterize the coal field, a particular formation, or a particular coal bed. From the Bacon Ridge Sandstone, Love, Antweiler, and Williams (1975) sampled two coal beds less than one foot thick and one coal bed two feet thick at locality 8 (Figure 1), and three coal beds of 5.5 feet, one foot, and less than one foot in thickness, respectively, from the vicinity of locality 6 (Figure 1). Evidently, complete standard proximate or ultimate analyses were not run on these samples, as only ash, total sulfur, and total carbon are reported. An average of seven samples from the above localities yielded 13.20% ash, 0.78% total sulfur, and 56.7% total carbon (presumably on an as-received basis). The heat value, (Btu content) of coal in the Bacon Ridge is not known, but a U.S. Bureau of Reclamation (1916) report states that coal from the Lava Creek area was "about 30" thick and is of good quality bituminous coal."

A single coal sample from a coal bed less than one foot thick in the Sohare sequence, 2½ miles north of the Pilgrim Creek locality, was reported by Love, Antweiler, and Williams (1975) to contain 27.80% ash, 1.02% total sulfur, and 51.3% total carbon. At an old coal mine on Pilgrim Creek, an average of two samples taken from a coal bed of unknown thickness in the Mesaverde Formation by Love, Antweiler, and Williams (1975) yielded 40.67% ash, 0.25% total sulfur, and 35.2% total carbon.

The only other published analyses of coal in the Jackson Hole Coal Field are for coals in the coal-bearing sequence in the Wind River Formation at Purdy Basin. A sample of unweathered, but perhaps abnormally dry, coal from a coal mine adit at locality 2 (Figure 1), which was analyzed over 40 years ago by the U.S. Bureau of Mines, yielded the following values on an as-received basis: moisture 10.3%, volatile matter 18.2%, fixed carbon 60.2%, ash 11.3%, sulfur 2.82%, and a heating value of 9,828 Btu/lb. (U.S. Minerals Management Service, unpublished file data; Rohrer, 1969). Unfortunately, four additional samples that were taken from outcrops in this same vicinity were weathered and had as-received heat values of 4,350, 5,170, 6,190, and 8,160 Btu/lb., respectively (Rohrer, 1969).

COAL RESOURCES

The only estimate of coal resources in the Jackson Hole Coal Field was made by Berryhill, et al., (1950): a total of 121,49 million short tons of subbituminous coal, all in the inferred category, was estimated for this coal field. This estimate, however, is far too conservative. At locality 2, using Figure 2 of Rohrer (1969) and an average total cost thickness of about 15 feet, there is over 220 million tons of coal in place in the Wind River Formation alone. Although some of this tonnage may be included in Berryhill, et al.'s estimate, it is obvious that their published estimate was far from inclusive. In short, there must be more coal than has been estimated, but, to date, no better estimate has been made.

COAL PRODUCTION

The Jackson Hole Coal Field has only produced a total of 11,382 tons of coal that can be documented since 1910. The following table summarizes coal production from the field:

The mining at Pilgrim Creek and Lava Creek between 1910 and 1916 was conducted by the U.S. Bureau of Reclamation to power construction equipment used in the building of the original Jackson Lake Dam (1910-1911) and its later reconstruction (1913-1916).

The mine at Lava Creek was at peak production during 1914 and 1915: over 3,000 tons were produced in 1914 and over 5,000 tons in 1915. Sixteen coal miners, 14 coal haulers, and three foremen were employed at this mine during 1915. The coal miners were paid at the rate of one dollar per ton of coal loaded on the cars in the mine; the cost of the coal used on the project was calculated at $7.105/ton delivered to the project site (U.S. Bureau of Reclamation, 1913-1916). A brief description of the coal mining operation in 1915 at the Lava Creek mine is given in the U.S. Bureau of Reclamation (1913-1916) report for 1915 (Chapter IV, p. 29-30):

> "In order to prevent encroachment on private property, it was necessary to establish a portion of the main entry on a steep grade against the direction of haul and branch off rooms about forty feet wide from that on practically a level grade. Coal cars were pushed by hand out of the rooms to the main entry and then hauled to the tipple by two burros working in tandem. By November 5 the incline had become so long that the burros could no longer handle the output of the mine and a twelve Horse Power hoisting engine was installed."

The report for 1916 (U.S. Bureau of Reclamation, 1913, 1916) describes the last mining that occurred in this mine and the deposition of the coal:

> "The fault that was encountered in December, 1914 was found to shut off progress in all directions and to obtain the required amount of coal, it was found necessary to go into the old workings and remove coal that had been previously abandoned on account of its inaccesibility. At the works, bins were constructed near where the coal was to be used and a portion of the coal deposited therein and the balance stored in discarded bunkhouses."

Locality (see Figure 1)	*Mine Name (Mine Method)*	*Years of Production*	*Production (tons)*	*Formation Produced From*
Pilgrim Creek	Pilgrim Creek (underground)	1910-1911	661	Mesaverde
Lava Creek	Lava Creek (underground)	1913-1916	9,447	Bacon Ridge Sandstone
5?	Buffalo (unknown)	1939	100	Bacon Ridge Sandstone
Lava Creek	Johnson (underground)	1939-1942	1,174	Bacon Ridge Sandstone
		TOTAL	11,382	

The Johnson and Buffalo coal mines were also located in the Lava Creek mine area. The exact location and mining method of the Buffalo mine is not known at the present time. The Johnson mine was an underground mine with two entries along the outcrop of a 2.2- to 2.5-feet-thick coal bed. The coal bed in the mine had a 20° dip to the northeast; the mine encountered two north-trending faults that displaced the coal from two to six feet and appeared to limit the extent of mining. The mining at these two mines was by private individuals and, evidently, the coal was sold for local consumption.

Two other underground coal mines in the Jackson Hole Coal Field have been reported: the Iter Gros Ventre mine at locality 4 (Figure 1), at which part of a total coal section 4.6 feet thick was mined (Love, et al., 1948) and the mine at Purdy Basin (locality 2, Figure 1), in the coal-bearing unit of the Wind River Formation. Coal production has not been documented for these two coal mines or any other coal prospects in the field.

The coal production from the Johnson and Buffalo mines is the only production in the coal field reported by the Wyoming State Inspector of Mines (annual, 1902-1981). An additional 10,138 tons of coal produced by the Bureau of Reclamation should be added to the total production for this field and for Teton County.

SUMMARY

Although much of the basic framework for the coal deposits in this field has been assembled, the continuity and correlation of coal beds has not been established, there is no adequate coal analysis for any coal bed in the field, and the coal resource estimate is not only over 30 years old, but it is also inadequate. Clearly, the Jackson Hole Coal Field remains one of the least known and least understood coal fields in Wyoming.

SELECTED REFERENCES

Berryhill, H. L., Jr., Brown, D. M., Brown, Andrew, and Taylor, D. A., 1950, Coal resources of Wyoming: U.S. Geological Survey Circular 81, 78 p.

Fogg, P. M., compiler, 1915, A history of the Minidoka Project, Idaho, to 1912 inclusive, Vol. I and II: U.S. Bureau of Reclamation Minidoka Project History (Archives).

Glass, G. B., 1981, Coal deposits of Wyoming, *in* Wyoming Geological Association Guidebook, 32nd Annual Field Conference, 1981: p. 181-236.

Hausel, W. D., Glass, G. B., Lageson, D. R., Ver Ploeg, A. J., and De Bruin, R. H., 1979, Wyoming mines and minerals map: Geological Survey of Wyoming Map MS-5, scale 1:500,000, colored.

Love, J. D., 1947, The Tertiary stratigraphy and its bearing on oil and gas possibilities in the Jackson Hole area, northwestern Wyoming: U.S. Geological Survey Oil and Gas Investigations Preliminary Chart 27.

Love, J. D., 1956, Geologic map of Teton County, Wyoming, *in* Wyoming Geological Association Guidebook, 11th Annual Field Conference, 1956: In pocket.

Love, J. D., 1973a, Preliminary geologic map of the Two Ocean Lake quadrangle, Teton County, Wyoming: U.S. Geological Survey Open-file Report, 1973.

Love, J. D., 1973b, Harebell Formation (Upper Cretaceous) and Pinyon Conglomerate (uppermost Cretaceous and Paleocene), northwestern Wyoming: U.S. Geological Survey Professional Paper 734-A, 54 p.

Love, J. D., 1974a, Geologic map of the south half of the Huckleberry Mountain quadrangle, Teton County, Wyoming: U.S. Geological Survey Open-file Report 74-54.

Love. J. D., 1974b, Geologic map of the south half of the Mount Hancock quadrangle, Teton County, Wyoming: U.S. Geological Survey Open-file Report 74-127.

Love, J. D., 1975a, Geologic map of the Whetstone Mountain quadrangle, Teton County, Wyoming: U.S. Geological Survey Open-file Report 75-138.

Love, J. D., 1975b, Geologic map of the Gravel Mountain quadrangle, Teton County, Wyoming: U.S. Geological Survey Open-file Report 75-220.

Love, J. D., Antweiler, J. C., and Williams, F. E., 1975, Mineral resources of the Teton Corridor, Teton County, Wyoming: U.S. Geological Survey Bulletin 1397-A, 51 p.

Love, J. D., and Christiansen, A. C., 1980, Preliminary correlation of stratigraphic units based on 1°x2° geologic quadrangle maps of Wyoming, *in* Wyoming Geological Association Guidebook, 31st Annual Field Conference, 1980: p. 279-282.

Love, J. D., Duncan, D. C., Bergquist, H. R., and Hose, R. K., 1948, Stratigraphic sections of Jurassic and Cretaceous rocks in the Jackson Hole area, northwestern Wyoming: Wyoming Geological Survey Bulletin 40, 48 p.

Love, J. D., Hose, R. K., Weitx, J. L., Duncan, D. C., and Bergquist, H. R., 1951, Stratigraphic sections of Cretaceous rocks in northeastern Teton County, Wyoming: U.S. Geological Survey Oil and Gas Investigations Chart OC-43.

Love, J. D., and Keefer, W. R., 1975, Geology of sedimentary rocks in southern Yellowstone National Park, Wyoming: U.S. Geological Survey Professional Paper 729-D, 60 p.

Love, J. D., Keefer, W. R., Duncan, D. C., Bergquist, H. R., and Hose, R. K., 1951, Geologic map of the Spread Creek-Gros Ventre River area, Teton County, Wyoming: U.S. Geological Survey Oil and Gas Investigations Map OM-118.

Love, J. D., Reed, J. C., Jr., Christiansen, R. L., and Stacy, J. R., 1973, Geologic block diagram and tectonic history of the Teton region, Wyoming-Idaho: U.S. Geological Survey Miscellaneous Geologic Investigations Map I-730.

Love, J. D., Weitz, J. L., and Hose, R. K., 1955, Geologic map of Wyoming: U.S. Geologic Survey Map, scale 1:500,000, colored.

Rohrer, W. L., 1969, Geologic map of the Sheridan Pass quadrangle, Fremont and Teton Counties, Wyoming: U.S. Geological Survey Open-file Map, 1969.

U.S. Bureau of Reclamation, 1913-1916, Annual project history, 1913-1916, Jackson Lake Enlargement Project: Bureau of Reclamation Minidoka Project History (Archives).

Wyoming State Inspector of Mines, (annual) 1902 through 1980, Annual report of the State Inspector of Mines of Wyoming: Rock Springs, Wyoming.

Thirty-Third Annual Field Conference — 1982
Wyoming Geological Association Guidebook

GEOCHEMICAL PROVENANCE OF ANOMALOUS METAL CONCENTRATIONS IN STREAM SEDIMENTS IN THE ASHTON 1:250,000 QUADRANGLE, IDAHO/MONTANA/WYOMING

BY SPENCER S. SHANNON JR.[1]

ABSTRACT

Stream-sediment samples from 1500 sites in the Ashton, Idaho/Montana/Wyoming 1:250,000 quadrangle were analyzed for 45 elements. Almost all samples containing anomalous concentrations (exceeding one standard deviation above the mean value of any element) were derived from drainage basins underlain by Quaternary rhyolite, Tertiary andesite or Precambrian gneiss and schist. Aluminum, barium, calcium, cobalt, iron, nickel, magnesium, scandium, sodium, strontium, and vanadium have an andesite provenance. Most anomalous manganese, europium, hafnium, and zirconium values were derived from Precambrian rocks. All other anomalous elemental concentrations are related to Quaternary rhyolite.

This study demonstrates that multielemental stream-sediment analyses can be used to infer the provenance of stream sediments. Such data are available for many parts of the country as a result of the National Uranium Resource Evaluation. This study suggests that stream-sediment samples collected in the Rocky Mountains can be used either as pathfinders or as direct indicators to select targets for mineral exploration for a host of metals.

INTRODUCTION

A geochemical stream-sediment survey for 45 elements was conducted by the Los Alamos National Laboratory in the Ashton 1:250,000 quadrangle in eastern Idaho, southwestern Montana, and northwestern Wyoming (Fig. 1). This study was undertaken as part of the nationwide Hydrogeochemical and Stream Sediment Reconnaissance (HSSR) Program, which was a major component of the U.S. Department of Energy's National Uranium Resource Evaluation (NURE).

A total of 1500 sediment samples was collected. These were obtained from 1433 stream channels, 52 springs, and 15 ponds. The samples from Montana were collected in August 1976, whereas those in Idaho and Wyoming were collected in the summer of 1978. All samples were collected at a nominal reconnaissance density of one sample location per 10 km^2. Standardized field, analytical, and data-base management procedures were used during both sampling programs.

All 67,500 analyses for elements in these samples (except for 3,000 As and Zr analyses) are tabulated in the Ashton quadrangle report (Shannon, 1980). These data plus the As and Zr data are also available on magnetic tape from GJOIS Project, UCC-ND Computer Applications Department, 4500 North Building, Oak Ridge National Laboratory, P.O. Box X, Oak Ridge, TN 37830. Additional HSSR data are available for small areas along the northern and eastern margins of Yellowstone National park in the Bozeman quadrangle (Bolivar, 1979; 1980a) and the Cody quadrangle (Bolivar, 1980b).

Field Procedures

Enough fine-grained, organic-rich, water-transported sediment to yield a composite sample of 25 g after processing (as indicated below) was taken from beneath the water level at three adjacent spots at each spring or stream loca-

[1]Los Alamos National Laboratory, Los Alamos, New Mexico 87545

tion. The sediment was put into a new, clean, and originally sealed, rip-top polyethylene bag which was then properly double labeled for delivery to the contractor's drying facility. After drying at <100°C, each sample was sieved through a 100-mesh stainless steel sieve. The minus 100-mesh fraction was put into a prewashed, 25-ml polyethylene vial which was then appropriately double-labeled and sealed for shipment to Los Alamos. All field observations and measurements were recorded on field data forms which were included with the samples.

The field observations represented the best judgment of the field sampler at a location and included general descriptions of the local bedrock, sediment, water, vegetation, terrain, weather, and possible contaminants. Because these observations were subjective and made quickly in the field, they should be held subordinate to formally documented information such as that provided by published topographic or geologic maps.

Each contractor was supplied field maps with the desired sample types and general locations symbolically premarked at Los Alamos. The field maps were generally 1:24,000 or 1:62,500 USGS topographic maps. As each location was sampled, a unique sample location number, preprinted on transparent adhesive labels that was provided with each identically numbered field data form, was placed on top of the precisely marked point representing the sample site on the field map. When a desired sample as specified could not be obtained, an identical or alternate sample type was picked, and the new sample type and location were marked on the field map and properly labeled as above. The latitude and longitude of each location was then computed by the sampling contractor within 48 h of taking

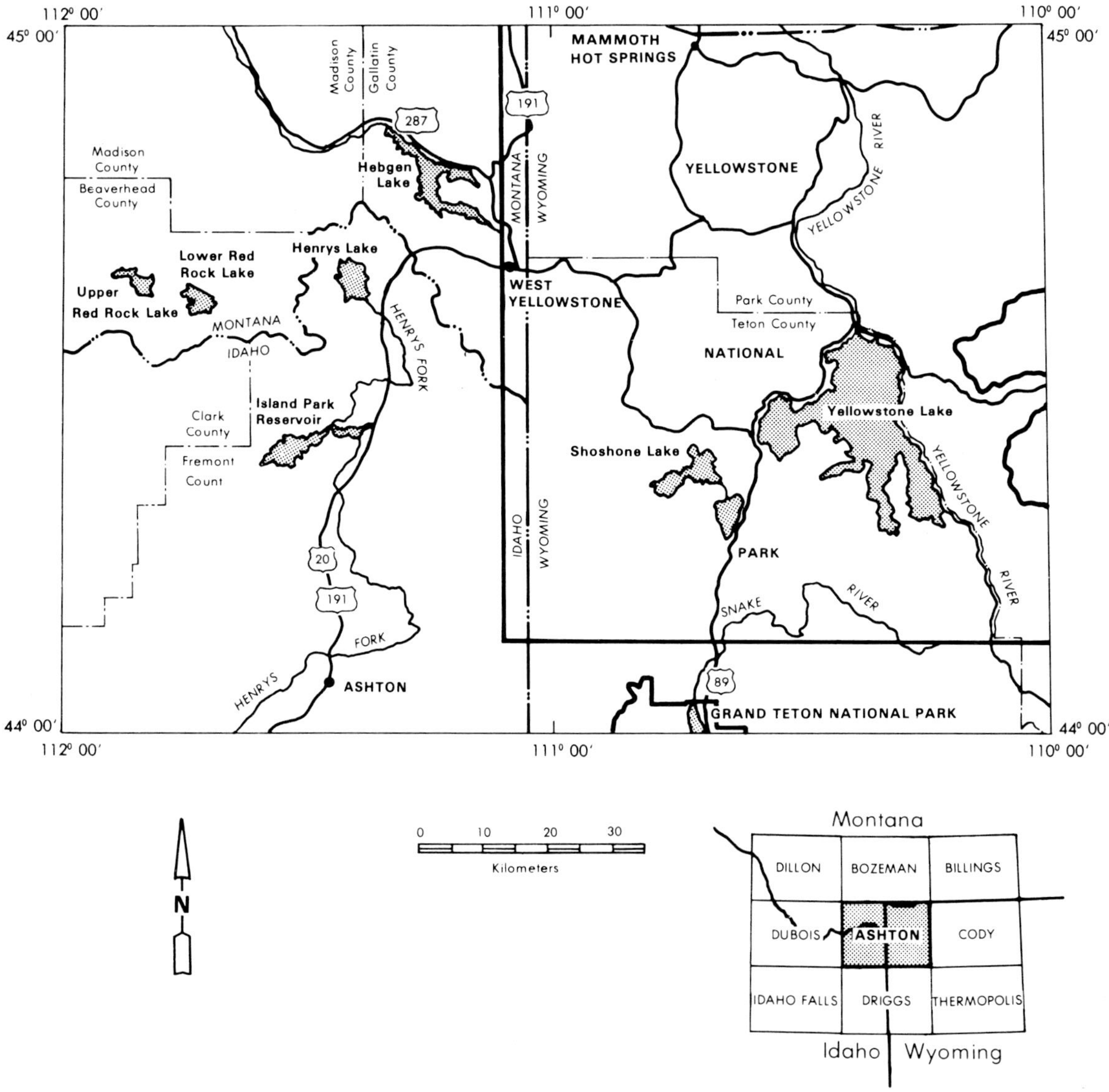

Fig. 1: Location map of the Ashton 1:250,000 quadrangle, Idaho/Montana/Wyoming.

each sample. Every location was later checked at Los Alamos by digitizing the sample locations on each map and comparing them to those computed in the field. The latitudes and longitudes were corrected if the field-computed locations were displaced by more than 300 m from the locations marked on the field maps. A final visual check of sample locations was made by overlying computer-produced location plots on the field maps used. The computer program for generating the Universal Transverse Mercator map projection overlys was described by Cheadle (1977).

Analytical Procedures

All sediment samples were analyzed for total uranium by delayed-neutron counting (DNC) (Minor et al., 1981). A split of each sample was transferred to a clean 4-ml rabbit, weighed, and its weight recorded along with the appropriate location number. These rabbits were then loaded into a 50-sample transfer clip. The reactor pneumatic-transfer system and background radiation levels were checked, and standards were run for calibration. The transfer clip was installed and the samples were cycled through the system. The uranium concentration was automatically measured, converted to ppm, and entered into the data base. The lower limit of detection of this method was 0.01 ppm uranium, far below the range or uranium concentrations in natural sediment samples. Above the 1 ppm level, the uranium values in sediment measured by DNC at Los Alamos had a one-sigma error of less than 4%. The specially designed delayed-neutron detectors, were described by Balestrini et al. (1976).

Immediately upon completion of the uranium analysis of sediment samples by DNC, the same 4-ml sediment splits were entered into the neutron-activation analysis (NAA) sequence. The concentrations of 33 additional elements were determined by this procedure. These elements were Al, As, Au, Ba, Ca, Ce, Cl, Co, Cr, Cs, Dy, Eu, Fe, Hf, K, La, Lu, Mg, Mn, Na, Rb, Sb, Sc, Sm, Sr, Ta, Tb, Th, Ti, V, Yb, Zn, and Zr. The full DNC/NAA timing sequence used at Los Alamos for each sediment sample was: 20-s irradiation, 10-s delay, 30-s DNC analysis, 20-min delay, 500-sγ-ray count for short-lived radionuclides, 96-s re-irradiation, 14-day delay, and finally a 1000-sγ-ray count for long-lived radionuclides. The γ-ray counting was done by lead-shielded Ge (Li) detectors; the 4096-channel γ-ray data were recorded and subsequently analyzed for each individual element by computer. The analytical data for each sample was automatically printed along with the associated statistical errors. The lower detection limits for the various elements as reflected by the "less than" values (denoted by a minus sign in front of a concentration) in the data listings were the values for the individual elements at which the statistical counting error approached 50%. Typical lower detection limits for the elements determined by NAA were reported in Nunes and Weaver (1978); however, the actual detection limit for an element depended upon the composition of the sample, and this limit may have been higher or lower than the typical value. At concentration values one order of magnitude above the lower detection limits, the relative errors were generally less than 10%.

A computer-controlled, energy-dispersive x-ray flourescence system was used to determine Ag, Bi, Cd, Cu, Nb, Ni, Pb, Sn, and W in sediments. The system consisted of an automatic 20-position sample changer, a lithium-drifted silicon detector, a pulsed molybdenum transmission-target x-ray tube, a multi-channel analyzer, and a minicomputer. The sediment samples were prepared for analysis by grinding 6 g of each minus 100-mesh sample to a minus 325-mesh powder. A computer program positioned the 6-g samples in the x-ray beam, unfolded overlapping peaks, determined peak intensities for each element, and calculated the ratio of the intensity of each peak to that of the molybdenum KαCompton peak. Concentrations of each element were then calculated using equations obtained by analyzing prepared standards. Detection limits were: 5 ppm for Ag, Bi, Cd, and Pb; 10 ppm for Cu and Sn; 15 ppm for Ni and W; and 20 ppm for Nb. When an analysis resulted in an elemental concentration that was below the detection limit, a minus sign preceding the value of the detection limit for that element was inserted in the data listings. The relative standard deviation was 10% or less at the 100-ppm level and 20% or less at the 20-ppm level. Details of the method and equipment used were described by Hansel and Martell (1977).

A 5-mg portion of the minus 325-mesh sample that had already been analyzed by x-ray flourescence was mixed with 10 mg of a buffer consisting of one part graphite and one part SiO_2. The sample/buffer mixture was placed into a graphite electrode that was used as the anode of a dc arc having a short circuit current of 6A for 10 s, then 17A for 50 s. Photo-multiplier tubes in a direct-reading spectrograph were used to measure the second order 313.0-nm line of Be, the first order 670.7- and 610.3-nm lines of Li, the background spectra near these lines, and the 327.6-nm line of V. the 670.7-nm Li line was used for Li concentrations up to 10 ppm and the 610.4-nm line of Li was used for concentrations above 10 ppm. The V line was used to correct the Be value when V is present. The signals from the photomultiplier tubes were read by a digital voltmeter and were processed by a desk-top calculator. The results were simultaneously printed on paper and written on cassette tape for later transmission to a computer data file. The elemental concentrations of Be and Li were determined from the spectra, based on the results of previously run calibration standards. The lower detection limit for both elements was 1 ppm. When an an analysis resulted in an elemental concentration that was below the detection limit, a minus sign preceding the value of the detection limit for that element was inserted in the data listings. Precision at the lower detection limit was ≈50% for both elements and improved to ≈25% at one order of magnitude above the lower limit.

Geology

A detailed discussion of the geology of the Ashton 1:250,000 quadrangle is beyond the scope of this paper. A brief summary of the lithology of the quadrangle is generalized geologic maps by the U.S. Geological Survey (1972) and IntraSearch (1979).

The Precambrian basement comprises dolomite, gneiss, schist, tremolite marble, and quartzite.

All Paleozoic and Mesozic rocks are sedimentary. Cambrian strata include limestone, shale, and sandstone. Dolomite is the dominant Ordovician and Devonian rock type, and limestone is the chief Mississippian rock type. Pennsylvanian rocks are principally sandstone and siltstone whereas Permian rocks are chiefly phosphatic shale and sandstone.

Triassic rocks consist of siltstone, shale, and sandstone. Jurassic strata include sandstone, shale, and limestone. Cretaceous rocks are predominantly sandstone and shale.

Almost all Tertiary rocks are volcanic. The chief exceptions are Paleocene sandstone and shale. Eocene rocks are chiefly andesite, basalt, trachyte and rhyodacite. Pliocene-Pleistocene volcanics comprise rhyolite flows and ash-flow tuffs. Pleistocene volcanics are comprised almost wholly of rhyolite flows and ash-flow tuffs plus local basalt. Pleistocene sediments are glacial deposits. Recent sediments include alluvium, fan and terrace deposits, landslides, and windblown sand. There are also local hot-spring and hydrothermal-explosion deposits.

Methodology

For each of the 45 elements, computer maps and tabulations were made showing the distribution of sites from which the elemental content of the sediment samples exceeded one (σ), two (2σ), or three (3σ) positive standard deviations above the arithmetic mean for that element. Values less than the detection limit or of zero were excluded from consideration.

Each individual element map of the 1° by 2° Ashton quadrangle (U.S. Geological Survey, 1972) was subdivided into 32 15-minute quadrangles numbered consecutively from the upper left corner as shown in Figure 2. The number of sample values for sites within each quadrangle between σ and 2σ, between 2σ and 3σ, and exceeding 3σ were counted separately and recorded. Each quadrangle was scored separately for each element by adding the number of values between σ and 2σ to twice the number of values between 2σ and 3σ and to thrice the number of values exceeding 3σ. Based on these scores, the four quadrangles containing the highest sums were taken to represent the peak geographic concentrations of sites in the Ashton quadrangle from which samples containing the most anomalous values of that element were obtained. Hence the one-eighth of the total area of the Ashton quadrangle, from which the largest number of anomalous values and the most strongly anomalous values for that element were reported, were identified as most likely to have a decipherable geologic relation to the provenance of the stream-sediment samples. This process was repeated, in turn, for each of the other 44 elements.

For each 15-minute quadrangle, the areas, which include clusters of samples containing anomalous values of one or more elements, were plotted on the Ashton topographic map as shown in Figure 2. The geology of these areas was noted by transferring information from the 1:250,000 geologic overlay (IntraSearch, 1979) included in the Ashton NURE report (Shannon, 1980). For areas within Yellowstone National Park, the 1:125,000 geologic map was used (U.S. Geological Survey, 1972). The resulting information was tabulated (see Appendix) to determine the association of anomalous values of each metal with specific rock types throughout the Ashton quadrangle. Each major rock type in the areas upstream from sites, from which samples having anomalous concentrations of various elements were obtained, was used to match such elements with the lithology of their probable derivation. Some samples may be recycled from terrace or glacial deposits, which ultimately had their provenance in other rock types cited.

Conclusions

The following groups of elements have a correlation of 1.00 (four out of four) between the four 15-minute quadrangles within which are located sites from which samples containing the most anomalous values of those elements were obtained. The quadrangle numbers are shown in parentheses below and also in Figure 2.

1. Zr and Hf (2, 3, 15, 23).
2. Sr, Sc, Na, Fe, Co, Ca, and Al (8, 16, 24, 32).
3. As and Sb (13, 14, 15, 21).
4. Ni and Mg (16, 24, 31, 32).
5. Yb, Sm, and La (23, 28, 29, 30).

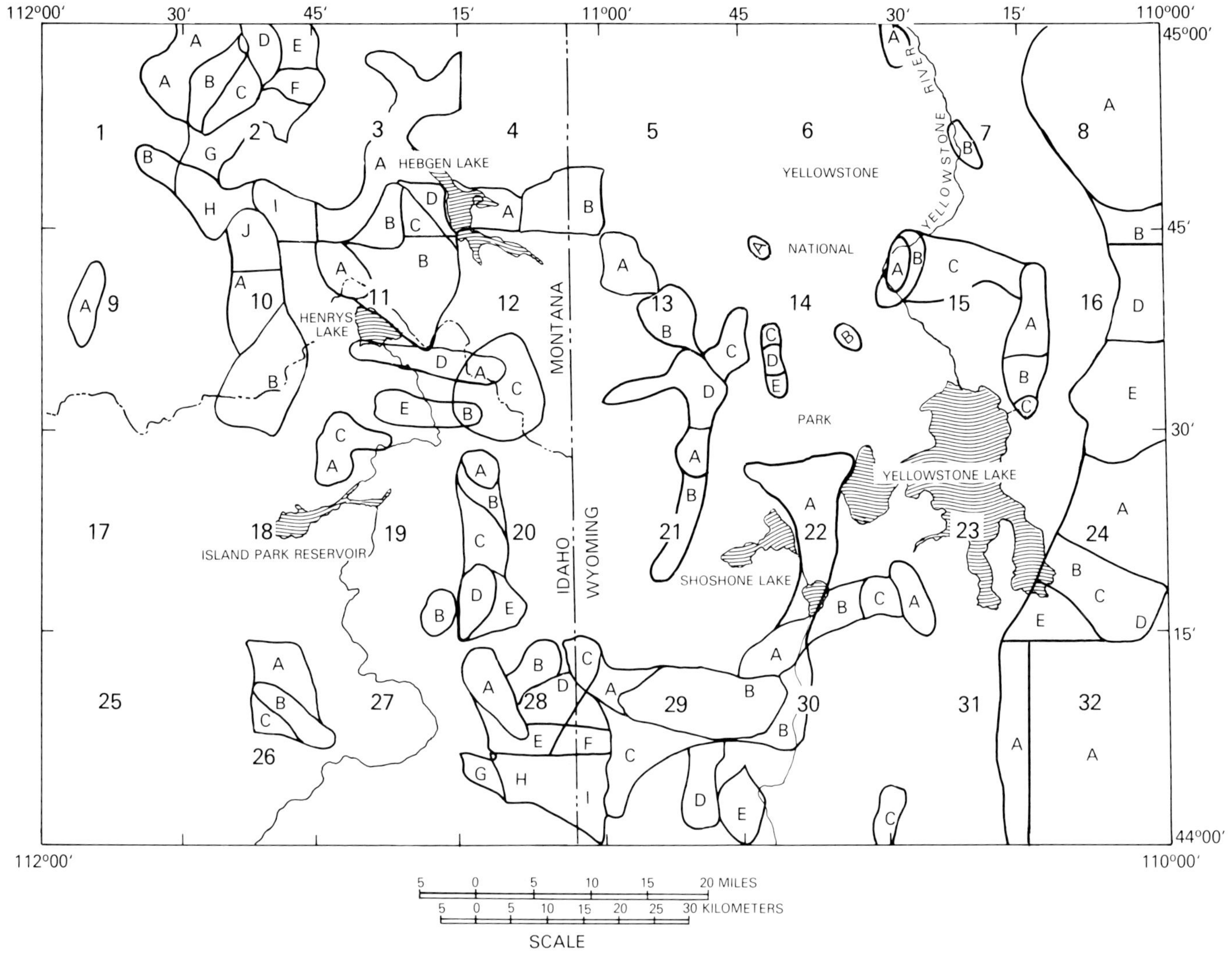

Fig. 2: Anomalous concentrations of elements in stream-sediment samples in the Ashton 1:250,000 quadrangle, Idaho/Montana/Wyoming.

6. Tb, Lu, and Dy (20, 28, 29, 30).

The following groups of elements have a correlation of 0.75 (three out of four) as explained above.

1. Zr and Eu (2, 3, 15).
2. Hf and Eu (2, 3, 15).
3. Mn and Cu (10, 11, 16).
4. Sn and Rb (20, 29, 30).
5. V and Sr, Sc, Na, Fe, Co, Ca, and Al (8, 16, 24).
6. W and As and Sb (14, 15, 21).
7. Ce and Th (4, 23, 29).
8. Ba, Ni, and Mg (16, 24, 32).
9. Ba and Sr, Sc, Na, Fe, Co, Ca, and Al (16, 24, 32).
10. Yb, Sm, and La versus Tb, Lu, and Dy (28, 29, 30).
11. U and Tb, Lu, and Dy (20, 28, 29).
12. K and Tb, Lu, and Dy (20, 28, 29).
13. Pb, Zn, and Be (13, 20, 29).

Most elements that occur in anomalous concentrations in stream-sediment samples have their provenance in Quaternary rhyolitic rocks. These include U. Th, seven rare earths, ore metals (Bi, Pb, Sb, Sn, W, and Zn), rare metals (Be, Cs, Li and Ta), and also K. Rb, As, and Cl. Those within Yellowstone National Park are associated chiefly with the Lava Creek and Huckleberry Ridge tuffs of the Yellowstone Group. In addition, 15 flows of the Central Plateau Member of the Plateau Rhyolite as well as the Mount Jackson and Lewis Creek rhyolites are also probable sources of anomalous metals in stream-sediment samples.

Stream-sediment samples containing anomalous amounts of major elemental constituents of andesite (Al, Ca, Fe, Mg, Na, and Ti) as well as Ba, Co, Ni, Sc, Sr, and V have a provenance in the andesitic rocks of the Absaroka Volcanic Supergroup of Eocene age along the eastern margin of the Ashton quadrangle. The chief source rocks are the Lamar River, Langford, Two Ocean and Wiggins Formations. Other probably contributors are the Wapiti and Mount Wallace Formations and the Trout Creek trachyandesite.

Most stream sediments containing anomalous amounts of Mn, Eu, Hf, and Zr have a provenance in Precambrian gneiss and schist. The Tanner manganese prospect (Mitchell et al., 1981) is upstream from one such anomalous-sample site.

For a few anomalous elements, the provenance is mixed. Cd and Cr seem to be equally associated with Tertiary andesite and Quaternary rhyolitic rocks. Some Eu, Hf, and Zr are also related to Quaternary rhyolite. Cu seems to be equally derived from Tertiary volcanics and Quaternary rhyolite while Nb has its probable provenance divided between Quaternary rhyolite and Quaternary basalt.

It is significant that erosion from other rock units in the Ashton quadrangle has not formed sediments containing appreciably anomalous values of any of the 45 elements in this study.

NURE stream-sediment samples originally were collected, analyzed, and the resultant data published to facilitate exploration for additional uranium resources in the United States. However, the interpretation of this array of multi-elemental analytical data may be far more important either directly in seeking new resources of many of the elements cited or indirectly as pathfinders to resources of other metals not readily analyzed in the NURE program.

Obviously, there is no intent to encourage mineral exploration in Yellowstone National Park. However, other uses of these geochemical data include the solution of purely scientific geologic problems related to the provenance of stream sediments within and adjoining the park. For example, the ultimate derivation of multicyclic sediments can be deduced. Also this is another tool for determining a more detailed Pleistocene geologic history of the park.

Acknowledgements

C. W. Myers encouraged the research pertaining to the interpretation of the Ashton geochemical data. A.C. Waters critically reviewed the manuscript.

REFERENCES CITED

Balestrini, S.J., Balagna, J.P., and Menlove, H.O., 1976, Two specialized delayed-neutron detector designs for assays of fissionable elements in water and sediment samples, Nuclear Instrum. and Methods, v. 136, pp. 521-524.

Bolivar, S.L., 1979, Uranium hydrogeochemical and stream sediment reconaissance of the Bozeman NTMS quadrangle, Montana, GJBX-8(79), US DOE, Grand Junction, CO, 98 p.

Bolivar, S.L., 1980a, Uranium hydrogeochemical and stream sediment reconnaissance of the Bozeman NTMS quadrangle, Montana, including concentrations of forth-two additional elements, GJBX-235(80), US DOE, Grand Junction, CO, 131 p.

Bolivar, S.L., 1980b, Uranium hydrogeochemical and stream sediment reconnaissance data release for the Cody NTMS quadrangle, Wyoming, including concentrations of forth-two additional elements, GJBX-233(80), US DOE Grand Junction, CO, 147 p.

Cheadle, J. III, 1977, Computer program for universal Transverse Mercator map projection, CJBX-54(77), US ERDA, Grand Junction, CO, 11 p.

Hansel, J.M. and Martell, C.J., 1977, Automated energy-dispersive x-ray determination of trace elements in streams, GJBX-52(77), US ERDA, Grand Junction, CO, 8 p.

IntraSearch, 1979, Geology of the Ashton quadrangle, map (1:250,000 scale) prepared for Bendix Field Eng. Corp. according to BFEC specifications 1125B, Grand Junction, CO.

Minor, M.M., Hensley, W.K., Denton, M.M., and Garcia, S.R., 1981, An automated activation analysis system, Modern Trends in Activation analysis 1981, Intl. Conf., U. of Toronto, Toronto, Canada.

Mitchell, V.E., Strowd, W.B., Hustedde, G.S., and Bennett, R.E., 1981, Mines and prospects of the Ashton quadrangle, Idaho, Idaho Bureau of Mines and Geology, 4 p.

Nunes, H.P. and Weaver, T.A., 1978, Hydrogeochemical and Stream Sediment Reconnaissance of the National Uranium Resource Evaluation program in the Rocky Mountain states of New Mexico, Colorado, Wyoming, and Montana and the state of Alaska, July-September 1977, GJBX-27(78), US DOE, Grand Junction, CO, 14 p.

Shannon, S.S. Jr., 1980, Uranium hydrogeochemical and stream sediment reconnaissance data release for the Ashton NTMS quadrangle, Idaho/Montana/Wyoming, including concentrations of 42 additional elements, GJBX 189(80), US DOE, Grand Junction, CO, 168 p.

U.S. Geological Survey, 1955 (limited revision 1972), Ashton, Idaho/Montana/Wyoming, topographic map NL 12-11 (1:250,000 scale), Denver, CO.

U.S. Geological Survey, 1972, Geologic map of Yellowstone National park, miscellaneous geologic investigations, map I-711 (1:125,000) scale), Denver, CO.

APPENDIX

ASSOCIATION OF ANOMALOUS CONCENTRATIONS OF ELEMENTS IN STREAM-SEDIMENT SAMPLES WITH GEOGRAPHIC AREAS AND LOCAL GEOLOGY, ORDERED BY QUADRANGLE

Quadrangle	Geologic Area	Elements	Geology
1	A. Horse Creek	Titanium	Upper Paleozoic strata (ss, ls, dol); fault parallels Horse Creek.
	B. Elk River	Titanium	Upper Paleozoic starata (ss, ls, dol).
2	A. Granite Mountain-Bobcat Creek	Titanium	Tertiary basalt and other volcanics, undivided Precambrian rocks and Upper Paleozoic strata (ss, ls, dol).
	B. Middle Horse Creek	Europium, hafnium, titanium, and zirconium	Undivided Precambrian rocks, Tertiary volcanics.
	C. Standard Creek-lower Horse Creek	Titanium	Undivided precambrian rocks, Tertiary volcanics.
	D. Moose Creek	Chromium and titanium	Quaternary terrace deposits and alluvium.
	E. Finger Mountain	Chromium	Precambrian gneiss, schist, and Quaternary terrace deposits.
	F. Squaw Creek	Chromium, europium, hafnium, tatanium, and zirconium	Precambrian gneiss and schist and Quaternary glacial and terrace deposits.
	G. Freezeout Mountain-Papoose Creek	Europium, hafnium, and zirconium	Precambrian gneiss and schist, Tertiary volcanics, and Quaternary glacial and terrace deposits.
	H. Elk River-Tepee Creek	Europium, hafnium, titanium, and zirconium	Precambrian gneiss and schist, Tertiary volcanics.
	I. Cliff Lake	Europium, hafnium, and zirconium	Tertiary volcanics, Quaternary terrace deposits.
	J. Meridian Creek	Copper	Tertiary volcanics.
3	A. Horn Creek-Earthquake Lake-Beaver Creek	Europium, hafnium, and zirconium	Precambrian gneiss and schist, Mesozoic strata (ss, sts, sh,) Tertiary volcanics, Paleozoic strata (ss, ls, sh), Quarternary alluvium and terace deposits.
	B. Sheep Point	Europium, hafnium, manganese, and zirconium	Precambrian gneiss and schist, Cambrian limestone and sandstone.
	C. Coffin Creek	Maganese	Precambrian gneiss and schist, Cambrian limestone and sandstone.
	D. Watkins Creek	Bismuth, europium, hafnium, magnese, and zirconium	Precambrian gneiss and schist.
4	A. Rainbow Point - Grayling Creek	Bismuth	Quaternary rhyolite (Huckleberry Ridge Tuff), Precambrian gneiss and schist, Tertiary basalt (Mt. Wallace Formation, Quaternary sediments.
	B. Gneiss Creek	Bismuth, cerium, lithium, and thorium	Quaternary rhyolite (Huckleberry Ridge and Lava Creek tuffs), Tertiary conglomerate (Sepulcher Formation), Precambrian gneiss and schist, Quaternary alluvium and glacial deposits.
5	None		
6	None		
7	A. Hellroaring Creek	Copper	Tertiary rhyodacite and andesite (Sepulcher Formation), Precambrian gneiss and schist, Quaternary rhyolite (Huckleberry Ridge Tuff) Quaternary sediments.
	B. Agate Creek	Copper	Quaternary rhyolite (Lava Creek tuff), Tertiary andesite (Lamar River Formation), Quaternary basalt (Junction Butte Basalt), Quaternary sediments.
8	A. Soda Butte Creek-Cache Creek	Alluminum, barium, calcium, cobalt, iron, magnesium, scandium, strontium, and vanadium	Tertiary andesite (Lamar River Formation), Tertiary andesite and basalt (Wapiti and Mount Wallace Formations), Tertiary trachyandesite (Trout Creek trachyandesite), Paleozoic strata (ls, dol), Quaternary rhyolite (Lava Creek Tuff), Quaternary sediments.

APPENDIX (Cont.)

Quadrangle	Geologic Area	Elements	Geology
	B. Chalfee Creek	Aluminum, barium, calcium, cobalt, iron, magnesium, nickel, scandium, strontium, and vanadium	Tertiary andesite and basalt (Wapiti Formation), Tertiary andesite (Lamar River Formation), Quaternary rhyolite (Lava Creek Tuff), Quaternary sediments.
9	A. Red Rock River	Chlorine	Quaternary alluvium, Tertiary volcanics, Permian phosphatic shale.
10	A. Elk Mountain-Hidden Lake	Copper	Tertiary volcanics, Precambrian gneiss and schist, Precambrian dolomite, Pliocene-Pleistocene rhyolite, Tertiary basalt, Quaternary fan deposits and alluvium.
	B. Nemesis Mountain	Maganese (includes Tanner Mtn. prospect (Mitchell et al., 1981)	Pliocene-Pleistocene rhyolite, Tertiary volcanics, Precambrian gneiss and schist, Precambrian dolomite, Precambrian quartzite, Paleozoic strata (ss, ls, dol, phosphatic sh), Triassic strata (ss, sh).
11	A. Raynolds Pass	Maganese and zinc	Precambrian gneiss and schist, Precambrian dolomite, Precambrian quartzite, Quaternary fan deposits and alluvium. Madison Range fault.
	B. Targhee Peak	Manganese	Precambrian gneiss and schist, Pliocene-Pleistocene rhyolite, Quaternary rhyolite, Precambrian Quaternary fan deposits and alluvium.
	C. Yale Creek	Chromium and copper	Quaternary rhyolite, Tertiary volcanic, Quaternary basalt, Quaternary alluvium.
	D. Jesse Creek	Chromium	Quaternary rhyolite, Tertiary volcanics, Quaternary sediments.
	E. Henrys Fork north of Macks Inn	Niobium	Quaternary rhyolite, Quaternary alluvium.
12	A. Reas Pass Siding	Chromium, rubidium, and uranium	Quaternary rhyolite.
	B. Thirsty Creek	Niobium and rubidium	Quaternary rhyolite, Quaternary alluvim and fan deposits.
	C. Reas Pass	Rubidium and uranium	Quaternary rhyolite, Quaternary alluvium and fan deposits.
13	A. Cougar Creek	Lithium and lutetium	Quaternary rhyolite (Lava Creek Tuff) Quaternary basalt (Madison River Basalt), Quaternary fan deposits.
	B. Madison Junction	Bismuth	Quaternary rhyolite (Lava Creek Tuff, West Yellowstone and Nez Perce Creek flows, Mount Jackson Rhyolite), Quaternary sediments.
	C. Nez Perce Creek	Bismuth, lead, and zinc	Quaternary rhyolite (Nez Perce Creek flow), Quaternary sediment.
	D. Lower Geyser Basin-Sentenel Creek	Antimony, beryllium, bismuth, chlorine, cesium, and lithium	Quaternary rhyolite (West Yellowstone and Elephant Back flows, Upper Basin Member), Quaternary hot spring deposits, Quaternary sediments.
14	West of Norris Junction	Antimony, arsenic, and tungsten	Quaternary rhyolite (Lava Creek Tuff), Quaternary hot-spring deposits, Quaternary sediments.
	B. Alum Creek	Antimony	Quaternary rhyolite (Hayden Valley flow), Quaternary sediments.
	C. Magpie Creek	Beryllium	Quaternary rhyolite (Nez Perce Creek and Mary Lakes flows), Quaternary sediments.
	D. Nez Perce Creek	Antimony, arsenic, beryllium, and tungsten	Quaternary rhyolite (Mary Lake and Elephant Back flows), Quaternary sediments.
	E. Juniper Creek	Beryllium	Quaternary rhyolite (Elephant Back flow), Quaternary sediments.
15	A. Artist Point	Antimony, arsenic, barium, cadmium, europium, hafnium, tungsten, and zirconium	Quaternary rhyolite (Upper Basin Member), Quaternary Sediments.

APPENDIX (cont.)

Quadrangle	Geologic Area	Elements	Geology
	B. Artist Point (outer ring)	Barium, europium, hafnium, zirconium	Quaternary rhyolite (Upper Basin Member, Hayden Valley and Solfatara Plateau flows), Quaternary sediments.
	C. Sour Creek-upper Broad Creek	Europium, hafnium, zirconium	Quaternary rhyolite (Upper Basin Member, Hayden Valley and Sofatara Plateau flows), Quaternary sediments.
16	A. Upper Pelican Creek	Aluminum, barium, europium, hafnium, and zirconium	Quaternary rhyolite (Lava Creek Tuff), Tertiary andesite (Langford Formation), Tertiary trachyandesite (Trout Creek Trachyandesite), and Quaternary sediments.
	B. Sedge Creek	Aluminum and barium deposits, Quaternary alluvium.	Quaternary rhyolite (Lava Creek Tuff), Quaternary hot-springs
	C. Tributaries to Turbid Lake	Silver	Quaternary rhyolite (Lava Creek Tuff), Quaternary hot-spring deposits, Quaternary sediments. Two northeasterly faults downdropped to the northwest.
	D. Cold Creek	Aluminum, barium, calcium, cobalt, iron, magnesium, scandium, sodium, strontium, titanium, and vanadium	Tertiary andesite and basalt (Wapiti and Mount Wallace Formations); Tertiary trachyandesite (Trout Creek Trachyandesite), Tertiary andesite (Landford Formation), Quaternary rhyolite (Lava Creek Tuff), Quaternary sediments.
	E. Jones Creek	Aluminum, barium, calcium, cobalt, copper, iron, magnesium, manganese, nickel, scandium, sodium, strontium, titanium, and vanadium.	Tertiary andesite (Langford Formation), Quternary sediments.
17	None		
18	None		
19	A. Hotel Creek	Chromium and copper	Quaternary rhyolite, Quaternary alluvium.
	B. Pineview Siding	Niobium	Quaternary basalt, Quaternary rhyolite. Northeasterly fault.
20	A. Moose Creek	Beryllium, lutetium, potassium, terbium, and uranium	Quaternary rhyolite, Quaternary alluvium.
	B. Chick Creek	Lutetium, potassium, terbium, and uranium	Quaternary rhyolite.
	C. Split Creek	Dysprosium, lead, lutetium, potassium, rubidium, uranium, and zinc.	Quaternary rhyolite.
	D. Partridge Creek	Beryllium, dysprosium, lutetium, potassium, rubidium, and terbium	Quaternary rhyolite.
	E. South Fork Partridge Creek	Lutetium, potassium, and terbium	Quaternary rhyolite.
21	A. Upper Geyser Basin	Antimony, beryllium, cesium, chlorine, and tantalum	Quaternary rhyolite (Mallard Lake Member, West Yellowstone flow, Upper Basin Member), Quaternary hot-spring deposits, Quaternary alluvium.
	B. Firehole River	Antimony, arsenic, chlorine, cesium, and tungsten	Quaternary rhyolite (Summit Lake, Grants Pass, and Spring Creek flows), Quaternary sediments.
22	A. Lewis Lake-Chickadee Lake	Lead and potassium	Quaternary rhyolite (Shoshone Lake Tuff Member, Aster Creek, Dry Creek, Elephant Back, and Pitchstone Plateau flows), Quaternary sediments.
	B. Aster Creek	Lithium	Quaternary rhyolite (Huckleberry Ridge Tuff, Aster Creek flow), Quaternary sediments.

APPENDIX (Cont.)

Quadrangle	Geologic Area	Elements	Geology
	C. Heart Lake Geyser Basin	Cesium and lithium	Quaternary rhyolite (Huckleberry Ridge Tuff, Aster Creek flow), Quaternary hot-spring deposits, Quaternary sediments.
23	A. Beaver Creek	Bismuth, cerium, hafnium, lanthanum, samarium, thorium, ytterbium, and zirconium	Quaternary rhyolite (Lava Creek Tuff, Mount Jackson Rhyolite, Aster Creek flow), Quaternary sediments.
24	A. Mount Langford	Aluminum, barium, calcium, cobalt, iron, magnesium, nickel, scandium, sodium, strontium, and vanadium	Tertiary andesite (Landgord Formation, Two Ocean Formation), Tertiary trachyandesite (Trout Creek Trachyandesite), Quaternary sediments.
	B. Beaverdam Creek	Aluminum, barium, cadmium, calcium, cobalt, iron, magnesium, nickel, scandium, sodium, strontium, and vanadium	Tertiary andesite (Langford Formation, Two Ocean Formation), Tertiary trachyandesite (Trout Creek Trachyandesite), Quaternary sediments.
	C. Trappers Creek	Aluminum, barium, cadmium, calcium, cobalt, iron, magnesium, nickel, scandium, sodium, strontium, titanium, and vanadium	Tertiary andesite (Langford Formation, Two Ocean Formation), Quaternary sediments.
	D. Mountain Creek	Aluminum, barium, cadmium, calcium, chromium, cobalt, iron, magnesium, nickel, scandium, sodium, strontium, titanium, and vanadium	Tertiary andesite (Langford Formation, Two Ocean Formation), Quaternary sediments.
	E. Trail Creek	Aluminum, barium, cadmium, calcium, cobalt, iron, magnesium, nickel, scandium, sodium, strontium, and vanadium	Andesite (Quaternary rhyolite) (Langford Formation), Quaternary sediments.
25	None		
26	A. Blue Creek	Bismuth	Quaternary rhyolite, Quaternary alluvium.
	B. Rattlesnake Creek	Bismuth and chlorine	Quaternary rhyolite, Quaternary alluvum.
	C. Spring Creek	Bismuth	Quaternary rhyolite, Quaternary basalt, Quaternary alluvium.
27	None		
28	A. Moose Creek-Snow Creek	Dysprosium, lanthanum, lithium, lutetium, niobium, potassium, samarium, tantalum, thorium, and uranium	Quaternary rhyolite, Quaternary basalt.
	B. Upper Snow Creek	Dysprosium, lanthanum, lithium, lutetium, potassium, tantalum, thorium, and uranium.	Quaternary rhyolite (Buffalo Lake flow), Quaternary glacial deposits.
	C. Lake Wyodaho	Berryllium, dysprosium, europium, lanthanum, lithium, lutetium, samarium, tantalum, thorium, tin, uranium, and ytterbium	Quaternary rhyolite (Summit Lake flow), Quaternary basalt (Falls River Basalt), Quaternary glacial deposits, Quaternary sediments.
	D. Horseshoe Lake	Lithium, potassium, and uranium	Quaternary rhyolite (Lava Creek Tuff), Quaternary basalt (Falls River Basalt), Quaternary glacial deposits, Quaternary sediments.
	E. Upper Rock Creek	Lithium, niobium, potassium, and uranium	Quaternary basalt (Falls River Basalt), Quaternary rhyolite.

APPENDIX (Cont.)

Quadrangle	Geologic Area	Elements	Geology
	F. ID-WY-YNP common point	Niobium	Quaternary basalt (Falls River Basalt), Quaternary rhyolite (Lava Creek Tuff), Quaternary sediments.
	G. Porcupine Creek	Titanium	Quaternary basalt, Quaternary rhyolite, Quaternary glacial deposits, Quaternary alluvium.
	H. Falls River Ridge	Lithium and titanium	Quaternary rhyolite, Quaternary basalt, Quaternary alluvium.
	I. Boone Creek	Titanium	Quaternary rhyolite, Quaternary rhyolite, Quaternary glacial deposits, Quaternary alluvium.
29	A. Lower Bechler River	Beryllium, dysprosium, eurpium, lanthanum, lutetium, samarium, terbium, tin, and ytterbium	Quaternary rhyolite (Pitchstone Plateau flow), Quaternary basalt (Falls River Basalt), Quaternary sediments.
	B. Mountain Ash Creek-Beula Lake	Beryllium, cerium, dysprosium, europium, lanthanum, lead, lutetium, potassium, rubidium, samarium, tantalum, terbium, thorium, tin, uranium, and ytterbium	Quaternary rhyolite (Pitchstone Plateau flow, Lewis Canyon rhyolite, Lava Creek Tuff, Huckleberry Ridge Tuff), Tertiary dacite, Quaternary basalt (Falls River basalt), Triassic strata (sts, sh), Tertiary andesite (Absaroka Volcanic Supergroup), Low Paleozoic strata (ls, dol, sh), Quaternary sediments. Numerous northerly faults in eastern part.
	C. Loon Lake	Cerium, dysprosium, europium, lanthanum, lutetium, potassium, rubidium samarium, terbium, and thorium	Quaternary rhyolite (Huckleberry Ridge Tuff, Lava Creek Tuff), Quaternary basalt (Falls River Basalt), Tertiary dacite, Upper Paleozoic strata (ss, ls, dol), Quaternary glacial deposits, Quaternary alluvium.
	D. Survey Peak	Europium	Quaternary rhyolite, Paleozoic strata (ls, dol, ss, sh), Tertiary basalt, undivided Precambrian rocks, Triassic shale, Quaternary glacial deposits.
	E. Lower Berry Creek	Chromium	Mesozoic strata (sh, ss), Quaternary rhyolite, Upper Paleozoic strata (ls, ss), Quaternary glacial deposits, Quaternary alluvium.
30	A. Upper Lewis River	Cesium, dysprosium, lanthanum, lead, lithium, lutetium, samarium, tantalum, and terbium	Quaternary rhyolite (Pitchstone Plateau flow).
	B. Lower Lewis River	Cerium, dysprosium, lanthanum, lithium, lutetium, samarium, and uranium	Quaternary rhyolite (Lava Creek Tuff, Lewis Canyon Rhyolite), Carboniferous strata (ls, ss), Triassic strata (sts, sh, ss), Quaternary sediments.
	C. Bobcat Ridge (partly Quadrangle #31)	Gold	Cretaceous claystones, Paleocene strata, Quaternary in rhyolite.
31	A. Eastern border	Same as all of Quadrangle #32.	
32	A. Entire quadrangle	Aluminum, barium, calcium, cobalt, iron, magnesium, nickel, scandium, sodium, and strontium	Tertiary andesite (Wiggins Formation, Two Ocean Formation), Paleozoic strata (ls, dol, ss), Quaternary basalt (Basalt of Mariposa Lake), Quaternary sediments, Quaternary glacial deposits.

Thirty-Third Annual Field Conference — 1982
Wyoming Geological Association Guidebook

ROAD LOG SECTION

COMPILED & EDITED BY NANCY M. DOELGER

The majority of the road logs included in this section of the guidebook were prepared for other meetings. They are included to provide as much coverage of the roadside geology of Yellowstone Park and nearby areas of geological interest as possible. One suggested reference for areas of the park not covered by this guidebook is:

Parsons, W.H., 1978, Field Guide to the
Middle Rockies and Yellowstone,
Kendall/Hunt Publications.

This book is for sale in the park. Another potential reference is *Roadside Geology of the Northern Rockies* by Donald Hyndman (1976, Mountain Press).

The road logs in this section are, in general, printed as they were originally written. Minor adjustments were made in cases where there had been changes in major highway routes or highway numbers, and the direction of travel was reversed on two logs (Tower Junction to Soda Butte Creek Valley and Madison Junction to Old Faithful). The geological portions of the logs were not changed.

There may be some overlap between the logs included in this guidebook and logs for the field trips associated with this conference. In the case of such overlap, the road logs prepared for field trips associated with this conference can be expected to be more current.

The logs included in this section of the guidebook are:

WEST YELLOWSTONE TO HEBGEN LAKE AREA

VIA U.S. HIGHWAY 191 AND U.S. HIGHWAY 287

W.J. McMANNIS AND D.L. KIEL

(reprinted by permission of the Billings Geological Society)

INTRODUCTION

This log leads along and across the differentially tilted faultblock of Hebgen Reservoir to Quake Lake where the upstream effects of the Madison Canyon landslide are displayed. This road log was prepared approximately one year after the earthquake which caused the slide for the 1960 Billings Geological Society Field Conference. Some of the post-earthquake phenomena may no longer be present as described in this log.

Mileage Cumulative	Mileage Increment	DESCRIPTIVE LOG
0.0	0.0	West Yellowstone, junction of Highway 191 and Highway 20. Leave town northward on U.S. 191, crossing the flat surface of the former lake beds of the West Yellowstone Basin, the so-called West Yellowstone Flats.
1.3	1.3	Side road to the left; leads along the south shore of Hebgen Lake.
2.6	1.3	The road descends from the upper surface of the West Yellowstone Flats onto the surface of the Madison River floodplain, losing about 70 feet in elevation.
3.5	0.9	Road to Madison Arm of Hebgen Lake on the left.
3.8	0.3	Highway bridge across the Madison River. The new asphalt approach-ramps on either end of the bridge were put in following the earthquake of 1959 because the fill approaches were so badly shaken away from the bridge itself.
4.1	0.3	The road climbs from the floodplain of the Madison River up to the surface of the West Yellowstone Flats. The roadcuts on either side contain sands with a very high proportion of black obsidian grains along with quartz.
4.5	0.4	Mountain peaks directly ahead are in the Madison Range.
5.1	0.6	Side road at the left leads to Horse Butte, a peninsula in Hebgen Lake. The prominent peak at 3 o'clock is Mount Holmes in Yellowstone Park. The road is still on the surface of the West Yellowstone lake beds. These sediments are very porous, and the drainage on them is poorly developed; however, they store quite a bit of moisture and support a dense stand of Lodgepole pine.
7.8	2.7	Descending from the surface of the West Yellowstone Flats to the floodplain of Cougar Creek.
8.0	0.2	Highway crosses Cougar Creek; just beyond the creek the highway was badly damaged by the earthquake of August 1959.
8.2	0.2	A small rise between the valleys of Cougar Creek and Duck Creek.
8.5	0.3	The Highway is cut by the new Red Canyon fault scarp (downthrown on the southwest, visible directly ahead and at 3 o'clock. The displacement on the scarp is approximately 10 to 13 feet between the Highway and the buildings of the Highway Maintenance Station at 3 o'clock. This fault scarp follows along a previously developed scarp several feet high, to which no particular significance had been attached prior to the earthquake of August 1959. This particular scarp dies out a few hundred feet to the northwest, while an en echelon scarp begins at the Duck Creek "Y" and continues to the northwest.
8.7	0.2	Duck Creek "Y" (now U.S. Highway 287). TURN LEFT. Small scarp crosses Highway 191 about 100 feet to the right. This scarp is en echelon with the one just crossed, southwest of Duck Creek bridge.
8.8	0.1	Low scarp at right lies a short distance west of a small graben which borders field just visible on upthrown side of near scarp. This graben is best viewed from Highway 191.
9.3	0.5	Road to ranch on right was offset some 15 feet by the prominent fault scarp. Mountains at 10 o'clock are underlain by Precambrian metamorphics and capped by Paleozoic strata. Cenozoic lava mantles this flank of the mountains.

9.9	0.6	TURN RIGHT, road to Blarneystone Ranch. Outcrops at mouth of Greyling Creek dead ahead are late Cenozoic rhyolites.
10.4	0.5	Greyling Creek.
10.5	0.1	SUGGESTED STOP 1. Blarneystone Ranch. Fault scarp developed across ranch buildings, offsetting one of cinder block structures in a peculiar fashion. The downthrown side of the building remained intact, whereas the upthrown part collapsed so that the roof of the latter is now lower than the roof of the former.
11.1	0.6	Return to main Highway. TURN RIGHT.
11.6	0.5	Sag in Highway developed during the earthquake of August 17, 1959.
12.3	0.7	Greyling Creek bridge. This bridge was badly shaken up during the earthquake, and fill on either side was loosened necessitating repairs. Just beyond bridge small reverse faults of about one-foot displacement developed in asphalt of the Highway. Large rhyolite boulders on right side of road.
12.7	0.4	Note series of small scarplets in slope at 1 o'clock; Paleozoic rocks in ridge beyond.
13.3	0.6	Old stream gravels in road cut.
13.5	0.2	Large spring emerges below road at left; pre-earthquake.
13.75	0.25	Note sine-curve buckling of white fence on left, a result of northwest-southeast shortening during the earthquake. Also note offset in white fence on right ahead. Highway centerline (now covered by new asphalt) was similarly offset.
14.1	0.35	Landslide topography at 3 o'clock.
14.3	0.2	Glimpse of Red Canyon scarp at 1:30 o'clock.
14.9	0.6	TURN RIGHT. Road to Red Canyon. Paleozoics cap mountain at 10 o'clock; remnant of rhyolite amongst trees at 9:30 o'clock. Prong of mountain extending toward lake is underlain by Precambrian metamorphics, which are overlain by Cambrian rocks.
15.1	0.2	Cattle guard. Paleozoic carbonates in ridge ahead lie on the east side of the Red Canyon scarp.
15.6	0.5	Landslide debris at 3 o'clock.
15.8	0.2	Larger, older landslide mass at 2 o'clock.
16.0	0.2	Cattle guard, stay on main road ahead.
16.05	0.05	Chimney on main cabin on right slightly disturbed by the earthquake.
16.8	0.75	The Red Canyon scarp skirts along the base of the Paleozoic outcrops high on the ridge at 3 o'clock.
17.2	0.4	Red Canyon Creek. Paleozoic limestones in roadside at left ahead.
17.6	0.4	Red Kootenai debris in road cut at left.
17.8	0.2	SUGGESTED STOP 2. Red Canyon. Walk about ¼ mile upstream to right to scarp. Displacement on scarp where it crosses the creek is about 20 feet. Falls produced by scarp are already being degraded by stream action. This fault has been active many times and has produced a sizeable stratigraphic displacement. This had been mapped prior to August 17, 1959, and the recent movement pinpointed the fault location.
20.6	2.8	Return to main Highway. TURN RIGHT.
21.6	1.0	The Narrows. From here to northwest along Hebgen Reservoir this side of the shore area was inundated by a differential tilting of the catchment basin.
22.0	0.4	Precambrian metamorphics in road cuts on right.
22.5	0.5	Middle Cambrian Flathead sandstone blocks on right lie close to the present outcrop of same.
22.6	0.1	Flathead sandstone outcrop on right.
22.8	0.2	Cambrian limestone at 1 o'clock.
22.95	0.15	Cabins on left were badly dislocated during the earthquake.
23.35	0.4	Rhyolite debris on shoulder of road cut to right.
24.1	0.75	TURN LEFT.
24.3	0.3	SUGGESTED STOP 3. Johnson Creek sand spouts. Several rather large "blowouts" of water occurred in this area as a consequence of settling of unconsolidated materials and squeezing out of water during and after the earthquake.
24.4	0.1	TURN LEFT on to main Highway.
24.6	0.2	Hebgen Lake Lodge. First view of Hebgen fault scarp at 3 o'clock on either side of group of cabins near mouth of Johnson Creek.
25.0	0.4	Note that cabins at 3 o'clock suffered very minor damage in spite of proximity to fault.

25.4	0.4	Dead willows on left were covered by reservoir waters after the earthquake.
25.9	0.5	SUGGESTED STOP 4. PARK ON FAR RIGHT SHOULDER OF ROAD. First major landslide along this side of Hebgen Reservoir. A large segment of old Highway 287 was removed by this slide. Shoo-fly road around slide to right was put in the day after the earthquake to facilitate rescue of people stranded down river.
26.1	0.2	On shoo-fly road. Slide directly below on left.
26.3	0.2	Back on old Highway 287.
26.5	0.2	Glimpse of Hebgen scarp at 2 to 4 o'clock.
27.3	0.8	This part of Highway was under water after the earthquake.
27.55	0.25	Kirkwood Ranch Motel.
27.8	0.25	Shoo-fly road around another large landslide area.
28.2	0.4	Good view of Hebgen scarp at 2 o'clock.
28.5	0.3	This bank of reservoir has moved perceptibly toward the lake.
28.9	0.4	SUGGESTED STOP 5. PARK ON FAR RIGHT SHOULDER OF ROAD. We are directly above the former position of Hilgard Lodge which slid into the lake during the earthquake, then floated about ¼ mile into an alcove formed by another large landslide.
29.0	0.1	Shoo-fly road around large landslide. This is the nearest large landslide to Hebgen Dam.
29.3	0.3	Slide area is directly below on the left.
29.9	0.6	SUGGESTED STOP 6. Hebgen Dam. PARK ON FAR RIGHT SHOULDER OF ROAD. The dam settled some 9 feet during the earthquake, and the earth fill on both sides of the concrete core slumped badly, more so on this side than on the other, so that the concrete core protruded above the downstream fill as much as 4 feet. The spillway was seriously cracked, the water ripping up large segments of it and carrying them down the spillway. The seiches, or waves, created by the readjustment of the water to a new basin shape topped the dam four times. Many of the more mobile survivors of the quake and slide-effects congregated here until rescue was effected.
30.6	0.7	TURN RIGHT.
30.65	0.05	SUGGESTED STOP 7. Cabin Creek Campground. The Hebgen scarp bisected the old campground, separating garbage cans from their fireplaces. Maximum displacement was about 15 feet, but this dies to nearly zero on the ridge to the north. A few minor horsetails extended across into Beaver Creek to the north and die there. The scarp waterfalls are being degraded rapidly by stream action. Note the well developed graben at toe of main scarp. Many huge boulders thundered down the mountain sides in this and nearby areas, but miraculously no one was killed here. A one-holer appears to have suffered a tremendous explosion.
30.7	0.05	TURN RIGHT on main Highway.
30.75	0.05	Cabin Creek.
31.4	0.65	Ridges to right and across river to left are Precambrian metamorphics. Many fresh rock-fall scars are noticeable.
31.5	0.1	Refuge Point. Many of survivors from the quake and slide gathered or were moved here by rescue workers.
31.8	0.3	Side road to Beaver Creek on right. Stay on Highway.
32.3	0.5	Abutments to Beaver Creek bridge were badly shaken, and small sand spouts formed on floodplain during the earthquake.
33.2	0.9	Dead trees mark high-water level of Quake Lake which extended almost to Cabin Creek.
33.45	0.25	SUGGESTED STOP 8. Quake Lake. Present water level of the lake is fairly stable. However, the level dropped rather rapidly from its high-level mark, shown by the dying limbs on the trees, when the water topped the Madison Canyon slide. The road log retraces the Highway back to Highway 191. For alternate return route, see note following road log.
50.8	17.35	Duck Creek "Y". TURN RIGHT. Proceed to West Yellowstone.
59.5	8.7	TURN RIGHT. West Yellowstone, junction of Highways 191 and 20.
60.2	0.7	Airport on left was utilized extensively in rescue operations following earthquake.
64.1	3.9	Madison Arm road to right. For last 3.9 miles road has been crossing West Yellowstone flats which are underlain by volcanics, Cenozoic sediments, and

		rhyolite-obsidian stream sands and gravels.
64.6	0.5	South Fork of Madison River.
67.1	2.5	TURN RIGHT. Road to south shore of Hebgen Reservoir. Lionhead ski area and Lionhead Mountain at 9 to 10 o'clock on Paleozoic rocks.
68.8	1.7	Basin Fire Guard Station.
69.7	0.9	Volcanics in slopes on left.
70.4	0.7	On right is former arm of Hebgen Reservoir which was drained as a result of differential tilt of the catchment basin during the earthquake.
70.7	0.3	Precambrian metamorphic knobs protrude through volcanics on both sides of the road.
70.75	0.05	TURN RIGHT. Lonesomehurst summer-homes area.
70.9	0.15	SUGGESTED STOP 9. Lonesomehurst summer-homes area on south arm of Hebgen Reservoir. Level of lake dropped some ten feet here during the August 17, 1959, earthquake. The water shifted rather rapidly and sloshed back and forth at least four times, topping dam that many times, and finally came to rest in its newly-shaped catchment basin. To the northeast the outcrops seen earlier along the opposite shore of Hebgen Reservoir are visible, and beyond on the skyline are the Paleozoic exposures forming Kirkwood Ridge along which lies the Red Canyon scarp. Far to the north are the peaks of the main part of the Madison Range. The beach sands here are composed of obsidian, rhyolite, and other heterogeneous rock types. RETURN TO MAIN ROAD, THENCE TO HIGHWAY 191, AND WEST YELLOWSTONE.
82.3	11.4	West Yellowstone, junction of Highways 191 and 20.

Note: State Highway 287, which is now U.S. Highway 287, was destroyed by the landslide beyond Stop 8. The highway has since been rebuilt, and it is now possible to drive along Quake Lake to the site of the Madison Canyon slide (stop 1 of the next road log), and to return to West Yellowstone via State Highway 87 and U.S. Highway 191-20. The next road log (West Yellowstone to Madison Canyon slide) may be followed in reverse for the return.

The condition of the secondary roads mentioned in this road log is not known.

WEST YELLOWSTONE TO MADISON CANYON SLIDE

VIA REYNOLDS PASS

via U.S. Highway 20-191, State Highway 87 and U.S. Highway 287

W.J. McMannis and D.L. Kiel

(Reprinted by permission of the Montana Geological Society)

INTRODUCTION

This log begins at West Yellowstone and proceeds over the lava-capped range to the west into the Henrys Lake Basin, and across the lava-choked Reynolds Pass to Missouri Flats at the upper end of the intramontane Madison Valley. From there the log leads to the Madison Canyon slide. The log was prepared approximately one year after the earthquake which caused the slide for the 1960 Billings Geological Society Field Conference. Some of the post-earthquake phenomena may no longer be present as described in the log.

Mileage Cumulative	Mileage Increment	DESCRIPTIVE LOG
0.0	0.0	West Yellowstone. Junction of Highways 191 and 20. Head out in direction of Idaho Falls on 191-20.
4.6	4.6	Madison Arm of Hebgen Reservoir road on right.
5.1	0.5	South Fork of Madison River.
5.8	0.7	Lion Mountain and Targhee Mountain dead ahead. State geologic map shows Madison limestone resting on Precambrian metamorphics in this range, as does the Idaho geologic map. The road log chairman doubts the accuracy of this relationship. For the last 5 miles you have been traveling across the very flat West Yellowstone Flats, an old river-cut surface underlain by rhyolite, rhyolite sands, and tuffaceous material.
6.5	0.7	Erosion surface above road at right capped by stream gravels consisting mainly of olivine-basalt cobbles and pebbles. Lion Mountain Ski Area dead ahead on open slopes near crest of main ridge.
7.9	1.4	Rhyolite in road cut on left and in low sage-covered hill at 3 o'clock across creek.
8.6	0.7	Rhyolite in road cut at right, large blocks of vesicular rhyolite to left, and also a few exotic blocks of basalt.
8.9	0.3	Glacial(?) boulders, including Paleozoic and Precambrian rocks, in flat area on either side of road.
9.3	0.4	Lion Mountain Ski Area road on right. More rhyolite.
9.9	0.6	Targhee Pass. Rhyolite floors pass. Elevation 7,079 feet. Entering Idaho.
10.3	0.4	Centennial Mountains dead ahead.
11.8	1.5	Road cut at left in Mississippian silty carbonates. To right is an old dissected stream terrace with gravels capping truncated carbonates.
12.4	0.6	More fossiliferous Mississippian carbonates in road cut on left. Paleozoic section in peaks at 4 to 5 o'clock.
13.4	1.0	Rhyolite boulders in talus pile on hill at 9 o'clock contain inclusions of scoriaceous basalt. Lava apparently caps ridges at 7 to 9 o'clock.
13.7	0.3	Shoulder at 3 o'clock cut on rhyolite and capped by stream gravels.
14.4	0.7	Junction of U.S. Highway 191-20 and State Highway 87. TURN RIGHT.
14.9	0.5	Henrys Lake at 9 o'clock, Centennial Mountains beyond. These mountains are comprised of Precambrian metamorphic block overlain by generally southward dipping Paleozoic and lower Mesozoic rocks. The latter are overlapped unconformably on the south by volcanics of the Snake River downwarp. At 11:30 o'clock are Precambrian metamorphics overlain by easterly dipping Paleozoic beds, modified fault facets form steep lower slopes. Henrys Lake Basin is floored by rhyolites equivalent to those capping ridges and filling floor of Targhee Pass. The latter are probably unfaulted.

15.7	0.8	Imbrication in gravels of road cut at right show current was to S. 20° W.
18.2	2.5	Henrys Lake Lodge at left. For last 2.5 miles you have been traveling along large alluvial fans at toe of old fault scarp.
20.0	1.8	Junction Centennial road (left) and State Highway 87 (right). Stay on 87. Note remnant of uplifted pediment surface at 1:30.
21.3	1.3	Cattle guard. Mountain slopes at 1 to 3 o'clock are Precambrian metamorphics. Note triangular facets on spur ridges. Recent fans are undisturbed by later fault movements. Hills at 10 o'clock are Precambrian also.
22.3	1.0	Driving on coarse alluvial fan debris. At 3 o'clock, a younger fan, or slide, is superimposed on older fan.
24.1	1.8	Low bench to left underlain by rhyolite, hills beyond are Precambrian.
24.5	0.4	Reynolds Pass; elevation 6,834 feet (Bench Mark 20 yards to left). Angular Precambrian debris mantles surface here. Outcrops to left and right are intensely crumpled and sheared Precambrian carbonates, similar to those forming prominent ledges at the Madison Canyon slide. Rhyolite underlies rounded hills and shoulders within pass itself, suggesting that entire pass may be floored by this volcanic ejecta. Teton Range and flat, lava-capped Yellowstone Plateau at 6 o'clock, Madison Range at 1 o'clock.
25.4	0.9	Cattle guard. Junction with Antelope Basin road on left. Continue ahead. Rounded knobs are underlain by rhyolite, high ridges to NW are Precambrian.
26.85	1.45	Junction, Horn Creek road on left. Stay RIGHT. The Horn Creek road goes down a canyon incised into rhyolite, which rests on Precambrian metamorphics. Hundreds of large boulders rolled into the road in the canyon on August 17 last year. This is also one way into Cliff and Wade lakes, where two people were killed by rolling boulders during the earthquake.
28.25	1.4	Precambrian metamorphics at left, strike N. 55° E., dip 60° NW. At 4 o'clock is a beautiful lateral moraine at mouth of canyon. Madison Range ahead.
29.55	1.3	At 3 o'clock is a well developed U-shaped valley with terminal moraine. Road on face to right of canyon goes up to an asbestos mine in the Precambrian.
30.05	0.5	Hill to left is Precambrian metamorphics, strike N. 60 E., dip 40° NW. Broad smooth surface ahead is called Missouri Flats and is coincident with the high erosional bench along the Madison Valley, probably the equivalent of the Cameron bench farther north.
32.05	2.0	A long-distance view of the Madison Canyon slide at 3 o'clock.
33.45	1.4	Junction of Highway 87 and Cliff Lake road. Stay on 87. Slopes to west are dip-controlled on rhyolite. Toward east these slopes are covered over by gravels of Missouri Flats, or are cut into by present streams.
33.75	0.3	At edge of Missouri Flats surface. This is the third prominent erosional level above the present Madison River, and probably corresponds to the high Cameron bench farther down river. View of big slide at mouth of Madison Canyon at 2 o'clock.
34.05	0.3	Leaving second erosional level above Madison River.
34.35	0.3	Leaving first erosional level (Ennis terrace?) above Madison River, dropping to present "floodplain".
35.05	0.7	Madison River bridge. DRIVE WITH CARE.
35.25	0.2	Rising from present "floodplain".
35.75	0.5	Junction State Highway 87 and U.S. Highway 287. Madison Canyon slide to RIGHT. Ennis to left. TURN RIGHT. Somewhat obscured recent scarplets along mountains on left. Small slump (August 17, 1959) at 11 o'clock.
37.75	2.0	Recent scarplet at heads of small fans at left. Reynolds Pass historical sign on right. Fine view at 2 o'clock of Missouri Flats terminating against triangular facets at base of mountains.
38.55	0.8	Forest Service maintained a barrier here to prevent automobile travel to slide area until visitor facilities could be constructed. A Bench Mark nearby subsided 6.8 feet relative to sea level during August 17 earthquake.
38.85	0.3	Outwash at right consists of slide debris built up since water topped the slide.

39.15	0.3	Recent undercutting of far bank of river on right.
39.25	0.1	Directly beneath us, river has undercut old Highway 287. Before water topped slide this portion of 287 was still intact. At 12:30 o'clock are dolomite cliffs adjacent to ones which collapsed on August 17, 1959.
39.45	0.2	Alluvial apron on right built of reworked slide debris.
39.65	0.2	Buried end of Highway 287, immediately after slide, was almost directly below this point.
39.95	0.3	High point of slide debris at 2 o'clock.
40.25	0.3	SUGGESTED STOP 1. Madison Canyon slide. Southwest across the canyon is a magnificent view of the headwall of the slide which broke off along foliation planes and cross fractures in mica schist. The steeply dipping foliation curves gradually to a more northwesterly trend and the ridge parallels same. Notice the loose material at south end and right center of upper slide. This debris is still moving appreciably down toward the river. Very little rock is falling from the headwall cliffs, although these cliffs dropped some 20 feet from their original position at the time of the slide along foliation planes on the back side of the ridge. On the back side of the ridge pre-slide talus from original crest has been beheaded by collapse of the top of the ridge. Edge of the slide at left is also controlled by foliation planes. At right edge of slide are dolomite and quartzite cliffs which originally extended across base of canyon wall. These massive rocks acted as a retaining wall until the night of August 17 when they gave way and the upper slopes came tumbling after. The slide measured approximately one-half mile across at the headwall, covered the old valley floor for about 4,400 feet, and extended a maximum of about a mile from headwall to upstream end. To the north, behind us, was the high point of slide debris, which was piled up to about 400 feet above original stream level. Note the size of some of the blocks in this northeast part of the slide, which is composed primarily of dolomite debris. The main mass of the slide from here to the headwall consisted of mica schist. Prior to Army Engineer activities and downcutting by the river this debris was 150-200 feet thick. The inner gorge is a combination of river and Army Engineers work.

WALK FROM HERE TO VANTAGE POINT NEAR QUAKE LAKE.

View of Quake Lake. Note old water level shown by dying timber. The water has dropped some 50 feet to present level, and originally backed up to near the mouth of Cabin Creek about 5 miles upstream. Prompt action by Army Engineers prevented water from backing up to toe of Hebgen dam. Note up-slide orientation of trees across inner gorge.

WALK TO LIP OF INNER GORGE.

Engineers cut and rip-rapped a spillway here and materially aided in reducing danger of sudden downcutting and flood in downstream areas. One huge block of folded schists still more of less intact on far side of gorge.

Return to West Yellowstone following same route or continue along Highway 287 and return to West Yellowstone via U.S. Highway 191. It was not possible to continue past stop 1 at the time the road log was written because the highway was destroyed by the landslide. The road has since been rebuilt, so it is now possible to drive along Quake and Hebgen Lakes, and return to West Yellowstone via Highway 191. The preceeding road log (West Yellowstone to Hebgen Lake Area) may be followed in reverse if this route is chosen.

The condition of the secondary roads mentioned in this road log is not known.

WEST YELLOWSTONE TO CANYON JUNCTION

DUNCAN FOLEY[1], DENNIS L. NIELSON[1], CLAYTON R. NICHOLS[2]

INTRODUCTION

This log is part of a road log prepared for a three day field trip taken by the Geothermal Resources Council in September of 1980. The log was adapted primarily from U.S. Geological Survey Geological and Surficial Geologic quadrangle mapping. The maps used in the log, and the portion of the road log each map pertains to are listed in Appendix A. Lithologic and outcrop descriptions were also drawn from these maps. They are included in Appendix B. The bibliography and appendix are located at the end of the road log section of the guidebook.

Potassium-Argon dates reported in this guide were calculated using pre-1977 decay constants. To compare with more recent dating, multiply the dates given herein by 1.025.

Glacial activity has been separated into Bull Lake and Pinedale times. Estimates for the ages of the ice are given in Table 1.

TABLE 1

Age estimates for Glacial Activity in Yellowstone (in thousands of years)

	(1971) Keefer	(1976) Richmond	(1979) Pierce
Pinedale	25-8	70-12	70-13
Bull Lake, late stade	125-45	98-85	160-130(?)
Bull Lake, early stade	125-45	120-110	160-130(?)

The W.G.A. would like to thank the authors for their permission to include this road log in the guidebook.

Mileage Dist-ance	Mileage Cumul-ative	
0.0	0.0	Beginning at the west entrance to the Park, drive across alluvial gravelly sand, deposited in the Pinedale-Bull Lake interglacial interval in the West Yellowstone Basin. The tremendous number of dead pine trees in this area is due to an infestation of the Mountain Pine Beetle, which kills the trees by cutting the circulation of sap. Many trees in the Island Park area to the south have been killed by the beetle, and the infestation is now moving northeast into Yellowstone National Park. Figure 1 illustrates the relations among the Quaternary volcanic units in the park. Although it is not comprehensive, the figure does illustrate some of the more important stratigraphic relations. More than 10,000 thermal features have been catalogued within Yellowstone Park. In addition to the heat phenomena associated with hot springs and geysers, the waters have a beneficial effect of increasing aquatic productivity. The Madison and Firehole rivers have some of the best trout fishing in the country. Many of the streams in the Park have normal productivity upstream from thermal features, but support a wider variety of aquatic life after the hot waters have mixed with the river waters. There are also a few sites where the opposite effect is true, thermal springs reduce stream productivity. Yellowstone is an area of general high elevation and high precipitation. This implies that, especially for areas to the west and south, it is an area of regional groundwater recharge.
2.2	2.2	The Madison River is visible on the north side of the highway. The river here flows along outwash sand and gravel from Pinedale glaciers. Pinedale outwash fans form the deposits away from the river.
1.9	4.1	The roadbed is over an outcrop of Madison River basalt.
1.9	6.0	The low ridge on the north side of the river is composed of Pinedale age till which is near the margin of Pinedale glaciation.
0.7	6.7	Approximate entrance to Madison River Canyon. The road east of here received ex-

[1]Earth Science Laboratory, University of Utah Research Institute, Salt Lake City, Utah.
[2]Department of Energy Division of Geothermal Energy, Idaho Falls, Idaho

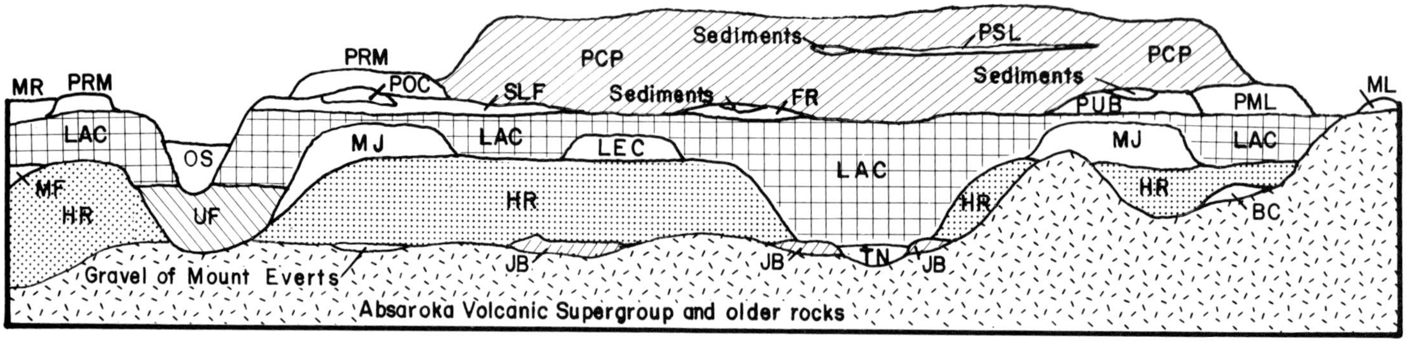

Second volcanic cycle

MF Mesa Falls Tuff (1.2)

First volcanic cycle

TN Sediments and basalts of The Narrows (1.6)
LEC Lewis Canyon Rhyolite
HR Huckleberry Ridge Tuff (2.0)
BC Rhyolite of Broad Creek
JB Junction Butte Basalt (2.0)

Third volcanic cycle

PRM Roaring Mountain Member (0.4-0.075(?))
PCP Central Plateau Member (02.-0.07)
PSL Shoshone Lake Tuff Member (0.18)
POC Obsidian Creek Member
PML Mallard Lake Member (0.6-0.53)
PUB Upper Basin Member (0.6-0.53)
OS Osprey Basalt
MR Madison River Basalt
SLF Swan Lake Flat Basalt
FR Falls River Basalt
ML Basalt of Mariposa Lake
LAC Lava Creek Tuff (0.6)
UF Undine Falls Basalt (0.7)
MJ Mount Jackson Rhyolite (0.79-0.64)

Figure 1: Diagrammatic relations among Quaternary volcanic units, Yellowstone National Park. Neither all stratigraphic units nor all possible stratigraphic relations are depicted. Numbers after units indicate approximate ages, in millions of years. Redrafted from Christiansen and Blank (1972), with modification of Christiansen and Blank (1974).

tensive damage from rockfalls and landslides during the August 1959 Hebgen Lake earthquake. The epicenter of the earthquake was 35 miles northwest of West Yellowstone; 26 people were killed near Hebgen and Earthquake lakes.

The rocks on the north side of the river are Member B of the Lava Creek Tuff (for lithologic description, see Appendix B). This tuff was emplaced immediately prior to the formation of the Yellowstone Caldera, and has been dated at 600,000 years (Christiansen and Blank, 1972). This was the third major tuff eruption in the history of caldera formation in this vicinity. The Huckleberry Ridge Tuff at 2 million years, and the Mesa Falls Tuff, at 1.2 million years, were the earlier eruptions.

The rocks on the south side of the river are flows of the Mount Jackson Rhyolite, which in this area has been separated into the Mt. Haynes flow and the Harlequin Lake flow. The Mt. Haynes flow forms the cliffs south of the river (for lithologic description, see Appendix B).

The Mt. Jackson Rhyolite has been dated at 790,000 and 640,000 years (Christiansen and Blank, 1972). Preservation of this unit outside the Yellowstone Caldera collapse area has helped define the caldera margin.

0.8 7.5 Bridge across the Madison River.

As we drive east, the Mount Jackson Rhyolite is exposed on the south side of the canyon. On the north slope, it is overlain by the upper part of Member A of the Lava Creek Tuff (for lithologic description, see Appendix B). Pinedale till, talus, rubble, and sand and gravel mantle the bedrock outcrops along the river.

2.8 10.3 Mt. Haynes is the prominent peak to the south.

1.9 12.2 About here you cross the northwestern fault scarp of the Yellowstone Caldera. As you travel east, the road generally follows the Caldera margin. Detailed seismic studies were undertaken by University of Utah personnel and others in July to determine whether any shallow magma may exist in the Caldera. Eaton and others (1975) concluded on the basis of gravity, aeromagnetic, earthquake epicenter and seismic wave patterns that: "a body composed at least partly of magma underlies the region of the rhyolite plateau, including the Tertiary volcanics immediately to its northeast."

Further geophysical evidence of the anomalous nature of the Yellowstone Caldera is presented in Iyer (1979), where it is suggested that a body of low seismic velocity material extends beneath the caldera to a depth of 200-250 km. Morgan and others (1977) conclude, on the basis of heat flow data, that a cooling batholith, perhaps with some re-supply of magma, exists beneath the caldera. Smith and others (1977) present seismic, aeromagnetic, and gravity data, and agree with the conclusion that a cooling batholith exists beneath the caldera. Smith and Cristiansen (1980) summarize all these data.

Heat supply for hydrothermal activity in the Park is undoubtedly related to the magmatic activity. Rhyolitic magmas typically occupy shallow chambers where their influence on the surface hydrothermal regime can be strong. Such magma, whether partially melted or solidified but still warm, could create the heat to drive Yellowstone hydrothermal systems. Basalts do not create large near-surface magma chambers; instead they come from much greater depth. This implies that basalts do not retain residual near-surface heat in the vast quantities found at Yellowstone (Fournier and others, 1976; Morgan and others, 1977), and a basalt eruption the same age as the Yellowstone rhyolite eruptions would not be able to maintain the hydrothermal systems.

The rhyolites exposed east of Mt. Haynes on the south side of the canyon are part of the Central Plateau rhyolite (see Appendix B for lithologic description).

The West Yellowstone flow, which is exposed from this point east to Madison Junction, has been dated at about 114,500 ± 7,300 years. North of the river, the Harlequin Lake flow of the Mt. Jackson Rhyolite is overlain by the lower part of Member A of the Lava Creek Tuff (see Appendix B for lithologic description).

Outcrops of Madison River Basalt, described as "gray to black containing abundant phenocrysts of plagioclase and olivine" have also been mapped on the north slope. Member B of the Lava Creek Tuff overlies member A on the hill north of the road.

1.7 13.9 Madison Junction, turn left (north). This is the confluence of the Gibbon and Firehole rivers, which form the Madison River. National Park Mountain, southwest of the intersection, commemorates the formation of the idea of a National Park to preserve Yellowstone's features. Cornelius Hedges, a journalist travelling with the Washburn expedition of 1870, is credited with suggesting, while camped at the site of the present campground, that the Yellowstone area should be set aside and preserved. The act setting aside the Park passed Congress on March 1, 1872.

0.8 14.7 Terrace Springs Complex. Bath Springs and Terrace Springs form the major outlets. Terrace Spring is 62°C, and has a flow of about 1220 gpm. The apparent boiling of this spring is due to CO_2 gas. Terrace Springs is one of four sites in the park where silica and travertine have both been deposited (Marler, 1973, p. 599-600). Fournier and Truesdell (1974) used Terrace Spring as a demonstration of mixing of hot and cold waters without enthalpy loss from the hot water component. The hills south of the river are composed of the Nez Perce Flow of the Central Plateau Rhyolite. This flow has been dated at about 160,000 years.

0.7 15.4 The Gibbon River flows along a bottom of humic alluvium, with Pinedale fan gravels, kame deposits, till and rubble forming the higher slopes.

0.3 15.7 The prominent cliff north of the road is composed of the lower and upper parts of member A of the Lava Creek Tuff. As you proceed east, you will be driving across the upper part of member A.

3.3 19.0 Gibbon Falls, which flows across the upper part of Member A of the Lava Creek Tuff, is east of the highway.

0.4 19.4 The small outcrop on the west side of the road has been mapped as part of the Lava Creek Tuff.

1.3 20.7 Bridge across the Gibbon River.

0.6 21.3 As you proceed north through Gibbon Canyon, the rocks on the west side of the river are member A of the Lava Creek Tuff, and the rocks on the east side of the river are the Gibbon River flow of the Plateau Rhyolites. The Gibbon River Flow has a reported K-Ar age of about 87,000 years; it erupted from a vent along the caldera rim, and flowed over pre-existing Nez Perce flows. The Nez Perce flows may be locally preserved on the west side of the canyon. They ponded a lake that may have extended as far north as the Norris Geyser Basin.

0.3 21.6 Bridge across the Gibbon River.

0.4 22.0 Sinter terraces on the east side of the river.

0.4 22.4 Beryl Springs. A stream vent blows behind the main springs. A large zone of alteration may be observed around the spring. Siliceous sinter deposits continue along the river to the north. These sinter deposits imply a base reservoir temperature above 180°C. The caldera margin is inferred through geologic mapping to be along or just east of the Gibbon River here.

As you travel through the Park, the stress placed on vegetation by thermal activity may be easily noted. White (1978), while including other possible factors such as depth to water table, concluded that normal tree growth occurs in areas of less than 200 HFU (heat flow units, 10^{-6} cal/cm^2/sec) conductive heat flows, stunted growth of trees occurs in areas of 200-400 HFU, and no growth occurs in areas with greater than 500 HFU.

0.4 22.8 Bridge.

0.6 23.4 Enter Gibbon Meadow, a popular place to observe elk. The surface deposits in the meadow are primarily diatomaceous silt. Surface thermal activity is most evident along the margins of the basin. The rocks surrounding the north, west, and south margins of the basin are members A and B of the Lava Creek Tuff. Outcrops of rhyolite adjacent to the east side of the road at the south end of the basin have been tentatively correlated with the Nez Perce Creek flow of the Central Plateau Member.

Paintpot Hill is the prominent hill on the southeast side of the basin. It is part of the Obsidian Creek Member of the Plateau Rhyolite (for lithologic description, see Appendix B). The Gibbon Hill and Geyser Creek domes occur east of the basin. The domes were probably emplaced shortly after the extrusion of the Lava Creek Tuff and the collapse of the caldera at 600,000 years ago.

1.4 24.8 Enter the small canyon associated with the Gibbon River Rapids. Bedrock is member A of the Lava Creek Tuff. Chocolate Pots hot springs are along both sides of the river. One small spring issues from gravel fill used as a road base.

0.7 25.5 Enter Elk Park. Bedrock here is covered by humic alluvium along the rivers, with diatomaceous silt and siliceous sinter deposits comprising the rest of the basin. The hills around the basin are composed of Member B of the Lava Creek Tuff. A few small hot springs are presently active in the park.

1.1 26.6 You are now driving along the south edge of the Norris Geyser Basin.

1.1 27.7 Intersection; turn right (east) toward Canyon. From here until the first suggested stop, the road traverses Pinedale till and sand deposits. The sand is either stream or outwash and stream deposits. Humic alluvium is restricted to the vicinity of present streams. Bedrock beneath the surficial deposits has been mapped as Member B of the Lava Creek Tuff.

1.6 29.3 SUGGESTED STOP 1. Turn off for the Virginia Cascades Road. The narrow road precludes parking at the outcrop, so it is necessary to park in a designated area and walk to the outcrop. The outcrops are Lava Creek Tuff, which show many of the features of ash flow tuff stratigraphy. The contact between members A and B of the tuff can be seen at this location.

2.0 31.3 Return to main highway and turn right (East).

0.6 31.9 Exposure of vitrophyre in new road cut. The rock is the Solfatara Plateau flow of

		the Central Plateau Member of the Plateau Rhyolite. This flow has a K-Ar age of about 100,000 years. The rhyolite is covered by Pinedale till. At approximately this point you are crossing the northern fault scarp of the Yellowstone Caldera.
3.6	35.5	Zone of hydrothermal alteration and hot ground.
1.3	36.8	Forest fire area on north side of road.
1.6	38.4	At the break in the forest, you enter Cascade Meadows. Outwash and alluvial deposits, with humic alluvium along stream courses, cover bedrock in the meadow.
1.0	39.4	Canyon Junction. End of this section of the road log.

CANYON JUNCTION TO MUD VOLCANO-SULPHUR CAULDRON AREA

DUNCAN FOLEY[1], DENNIS L. NIELSON[1], CLAYTON R. NICHOLS[2]

INTRODUCTION

This log is part of a road log prepared for a three day field trip taken by the Geothermal Resources Council in September of 1980. The log was adopted primarily from U.S. Geological Survey Geologic and Surficial Geologic quadrangle mapping. The maps used in preparing the log, and the portion of the road log each map pertains to are listed in Appendix A. Lithologic and outcrop descriptions were also drawn from these maps. They are included in Appendix B. The bibligraphy and appendix are located at the end of the road log section of this guidebook.

Potassium-Argon dates reported in this road log were calculated using pre-1977 decay constants. To compare with more recent dating, multiply the dates given herein by 1.025.

Glacial activity has been separated into Bull Lake and Pinedale times. Estimates for the ages of the ice are given in Table 1.

TABLE 1.
Age estimates for Glacial Activity in Yellowstone
(in thousands of years)

	Keefer (1971)	Richmond (1976)	Pierce (1979)
Pinedale	25-8	70-12	70-13
Bull Lake, late stade	125-45	98-85	160-130(?)
Bull Lake, early stade	125-45	120-110	160-130(?)

The W.G.A. would like to thank the authors for their permission to include this road log in the guidebook.

Mileage Distance	Cumulative	
0.0	0.0	Starting at Canyon Junction, drive south. Immediately south of the intersection the road travels on Pinedale sand and till. The bedrock to the south is composed of the Canyon flow of the Upper Basin Member and the Hayden Valley flow of the Central Plateau Member; these are separated by sediments of the Upper Falls. The sediments are chiefly lacustrine sands and silts, with local gravels and pumiceous tuff, formed during the intra-Bull Lake interval.
2.5	2.5	Turn left (east) for Artists Point. The oldest known sinter deposits in the park are exposed in roadcuts on the way to the parking area; they are interbedded with caldera fill sediments (D.E. White, pers, comm., 1980).
1.7	4.2	SUGGESTED STOP 1. Artists Point parking area. Walk to the overlook of the Grand Canyon of the Yellowstone.

The canyon is cut in the Canyon Flow of the Upper Basin Member of the Plateau rhyolite. The Canyon Flow has been K-Ar dated at about 590,000 years. Christiansen and Blank (1972, p. 13) say:

> "In our interpretation, the rhyolite section in the upper Grand Canyon area consists of a basal, largely agglutinated air-fall tuff overlain successively by the Canyon flow proper and by a younger flow exposed near the Dunraven Pass Road."

The Canyon flow probably erupted into a lake ponded by the collapse of the caldera following the eruption of the Lava Creek Tuff (Richmond, 1976, p. 6)

The colors of the rocks are primarily due to the formation of iron oxides and hydroxides from hydrothermal activity. Hot springs may still be seen near river level both upstream and downstream from Artist's Point.

The canyon itself has been cut in successive stages, related to interglacial meltwaters, floods produced by volcanic eruptions into lakes, and catastrophic lowering of lakes by sudden melting and hydrothermal explosions (Richmond, 1976).

[1]Earth Science Laboratory, University of Utah Research Institute, Salt Lake City, Utah.
[2]Department of Energy, Division of Geothermal Energy, Idaho Falls, Idaho.

1.7 5.9 Return to main road and proceed right (south) toward Mud Volcano.

1.3 7.2 The Solfatara Plateau flow forms the hill west of the river; the Hayden Valley flow forms the hill east of the river.

0.7 7.9 Enter Hayden Valley. Glacial lake, kame, and till deposits form the valley floor. The earliest lake was older than the 100,000 year old Hayden Valley flow of the Central Plateau Member of the rhyolites. Three younger lakes also formed during Bull Lake late stage, Bull Lake Pinedale interglacial, and Pinedale glacial times. These lakes also existed in the present area of Yellowstone Lake. The youngest glaciation in Hayden Valley was probably about 12,600 to 12,000 years ago (Richmond, 1976).

1.0 8.9 The road climbs a small hill of Hayden Valley flow, Pinedale till, Bull Lake sediments, and flood deposits.

0.6 9.5 SUGGESTED STOP 2. Turn out for a view of the Sour Creek Dome, Hayden Valley, the Absaroka and Washburn Ranges. This stop provides a view of the Sour Creek resurgent dome to the east across the Yellowstone River. The resurgent phase of the caldera followed collapse and is produced by magma pressure from beneath. This is a structural dome in contrast to extrusive lava domes exposed elsewhere in the park. As a structural uplift, older portions of the stratigraphy are exposed with both members A and B of the Lava Creek Tuff and small areas of Huckleberry Ridge Tuff present. The surface of the resurgent dome is broken into discrete blocks by numerous normal faults forming an orthogonal pattern. An apical graben often forms as part of this faulting which is produced by extension associated with uplift. The topographic expressions of the faults can be observed although the faults themselves are mantled by till. Numerous hot springs and areas of altered ground can be found on the margins of the resurgent dome although the central portion of the structure contains no thermal manifestations.

1.1 10.6 Hydrothermally altered sediments and landslide topography are visible on the west side of the highway.

2.7 13.3 SUGGESTED STOP 3. Mud Volcano-Sulphur Cauldron area. Park in the Mud Volcano parking lot and walk the trail to the southwest. You may also wish to walk the several hundred yards north along the highway to the overlook of Sulphur Cauldron. This stop will provide the opportunity to observe the surface manifestations of a vapor-dominated system. Numerous acid sulfate springs and mud pots are present. Note the small amount of actual run off from the entire area. The sulfur cauldron area has been mapped in detail by White and others (1971).

Christensen and Blank (1975) have mapped siliceous hot spring deposits along the Yellowstone River indicating that the geothermal system at one time was water dominated. The change from a water dominated to a vapor dominated system is thought to result from the development of impermeable margins, which restricts recharge of water and leads to the lowering of the water table through boiling.

The transitory nature of thermal phenomena is illustrated at Mud Volcano. A series of earthquakes in early 1979 led to greatly increased thermal activity here. Many trees have since been killed, and new mudpots are developing. For chemical analyses of waters from this area, see Rowe, et. al, 1973 and Thompson and Yadav, 1979.

Smith (pers. comm., 1980) reports uplift rates in the Mud Volcano area of approximately 14 mm/year. It is important in thermal areas to remain on the walkways provided. In particular, hydrothermal sinter deposits often form a thin crust on the edges of pools, which will collapse under the weight of a person. Many people have been burned by breaking through such thin crusts.

End of this section of road log.

CANYON JUNCTION TO TOWER JUNCTION

DUNCAN FOLEY[1], DENNIS L. NIELSON[1], CLAYTON R. NICHOLS[2]

INTRODUCTION

This log is part of a road log prepared for a three day field trip taken by the Geothermal Resources Council in September of 1980. The log was adapted primarily from U.S. Geological Survey Geologic and Surficial Geologic quadrangle mapping. The maps used in preparing the log, and the portion of the road log each map pertains to are listed in Appendix A. Lithologic and outcrop descriptions were also drawn from these maps. They are included in Appendix B. The bibliography and appendix are located at the end of the road log section of this guidebook.

Potassium-Argon dates reported in this road log were calculated using pre-1977 decay constants. To compare with more recent dating, multiply the dates given herein by 1.025.

Glacial activity has been separated into Bull Lake and Pinedale times. Estimates for the ages of the ice are given in Table 1.

TABLE 1
Age estimates for Glacial Activity in Yellowstone
(in thousands of years)

	Keefer (1971)	Richmond (1976)	Pierce (1979)
Pinedale	25-8	70-12	70-13
Bull Lake, late stade	125-45	98-85	160-130(?)
Bull Lake, early stade	125-45	120-110	160-130(?)

The W.G.A. would like to thank the authors for their permission to include this road log in the guidebook.

Mileage Dist-ance	Cumul-tive	
0.0	0.0	Starting at Canyon Junction, drive north toward Mt. Washburn and Tower Junction. As you drive north from the intersection, the road is on the Canyon Flow of the Upper Basin Member of the Plateau Rhyolites (see Appendix B for lithologic description). This flow is exposed above the upper falls of the Yellowstone River at Canyon. Pinedale till forms the surficial deposits.
1.6	1.6	Cross the mapped trace of the northern fault scarp of the Yellowstone Caldera. The bedrock changes from the Canyon flow to the Dunraven Road flow of the Upper Basin Member of the Plateau Rhyolites (for lithologic description, see Appendix B).
0.5	2.1	The rock on the south side of the road is Dunraven Road flow, with andesite flows of the Lamar River Formation on the north side of the road. The Lamar River Formation is part of the Absaroka Volcanic Supergroup, which is preserved outside the caldera. You can view the Absaroka volcanic field from Mt. Washburn. Smedes and Prostka (1972) describe the Lamar River Formation as: ". . . well-bedded coarse alluvial facies volcanic conglomerates, breccias, and tuffs." The andesite flows are described in Appendix B. Smedes and Prostka (1972) considered the Lamar River Formation to be late Wasatchian or early Bridgerian age; this is about 49 m.y.
0.8	2.9	The road here crosses onto the Sepulcher Formation (for lithologic description, see Appendix B). At this point, the road climbs out of plateau rhyolites and will remain in various facies of the Absaroka Volcanic Supergroup to Dunraven Pass.
0.6	3.5	View east toward Washburn Hot Springs. These springs have the characteristics of a vapor dominated system. D.E. White (written comm., 1980) points out, however, that the discharge of high chloride waters through the hot springs at Seven Mile Hole in the Grand Canyon of the Yellowstone indicates an underlying water dominated part of this hydrothermal system.

[1]Earth Science Laboratory, University of Utah Research Institute, Salt Lake City, Utah
[2]Department of Energy, Division of Geothermal Energy, Idaho Falls, Idaho

1.5 5.0 SUGGESTED STOP 1. Dunraven Pass Picnic area. You can hike the Chittenden Road to the overlook of the Yellowstone Caldera and the Absaroka Range. Schultz (1962) summarized the topography and volcanic history of Mt. Washburn by saying:

"The cone associated with the vent of the Washburn volcano was broad and erosion-scarred, with dips up to 25° near the vent. Generally, slopes varied between 6° and 12° away from the vent area, but were only 3° to 4° in outlying areas. The cone was approximately 25 to 30 miles in diameter at the base and stood 3,000 feet above the general terrain. In relation to the present general elevation of the region and the height of Mt. Washburn (10,243), the maximum elevation of the cone was probably nearly 12,000 feet. Eruptions from the vent were spasmodic, and were separated by relatively long periods of quiescence. During these periods, rapid erosion attacked the cone, causing the paleotopography and the cut-and-fill structures that occur in breccia outcrops of Mt. Washburn."

"It is proposed that the vent or vents for the volcanic rocks in the Mt. Washburn region is [are] located two to three miles southeast to south-southeast of Mt. Washburn. The vent is now buried beneath the plateau formed by the Canyon (Plateau) flows and probably is situated on the northwest side of the Grand Canyon."

Mt. Washburn has many of the characteristics, such as extensive mud flow deposits, that are found on modern volcanos like Mt. St. Helens.

The rocks exposed along the trail are part of the vent facies of the Lamar River Formation (for lithologic description, see appendix B). For the next several miles toward the north, the breccias of this facies are well exposed on the east side of the highway. The high peaks west of the pass are also composed of Absaroka volcanic rocks.

1.9 6.9 The valley of Tower Creek is visible to the north and northwest. The northeast-sloping benches in the valley are Member B of the Lava Creek Tuff. As you proceed, the road passes from vent facies to alluvial facies of the Lamar River Formation. The alluvial facies of the Langford Formation is also present along the road (for lithologic description, See Appendix B). The Langford Formation has been dated at about 48 m.y. (Smedes and Prostka, 1972).

3.1 10.0 Turn off to Chittenden Road parking area. Proceed along main highway. The tight switchback immediately east of the turn off is in the alluvial facies of the Lamar River Formation (for lithologic description, see Appendix B). Pinedale till and rubble mantle the slopes.

1.1 11.1 View toward the northeast of the Absaroka volcanic pile. The near hills across the river are Specimen Ridge, where extensive Eocene forests have been preserved. Dorf (1960) described these forests as covering more than 40 square miles, having hundreds of upright trees, identifiable parts of over 100 plants, and a vertical succession of up to 27 forests. Fritz (1980a, 1980b) has re-examined these concepts, and prefers a two-part model. Trees from high Absaroka volcanic slopes could have been transported downslope by mud flows, which then would have been reworked by braided streams. This model could explain the mixing of cooler and warmer climate trees found on Specimen Ridge. The forests are in part of the Lamar River Formation.

1.7 12.8 Lamar River alluvial facies crop out on the west side of the road. Pinedale kame deposits form the slope toward Antelope Creek.

1.3 14.1 Member B of the Lava Creek Tuff is exposed in the outcrop along the road.

1.4 15.5 Junction Butte Basalt (approximately 2 m.y. old) exposed in road cut. (For lithologic description, see appendix B.) The road passes better outcrops between Tower Falls and Calcite Springs.

0.7 16.2 Tower Falls turn out, proceed north on main road.

0.1 16.3 Cross Tower Creek. The falls and creek are in the Sepulcher Formation. Off-trail hiking in this area is particularly dangerous as the rocks of the Sepulcher Formation are loose.

0.2 16.5 The overhanging cliff flow of the Junction Butte Basalt is exposed in the spectacular roadcut. The basalt here is about 30 m thick, and shows two-tiered jointing (Christiansen and Blank, 1972, p. B9).

0.5 17.0 SUGGESTED STOP 2. Turnout for Calcite Springs overlook. Good exposures of the sediments and basalts of the Narrows are present in the canyon walls.

The mountains to the north are composed of Precambrian rocks, while those to the east are Absaroka volcanic rocks. Calcite Springs is visible along the Yellowstone River at the base of Bumpus Butte. Pinedale till mantles the ridgetops on both sides of the river.

Proceed north toward Tower Junction. As you drive north, Junction Butte Basalt forms the cap on Bumpus Butte, Member B of the Lava Creek Tuff forms the hill behind Rainey Lake, and hot springs travertine deposits form the last hill before dropping down to the junction.

1.6 18.6 Tower Junction. The surficial deposits at the intersection are Pinedale kame and fan gravels, with recent gravel and fine-grained alluvium along stream courses.

End of this section of the road log.

TOWER JUNCTION TO SODA BUTTE CREEK VALLEY

H.J. PROSTKA

INTRODUCTION

This log was prepared for a 1968 meeting of the Rocky Mountain Section of the Geological Society of America. At the time the road log was written, the volcanic units in the park had not been formally named. Since that time, the informal names used in this road log have been superseded by formal names.

The W.G.A. would like to thank the author for permission to include this road log in the guidebook.

Mileage Cumulative	Distance	
0.0	0.0	Tower Junction. Drive east-northeast toward Cooke City.
0.2	0.2	Pleasant Valley. Amethyst Mountain and Specimen Ridge at 3 o'clock. At 6 o'clock is Roosevelt Lodge with a timber-covered cliff of Yellowstone Tuff behind it. Ledge of Lost Creek Tuff with abundant talus is at 8 o'clock. Beyond it on the skyline is Crescent Hill. The road from here to Crescent Hill is built on the till-mantled bench of Lost Creek Tuff which rises toward a structural high at Crescent Hill.
0.4	0.2	Small exposure of Tower Creek Conglomerate (probably Quaternary) right of road, and upper Cenozoic basalt on the left.
0.7	0.3	Cross Yellowstone River. Note the hot-spring deposits. Greenish conglomerate in cliffs near river at 10 o'clock is basal Cathedral Cliffs Formation.
2.2	1.5	Subdued saddle at 3 o'clock is the blocked preglacial channel of Yellowstone River. For the past mile the road approximately followed the debris-filled preglacial channel of Yellowstone River.
3.4	1.2	Large erratic blocks in Pinedale till deposited by the glacier from Slough Creek valley.
4.2	0.8	North end of Specimen Ridge at 2 o'clock. Ledge on ridge crest at 5 o'clock is Lost Creek Tuff of Brown (1961). Small mesa in foreground at 7 o'clock is Junction Butte — an erosional remnant of upper Cenozoic basalt similar to the basalt across the Lamar River from the Buffalo Ranch.
5.3	1.1	Cross Lamar River.
5.6	0.3	Roadcut in sands and gravels of outwash plain.
6.6	1.0	Enter Lamar Canyon. Directly behind, about 25 miles away on the skyline, is Electric Peak, the highest peak in the Gallatin Range, located in the northwestern part of the park. At 5 o'clock is Prospect Peak, the northernmost peak in the Washburn Range. The broad valley directly behind is near the confluence of Slough Creek (not visible) and Lamar River. These flats are pitted outwash terraces formed around the snout of a glacier from Slough Creek valley during the waning stages of Pinedale Glaciation. At the Pinedale maximum the ice was as much as 3,000 feet thick in this valley (K.L. Pierce, oral communication, 1967).
7.8	1.2	Leave Lamar Canyon. The volcanics of the Absaroka field overlap this basement high of Precambrian granitic gneiss (age about 2.7 billion years). Pinedale till and kame gravels within the canyon indicate that it was cut in pre-Pinedale time.
10.6	2.8	Lamar Ranger Station, formerly the Buffalo Ranch, where buffalo or bison were herded from 1902 to 1922 to save them from extinction. Behind the Ranger Station is the Rose Creek landslide and the associated scarps which expose Cathedral Cliffs Formation. Across the river, at 3 o'clock, are two prominent benches underlain by upper Cenozoic volcanics. The upper one is Yellowstone Tuff and the lower one is upper Cenozoic basalt. The ancestral valleys of Lamar River and Yellowstone River were repeatedly inundated by upper Cenozoic volcanic rocks such as these. When the valleys were recut, many benchlike remnants of these flows were left.

12.5	1.9	Fossil Forest exhibit. Behind on the skyline is the Buffalo Plateau which consists mainly of Precambrian metamorphics overlain partly by gently-dipping Paleozoic rocks. At 8 o'clock is Druid Peak with its capping of "early basalt sheets". The two ledges in the lower grassy slopes at 9-10 o'clock are Elk Creek Basalt flows in the lower part of the Cathedral Cliffs Formation. At 12 o'clock, about 15 miles away, is Saddle Mountain and other peaks near the headwaters of Lamar River. Mirror Plateau, at 2 o'clock, is capped by several hundred feet of Early basalt sheets. Across the river, at 3-4 o'clock, is the famous Fossil Forest. The petrified remains of 27 successive buried forests of late early Eocene to middle Eocene age are exposed in these cliffs. Erling Dorf (1960), who studied this flora in detail, identified more than 100 species of fossils plants from this locality. The Cathedral Cliffs strata containing the Fossil Forest are of an intermediate facies, composed of pyroxene-rich fragments characteristic of the dark facies mixed with the lighter-colored, finer-grained, dominantly hornblendic and biotitic material more typical both of the type section and of the sequence farther west.
13.3	0.8	Madison in roadcut.
14.2	0.9	Madison exposed along road; overlain by Cathedral Cliffs Formation with ledge formed by an interbedded flow of Elk Creek Basalt.
14.8	0.6	Parking area for trail head. Good view of Mirror Plateau at 4 o'clock, and Amethyst Mountain and Specimen Ridge directly behind.
15.7	0.9	Soda Butte — a postglacial travertine cone deposited by a hot spring which is now almost extinct. The original hummocky topography on the landslide deposits at 7-8 o'clock has been modified by the overriding of Pinedale ice.
16.0	0.3	Exposure of Madison at 9 o'clock.
21.2	5.2	Ice Box picnic area. Excellent views ahead of The Thunderer with its capping of Early basalt sheets.
21.5	0.3	SUGGESTED STOP 1. Ice Box Canyon — park alongside road. Here, volcaniclastic rocks in the lower part of the dark facies of Cathedral Cliffs Formation can be examined close at hand. Note the varieties of andesitic rock fragments and the scattered pebbles of Precambrian metamorphic and Paleozoic sedimentary rock. The sorting and bedding characteristics of these strata imply a fluvial and/or debris-flow origin.
21.8	0.3	Roadcut in Pinedale kame silt.
23.0	1.2	Madison exposed in roadcuts.
23.2	0.2	Cross Pebble Creek.
23.8	0.6	Landslide blocks of volcanics at 3 o'clock. At 6 o'clock, in canyon of Pebble Creek, is cliff of limestone of the Madison Group.
24.1	0.3	Round Prairie — excellent view of Barronette and Abiathar Peaks, Amphitheater Mountain, and The Thunderer. The dark facies of Cathedral Cliffs Formation with interbedded flows (Elk Creek Basalt) in its lower part is a few hundred feet above the road on the flanks of The Thunderer, Barronette Peak, and Mount Hornaday.
24.4	0.3	Landslide blocks of volcanics of the Absaroka field of Eocene volcanic rocks overlying Pinedale kame gravels, sands, and silts in roadcut.
25.8	1.4	SUGGESTED STOP 2. Valley of Soda Butte Creek 0.4 mi. NE of Soda Butte. Park along road. Looking generally northeastward, the prominent mountains seen clockwise starting at 8 o'clock are Druid Peak, Mount Hornaday, Barronette Peak, Abiathar Peak at 11:30, Amphitheater Mountain (double peak), The Thunderer at 1:30, and Mount Norris at 2. At 6 and 7 o'clock are Amethyst Mountain and Specimen Ridge, with the small travertine cone of Soda Butte along the highway in the valley. These mountains are made up largely of gently dipping Eocene volcaniclastic rocks and lava flows.

The valley bottom and lowermost slopes are underlain by Paleozoic carbonate rocks (chiefly Madison Group) mantled by a variety of surficial deposits. Directly overlying the Paleozoic rocks is about 2,000 feet of a dark, pyroxenic facies of Cathedral Cliffs Formation — medium-brown, well-bedded, volcanic breccias, conglomerates, and sandstones — well exposed in the ridge at 8 o'clock and in the lower cliffs of The Thunderer and Mount Norris. Cathedral Cliffs Formation also makes up approximately the lower two-thirds of Barronette and Abiathar Peaks and Ampitheater Mountain. The top of the Cathedral Cliffs Formation commonly forms a grassy bench. Unconformably above the Cathedral Cliffs Formation are somewhat darker brown, more resistant, coarse pyroxene andesite breccias which comprise the middle one-third of The Thunderer and the uppermost parts of Barronette and Abiathar Peaks and Amphitheater Mountain. This contact was not mapped by earlier workers who lumped both the dark facies of Cathedral Cliffs Formation and the overlying breccias as Early basic breccia (Hague and others, 1896; Rouse, 1937; Brown, 1961). Conformably overlying the volcaniclastic rocks are trachybasalt and trachyandesite lava flows and flow breccias called the Early basalt sheets. These lavas cap Druid Peak and The Thunderer. On the southeast face of Mount Hornaday the lavas and the underlying breccias are seen to interfinger and intermix with light-colored pumiceous volcaniclastic rocks of the western facies.

Most of the large cliffs are landslide scarps. The landsliding occurred both before and after Pinedale Glaciation about 10,000 to 25,000 years ago.

End of this section of road log.

TOWER JUNCTION TO MAMMOTH HOT SPRINGS

DUNCAN FOLEY[1], DENNIS L. NIELSON[1], CLAYTON R. NICHOLS[2]

INTRODUCTION

This log is part of a road log prepared for a three day field trip taken by the Geothermal Resources Council in September of 1980. The log was adapted primarily from U.S. Geological Survey Geologic and Surficial Geologic quadrangle mapping. The maps used in preparing the log, and the portion of the road log each map pertains to are listed in Appendix A. Lithologic and outcrop descriptions were also drawn from these maps. They are included in Appendix B. The bibliography and appendix are located at the end of the road log section of this guidebook.

Potassium-Argon dates reported in this road log were calculated using pre-1977 decay constants. To compare with more recent dating, multiply the dates given herein by 1.025.

Glacial activity has been separated into Bull Lake and Pinedale times. Estimates for the ages of the ice are given in Table 1.

TABLE 1

Age estimates for Glacial Activity in Yellowstone (in thousands of years)

	Keefer (1971)	Richmond (1976)	Pierce (1979)
Pinedale	25-8	70-12	70-13
Bull Lake, late stade	125-45	98-85	160-130(?)
Bull Lake, early stade	125-45	120-110	160-130(?)

The W.G.A. would like to thank the authors for their permission to include this road log in the guidebook.

Mileage Distance	Cumulative	
0.0	0.0	Tower Junction. Proceed west. As you do so you are driving over various facies of Absaroka volcanic rocks which are covered by Pinedale till and rubble.
0.6	0.6	The road climbs the hill approximately along the contact of the main body of the Sepulcher Formation in the hills south of the road and the Lost Creek Tuff member of the Sepulcher Formation which covers the flats north of the road. The main body of the Sepulcher is described in Appendix B. Good exposures of the Lost Creek Tuff Member will be mentioned below.
0.9	1.5	Road to Petrified Tree. The tree is in the Sepulcher Formation, immediately below strata mapped as interbedded Lamar River Formation alluvial facies. Small exposures of Pliocene siltstone and sandstone are also present on the hill with the Petrified Tree (Richmond and others, 1978).
1.0	2.5	The ridge east of the highway is formed of the Lost Creek Tuff member of the Sepulcher Formation (see Appendix B for lithologic description).
0.7	3.2	Floating Island Lake. The main body of the Sepulcher Formation with interbedded Lamar River Formation forms the hillside south of the lake. The Lost Creek Tuff forms the hills north and west of the lake.
0.6	3.8	Elk Creek Basalt Member of the Lamar River Formation is the bedrock at the trailhead road. (See Appendix B for lithologic description).
1.9	5.7	The road climbs along the contact between the Lost Creek Tuff on the north side and the main body of the Sepulcher Formation on the south side.
1.3	7.0	Exposures of Crescent Hill Basalt on both sides of the road. The Crescent Hill Basalt: "... consists of two flows of scoriaceous, partly pillowed trachybasalt which have an aggregate thickness of about 250 feet." (Smedes and Prostaka, 1972, p. C23)

Smedes and Prostka (1972) consider the Crescent Hil Basalt to have an early Bridgerian Provincial age (approx. 48 m.y.).

[1]Earth Science Laboratory, University of Utah Research Institute, Salt Lake City, Utah
[2]Department of Energy, Division of Geothermal Energy, Idaho Falls, Idaho

1.0	8.0	Phantom Lake on the south side of the highway.
0.4	8.4	The Undine Falls Basalt is exposed along the road. This basalt is estimated to be between 700,000 and 600,000 years old; it conformably underlies the Lava Creek Tuff (Christiansen and Blank, 1972, p. B12).
0.5	8.9	Blacktail Plateau Drive.
1.3	10.2	As you round the corner, the Gallatin Range becomes visible on the skyline to the west. The road at this point has traveled near Mississippian Mission Canyon Limestone and Pennsylvanian Quadrant Sandstone, which are brought to the surface along Gardner Fault Zone (Ruppel, 1972). The Gardiner Fault Zone is discussed in detail by Fraser and others (1969). The fault has been mapped from Cinnabar Mountain, north of the park, southeastward to Mt. Washburn, but Fraser and others (1969) cite evidence that the fault may be 140 miles long, from the vicinity of Cody, Wyoming to the Madison Range in Montana. It is a high-angle reverse fault, with more than 10,000 feet of displacement. Most of the movement took place during Laramide tectonism, but Pliocene(?) basalt has been offset 400 feet, and Pleistocene deposits have been offset a few tens of feet along the fault. The Gardiner Fault forms the southwest margin of the Beartooth Block; in the vicinity of Gardiner, Precambrian rocks are adjacent to Paleozoic and Mesozoic rocks. Isolated exposures of Huckleberry Ridge Tuff, which will be discussed below, are also preserved. The Gallatin Range is an uplifted horst of Precambrian to Eocene metamorphic, sedimentary, and intrusive rocks. The intrusions in the Gallatin Range are Eocene and are a postulated source for some of the Absaroka volcanic rocks. (Smedes and Prostka, 1971). As you drive west, the road crosses Pinedale till, with kame deposits near the creeks.
1.8	12.0	The low hills to the north are Lava Creek Tuff.
0.4	12.4	Outcrop of Lava Creek Tuff on the southside of the road. A low outcrop of this tuff continues west into the valley on the north side of the highway.
0.2	13.6	Turn off for Wraith Falls where Lava Creek Tuff is well exposed. The Swan Lake Flat Basalt is found to the west. This unit is younger than the Lava Creek Tuff, but was erupted soon enough after the emplacement of the Lava Creek Tuff that in many places it preserves the non-welded ashy top of the Lava Creek cooling unit. The Swan Lake Flat Basalt is described in Appendix B.
0.4	14.0	Lupine Creek
0.5	14.5	Turn out for Undine Falls parking area.
0.1	14.6	Exposure of Swan Lake Flats Basalt.
0.1	14.7	Exposure of Undine Falls Basalt. For the next several miles, Mt. Everts is visible north of Lava Creek Canyon. Prevolcanic rocks on Mt. Everts are, in ascending order, the Cretaceous Mowry Shale, Frontier Sandstone, and Cody Shale. Further west, the Telegraph Creek Formation, Eagle Sandstone, and Everts Formation are also exposed (Ruppel, 1972).
0.1	14.8	Mammoth Hot Springs is visible to the west.
0.2	15.0	The cap of Huckleberry Ridge Tuff is easily visible on Mt. Everts. This unit is the oldest of the ash flow tuffs in the Yellowstone group, at 2 m.y. It was erupted in the first caldera cycle. Smith and Christiansen (1980) draw the southern boundary of the caldera from which this tuff erupted. The northern and eastern boundaries of the caldera have been obscured by later volcanic activity. The Huckleberry Ridge Tuff has been described and separated into three members (Christiansen and Blank, 1972; Christiansen and others, 1978). See Appendix B for lithologic description of the Huckleberry Ridge Tuff. Christiansen (1979) has estimated the combined volume of the three members of the Huckleberry Ridge Tuff at 2,450 km^3.
1.6	16.6	View toward northwest of Gardiner River Canyon. This area was the site of a wikiup village occupied by Sheepeater Indians and one of the last native American inhabitations in Yellowstone Park.

0.1 16.7 Gardiner River. From here to Mammoth, the road travels on Pinedale Till. Travertine deposits of Mammoth Hot Springs and on Terrace Mountain are visible to the west.

2.3 19.0 Enter Mammoth, the park headquarters. Continue straight on road; turn left at hotel, and proceed south.

0.6 19.6 SUGGESTED STOP 1. Parking area at Lower Terrace of Mammoth Hot Springs. Walk through the Main Terrace and Upper Terrace.

Mammoth Hot Springs is a carbonate-depositing thermal system, and as such is different from the vapor-dominated system at Mud Volcano and the silica-depositing systems at Norris and the Upper and Lower Geysers basins.

Individual thermal features at Mammoth vary in their level of activity from year to year. Carbonate-depositing hydrothermal systems are transitory in their surface expression as are the vapor-dominated systems, such as at the Mud Volcano, and the siliceous sinter-depositing systems, as at Norris Geyser Basin and Old Faithful.

A wide variety of features may be seen on the walk through Mammoth. These include hot-spring cones, terracettes, collapse features, tension fractures, fissure ridges, caves, calcite ice, and micro-organism deposited travertine (Bargar, 1978).

Although Mammoth Hot Springs is lower temperature than some other springs in the Park, the waters are postulated to have circulated through limestones, and picked up their carbonate load by reacting with the bedrock. If the waters had not circulated through limestones, their mineral load would be much different. Bargar (1978) quotes Allen (1934) as recording rates of travertine deposition that range from 2.8-56.5 cm/yr, with an average of 21.1 cm/yr. Color variations in hot spring pools are due to not only the reflection of the sky, but also to several kinds of bacteria and algae. Two bacteria found in thermal waters in the park, *thermoanerobacter ethanolicus* and *thermoanerobium brockii*, have been found to be very efficient at producing ethanol from carbohydrates. Iron and manganese oxides also strongly color some hot springs.

Travertine deposits on the top of Terrace Mountain, southwest of the present hot springs, have been dated at 63,000 ± 9,000 years (Rosholt, 1976). Travertine on Terrace Mountain is overlain by Pinedale till.

Research drilling by the U.S. Geological Survey encountered 253.5 feet of travertine, which was porous near the surface but cemented at depth (White and others, 1975). Jurrassic (?) rocks were encountered beneath the travertine. A maximum temperature of 73°C was encountered, and White and others (1975, p. 64) conclude:

> "Evidently all of the water of the system is heated to this temperature or only a little higher, and it rises as liquid water and dissolved gasses with no significant change until total vapor pressure has decreased to a few atmospheres."

Truesdell and Fournier (1976) suggest that Mammoth Hot Springs waters may be waters from Norris Geyser Basin that have moved north and reacted with limestone.

Corwin Springs, LaDuke Hot Springs, and Bear Creek Warm Springs are other travertine depositing springs north of Mammoth Hot Springs. Water circulation for these hydrothermal systems is postulated to involve deep circulation in the Madison Limestone, heating by the earth's thermal gradient, and circulation to the surface along the Gardiner Fault zone (Struhsacker, 1976).

End of this section of road log.

MAMMOTH HOT SPRINGS TO NORRIS JUNCTION

DUNCAN FOLEY[1], DENNIS L. NIELSEN[1], CLAYTON R. NICHOLS[2]

INTRODUCTION

This log is part of a road log prepared for a three day field trip taken by the Geothermal Resources Council in September of 1980. The log was adapted primarily from U.S. Geological Survey Geologic and Surficial Geologic quadrangle mapping. The maps used in preparing the log, and the portion of the road log each map pertains to are listed in Appendix A. Lithologic and outcrop descriptions were also drawn from these maps. They are included in Appendix B. The bibliography and appendix are located at the end of the road log section of this guidebook.

Potassium-Argon dates reported in this road log were calculated using pre-1977 decay constants. To compare with more recent dating, multiply the dates given herein by 1.025.

Glacial activity has been separated into Bull Lake and Pinedale times. Estimates for the ages of the ice are given in Table 1.

TABLE 1
Age estimates for Glacial Activity in Yellowstone
(in thousands of years)

	Keefer (1971)	Richmond (1976)	Pierce (1979)
Pinedale	25-8	70-12	70-13
Bull Lake, late stade	125-45	98-85	160-130(?)
Bull Lake, early stade	125-45	120-110	160-130(?)

The W.G.A. would like to thank the authors for their permission to include this road log in the guidebook.

Mileage Dist-ance	Mileage Cumul-ative	
0.0	0.0	Upper Terrace parking lot, Mammoth Hot Springs. Go to main road and turn south (right) The road is on Pinedale till, kame, and landslide deposits to the south. Some landsliding is still active, as evidenced by the breaks in the highway.
1.7	1.7	The Hoodoos. This is a landslide of travertine springs deposits which has slumped to its present position from the top of Terrace Mountain. Bunsen Peak is the mountain east of Golden Gate Canyon. It is composed of an intrusive dacite considered to be related to the Absaroka Volcanic Supergroup, but older than the Sepulcher Formation.
1.6	3.3	Entering Golden Gate. Excellent exposures of the Huckleberry Ridge Tuff. The 1959 Hebgen Lake earthquake caused much damage to the bridge and road here.
0.5	3.8	Enter Swan Lake Flat. The Gallatin Range is the prominent mountain chain to the west. Pinedale kame, till, and fan deposits, and recent alluvium cover bedrock in the flats. An extensive drumlin field covers the western portion of the flats (Pierce, 1979). The Sepulcher Formation is mapped west of Swan Lake, and the Swan Lake Flat Basalt is mapped east of the highway. You will drive south, along the Norris-Mammoth fracture zone which lies east of the Gallatin Horst. The Norris-Mammoth zone is an approximately 22-mi long zone of north-south trending faults, rhyolite and basalt vents, and gravity and magnetic anomalies (Eaton and others, 1975). The zone has much hydrothermal activity.
3.2	7.0	Road to Sheepeater Cliffs. Rocks that are the products of mixing of basalt and rhyolite magmas are preserved in these cliffs (not visible from the highway). Struhsacker (1978) presents a detailed summary of the history of scientific investigation of these rocks, and interprets the mixed magmas in a regional context.
0.1	7.1	Gardiner River. The road crosses Swan Lake Flats Basalt to the south.
0.4	7.5	Indian Creek Campground. The hill on the east side of the highway south of the campground is composed of Lava Creek

[1]Earth Science Laboratory, University of Utah Research Institute, Salt Lake City, Utah
[2]Department of Energy, Division of Geothermal Energy, Idaho Falls, Idaho

Tuff, with a mantle of Pinedale till and scree.

2.7 10.2 Appolinaris Spring.

0.9 11.1 Outcrops of Lava Creek Tuff on the west side of the highway.

0.6 11.7 Obsidian Cliffs on east side of road; Lava Creek Tuff on west. Obsidian Cliffs are formed of the Obsidian Cliff flow of the Roaring Mountain Member of the Plateau Rhyolite (see Appendix B for lithologic description).

Obsidian quarried by indians from this locality was both worked locally and traded extensively throughout the region.

3.0 14.7 Clear Water Springs.

0.2 14.9 Hydrothermally altered alluvium.

0.4 15.3 SUGGESTED STOP 1. Roaring Mountain. Since early observation Roaring Mountain has declined in the intensity of its fumarolic activity. It is impressive on cool mornings when the condensate is most apparent. The decline of the "roar" is typical of the everchanging nature of surface hydrothermal phenomena.

Roaring Mountain is a hydrothermal explosion deposit overlying the Lava Creek Tuff. The explosion took place during deglaciation in Middle Pinedale time. Roaring Mountain shows the character of a near-surface vapor-dominated system. Manifestations of an active water-dominated system, which probably represents the associated boiling water table can be seen along Obsidian Creek to the west of the road.

Between Roaring Mountain and Norris, the amount of hydrothermal activity and alteration increases. Bedrock is member B of the Lava Creek Tuff.

4.1 19.4 Gibbon River.

0.8 20.2 SUGGESTED STOP 2. Norris Junction. Turn right (west) into parking area (loop road not included in mileage log). Norris Basin is a large seismically active system of silica depositing hot springs and geysers. You can walk the Back Basin, where Steamboat Geyser, the largest geyser in the world (recorded eruptions to 380 feet; Byron, 1979) and Echinus Geyser, which in June 1980 was erupting every 40 minutes, can be seen.

Echinus Geyser presents a good opportunity to observe the full cycle of geyser activity described by White (1867). In particular, the draining of the geyser after an eruption, as water levels in the geyser tube equilibrate with the surrounding ground water, a rise in water level prior to the eruption, initiation of eruption with the unloading of cooler water from the top of the pool, an early water phase to the eruption, which is followed by a predominantly steam phase, which tapers as the eruption dies out, can be seen. White (pers. comm., 1980) reports that pyrite and marcasite are among the materials ejected during vigorous eruptions of Echinus.

End of this section of road log.

(Note: the road log from West Yellowstone to Canyon Junction covers the geology between Norris Junction and Madison Junction.)

MADISON JUNCTION TO OLD FAITHFUL

DUNCAN FOLEY[1], DENNIS L. NIELSON[1], CLAYTON R. NICHOLS[2]

INTRODUCTION

This log is part of a road log prepared for a three day field trip taken by the Geothermal Resources Council in September of 1980. The log was adapted primarily from U.S. Geological Survey Geologic and Surficial Geologic quadrangle mapping. The maps used in preparing the log, and the portion of the road log each map pertains to are listed in Appendix A. Lithologic and outcrop descriptions were also drawn from these maps. They are included in Appendix B. The bibliography and appendix are located at the end of the road log section of this guidebook.

Potassium-Argon dates reported in this road log were calculated using pre-1977 decay constants. To compare with more recent dating, multiply the dates given herein by 1.025.

Glacial activity has been separated into Bull Lake and Pinedale times. Estimates for the ages of the ice are given in Table 1.

TABLE 1
Age estimates for Glacial Activity in Yellowstone
(in thousands of years)

	Keefer (1971)	Richmond (1976)	Pierce (1979)
Pinedale	25-8	70-12	70-13
Bull Lake, late stade	125-45	98-85	160-130(?)
Bull Lake, early stade	125-45	120-110	160-130(?)

The W.G.A. would like to thank the authors for their permission to include this road log in the guidebook.

Mileage Distance	Cumulative	
0.0	0.0	Madison Junction. Proceed South.
0.1	0.1	Picnic area entrance.
0.2	0.3	Gibbon River.
0.2	0.5	Firehole River Drive entrance. The Nez Perce Creek flow is on the east side of the river, and the West Yellowstone flow is on the west side. Spectacular rhyolite breccias are exposed in a road cut part way up the canyon.
0.3	0.8	Cliffs of the west Yellowstone flow are visible across the Firehole River.
1.5	2.3	Firehole River Drive exit.
1.0	3.3	The Nez Perce Creek flow is on both sides of the highway. This spot is marked by a conspicuous talus slope of West Yellowstone flow on the west side of the river.
0.6	3.9	The Nez Perce Creek flow is east and the West Yellowstone flow is west of the road. The level area north of this point is composed of Pinedale alluvial sands and gravels.
2.2	6.1	Nez Perce Creek. The creek flows a course toward the northeast which marks a zone associated with the ring fracture system of the Yellowstone Caldera. The Elephant Back flow is present south of the creek, and the Nez Perce flow is found north of the creek.
1.3	7.4	The road travels across a flat of largely diatomaceous silt. The hills to the north are composed of Bull Lake and Pinedale Kame deposits. The Bull Lake deposits are described in Appendix B.

Twin Buttes are the hills to the southwest, in front of the West Yellowstone flows as viewed from the highway. These hills are also composed in part of cemented kame gravels. The lower eastern and southern slopes of Twin Buttes, as well as the low hills surrounding Pocket Basin, which is just west of the highway at this point, are composed of hydrothermal explosion deposits. These deposits have been described as:

> "Rubble formed by steam explosion in surficial deposits triggered by abrupt decrease in confining pressure."

An extensive discussion of the Pocket Basin area, as well as other hydrothermal explosion features, is contained in Muffler and others (1971).

[1]Earth Science Laboratory, University of Utah Research Institute, Salt Lake City, Utah
[2]Department of Energy, Division of Geothermal Energy, Idaho Falls, Idaho

0.1	7.5	The road descends a small hill of the Elephant Back flow.
0.6	8.1	North junction, Firehole Lake Drive.
0.0	8.1	SUGGESTED STOP 1. Fountain Paint Pots parking area. This area presents a small but vivid demonstration of the effect of topographic control on water- and vapor-dominated hydrothermal systems. The surficial deposits along the road from this point to the south junction of the Firehole Lake Drive are composed of siliceous sinter.
1.1	9.2	South junction, Firehole Lake Drive. Biscuit Basin flow forms the hill northeast of the intersection. East of the Highway, the Elephant Back flow of the Central Plateau Member of the Plateau Rhyolite forms the hill. The Mallard Lake Dome is the hill to the southeast of the intersection.
1.0	10.2	Midway Geyser Basin. The Biscuit Basin flow is exposed on the east side of the highway. From the Midway Geyser Basin north to the north edge of the Lower Geyser Basin, the West Yellowstone flow forms the western skyline.
3.7	13.9	SUGGESTED STOP 2. Biscuit Basin. Note the thermal activity in the parking lot. The lower slopes to the west are composed of the Biscuit Basin flow of the Upper Basin Member of the Plateau Rhyolite (see Appendix B for lithologic description.) The upper hills to the southwest are formed of the Summit Lake Rhyolite flow, and the hills to the northwest are composed of the West Yellowstone flow. These units are both members of the Central Plateau rhyolites. Ubiquitous Pinedale till covers the bedrock.
1.6	15.5	Black Sand Basin on west side of Highway. This is a group of about 12 geysers and springs.
0.3	15.8	Old Faithful intersection. Interchange springs, at the turn off for Old Faithful, was created during road construction. Fournier and Truesdell (1974) used this spring as a demonstration of their model of mixing after steam loss. 200-208°C water apparently rises, cools adiabatically giving off steam to approximately 100°C; this cooled water then mixes with ground water to give the 76°C spring water.

SUGGESTED STOP 3. Old Faithful Visitor Center. The Upper Geyser Basin contains more than 25% of the world's geysers in a one square mile area. Marler (1973) contains an extensive catalogue of geyser and spring features and cycles. Bryan (1979) presents a popularized edition of geyser descriptions.

Geyser cycles, a variety of sinter morphologies, siliceous deposits, a range of hot spring conditions, and thermophyllic algae can be seen in this area.

The walk from the Upper Geyser Basin to Biscuit Basin is suggested (2.5 miles). Along the way, you can visit Seismic Geyser, which has been extensively discussed (Marler and White, 1975; Marler, 1964), and Artemesia Geyser, which has a large crater and intricate sinter deposits.

The U.S. Geological Survey and the Carnegie Institute have drilled in the vicinity of the Upper Basin. Results of these holes are summarized on Table 2.

Lone Star Geyser is about three miles south of Old Faithful, and may be part of the same convection system. Black Sand Basin is approximately one mile west of Old Faithful Inn.

Looking northeast from the visitor center, extensive deposits of sinter form the foreground. The hill in the background is the Mallard Lake resurgent dome, composed of the Mallard Lake Flow of the Plateau Rhyolites. It is mantled by Pinedale till. The Mallard Lake Dome is a resurgent dome of the caldera similar to the Sour Creek Dome.

End of this section of road log.

TABLE 2

Drilling Results, Upper Geyser Basin and Vicinity
(After White and others, 1975; Keith and others, 1978)

SITE	Total Depth (in ft)	Max. Temp. (°C)	Interval (in ft)	Description
USGS Lone Star (Y-6)	500	181	0-69.5	uncemented sand and gravel
			69.5-500	Scaup Lake Rhyolite Flow

USGS Black Sand (Y-1)	215	171	0-11.5	sinter
			11.5-211.5	zeolitized and cemented obsid. rich seds.
			211.5-215	Biscuit Basin rhyo flow
USGS Biscuit (Y-7)	242	143	0-5.5	sinter
			5.5-173	partly altered and cemented sand and gravel
			173-242	Biscuit Basin pumiceous flow breccia
USGS Biscuit (Y-8)	503	170	0—5	sinter
			5-181	cemented sand and gravel
			181-206	Biscuit Basin flow breccia
			206-503	Biscuit Basin pumiceous tuff
Carnegie I (Myriad)	406	180	0—7	sinter
			7-220	cemented sand and gravel
			220-406	Biscuit Basin rhyolite flow

CODY TO EAST ENTRANCE YELLOWSTONE PARK

via U.S. Highway 14-20/U.S. Highway 16

NORRIS REQUIST

INTRODUCTION

This road log was prepared for the 1964 Wyoming Geological Association publication *Highway Geology of Wyoming.* Some of the landmarks described in the log may have changed since it was written, but the route appears to be the same. It is also probable that the understanding of the geology of some areas and the dating of some units have been refined since 1964. This is particularly true in the case of the Absaroka volcanics.

Mileage		
Dist- ance	Cumul- ative	
0.0	0.0	West end of Sheridan Street, one block east of Buffalo Bill statue and museum complex. Follow U.S. Highway 14-20/ U.S. Highway 16.
0.6	0.6	Curve. Driving on Cody alluvial terrace. Powell terrace is 60 feet above Cody terrace.
0.2	0.8	Muddy sandstone and bentonites and concretion zones of Thermopolis Shale (Cretaceous 650 feet) outcropping on banks of Sulphur Creek.
0.3	1.1	Riverside Cemetary.
0.5	1.6	Heart Mountain to north. Cedar Mountain to southwest. Rattlesnake Mountain to west, Red Butte to northwest with Pat O'Hara Mountain beyond in the distance. Grass-covered slope is Phosphoria dip slope on the east flank of the anticline; softer Chugwater and younger Mesozoic rocks have been removed by erosion. Dips average 14°.
0.2	1.8	Continue straight ahead. State Highway 291 to south leads around south end of Cedar Mountain and up South Fork of Shoshoni River.
0.9	2.7	Wyoming Historical Marker — "Colter's Hell" — John Colter, member of the Lewis and Clark expedition, mountain man, trapper, early explorer, was the first known white man to see this area. He crossed this Shoshoni River — "Stinking Waters" at this point in 1807. Later explorers believing this spot to be what Colter had found and described — the present Yellowstone geyser basins - called it "Colter's Hell". Road on travertine terrace deposits. Travertine and terrace gravels rest on Chugwater (Triassic 800 feet). In a distance of six miles northeastward from this point the entire Mesozoic section is exposed along the banks of the Shoshoni River.
0.1	2.8	Travertine cones and abandoned sulfur mines. Mines recovered sulfur from travertine deposits during and prior to World War I.
0.1	2.9	Leached, slumped and brecciated Dinwoody (Triassic) and Phosphoria (Permian) in canyon below and to north. Light colored beds supporting dip slope at top of canyon walls in large canyon above are also Phosphoria 143+ feet). Darker rocks below are Tensleep Sandstone (Pennsylvanian - 177 feet). Purple clays and underlying red-stained limestones and cherts are the Amsden Formation (Pennsylvanian - 250 feet). Resistant gray cliff at base is Madison Limestone, (Mississippian - 700 feet). The Phosphoria, Tensleep and Madison produce oil at Oregon basin and other Big Horn basin fields. (Formation thicknesses taken from Stipp, 1947).
0.1	3.0	Sulfur in travertine deposits.
0.3	3.3	Road left leads up Cedar Mountain to Spirit Mountain Caverns in Madison Limestone.
0.1	3.4	Wyoming Historical Marker — "Shoshoni Canyon is a gorge cut across Rattlesnake Mountain by the wear-action of the Shoshoni River over a long period of time. The Mountain is an upfold in the earth's crust. Beds of sedimentary rock on the east flank slope eastward beneath the plains. The same units bend up and over the crest of the mountain and stand vertical along the west flank. Granite over

		two billion years old is exposed around Buffalo Bill Dam. South of the Dam is a vertical fracture, or fault, in the rocks along which the mountain was uplifted over 2,000 feet".
0.2	3.6	Heart Mountain Siphon. A lined tunnel along the south side of the river brings irrigation water from the Buffalo Bill Reservoir down to this point where the siphon then carries it across the river, through the north canyon wall and into the Heart Mountain Canal. The Canal takes the water eastward along the north side of the Shoshone River Valley to irrigate 26,000 acres downstream.
0.1	3.7	Madison Limestone just above road level. Occasional patches of travertine terraces are preserved.
0.1	3.8	Shoshone Canyon mouth. Canyon is cut through Rattlesnake anticline by superimposed Shoshone River.
0.3	4.1	Spoil pile at left is from excavation of tunnel bringing water from Shoshone Reservoir down to Heart Mountain Siphon.
0.4	4.5	Terrace gravels at level of Cody terrace.
0.2	4.7	Gallatin (Upper Cambrian - 515 feet) green shales and flat-pebble intraformational conglomerates along old road in canyon below. Base of massive Bighorn Dolomite (Ordovician - 410 feet) at west end of old bridge. Bighorn Dolomite is overlain by Jefferson Limestone (Devonian — 20+ feet) and less resistant Three Forks Formation (Devonian - 200 feet) composed of greenish-gray shale and buff limestone. Madison Limestone (Mississippian - 700 feet) forms upper cliff. (Formation thicknesses taken from Stipp, 1947).
0.3	5.0	Turnout.
0.1	5.1	Bridge. Flathead Sandstone (Middle Cambrian - 115 feet) lies on Precambrian near old road below. Overlying Gros Ventre (Middle Cambrian - 655 feet) and Gallatin (Upper Cambrian - 515 feet) are also exposed. (Formation thickness taken from Stipp, 1947).
0.5	5.6	Turnout.
0.2	5.8	Flathead Sandstone.
0.2	6.0	Tunnel. Contact of Precambrian granite and basal Cambrian Flathead Sandstone above east end of tunnel.
0.05	6.05	Tunnel.
0.05	6.10	Enter long tunnel in Precambrian.
0.7	6.8	Leave long tunnel. Parking area. Tunnel on old road is now used as foot path to top of Buffalo Bill Dam. Dam is 325 feet high and was completed in 1910. Shoshone Reservoir behind dam has capacity of 439,850 acre-feet of water. Power generating facilities are below the Dam at the 5,600 kilowatt Shoshone Power Plant and the 5,000 kilowatt Heart Mountain Power Plant.
		"Tunnel is cut along a basaltic sill which contains angular granulitic blocks in base and shows minor mylonitic layer at top. One feeder to sill is visible above road gutter."
		"Granite fault scarp shows curving rupture surfaces at the road level at the west end of the tunnel. North of the highway the granite comprises the hanging wall in fault contact with Upper Cambrian shales, and the Ordovician Bighorn Dolomite. South of the reservoir in Cedar Mountain this fault extends upward into but not through the Madison Limestone."
0.4	7.2	Tunnel.
0.1	7.3	Tunnel cutting horizontal travertine deposits and dipping Madison Limestone on west limb of Rattlesnake Mountain anticline.
0.3	7.6	Carter Mountain capped by Eocene volcanic rocks to the south across Shoshone Reservoir. Valley of South Fork of Shoshone River to the southwest.
2.0	9.6	Cody Shale (Cretaceous - 1800+ feet).
1.0	10.6	Horizontal Paleozoic rocks on crest of Rattlesnake Mountain anticline and flatirons of the same rocks on the eroded west limb.
0.6	11.2	Half Moon Motel and Cafe.
0.1	11.3	Bridge over Rattlesnake Creek.
0.3	11.6	Cody Shale.
2.0	13.6	Logan Mtn. north of highway; Sheep Mtn. south of highway. Both are largely Cody Shale. The lower slopes are capped by Paleozoic Madison and Bighorn Limestone remnants of the Heart Mtn. thrust sheet. A thin unit of Willwood ("Wasatch") Formation (Late Early Eocene) is present beneath the fault plane on the south side of Logan Mtn. indicating the fault is at least that young. Farther

		west volcanic rocks of the Early Basic Breccia (late Early or Middle Eocene) onlap the Paleozoic limestones of the thrust showing that the thrust emplacement was no later than Middle Eocene. The movement of the Heart Mtn. thrust masses therefore probably took place near the close of the Early Eocene.
1.2	14.8	Hummocky topography in lower slopes due to landslides (?) in Cody Shale.
1.2	16.0	Cross Trout Creek. Dark brown rocks on hill to north on the west side of Creek are several hundred feet of volcanic agglomerate, conglomerate and breccia of the Early Basic Breccia (Late Early Eocene). These rocks are at the same elevation as remnants of the Heart Mtn. thrust sheet on Logan Mtn. on the east side of the creek. The Early Basic Breccia was deposited upon an erosional surface on which were scattered residual limestone blocks from the Heart Mountain thrust sheet.
1.1	17.1	Cody Shale.
1.6	18.7	North of highway Cody Shale (Cretaceous - 1800+ feet) is at base of cliff. Bedded sandstone and claystones are Willwood (Eocene) and are overlain by Early Basic Breccia which continues up to summit of Black Mtn. in the distance. Note dike cutting sediments. Southeast from this point at the west end of Sheep Mtn. the volcanics rest against a cliff-like face of the Paleozoic limestones. The volcanics and limestones rest in turn on the underlying Willwood and Cody Shale.
1.3	20.0	Cross Jim Creek. Willwood Formation east of Creek.
0.5	20.5	Bridge over Shoshone River.
0.2	20.7	Wapiti Lodge and Post Office.
1.2	21.9	Willwood Formation. This typical section is interbedded coarse limonitic sands and pastel-colored claystones.
0.1	22.0	Wyoming Historical Marker — "Absoraka Volcanic Field — passes through a series of volcanic rocks over 9,000 feet thick covering 3,000 square miles. The rocks include lava, volcanic ash and other sorts of volcanic material ——". Numerous dikes which were feeders for lava flows show in the canyon walls as thin, narrow bands resembling stone walls. The peculiar castle-like forms are the result of weathering and removal of softer material by water."
1.3	23.3	Green Creek. The Chinese Wall, a basic igneous dike, crosses from northwest to southeast.
0.8	24.1	Willwood Formation (Eocene) in lower Shoshone River valley below dissected alluvial fan.
0.8	24.9	Canyon Creek.
0.2	25.1	Mountain View Lodge. 5,000 feet of volcanic rocks of the Early Basic Breccia are present around this broadening of the river valley.
0.2	25.3	Enter Shoshone National Forest. The Shoshone National Forest was set aside by proclamation of President Benjamin Harrison as the Yellowstone Park Timberland Reserve on March 30, 1891. It was the first unit of its kind created after passage of the Forest Reserve Act.
0.6	25.9	Southwest dip in volcanic rocks. Appears to be due to local subsidence.
1.0	26.9	Road to Nameit Creek Lodge.
0.6	27.5	Hanging Rock Camp Ground. Good exposures of volcanic agglomerate at Hanging Rock.
0.6	28.1	Clocktower Creek, Ptarmigan Mtn. to south composed of Eocene volcanics of the Early Basic Breccia.
0.4	28.5	Pagoda Creek.
0.3	28.8	Horse Creek Picnic Ground.
0.6	29.4	Volcanic agglomerate.
0.5	29.9	Big Game Campground.
0.4	30.3	Wyoming Historical Marker — "Wapiti Ranger Station — This is the oldest national forest ranger station in the United States. It was built in 1903 as a supervisor and ranger headquarters for the Shoshone division of the Yellowstone Timberland Reserve."
0.4	30.7	Several dikes north of road. Discordant dips south of road suggest water-lain continental cut-and-fill deposition of the volcanic rocks in this area.
0.3	31.0	Elk Fork Creek and Campground. One mile south and up Elk Fork Creek an outcrop of Willwood Formation (Eocene) is present at an elevation considerably higher than the Shoshone River whose valley is cut in younger volcanics. The older Willwood outcrop at a higher eleva-

tion probably is due to unevenness of the pre-volcanic erosional surface rather than to post-volcanic uplift since no strong north dips are present in the volcanics.

There are no known pre-volcanic sedimentary rock outcrops between this point and Yellowstone Lake in Yellowstone National Park.

1.8	32.8	Bridge over Shoshone River. Cliff face south of road shows continental channel-fill type deposition.
0.6	33.4	Clearwater Creek.
1.2	34.6	Dike.
0.4	35.0	Aspen Creek.
1.2	36.2	Moss Creek.
1.3	37.5	Rex Hale Campground. Volcanics in the area are even-bedded. Northwest of this point in a high cave the mummified body of a man was excavated in August, 1963. Radio-carbon dating of a fragment of his mountain sheep skin garment indicates that he died 1300 years ago (c. 680 A.D.). The mummy and artifacts from the cave are on display in the Whitney Galley of Western Art in Cody.
0.6	38.1	Fire Fighters Memorial. Fifteen men lost their lives fighting a forest fire five miles south of here on Blackwater Creek in August, 1937.
0.7	38.8	Newton Creek
2.3	41.1	Gunbarrel Creek and Absoraka Lodge.
0.6	41.7	Elephant Head Lodge.
1.8	43.5	Goff Creek and Goff Creek Lodge.
1.7	45.2	Libby Creek.
2.2	47.4	Mormon Creek.
2.3	49.7	Sleeping Giant Campground.
0.3	50.0	Grinnell Creek.
0.1	50.1	Sleeping Giant Winter Sports Area and Shoshone Lodge.
0.6	50.7	Three Mile Campground.
0.8	51.5	Bridge over Shoshone River.
0.2	51.7	Pahaska Tepee. Buffalo Bill's old hunting lodge stands east of the group of buildings erected later to serve tourists.
2.3	54.0	East Entrance, Yellowstone National Park.

End of this section of road log.

ROAD LOGS REFERENCES CITED

Allen, E. T., 1934, The agency of algae in the deposition of travertine and silica from the thermal waters: Am. Jour. Sci., v. 28, p. 373-389.

Allen, E. T., and A. L. Day, 1935, Hot Springs of the Yellowstone National Park: Carnegie Inst. Wash. Publ. 466.

Bargar, K. E., 1978, Geology and thermal history of Mammoth Hot Springs, Yellowstone National Park, Wyoming: U.S. Geol. Surv. Bull. 1444, 55 p.

Boyd, F. R., 1961, Welded tuffs and flows in the Rhyolite Plateau of Yellowstone National Park, Wyoming: G.S.A. Bull., v. 72, p. 387-426.

Brown, C. W., 1961, Cenozoic stratigraphy and structural geology, northeastern Yellowstone National Park, Wyoming and Montana: Geol. Soc. America Bull., v. 72, p. 1173-1194.

Bryan, T. S., 1979, The Geysers of Yellowstone: Colo. Assoc. Univ. Press, 225 p.

Christiansen, R. L., 1975, Geologic map of the Norris Junction Quadrangle, Yellowstone National Park, Wyoming: U.S. Geol. Surv. Map GQ-1189.

Christiansen, R. L., 1979, Cooling units and composite sheets in relation to caldera structure, *in* Chapin, C. E. and Elston, W. E., (eds.), Ash-Flow Tuffs: Geol. Soc. America Special Paper 180, p. 29-42.

Christiansen, R. L., unpub., Quaternary and Pliocene volcanism of the Yellowstone Plateau volcanic field of Wyoming, Idaho, and Montana: Unpub. manuscript, 7 p.

Christiansen, R. L., and H. R. Blank, 1972, Volcanic Stratigraphy of the Quaternary Rhyolite Plateau in Yellowstone National Park: U.S. Geol. Surv. Prof. Paper 779-B, 18 p.

Christiansen, R. L., and H. R. Blank, 1974a, Geologic Map of the Madison Junction Quadrangle, Yellowstone National Park, Wyoming: U.S. Geol. Surv. Map GQ-1190.

Christiansen, R. L., and H. R. Blank, 1974b, Geologic Map of the Old Faithful Quadrangle, Yellowstone National Park, Wyoming: U.S. Geol. Surv. Map GQ-1189.

Christiansen, R. L., and H. R. Blank, 1975, Geologic Map of the Canyon Village Quadrangle, Yellowstone National Park, Wyoming: U.S. Geol. Surv. Map GQ-1192.

Christiansen, R. L., H. R. Blank, Jr., J. D. Love, and J. C. Reed, Jr., 1978, Geologic Map of the Grassy Lake Reservoir Quadrangle, Yellowstone National Park and Vicinity, Wyoming: U.S. Geol. Surv. Map GQ-1459.

Dorf, E., 1960, Tertiary fossil forests of Yellowstone National Park Wyoming, in West Yellowstone-Earthquake area, Billings Geological Society Guidebook, 11th Annual Field Conference, 1960, p. 253-260.

Eaton, G. P., R. L. Christiansen, H. M. Iyer, A. M. Pitt, D. R. Mabey, H. R. Blank, Jr., I. Zietz, and M. E. Gettings, 1975, Magma beneath Yellowstone National Park: Science, v. 188, p. 787-796.

Fournier, R. O., and A. H. Truesdell, 1974, Geochemical indicators of subsurface temperature — Part 2: estimation of temperature and fraction of hot water mixed with cold water: Jour. Res. U.S. Geol. Surv., v. 2, no. 3, p. 263-270.

Fournier, R. O., D. E. White and A. H. Truesdell, 1976, Convective heat flow in Yellowstone National Park: Second United Nations Symposium on the Development and Use of Geothermal Resources, p. 731-739.

Fraser, G. D., H. A. Waldrop, H. J. Hyden, 1969, Geology of the Gardiner Area, Park County, Montana: U.S. Geol. Surv. Bull. 1277, 118 p.

Fritz, W. J., 1980a, Stratigraphic framework of the Lamar River Formation in Yellowstone National Park: Northwest Geology, v. 9, p. 1-8.

Fritz, W. J., 1980b, Reinterpretation of the depositional environment of the Yellowstone "fossil forests": Geology, v. 8, no. 7, p. 309-313.

Hague, A., J. P. Iddings, and W. H. Weed, 1899, Geology of the Yellowstone National Park, Pt. II, Descriptive geology, petrography, and paleontology: U.S. Geol. Survey Mon. 32, p. 1-439.

Hamilton, W., 1960, Late Cenozoic tectonics and volcanism of the Yellowstone region, Wyoming, Montana, and Idaho: Billings Geol. Soc. 11th Annual Field Trip Guidebook, p. 92-106.

Iyer, H. M., 1979, Deep continental structure under Yellowstone National Park, U.S.A.: A continental "hot spot": Tectonophysics, v. 56, p. 165-197.

Keefer, W. R., 1971, The geologic story of Yellowstone National Park: U.S. Geol. Surv. Bull 1347, 92 p.

Keith, T. E. C., D. E. White, and M. H. Beeson, 1978, Hydrothermal alteration and self-sealing in Y-7 and Y-8 drill holes in northern part of Upper Geyser Basin, Yellowstone National Park, Wyoming: U.S. Geol. Surv. Prof. Paper 1054-A, 26 p.

Marler, G. D., 1951, Exchange of function as a cause of geyser irregularity: Am. Jour. Sci., v. 249, p. 329-342.

Marler, G. D., 1964, Effects of the Hebgen Lake earthquake of August 17, 1959, on the hot springs of the Firehole Geyser Basin Yellowstone National Park: U.S. Geol. Surv. Prof. Paper 435-Q, p. 185-197.

Marler, G. D., 1973, Inventory of thermal features of the Firehole River Geyser Basins and other selected areas of Yellowstone National Park, NTIS PB-221 289, 639p.

Marler, G. D., and D. E. White, 1975, Seismic Geyser and its bearing on the origin and evolution of geysers and hot springs of Yellowstone National Park: Geol. Soc. of America Bull., v. 86, p. 749-759.

Morgan, P., D. D. Blackwell, and R. E. Spafford, 1977, Heat flow measurements in Yellowstone Lake and the thermal structure of the Yellowstone Caldera: Journ. Geophys. Res., v. 82, p. 3732.

Muffler, L. J. P., D. E. White, and A. H. Truesdell, 1971, Hydrothermal explosion craters in Yellowstone National Park: Geol. Soc. of America Bull., v. 82, no. 3, p. 723-740.

Parsons, 1958, Origin, age, and tectonic relationships of the volcanic rocks in the Absaroka-Yellowstone-Beartooth region, Wyoming-Montana: Billings Geol. Soc. 9th Annual Field Conference Guidebook, p. 36-43.

Pierce, W. G., 1963, Cathedral Cliffs Formation, the early acid breccia unit of northwestern Wyoming: Geol. Soc. America Bull., v. 74, p. 9-22.

Pierce, K. L., 1973, Surficial geologic map of the Mammoth Quadrangle and part of the Gardiner Quadrangle, Yellowstone National Park, Wyoming and Montana: U.S. Geol. Surv. Map I-641.

Pierce, K. L., 1974, Surficial Geologic Map of the Tower Junction Quadrangle and part of the Mount Wallace Quadrangle, Yellowstone National Park, Wyoming and Montana: U.S. Geol. Surv. Map I-647.

Pierce, K. L., 1979, History and dynamics of glaciation in the northern Yellowstone National Park Area: U.S. Geol. Surv. Prof. Paper 729F, 90 p.

Prostka, H. J., H. R. Blank, Jr., R. L. Christiansen, and E. T. Ruppel, 1975, Geologic map of the Tower Junction Quadrangle, Yellowstone National Park, Wyoming and Montana: U.S. Geol. Surv. Map GQ-1247.

Richmond, G. M., 1976, Surficial geologic history of the Canyon Village Quadrangle, Yellowstone National Park, Wyoming, for use with Map I-652: U.S. Geol. Surv. Bull. 1427, 35 p.

Richmond, G. M., 1977, Surficial geologic map of the Canyon Village Quadrangle, Yellowstone National Park, Wyoming: U.S. Geol. Surv. Map I-652.

Richmond, G. M., and H. A. Waldrop, 1975, Surficial geologic map of the Norris Junction Quadrangle, Yellowstone National Park, Wyoming: U.S. Geol. Surv. Map I-650.

Richmond, G. M., W. Mollenders, and M. Coremans, 1978, Climatic implications of two pollen analyses from newly recognized rocks of latest Pliocene age in the Washburn Range, Yellowstone National Park, Wyoming: U.S. Geol. Surv. Bull. 1455, 13 p.

Rosholt, J. N., 1976, $^{230}Th/^{234}U$ dating of travertine and caliche rinds, Geol. Soc. of Am. Abs. with Progs., v. 8, no. 6, p. 1076.

Rouse, J. T., 1937, Genesis and structural relationships of the Absaroka volcanic rocks, Wyoming: Geol. Soc. America Bull., v. 48, p. 1257-1296.

Rowe, J. J., R. O. Fournier, and G. W. Morey, 1973, Chemical analysis of thermal waters in Yellowstone National Park, Wyoming, 1960-65: U.S. Geol. Surv. Bull. 1303, 31 p.

Ruppel, E. T., 1972, Geology of pre-Tertiary rocks in the northern part of Yellowstone National Park, Wyoming: U.S. Geol. Surv. Prof. Paper 729-A, 66 p.

Schultz, C. H., 1962, Petrology of Mt. Washburn, Yellowstone National Park, Wyoming: unpub. PhD dissertation, The Ohio State University, 225 p.

Smedes, H. W., and H. J. Prostka, 1972, Stratigraphic framework of the Absaroka volcanic supergroup in the Yellowstone National Park Region: U.S. Geol. Surv. Prof. Paper, 729-6, 33 p.

Smith, R. B., R. T. Shuey, J. R. Pelton, and J. P. Bailey, 1977, Yellowstone Hot Spot: contemporary tectonics and crustal properties from earthquake and aeromagnetic data: Jour. Geophys. Res., v. 82, p. 3665-3676.

Smith, R. L., H. R. Shaw, R. G. Luedke, and S. L. Russell, 1978, Comprehensive tables giving physical data and thermal energy estimates for young igneous systems of the United States: U.S. Geol. Surv. Open File Report 78-925, 15 p.

Smith, R. B., and R. L. Christiansen, 1980, Yellowstone Park as a window on the earth's interior: Scientific American, v. 242, p. 104-117.

Stipp, T. F., 1947, Paleozoic formations near Cody, Park County, Wyoming: A.A.P.G. Bull., v. 31, p. 274.

Struhsacker, D. W., 1978, Mixed basalt-rhyolite assemblages in Yellowstone National Park: the petrogenetic significance of magma mixing: unpub. M.S. Thesis, Montana State Univ., Bozeman, Montana, 112 p.

Struhsacker, E. M., 1976, Geothermal systems of the Corwin Springs-Gardiner Area, Montana: possible structural and lithologic controls: unpub. M.S.thesis, Montana State Univ., Bozeman, Montana, 93 p.

Thompson, J. M., and S. Yadav, 1979, Chemical Analysis of Waters from geysers, hot springs and pools in Yellowstone National Park, Wyoming from 1974 to 1978: U.S. Geol. Surv. Open File Report 79-704.

Truesdell, A. H., and R. O. Fournier, 1976, Conditions in the deeper parts of the hot springs systems of Yellowstone National Park, Wyoming: U.S. Geol. Surv. Open File Report 76-428.

U.S. Geological Survey, 1972a, Geologic map of Yellowstone National Park: U.S. Geol. Surv. Map I-711.

U.S. Geological Survey, 1972b, Surficial geologic map of Yellowstone National Park: U.S. Geol. Surv. Map I-710.

Waldrop, H. A., 1975a, Surficial geologic map of the West Yellowstone Quadrangle, Yellowstone National Park and adjoining area, Montana, Wyoming, and Idaho: U.S. Geol. Surv. Map I-648.

Waldrop, H. A., 1975b, Surficial geologic map of the Old Faithful Quadrangle, Yellowstone National Park, Wyoming: U.S. Geol. Surv. Map I-649.

Waldrop, W. A., and K. L. Pierce, 1975, Surficial geologic map of the Madison Junction Quadrangle, Yellowstone National Park, Wyoming: U.S. Geol. Surv. Map I-651.

White, D. E., 1967, Some principles of geyser activity, mainly from Steamboat Springs, Nevada: Am. Jour. Sci., v. 265, p. 641-684.

White, D. E., L. J. P. Muffler, and A. H. Truesdell, 1971, Vapor-dominated hydrothermal systems compared with hot-water systems: Econ. Geol. v. 66, p. 75-97.

White, D. E., R. O. Fournier, L. J. P. Muffler, A. H. Truesdell, 1975, Physical results of research drilling in thermal areas of Yellowstone National Park, Wyoming: U.S. Geol. Surv. Prof. Paper 892, 70 p.

White, D. E., 1978, Conductive heat flows in research drill holes in thermal areas of Yellowstone National Park, Wyoming: Jour. Research, U.S. Geol. Survey, v. 6, p. 765-774.

Wilson, W. H., 1963, Correlation of volcanic rock units in the southern Absaroka Mountains, Northwest Wyoming: Univ. of Wyo. Contrib. to Geol., v. 2, p. 13-20.

APPENDIX A

Geologic and Surficial Geologic Map References

MILES	QUADRANGLE name	BEDROCK	SURFICIAL
all	——	USGS 1972a	USGS 1972b
West Yellowstone to Canyon			
0.0-4.1	West Yellowstone	——	Waldrop, 1975a
6.0-19.4	Madison Junction	Christiansen and Blank, 1974a	Waldrop and Pierce, 1975
20.7-38.4	Norris Junction	Christensen, 1975	Richmond and Waldrop, 1975
39.4	Canyon Village	Christiansen and Blank, 1975	Richmond, 1977
Canyon to Mud Volcano			
0.0-13.3	Canyon Village	see above	see above
Canyon to Tower			
0.0-1.6	Canyon Village	see above	see above
1.7-18.6	Tower Junction	Prostka and others, 1975	Pierce, 1974
Tower to Mammoth			
0.0-5.4	Tower Junction	see above	see above
5.5-19.6	Mammoth	—	Pierce, 1973
Mammoth to Norris			
0.0-15.3	Mammoth	see above	see above
15.3-20.2	Norris Junction	see above	see above
Madison to Old Faithful			
0.0-13.0	Madison Junction	see above	see above
13.0-15.8	Old Faithful	Christiansen and Blank, 1974b	Waldrop 1975b

APPENDIX B

Lithologic Descriptions in Alphabetic Order, Labeled as They Appear in Text

Bull Lake Kame Deposit

"Gray to brownish-gray sandstone and conglomerate composed almost entirely of rhyolite clasts. Firmly cemented with opal and zeolites by hydrothermal action. At Porcupine Hills in Lower Geyser Basin the conglomerate is characterized by great variation in texture, sorting, stratification, and local post-depositional slumping typical of ice-contact deposition."

Central Plateau Rhyolite

Large rhyolitic lava flows, each containing a wide variety of lithologies and distinguishable from one another only by topographic forms and zonal development of various textural features and emplacement structures. Most exposures are dark pitchstone or obsidian vitrophyres of flow-layered or flow-brecciated rhyolite, but gray to red pumiceous rhyolite and gray, brown, or varicolored crystalline rhyolite also are common; all varieties can occur within a single flow. All flows contain abundant 1-3 mm phenocrysts of quartz, sanidine, opaque oxides, and, in glassy zones, sparse clino-pyroxene and fayalitic olivine; a few flows also contain minor sodic plagioclase phenocrysts.

Huckleberry Ridge Tuff

Compound cooling unit of ash-flow tuff. Gray to brown generally densely welded and devitrified but locally glassy or partly welded. Most parts contain abundant phenocrysts of quartz, sanidine, and sodic plagioclase; sparse opaque oxides, clinopyroxene, and fayalitic olivine.

Member C — Entirely devitrified; commonly shows strong lineation of stretched welded pumice and aligned phenocrysts; compaction foliation commonly deformed by flowage folding. Abundant lithic inclusions. Phenocrysts abundant but generally smaller (less than 1 mm) than in members A and B. Thickness, 0-300 feet (0-90 m).

Member B — Entirely devitrified. Phenocrysts abundant and particularly large (as much as 5 mm) in upper part, sparse in lower part. Two types of welded pumice in upper part: one very dark and scoriaceous, the other light-colored and compact. Thickness 0-500 feet (0-150 m).

Member A — Mainly devitrified, but black vitrophyre at base. Phenocrysts, abundant in lower part, become progressively less abundant upward. Thickness, 0-700 feet (0-210 m).

Junction Butte Basalt

Dark-gray basalt lava flows characterized by well-developed two-tiered columnar jointing. Basalt is dense, aphanitic, and contains sparse plagioclase phenocrysts; locally overlies gravel, sand, and silt that are mapped with the basalt.

Lamar River Formation — Alluvial Facies

Medium- to light-brown, yellow, and green, well-bedded alluvial-facies volcanic conglomerate sandstone, tuff and well sorted breccia; fragments are subangular to subrounded clasts of pyroxene andesite and pyroxene-hornblende andesite; lower 100 feet (30 m) of unit locally contains sporadic Precambrian metamorphic rock fragments. Fossil trees, common throughout this unit, are especially abundant in the finer grained beds. This unit grades into and interfingers with the Sepulcher Formation to the west, and with the vent facies of the Lamar River Formation to the southwest. 80-1,400 feet (24-427 m) thick.

Lamar River Formation — Andesite Flows

Medium-gray and brown platy-jointed lava flows and flow breccias of hypersthene-hornblende andesite and dacite. This unit forms a small volcanic shield in the lower part of the Mount Washburn composite volcanoe. 0-1,400 feet (0-427 m) thick.

Lamar River Formation — Elk Creek Basalt Member

Medium- to dark-gray and brown-columnar-jointed lava flows and reddish-brown scoriaceous flow breccias of shoshonite. Phenocrysts, pyroxene, and olivine occur in highly variable amounts and sizes from flow to flow. Unit is at different localities a member of the Sepulcher or the Lamar River Formation and is interlayered with lower parts of both formations. 0-400 feet (0-122 m) thick.

Lamar River Formation — Vent Facies

Medium- to dark-gray and brown laharic breccias, autoclastic flow breccias, thin discontinuous lava flows, and tuffs; all variably altered to shades of red, purple, and light yellowish gray. Lava flows and rock fragments are of pyroxene andesite and pyroxene-hornblende andesite. These deposits have primary dips of as much as 30 degrees and form part of the Mount Washburn composite volcano centered in the southwestern corner of the quadrangle. This unit grades into and interfingers with the alluvial facies northward and northeastward away from the volcanic center. 0-3,000 feet (o-914 m) thick.

Langford Formation — Alluvial Facies

Light-gray to medium-gray well-sorted, massively bedded alluvial-facies volcanic breccia and conglomerate consisting of dark-gray subangular to subrounded clasts of pyroxene andesite and hornblende andesite in a light- to medium-gray ash-rich matrix. 2,000 feet (610 m) thick.

Lava Creek Tuff — Member A — Upper Part

Brown to bright pink. Densely welded throughout. Devitrified except for minor vitrophyric lenses in lower part. Contains abundant phenocrysts as much as 5 mm and larger of quartz, sanidine, and sodic plagioclase, moderately abundant opaque oxides and hornblende, and sparse allanite.

Lava Creek Tuff — Member A — Lower Part

Generally pinkish gray to light gray nonwelded to partly welded ash-flow tuff. Glassy to vapor-phase zone. Contains abundant lithic inclusions and some very large pumice blocks (some nearly 1 m). Contains moderately abundant small (generally 1 mm or less) phenocrysts of quartz, sanidine, and sodic plagioclase, moderately abundant hornblende and opaque oxides, and sparse allanite.

Lava Creek Tuff — Member B

Gray, brown, or pinkish-gray ash-flow tuff. Devitrified throughout; generally densely welded except for partially welded vapor-phase zones at top and bottom. Lithic inclusions are locally abundant. Abundant 1-5 mm phenocrysts of quartz, sanidine, and sodic plagioclase; some sanidine phenocrysts as much as 1 cm; sparse opaque oxides, clinopyroxene, and fayalitic olivine phenocrysts. Thin basal zone of phenocryst-poor tuff grades downward from densely welded to partly welded.

Mt. Jackson Rhyolite — Mt. Haynes Flow

Mainly gray crystalline rhyolite with local light-gray vitrophyric margins. Contains 1-3 mm phenocrysts of quartz, sanidine, plagioclase, and opaque oxides.

Plateau Rhyolite — Obsidian Creek Member

Extrusive rhyolitic domes, each with a core of gray crystalline rhyolite and a shell of black vitrophyre or gray pumiceous flow breccia. Contains 1-2 mm phenocrysts of quartz, sanidine, sodic plagioclase, opaque oxides, and sparse clinopyroxene. Age relations between individual domes or between the Obsidian Creek and Upper Basin Members are uncertain.

Plateau Rhyolite — Roaring Mountain Member — Obsidian Cliff Flow

Phenocryst-free or very phenocryst-poor rhyolite which contains abundant fresh black obsidian as well as crystallized and partly crystallized material. Obsidian Cliff flow yielded a young apparent age (75,000 years); the reliability of the date is uncertain (Christianson and Blank, 1972, p. B15).

Plateau Rhyolite — Upper Basin Member — Biscuit Basin Flow

Generally exposed as black perlitic vitrophyre of flow-brecciated rhyolite. Contains abundant phenocrysts, as much as 5 mm, of deeply embayed and sieved plagioclase, moderately abundant quartz and clinopyroxene, and sparse sanidine and opaque oxides.

Plateau Rhyolite — Upper Basin Member — Canyon Flow

Rhyolitic lava flow. Most exposures are brown or dark-gray crystallized rhyolite, but black pitchstone vitrophyre and gray pumiceous rhyolite occur at the margins. Contains abundant 1-5 mm phenocrysts of quartz and sodic plagioclase, and, in glassy zones, clinopyroxene. No sanidine phenocrysts.

Plateau Rhyolite — Upper Basin Member — Dunraven Road Flow

Similar to rhyolites of Central Plateau Member but contains abundant phenocrysts of sodic plagioclase; clinopyroxene phenocrysts are more abundant.

Sepulcher Formation — Lost Cabin Tuff Member

Gray, green, purple, and yellow welded ash-flow tuff of trachyte-rhyodacite, locally contains fragments of Precambrian metamorphic rocks and pieces of charred wood. Central densely welded zone has platy and columnar joints. Unit is interlayered with lower part of Sepulcher Formation and thins eastward and pinches out near west end of Lamar Canyon.

Sepulcher Formation — Main Body

Light-brown to yellowish-gray and greenish-gray epiclastic volcanic breccia, conglomerate, sandstone, and tuffs. Clasts are a variety of colors — gray, green, purple, red, brown, and yellow — of hornblende- and biotite-bearing pyroxene andesite, quartz latite, and dacite. Fragments of Precambrian metamorphic rocks occur locally in lower 100 feet (30 m) of unit. Fossil trees are especially abundant in some horizons, and fossil leaves occur in thin porcelaneous tuff beds. Proportion of coarser beds increases eastward and southward as the unit grades into and interfingers with the alluvial facies of the Lamar River Formation. 0-1,400 ft (0-427 m) thick.

Swan Lake Flat Basalt

". . . Generally . . . light gray to moderate gray and contains sparse to abundant large phenocrysts of plagioclase, as much as 1 cm across, and locally some olivine phenocrysts." (Christiansen and Blank, 1972, p. B15).

CODY TO CLARKS FORK CANYON, DEAD INDIAN HILL, AND RETURN

W.L. ROHRER, R.L. BULLOCK, J.J. COOPER,

M.D. JOHNSON, AND T.T. SCHOWALTER

MILEAGE Cumulative	Increment	
0.0	0.5	Parking lot at Holiday Inn, Cody, Wyoming. Proceed west on U. S. Highway 14 to Wyoming 120 intersection and turn north. This part of Cody is built on the Cody Terrace, 160 feet above the Shoshone River, which correlates with Pinedale Glaciation (Moss & Bonini, 1961)
0.5	0.3	Junction Wyoming 120 and U. S. Highway 14A; turn left.
0.8	0.1	Now traveling on the Lower Cody Terrace, about 100 feet above stream level.
0.9	0.7	Bridge over Shoshone River. Cody Shale-Frontier Formation contact crosses river ¼-mile upstream. The Cody Shale and Frontier Formation are 2,000±200 and 475±25 feet thick, respectively. Gypsum plant and Husky Oil Refinery are located downstream on left bank. Headquarters for Husky Oil are in Cody, a stone's throw from the refinery. From a small skimming plant with 19 employees and a production capacity of 900 barrels a day, the Cody refinery now employes 250 during its peak season and has a crude capacity of 11,000 barrels a day. The plant refines gasoline, diesel and stove fuels, and all grades of asphalt. An emulsion plant is presently under construction to further complement refining capabilities.
1.6	0.6	Traveling on Cody Terrace.
2.2	2.9	Traveling on Powell Terrace. In this vicinity the Powell Terrace averages about 290 feet above the Shoshone River. The Powell Terrace correlates with the late Bull Lake Glaciation (Moss & Bonini, 1961).
5.1	0.7	On Cottonwood Creek alluvium. Ridge east of Cottonwood Creek is of Cretaceous Mowry Shale. Disturbed material on the ridge is where bentonite was mined in the past.
5.8	1.2	Looking west at 9:00 o'clock is the east flank of Cottonwood anticline. Location: N½SW¼ Section 11, T53N, R102W. Cottonwood anticline has no production. The deepest hole, Texaco No. 1 Govt. McAlpin, topped Madison Limestone at 2,315 feet, was a duster. A small fault mapped here (Pierce, GQ-542, 1966) is shown as having right-lateral strike-slip displacement. Red shales, in road cut, and the sandstone ridge, immediately east of the highway, are in the undifferentiated Jurassic Morrison and Cretaceous Cloverly Formations. A narrow bed of kaolinitic potter's clay occurs in this unit on the east side of the ridge.
7.0	1.0	Bridge. Indian artifacts and buffalo skulls have been found on and in the Cottonwood Creek alluvium in this area.
8.0	1.1	Buildings east of highway are the old fish hatchery. The fish hatchery springs discharge from the Mowry Shale and Frontier Formation sandstones along the synclinal axis between the Cottonwood and Fish Hatchery anticlines. The Fish Hatchery anticline, east of the buildings, has no production although several holes have been drilled on the strucutre.

*Reprinted from the Twenty-Seventh Annual Field Conference — 1975 Wyoming Geological Association Guidebook.

9.1 3.5 Heart Mountain to the north. Looking across the north plunge of Fish Hatchery anticline. Figure 1, by W. G. Pierce, was taken from the E½SE¼ Sec. 22, T54N, R102W, about 1.75 miles northeast of the highway. Rattlesnake Mountain access road on west side; Shoshone National Forest is 12 miles distant.

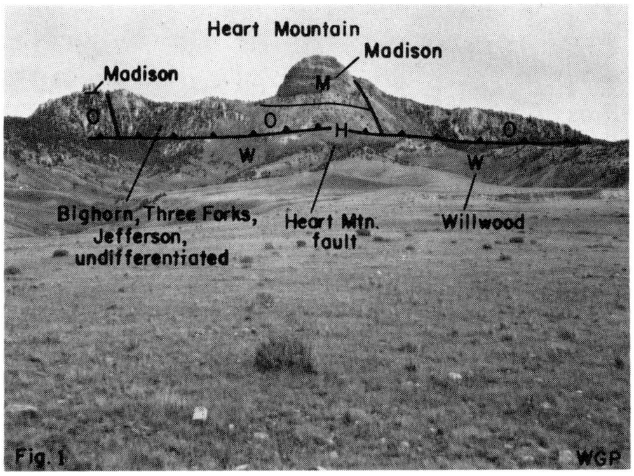

12.6 1.2 View north (12 o'clock) of syncline in Mesaverde Formation and Cody Shale located west of the Heart Mountain anticline. The picture, Figure 2, shows the re-entrant along the anticlinal axis. The Heart Mountain structure has cumulative production of 37,858 MMCF gas and 12,434 barrels oil from the Frontier Formation as of December, 1974.

13.8 1.4 Coal mine in lower Mesaverde in small draw to northeast.

15.2 2.2 Hill of Mesaverde to NNE, at 2 o'clock, is capped by heterogeneous colluvial deposits (Pierce, GQ-542, 1966), some of which have a reddish color suggestive of the variegated Eocene Willwood Formation. Traveling on Cody Shale.

17.4 0.2 Picture of Cody-Mesaverde contact along east side of highway; Figure 3. For mapping purposes the contact is at the base of the lowest massive sandstone. The transitional zone in the upper part of the Cody Shale is a classic regressive sequence of alternating sandstone and shale wherein the sandstone beds increase in thickness and dominance progressively upward to massive sandstone.

17.6 0.8 Junction with Dead Indian Hill road. Continue northward on State 120. Mesaverde forms bluffs on both banks of Pat O'Hara Creek.

18.4 0.8 Mesaverde-Meeteetse contact in short re-entrant to east. Uphill to northeast is the Meeteetse overlain unconformably by the Eocene Willwood Formation; Figure 4.

19.2 0.3 Willwood variegated beds along east side of highway. The Willwood Formation is of Early Eocene age. Many mammalian fossils from the Willwood have been catalogued in several museums. All three Early Eocene faunal zones (Greybullian, Lysitean, and Lost Cabinian) are represented in the formation at Tatman Mountain (several miles east of Meeteetse, Wyoming).

19.5 1.2 Traveling on Chapman Bench and terrace deposits which are approximately equivalent to the Powell Terrace along Shoshone River (Pierce, GQ-477, 1965).

20.7 1.9 Clarks Fork Canyon at 10 o'clock and Beartooth uplift just beyond. The surface on top of the Precambrian rocks in the Beartooth block has been called the subsummit peneplain. These features and the north end of Dead Indian Hill are shown in the accompanying picture which was taken 2.7 miles down the road; Figure 5. Badger Basin oil field is several miles beyond the small point at 12:30 o'clock. The Pryor Mountains are on the skyline at 1 o'clock. In the foreground, same direction, are the Badland Hills, an area where Early Eocene vertebrates have been collected from the Willwood Formation at various times since the turn of the century.

22.6 7.7 Pat O'Hara Creek in left foreground. A small remnant of the Chapman Terrace is at 9 o'clock and just beyond is the surface of the Kimball Bench which is a somewhat higher and older terrace level. A remnant of a still older terrace deposit occurs 3.5 miles to the southwest.

30.3 0.6 Bridge over Clarks Fork River. Approximate contact of Fort Union and Willwood Formations is buried under alluvium here. All strata from the river bend, to the east, and for several miles downstream are Ft. Union (Polecat Bench) Formation. The dip is 3 to 6 degrees south.

30.9 2.6 Junction with Wyoming 292. Turn left toward Clark, Wyoming. Now traveling on undifferentiated Quaternary alluvium. Note the boulder field for the next several miles. The boulder field results from the bursting of a morainal or glacial dam either at the mouth of the Clarks Fork Canyon or somewhere upstream.

33.5 0.2 Bennett Creek crossing.

33.7 0.8 Clark cemetary hill and site of Miles battlefield. The hill is composed of Willwood Formation capped by remnants of Chapman Bench gravel.

34.5 4.4 Junction. Turn left to Clarks Fork Canyon. Right fork goes to Clark.

38.9 2.1 STOP 1. Late Bull Lake terminal moraine on south side of road. This breached moraine extends around the west sides of high hills of steeply dipping (17-40 degrees E) Willwood sandstones and conglomerates north and south of the road. There are patches of the moraine surrounded by alluvium 4 miles south of our position. If these morainal remnants represent the breached dam that disgorged the boulders seen several miles eastward, the lake could have been as much as 500 feet deep and extended 8 miles upstream. A patch of morainal debris is plastered on the hillside immediately north of the road; Figure 6. The Willwood conglomerates mainly consist of chert and carbonates derived from the Paleozoic rocks.

41.0 0.4 Junction. Turn right to Singer Ranch (Martins Ranch on GQ-478).

41.4 0.3 STOP 2 at Singer Ranch house. Turn around. Permission has been granted to us to park and examine the Mowry Shale and Frontier Formation. The Mowry consists of gray and brown shales, in part siliceous, and containing numerous bentonite beds and abundant fish scales; about 450 thick. The Frontier is thick, lenticular, gray sandstone, gray and brown carbonaceous shale, and bentonite; about 450 thick. The formations are nearly vertical and well-exposed along the ridge; Figure 7. W. G. Pierce has noted that the middle or upper part of the Frontier contains unusual conglomerate beds. At this locality there is a 3-foot bed of white conglomerate composed mostly of felsite porphyry pebbles. Three miles to the north there are 2 similar conglomerate beds. One of these beds, near the middle of the Frontier, is composed of black, gray, white, and mottled chert pebbles up to 3 inches across and some light-colored felsite porphyry pebbles; and, the other, about 50 feet below the top of the Frontier is similar but contains pebbles up to 6 inches across. Refer to the articles in this guidebook by Charles Seimers, Al Merewether, and Bill Cobban for detailed discussions of the Frontier. Return to main road.

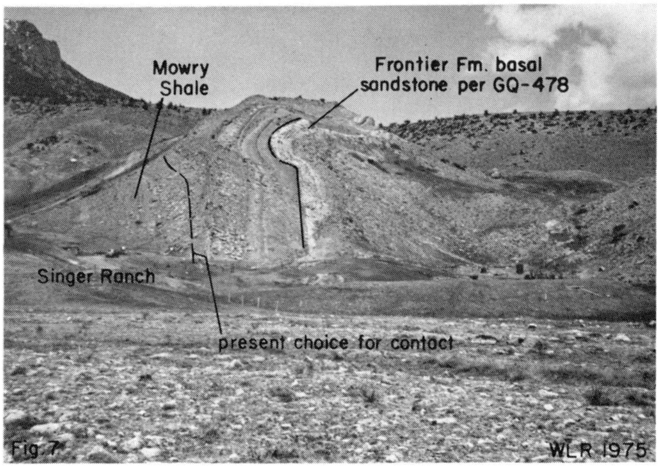

41.7 0.6 Junction. Turn right toward Clarks Fork Canyon.

42.3 1.0 View northward of Triassic, Jurassic and Lower Cretaceous rocks; Figure 8. The Triassic Dinwoody Formation is not well exposed here. The Chugwater Formation consists of red siltstone, shale and fine-grained sandstone; 650 ± 50 feet thick. Lower Jurassic rocks are absent; however the later Jurassic is represented. The Gypsum Spring Formation consists of red and gray shale, fossiliferous limestone and gypsum. The white band just to the right of the Chugwater represents massive gypsum in the basal part of the Gypsum Spring. The Sundance Formation consists of green and gray shale, glauconitic limy sandstone and fossiliferous limestone. The formation generally forms a valley with only the limestone cropping out as low ridges. Combined thickness of the Sundance and Gypsum Spring Formation is 425 to 500 feet. Uppermost Jurassic rocks in this area are represented by the Morrison Formation which is variegated claystone and gray silty sandstone. The Morrison is commonly combined with the overlying Early Cretaceous Cloverly Formation because of poor exposures. The Cloverly is gray and variegated shale, light gray sandstone, and lenticular chert conglomerate. Combined thickness of the Morrison and Cloverly is about 550 feet. The Thermopolis Shale consists of black shale with numerous bentonite beds and contains the Muddy Sandstone Member; about 550 feet thick (Pierce, GQ-478, 1965).

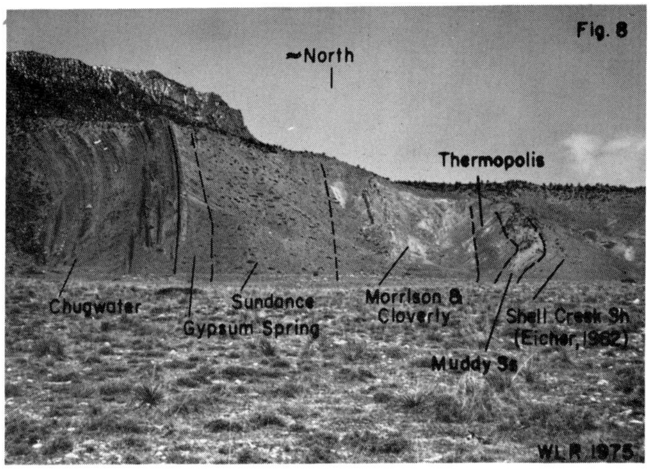

43.3 1.6 STOP 3. Road turn off. Clark Fork Canyon turn-around. View north of lower Paleozoic formations and of the Precambrian granite and granite gneiss; Figure 9. Note: The Gallatin has been raised to group rank and consists of the Pilgrim Limestone, Snowy Range, and Grove Creek Formations. The Flathead Sandstone is a hard, ledge-forming, quartzitic sandstone 100 feet thick. The overlying Gros Ventre consists of green micaceous shale, thin-bedded, gray limestone, and limestone-pebble conglomerate, 625 to 700 feet thick. The lowest formation of the Gallatin Group is the Pilgrim Limestone consisting of massive, light-gray, mottled, oolitic limestone which forms a prominent ledge; 100 to 125 thick. The Snowy Range is gray-green shale and greenish, flat-pebble conglomerate about 300 feet thick. The uppermost Gallatin formation is the Grove

Creek, consisting of gray, buff and orange limestone and dolomite; green shale, and gray-green limestone pebble conglomerate; 50 feet thick. The Upper Ordovician Bighorn Dolomite forms the lower massive cliff. It is gray dolomite and dolomitic limestone ranging from 400 to 475 feet thick. There is some uncertainty as to whether the Lower Devonian Beartooth Butte Formation can be seen from this vantage point. Charlie Sandberg, one of our trip leaders, will clarify this in his discussion of the Devonian and Early Mississippian rocks. The Beartooth Butte is a stream-channel deposit of red calcareous siltstone, red and yellowish-gray silty limestone and siltstone, and siltstone and limestone conglomerate and breccia; 0 to 75 feet thick. The overlying Jefferson Formation consists of fetid brown dolomite and light gray and tan limestone; uppermost part is mottled yellowish-orange dolomite and yellowish-gray siltstone; about 300 feet thick. The Devonian-Mississippian Three Forks Formation is yellow, greenish-gray and dark gray dolomitic siltstone, black fissile shale, and silty dolomite; about 90 feet thick. For a detailed description of the Devonian rocks in this area see: 1967 Wyoming Geological Survey Bull. 52. A ridge of Madison Limestone is present on the north skyline. Return to Dead Indian Hill road.

44.9 6.5 Singer Ranch road to north.

51.4 3.5 Junction with road to Clark, Wyoming. Continue eastward.

54.9 11.6 Junction with Wyoming 120. Turn right toward Cody.

66.5 0.8 Pat O'Hara Mountain, a large anticline, is on the skyline at 12 o'clock.

67.3 0.8 Contact of Meeteetse and Willwood Formations in draw on east side of road; refer to Figure 4 at mileage point 18.4.

68.1 0.6 Road junction. Turn west (right) onto Dead Indian Hill road.

68.7 0.4 On Two Dot Flats. This is the Chapman Bench Terrace level (see mileage point 95.9 for picture looking east).

69.1 1.2 Cattle guard.

70.3 0.2 Boulders along road are let down (lag) from the Heart Mountain fault.

70.5 0.5 Crossing contact of Cody Shale and Frontier Formation. Blaine Creek is along north side of road; a poor exposure of the Frontier is in the hill just beyond the creek.

71.0 0.7 Crossing contact of Mowry and Thermopolis Shales which is not exposed here. Cliffs on skyline at 11 o'clock are Madison Limestone, Three Forks Formation, Jefferson Formation and Bighorn Dolomite in the upper plate of the Heart Mountain fault; Chugwater red beds underlie these rocks.

71.7 0.7 Contact of Thermopolis Shale and Cloverly Formation; not exposed.

72.4 0.8 Unexposed contact of Morrison and Sundance Formations. Sandstones across creek toward the south are in the basal Morrison.

73.2 0.5 STOP 4. Excellent exposure of contact of Gypsum Spring and Chugwater Formations in hillside on north (right) side.

73.7 0.7 Picture of Chugwater and Gypsum Spring Formations; Figure 10. The accompanying composite graphic section

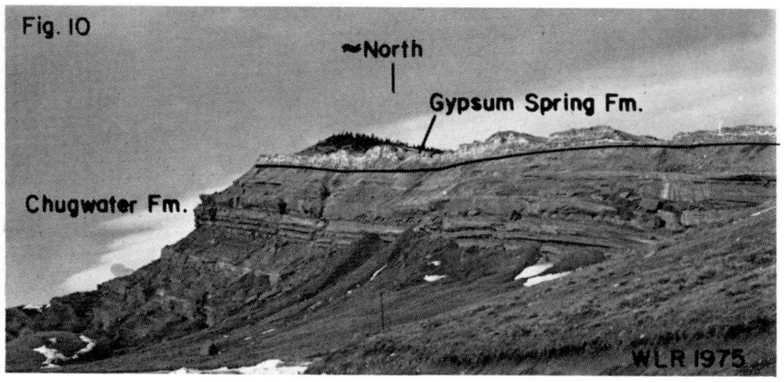

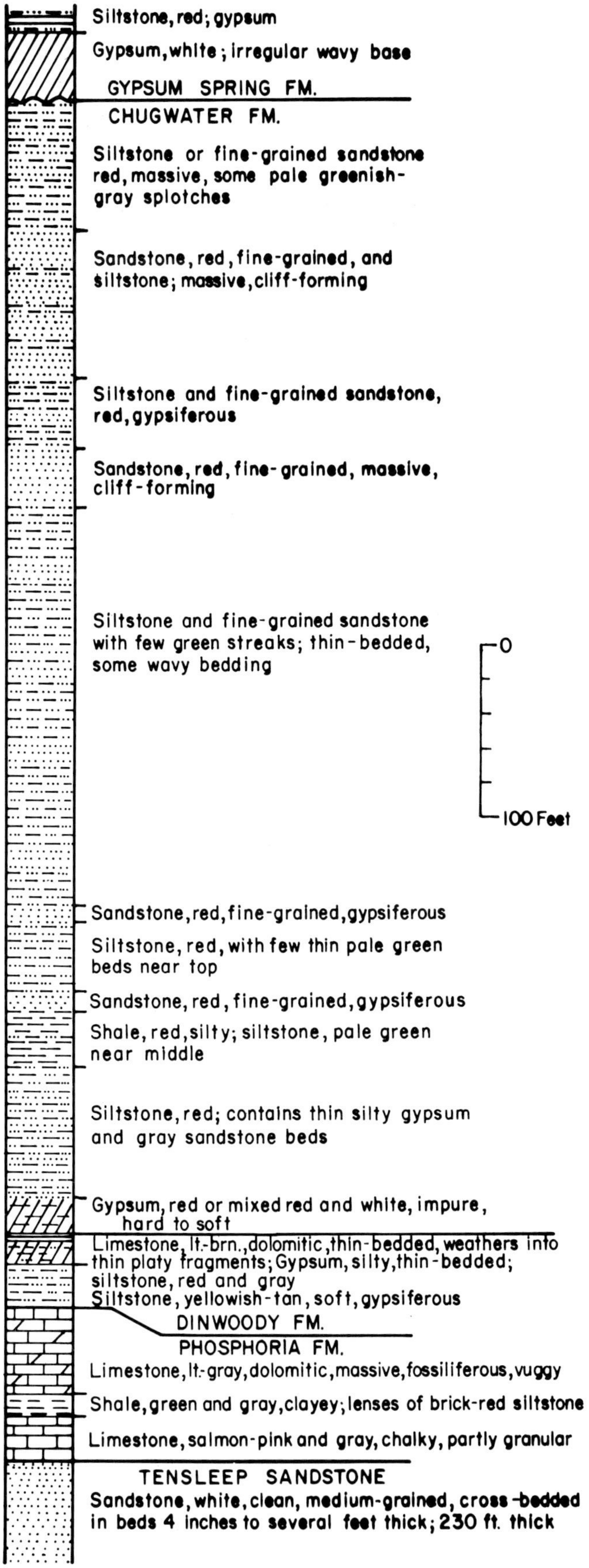

Fig. 11 - Composite section. Tensleep and Phosphoria measured at Clarks Fork Canyon; Dinwoody, Chugwater and Gypsum Spring measured on Dead Indian Hill. By W. G. Pierce.

was measured by W. G. Pierce in this area; Figure 11.

74.4 1.3 On Dinwoody Formation at crossing in draw. At the next bend in the road to the right, at cut bank, is a highly fossiliferous outcrop of the Phosphoria Formation.

75.7 0.2 The Phosphoria Formation and Tensleep Sandstone are exposed in the canyon wall to the north. Note: because the name, Phosphoria Formation, is used by most geologists in the petroleum industry, its use is continued here. The USGS prefers to use the name Park City Formation for Permian rocks in this area. The Phosphoria consists of siliceous limestone and dolomite, nodular chert and tan and gray shale; 70 to 110 feet thick. No attempt has been made here to recognize the formational members. The reader is directed to the article by Ed Maughan in this guidebook for additional information. The underlying Tensleep Sandstone is light-gray, well-sorted, crossbedded, massive sandstone. Thin beds of limestone and dolomite occur in the lower part of the Tensleep; 170 to 220 feet thick.

75.9 0.3 Heart Mountain on the skyline at 9 o'clock. At 10 o'clock, in the middle distance, are dip slopes of the Cloverly (Greybull Sandstone Member), easternmost; Sundance, middle slope; and Gypsum Spring.

76.2 0.3 STOP 5. The Triassic Dinwoody Formation is well-exposed in the road cut to the north. The Dinwoody is tan, gray, and red siltstone, gypsum and dolomite; 20 to 50 feet thick. A short dip slope on top of the Phosphoria is immediately uphill from the roadcut. The Chugwater caps the small hill north of the Dinwoody exposure; Figure 12. Refer to the composite section, Figure 11.

76.5	1.2	Cattle Guard.
77.7	1.5	Phosphoria Formation in road cut; poor exposure.
79.2	0.6	Phosphoria Formation and Tensleep Sandstone are exposed across canyon north of road. The contact is at the top of the cliff. The Mississippian and Pennsylvanian Amsden Formation is partly exposed in the bottom of the canyon. The Amsden consists of red shale, dolomitic limestone, some chert and hematite nodules; siltstone or sandstone are commonly present in the lower part; 250 to 300 feet thick. It is interesting to note that Pierce and Nelson (GQ-755, 1968) observed hematite nodules in the formation. Pisolitic and nodular hematite has been observed in the Amsden in the Wind River Mountains and Wyoming Range. What conditions so ubiquitous lead to the wide-spread occurrence of the Amsden hematite? Some of the hematite runs as high as 50 per cent iron, albeit as thin pods.
79.8	1.7	The road crosses a NNW-trending normal fault, downthrown on the west side, upon exiting from the upper end of the previously mentioned canyon; now traveling on Chugwater Formation.
81.5	1.0	Road crosses Heart Mountain fault.
82.5	13.4	STOP 6. Top of Dead Indian Hill and Sunlight Basin Overlook. Turn around. This is a lunch stop. The best view is from a point along the ridge north of the road; about a 15-minute hike. The accompanying panorma, Figure 13, is by W. G. Pierce. Pierce will discuss the Heart Mountain fault. For a discussion of the Sunlight mineralized area, refer to the article by Max Bergendahl in this guidebook. Jim Elliot may be on hand to discuss the Cooke City mining district. This is one of the most spectacular views in Wyoming. To the north and northwest is the Beartooth Uplift with the Clarks Fork fault bounding the southeast side of the uplift and the Beartooth fault (20,000 feet of displacement) on the east side. The bedding plane phase of the Heart Mountain fault can be seen extending over many square miles to the west, from Pilot Peak to the foot of the west slope of Dead Indian Hill. The transgressive phase can be seen as it follows up the west side of Dead Indian Hill to the summit, where it passes onto the former land surface. This land surface phase occurs at Heart Mountain and McCulloch Peaks. Here at

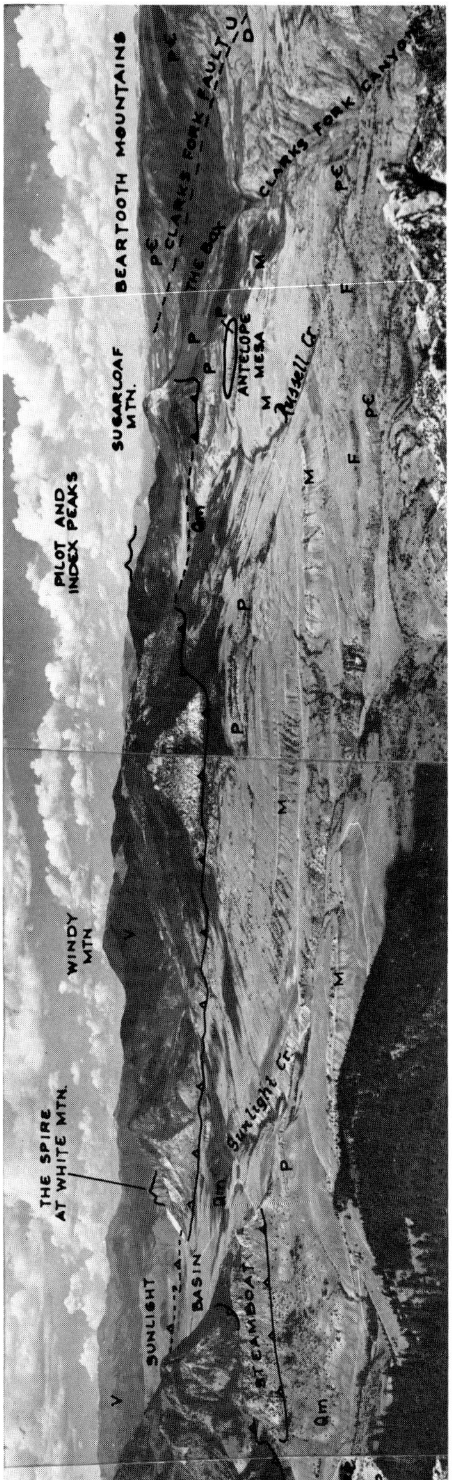

Symbols: Qm, glacial moraine; V, volcanic rocks; P, Pilgrim Limestone; M, Meagher Limestone; F, Flathead Sandstone; P€, Precambrian.

the summit of Dead Indian Hill the Phosphoria rests on silicified Tensleep Sandstone. Return to Wyoming 120.

95.9 0.6 Now traveling on Two Dot Flats of the Chapman Bench Terrace. The view east is of the Mesaverde Formation, tree-covered ridge; the overlying Meeteetse, clay shales; and the variegated Willwood Formation, which is resting unconformably on the Meeteetse and on the Lance, small wedge below the triangulation station (Figure 14).

96.5 1.1 Junction with Wyoming 210. Turn south.

97.6 1.2 Abandoned coal mine in lower Mesaverde Formation east of highway.

98.8 8.5 Two Dot Ranch road on west side.

107.3 9.1 STOP 7. Picture stop for afternoon lighting on Heart Mountain; refer to Figure 1 at mile point 9.1.

116.4 Intersection of U. S. 14 and Wyoming 120 in Cody.

(Turn to next road log for trip continuation)

DEAD INDIAN HILL TO YELLOWSTONE NATIONAL PARK, NORTHEAST ENTRANCE

BY THOMAS E. TUCKER[1]

This road log has been compiled to provide a general geologic description of an area of scenic beauty and unusual and well exposed geology. It is largely an up-date of a log appearing in the Billings (Montana) Geological Society Ninth Annual Field Conference Guidebook, Beartooth Uplift and Sunlight Basin, 1958, apparently written by W.G. Pierce and others, and now out of print.

(It is recommended that the road log from Cody to Dead Indian Hill, Wyoming Geological Association Guidebook, Big Horn Basin, 1975, P 272-280, by W.H. Rohrer, et al, be used to complete ones trip through this area.)

Mileage West	Mileage East	Increment	
0.0	52.1	.3	Dead Indian Pass (elev. 8048). Small remnants of Phosphoria Fm. lie on silicified Tensleep Sandstone. Historical marker on the south side at the crest commemorates early settlers and their efforts to build this road: DEAD INDIAN HILL SUMMIT Altitude 8000 feet "This pass is the summit of Dead Indian Hill. Through this portal, great herds of wild game seasonally migrated from the mountains to the plains. This high pass was the gateway for countless Indian hunting and war parties. Through this portal, Chief Joseph, in 1877, let his Nez Perce Indians in a strategic and defensive retreat pursued by the U.S. Army soldiers. Over this one and only opening to the valleys of the west, travelled a vast army of miners to seek the wealth of Cooke City and down this steep hill the early settlers of Sunlight Basin braved dangers. The first road improvement was made possible in 1909 by dwellers of Sunlight Valley whose names are here inscribed . . ., Dedicated 1940." Just beyond, a Forest Service rest area and overlook provides a panoramic view of the region to the west.
0.3	51.8	.6	Amsden Limestone outcrop on the right, below large talus slope of Tensleep Sandstone, contains intraformational chert and limestone breccia. Tensleep float is found quite some distance down the hill. Contacts with overlying Tensleep and underlying Madison Fms. are obscure.
0.9	51.2	1.0	Switchback to the left. Roadcut on the right is in Madison Limestone. Good view to the right of the Clarks Fork Canyon in Precambrian Granite.
1.9	50.2	.5	Switchback to the right. Vertical beds of purple to orange shale, and orange to tan dolomite, some with petroliferous odor and "pseudobreccia" (?) outcrop on the right. This is mapped as Madison Fm., perhaps the basal Cottonwood Canyon Shale member, but may be the Devonian Three Forks Formation.
2.4	49.7	.1	The road turns to the left in poor outcrops of Three Forks and Jefferson Formations.
2.5	49.6	.1	Steeply dipping Three Forks Shale (?), the same formation as at the last switchback, in the right shoulder of the road. Steeply dipping mass of limestone above the road to the right is intensely fractured Madison Limestone in a klippe above the transgressive phase of the Heart Mountain Fault.
2.6	49.5	.3	The road turns to the right around an outcrop of the Madison Limestone klippe.
2.9	49.2	.2	Switchback to the left, probably Jefferson and Three Forks formations.
3.1	49.0	.4	The road turns progressively to the left and here passes a series of small dikes and sills of basalt cutting Big Horn Dolomite.
3.5	48.6	.2	Switchback to the right in poor outcrop of Jefferson Fm., and then crossing obscure contact with Big Horn Dolomite.
3.7	48.4	.5	Gradual turn to the left in Big Horn Dolomite.
4.2	47.9	.3	Turn to the left and pass glacial morainal deposits approximately 1000 feet above the present valley floor.
4.5	47.6	.1	Switchback to the left near base of Snowy Range Formation, green shale and flat

[1]Mapco Prod. Co.
Billings, Montana

			pebble conglomerate.
4.6	47.5	.4	Outcrop of grey, dense limestone with limestone pebble conglomerate to the left, Snowy Range Formation.
5.0	47.1	.3	Switchback to the right in glacial morainal deposits.
5.3	46.8	.3	Cross cattle guard and outcrop of Pilgrim Limestone at small fault (?). The Pilgrim forms nearly everywhere a prominent escarpment of from 50-200 feet, and is readily recognized throughout the region.
5.6	46.5	.6	Park Shale below Pilgrim Limestone to the right.
6.2	45.9	.2	Outcrop of Park shale and limestone pebble conglomerate on the right at the last hairpin curve to the left. The outcrop exhibits several small chevron folds. Excellent view of the Clarks Fork Canyon to the north.
6.4	45.7	.3	Outcrop of Meagher Limestone on the left, dipping more steeply at the north end due to proximity of a down to the north fault.
6.7	45.4	.3	Local earthslides in the Park Shale and Wolsey Shale have cut into the road at several points. They are in places "stabilized" with a rip rap cover of Tensleep Sandstone. Mining claim corners seen in the area are signs of exploration for Uranium in underlying Flathead Sandstone. Note strong joint pattern in the granite in the valley floor to the right and behind us.
7.0	45.1	.7	View left ahead of steeply dipping Madison and Big Horn formations above nearly horizontal Pilgrim Limestone. They are separated by the Heart Mountain Fault.
7.7	44.4	.1	Cross Dead Indian Creek. Forest Service campground to the right.
7.8	44.3	1.0	Meagher Limestone forms prominent outcrop at left side of the road. Flathead Sandstone outcrops below the road to the right. Road is traversing Wolsey Shale.
8.8	43.3	.5	Road turns to the left into the valley of Sunlight Creek, still in Wolsey Shale. To the right is the Clarks Fork Canyon cut into Precambrian granite.
9.3	42.8	.2	Small outcrops of Flathead Sandstone to the left above a small normal fault. View up Sunlight Basin at 12:30. The deep cut of Sunlight Creek through glacial morainal deposits can be seen.
9.5	42.6	.2	Junction of Sunlight Basin Road, straight ahead, with road to Cooke city, to the right. Turn Right. The road into Sunlight Basin can be traversed by car for approximately 15 miles, providing spectacular views of the Absaroka Volcanics, Heart Mountain Thrust blocks, and glacial moraine features. Beyond lies the Sunlight mining region, a remote silver-copper district with minor production in the past, and the native sulphur prospects at Lee City and Sulphur Camp.
9.7	42.4	.7	Descend rapidly through morainal deposits.
10.4	41.7	.4	Cross Sunlight Creek. The road at both ends of the bridge is flanked by outcrops of Flathead Sandstone. To the left, Elk Creek cascades over extensive travertine deposits. Sunlight Creek picnic area to the left.
10.8	41.3	.1	The road climbs rapidly out of the Sunlight Creek Canyon through glacial moraine deposits. At the top of the hill is a particularly hazardous curve when coming from the other direction. Last October four vehicles missed the curve in one day, at least one of them making the trip all the way down to the creek. Careful when wet or icy. Good view behind us up into Sunlight Basin.
10.9	41.2	.5	Begin newly paved road. For the next several miles we travel on Wolsey and Park Shale, together, the Gros Ventre Formation, which is typically poorly exposed, even in new road cuts, as at Russell Creek. Pilgrim Limestone forms an escarpment which can sometimes be seen to our left.
11.4	40.7	1.6	Excellent views of the entire Clarks Fork Valley and Sunlight Basin region. To the north are the Beartooth Mountains, composed of Precambrian granite and granite gneiss, and reaching elevations of over 12,000'. The old Morrison Ranch can be seen on a bench of Gros Ventre Formation cut off to the north by the Clarks Fork Fault. The ranch can be reached by arduous jeep ride from the mouth of the Clarks Fork Canyon, or by airplane. To the northeast is the canyon of the Clarks Fork River, exposing on the southeast side

			in a palisade the entire section from Madison Limestone through Precambrian Granite. Bald Peak (elev. 8630) is some 2200 feet above the mouth of the canyon. To the east, is Dead Indian Hill which we have just descended. The Pilgrim Limestone forms perhaps the best outcrops, a third of the way up the slopes. To the south of Dead Indian Pass are erratically dipping blocks of Paleozoic formations, lying above the Heart Mountain Fault. To the south is Sunlight Basin, generally surrounded by volcanic rocks of the Absaroka Volcanic series.
13.0	39.1	.2	Cross Russell Creek. Ahead to the right Antelope Mountain is a prominent mesa supported by Pilgrim Limestone. An up to the west normal fault forms its western margin.
13.2	38.9	.9	Dirt road to the right leads 100 yards to the Frank N. Hammitt Monument, memorializing a ranger with the Yellowstone National Park Forest Reserve, 1898-1903.
14.1	38.0	1.1	Outcrops of slumped Gros Ventre Formation to left of road. Ahead the road has slumped the last two Springs. A half inch crack was apprarent along the left shoulder of the road in June. This is a characteristic problem in the Gros Ventre Formation throughout the region.
15.2	36.9	2.0	Above the road to the left is Sugarloaf Mountain, elev. 8780, a klippe above the Heart Mountain Fault. Outcrops along the road for the next several miles are in the Gros Ventre Formation.
17.2	34.9	.6	Camp Creek.
17.8	34.3	.4	To the left the Meagher Limestone underlies both the flats in the foreground and the higher bench in the distance. They are separated by a normal fault which we cross at 18.0.
18.2	33.9	1.4	Meagher Limestone (?) outcrop to the left.
19.6	32.5	.6	Scenic view turnout to the right. In early summer, Canyon Creek forms an impressive cascade across the valley down the face of the Beartooth Mountains.
20.2	31.9	.8	Road crosses approximate Cambrian Flathead-Precambrian granite contact. Gneiss and granite outcrop along the road for the most of the next 25 miles.
21.0	31.1	1.0	Scenic view pull out to the right.
22.0	30.1	2.2	Cross Reef Creek. Forest Service Campground to the left.
24.2	27.9	1.5	Scenic View pull-out to the left. Excellent view of Windy Mtn. elev. 10,262, and the Cathedral Cliffs, another Heart Mountain fault block, composed of Mississippian Madison through Ordovician Big Horn Formations above the Heart Mountain bedding fault. The fault block is in turn partially covered by post-faulting Eocene Wapiti Formation composed of andesitic volcanic conglomerate and the Cathedral Cliffs Formation, a volcanic sandstone and tuff, composed primarily of latite and andesite fragments. The entire mass is cut by numerous latite and basalt dikes. To the north across the valley of the Clarks Fork of the Yellowstone River, is Table Mountain, composed of Pilgrim and Gros Ventre Formations. The up dip side of the mountain is truncated by the Clarks Fork Fault. Beyond Table Mountain, Beartooth Butte can be seen on the horizon, an autochthonous block of Devonian through Cambrian carbonates in a down-faulted position on the Beartooth Plateau (elev. 10,514).
25.7	26.4	.3	Crandall Ranger Station road to the left.
26.0	26.1	.5	Cross Crandall Creek.
26.5	25.6	1.0	Power line crosses the road. Above the left side of the road is a small outcrop of Flathead Sandstone. Straight ahead on the horizon are Clay Butte and Beartooth Butte. Behind us at 4:00 is a good view of Cathedral Cliffs.
27.5	24.6	.5	Cross Clarks Fork of the Yellowstone River. It was down this valley that Chief Joseph led his band of Nez Perce Indians in 1877 in a futile attempt to escape the pursuing U.S. Cavalry. Various versions of the story have the party traversing the spectacular Clarks Fork Canyon, or climbing out of Sunlight Basin over Dead Indian Hill, either way, gaining several days on the less mobile cavalry.

28.0	24.1	.9	Hunter Peak, elev. 9034', to the left, a Paleozoic klippe.
28.9	23.2	2.5	Forest Service campground to the left, Hunter Peak.
31.4	20.7	1.2	Good outcrop of Precambrian granite and gneiss on the right. The west end of the cliff has a large pegmatite vein composed of quartz and microcline feldspar.
32.6	19.5	1.2	Cross Lake Creek. Forest Service campground to the right.
33.8	18.3	.3	Junction with U.S. 212. Turn left toward Cooke City. To the right the Beartooth highway leads up and over the Beartooth Mountains to Red Lodge, Montana, providing the motorist some of the most spectacular mountain scenery in the country.
31.4	18.0	.6	Begin winding descent to the floor of the Clarks Fork Valley as the road narrows. Glacial moraine deposits along the road.
34.7	17.4	.2	Single car pull-out on the left side of the road (very narrow) provides view up the Clarks Fork Valley to Pilot and Index Peaks (elevs. 11,708 and 11,313). Pilot Peak is to the south and is a sharp glacial horn, composed of Eocene Wapiti, Lamar River, and Cathedral Cliff andesitic volcaniclastic rocks, above the Heart Mtn fault. Pilot and Index were first climbed in 1932 & 1939 respectively. These peaks and the surrounding country were first seen by a white man in the winter of 1807-08 when John Colter made his famous trek eastward through the Yellowstone Park Area. Colter called them "Pilot" and "Finger" peaks, and states that "one of them derives its name from its shape, like a closed hand with the index-finger extending upward." Much confusion existed in the literature between 1872-1952 as to which is actually Index Peak and which is Pilot Peak. Library research forces one to conclude that Index Peak is to the north and Pilot Peak to the south. The south wall of the valley is composed of Eocene Wapiti andesitic volcanic rocks.
34.9	17.2	1.7	Cross Gilbert Creek in tight turn to the left.
36.6	15.5	1.2	Cross Crazy Creek. Campground to the left. Impressive cascade, up Crazy Creek 100 yards to the right.
37.8	14.3	.7	Jim Smith Peak, elev. 10,312, across the valley to the southwest.
38.5	13.6	1.6	Cross Clarks Fork of the Yellowstone River. Good views across the valley of Jim Smith Peak. Several spectacular waterfalls can be seen in early summer. Peak is composed of Wapiti Formation, above the Heart Mountain fault. The Pilgrim Limestone makes the prominent and continuous outcrop just below the fault.
40.1	12.0	.3	Flathead Sandstone outcrop on left side of road contains an abundance of burrows and trails in micaceous shale and sandstone.
40.4	11.7	.7	Fox Creek Campground to the right.
41.4	11.0	.5	Fox Creek. Good, though brief, views of the Heart Mountain fault zone straight ahead above the road in the next .7 miles.
41.6	10.5	.4	Glacial morainal deposits to the left.
42.0	10.1	.4	Index Creek.
42.4	9.7	.4	Good view to the right down the Clarks Fork Valley with Precambrian rocks of the Beartooth Mountains to the north and Eocene volcanic rocks above the Heart Mountain Fault to the south.
42.8	9.3	.3	Cross 45° North latitude, halfway between the equator and north pole. Montana-Wyoming state line was missurveyed a quarter mile to the north in 1879-1880.
43.1	9.0	.5	Road narrows at Montana-Wyoming state line. Talus slope on the left below Paleocene trachyandesite porphyry sill intruded into Park Shale.
43.6	8.5	.5	More tracyandesite prophyry talus.
44.1	8.0	.4	Cooke Pass, elev. approximately 8040.
44.5	7.6	.6	Chief Joseph campground to the left.
45.1	7.0	.5	Crooke Pass community, inhabited by a few hardy souls all year long.
45.6	6.5	.1	Colter Pass, elev. approx. 8005.
45.7	6.4	.1	Eocene andesite porphyry in roadcut on the left.
45.8	6.3	.3	Colter campground to the right.
46.1	6.0	1.0	Dirt road to right to Lulu Pass and the New World mining district, passable to cars for about two miles, then a jeep road.

47.1 5.0 Cooke City cemetery on the left.
.1 Historical sign reads:

COOKE CITY CEMETERY

Here within the shadows of the mountain they loved and lived among, rest a few of the rugged pioneers who discovered in 1870 the mineral wealth these mountains held. They established Cooke City and the New World mining district even before the country to the west became Yellowstone National Park. They lived and died for the dream their mountains had given them and the tribute from the earth that they so loved. The wild forget-me-not blooms each spring.

Few graves are marked, but at least one recent mine cave-in victim lies here.

47.2 4.9 Precambrian granite and granite gneiss
.1 outcrop on the right shoulder of the road. This is the last Precambrian outcrop along the road until Lamar Canyon, approximately 20 miles to the west.

47.3 4.8 Dirt road to right to Daisy Pass, where
.1 impressive views of the surrounding countryside, including much of the New World mining district, may be held. Passable by cars with care.

47.4 4.7 Small mine dump and collapsed adit
.1 above road to the right in Eocene rhyodacite prophyry. The only mineralization observed on the dump was disseminated pyrite.

47.5 4.6 Cross Miller Creek. Cleared area below
.5 road to the left is the site of a smelter to process ores from the New World district. The abundant pyrite in the waste rock is now creating an acid drainage problem.

48.0 4.1 Cooke City, elev. 7573, permanent
.4 population about 100, elementary school has one pupil.

Cooke was settled by prospectors and trappers in the late 1870's, following the discovery of gold and silver in the surrounding area, primarily on Miller and Republic Mountains, north and south of town respectively. Later development centered around Henderson Mountain two to four miles north of town. Activities ceased between 1886 and 1893. Oxidized gold ore from Henderson Mountain revived activity for a year or two, and then no major operations were carried on until about 1904. From 1904 until about 1925, mining activities continued at a moderate pace, however. By this time "tourists" had started to become important to the economics of Cooke City. Ore minerals include gold, silver, argenite, pyrargyrite, galena, sphalerite, chalcopyrite, pyrite, with associated cerussite, anglesite, calcite, siderite, maganous ankerite, rhodochrosite, quartz, hematite. Metamorphic skarn minerals include garnet, epidote, and temolite. Approximately 45 mines and prospects produced an estimated $215,000 by 1923 from approximately 650 ounces of gold, 100,000 ounces of silver, and 1,300,000 pounds of lead. Production since then is not known, however, the Irma-Republic Mine was active until the late 1950's. Estimated total value of production is $4-5,000,000.

Recent exploration activity has centered around Daisy and Lulu Passes. Rancher's Exploration had announced plans for a mine on 500,000 tons of gold reserves at Lulu Pass, but that is apparently now on hold.

48.4 3.7 Historical Marker on left: "In 1870 a
.4 party of prospectors came into this country by way of Soda Butte Creek. They found rich float but were set afoot by Indians. Cacheing their surplus supplies on a stream now called Cache Creek they made it back to the Yellowstone and reported their find. In the next four years many prospectors combed these mountains. The first real development began about 1876. Chief Joseph's band of fugitive Nez Perce Indians came through here in 1877. In 1883, there were 135 log cabins in the settlement, two general stores and 13 saloons. Cooke City has been waaiting for years for reasonable transportation connections to the outside world, so that her promising ore deposits may be profitably mined. She is no blushing maiden, but this highway is the answer to her prayers."

48.8 3.3 Single car pull out to left provides view of
.5 Republic Mountain and a single mine adit. The mountain is composed of Eocene Lamar River and Cathedral Cliffs andesitic volcaniclastic rocks. Apparent

dips in beds range up to 30°. Pilgrim Limestone is prominent just below the Heart Mountain bedding fault.

49.3 2.8 .5 Cross Sheep Creek.

49.8 2.3 1.0 Mount Republic Chapel of Peace on left with large parking area and good view of Republic Mountain.

The Heart Mountain fault is well exposed in the south and southwest parts of the Cooke City area. Features of the Heart Mountain fault have been described by Pierce (1957, 1960). Many features crucial crucial to the interpretation of the fault mechanism are exposed in and near the Cooke City area. The "break-away" fault is exposed a short distance west of the southwest corner of the map area and the trace of this fault to the north parallels 110° 00' to the northernmost exposures of it just west of the upper Stillwater River. Two large and several small Heart Mountain fault blocks are present in this area. One is well exposed at the north end of Republic Mountain; it contains Bighorn Dolomite, Jefferson Formation, Three Forks Formation, and Madison Limestone. Another probable Heart Mountain fault block with the same units exposed crops out both north and south of the divide between Sheep Creek and the upper Stillwater River. Throughout most of the Cooke City area the volcanic rocks, principally Lamar River formation, rest on the tectonically denuded surface of the Heart Mountain fault.

The major displacement along the Heart Mountain fault occurred during the deposition of the Lamar River Formation. Volcanic rocks, including all of the Cathedral Cliffs Formation and most of the Lamar River Formation, and Paleozoic sedimentary rocks east of the break-away fault were transported southwestward on the Heart Mountain fault. The youngest part of the Lamar River Formation was deposited after Heart Mountain faulting (Pierce and others, 1973).

From: Elliott, J.E., 1979, U.S.G.S. Map I-1084

50.8 1.3 .2 Small earthslide in landslide debris beneath cabin to right.

51.0 1.1 .2 Silver Gate, created in mid-1930's as a tourist road camp. Winter population, about 6. Wallrock Mountain to the south, Lamar River volcaniclastics. Surrounding mountains shed spectacular avalanches in April and May which contribute to significant mass wasting. Interior of Range Rider Lodge, right, exhibits interesting Swiss log architecture.

51.2 .9 .5 Soda Butte Creek to the left. Conglomeratic andesitic volcaniclastic rocks in landslide debris along road. During Pinedale glacation, 20,000-40,000 years ago, glacial ice flowed from the Beartooth Plateau area, down Soda Butte Creek, Lamar River and Yellowstone River valley to near Emigrant, Montana. With a thickness of over 3000 feet and a length of 91 miles this formed the longest glacier in the U.S., excluding those originating in the Canadian and Alaskan icecaps.

51.7 .4 .4 Enter Yellowstone National Park.

52.1 0.0 Northwest Entrance Gate and Ranger Station. Entrance fee $2.00. The road from here to Mammoth Hot Springs, and Gardiner, Montana is the only Yellowstone National Park road kept open all year. Wildlife is plentiful, especially elk, buffalo, moose, and coyote. This section of the park contains the most spectacular mountains, and better geological exposures.

(The following general log will complete the route to Soda Butte and there tie with the log by H.J. Prostka which appears elsewhere in this volume.)

0.0 12.3 1.2 Northeast entrance of Yellowstone National Park.

1.2 11.1 6.6 Warm Creek Picnic area. From this general area, there are numerous views of the surrounding ice and snow sculptured peaks. Abiathar Peak, elev. 10,928, to the left, was named after paleontologist Abiathar White, 1885. It's the first major peak to the west of the Heart Mountain detachment fault, and thus exhibits a normal section of Eocene volcanic rocks over an undisturbed Paleozoic section from

Mississippian Madison Limestone through Cambrian Pilgrim Limestone and Snowy Range Formation.

Ahead of us and to the right, is Barronette Peak, elev. 10,404, which contains the same stratigraphic sequence.

7.8	4.5 1.3	Icebox Canyon, contains some ice most of the year. It is cut through Lamar River andesitic conglomerate, which is well exposed along the roadside.
9.1	3.2 2.7	Cross Pebble Creek, campground to the right. Madison Limestone outcrops poorly to the right in this area.
11.8	0.5 .5	Outcrops to the right of Ordovician Big Horn Dolomite.
12.3	0.0	Soda Butte, a conical mound of travertine deposited in the recent past by a warm spring which today emits the hydrogen sulfide (rotten eggs odor) one smells in the area. Excellent views from here of the surrounding peaks, and back up the valley toward Cooke City.

BIBLIOGRAPHY

Bonney, Orrin H. & Bonney, Lorraine, 1963, Field Book, The Absaroka Range, Yellowstone Park: Denver, CO, Sage Books, p 233-302.

Elliott, James E., 1979, Geologic map of the southwest part of the Cooke City quadrangle: U.S.G.S., Misc. Investigations, I-1084.

Glidden, Ralph, 1976, Exploring the Yellowstone high country: by the author, Cooke City, Montana, 78 p.

Hewett, D.F., 1914, Sulphur deposits of Sunlight Basin, Wyoming: U.S.G.S. Bulletin 530-H, p. 350-362.

Lovering, T.S., 1929, The New World or Cooke City mining district, Park County, Montana: U.S.G.S., Bulletin 811 A, 1-87 p.

Parson, Willard H., 1978, Middle Rockies and Yellowstone Field Guide: Dubuque, Iowa, Kendall/Hunt Publishing Co., 233 p (excellent general guide to the area)

Pierce, Kenneth L., 1979, History and Dynamics of Glaciation in the northern Yellowstone Park area: U.S.G.S. Professional Paper 729-F, 90p.

Pierce, W.G., 1973, Principal features of the Heart Mountain fault and the mechanism problem in: Gravity and Tectonics, Eds. Delong, K.A., and Scholten, R., John Wiley & Sons Inc., New York; and reprinted in:Wyoming Geological Association Guidebook, Bighorn Basin, 1975.

Pierce, W. G., 1965, Geologic map of the Deep Lake Quadrangle: U.S.G.S. quadrangle map, GQ 478.

Pierce, W.G.(?), et al, 1958, Road Log, Cooke City through Sunlight Basin to Heart Mountain; Billings Geological Society Guidebook, Beartooth Uplift and Sunlight Basin, p 90-93. (out of print)

Pierce, William G., and Nelson, Willis H., 1971, Geologic map of the Beartooth Butte quadrangle; U.S.G.S. Geological Quadrangle GQ-935.

______, et al, 1973, Geologic map of the Pilot Peak Quadrangle: U.S.G.S., Miscellaneous Geologic Investigations, I-816.

Pierce, W. G. and Nelson, W. H., 1968, Geologic map of the Pat O'Hara Mountain Quadrangle: U.S.G.S. Geological Quadrangle, GQ-755.

Prostka, H.J., et al, 1975 Geologic Map of the Abiathar Peak Quadrangle: U.S.G.S., Geological Quadrangle, GQ-1244.

Sands, William J., 1975, Stratigraphic sections of the Madison Limestone and Amsden Formation, west flank of the Bighorn Basin: Wyoming Geological Association Guidebook, Geology and Mineral resources of the Bighorn Basin, p 129-137.

Sandberg, Charles A., 1967, Measured Sections of Devonian rocks in northern Wyoming, Geological Survey of Wyoming, Bulletin 52, 93 p.

U.S. Geological Survey, 1972, Geologic Map of Yellowstone Park: U.S.G.S., Misc. Geologic Investigations, I-711.

Urbanek, Mae, 1974, Wyoming Place Names: Boulder, CO, Johnson Publishing Co., 236 p.

LIST OF ADVERTISERS